Floodplains: Interdisciplinary Approaches

Geological Society Special Publications

Series Editors

A. J. Fleet

R. E. Holdsworth

A. C. Morton

M. S. Stoker

GEOLOGICAL SOCIETY SPECIAL PUBLICATION NO. 163

Floodplains: Interdisciplinary Approaches

EDITED BY

SUSAN B. MARRIOTT
University of the West of England, UK

JAN ALEXANDER
University of East Anglia, UK

1999
Published by
The Geological Society
London

THE GEOLOGICAL SOCIETY

The Geological Society of London was founded in 1807 and is the oldest geological society in the world. It received its Royal Charter in 1825 for the purpose of 'investigating the mineral structure of the Earth' and is now Britain's national society for geology.

Both a learned society and a professional body, the Geological Society is recognized by the Department of Trade and Industry (DTI) as the chartering authority for geoscience, able to award Chartered Geologist status upon appropriately qualified Fellows. The Society has a membership of 8600, of whom about 1500 live outside the UK.

Fellowship of the Society is open to persons holding a recognized honours degree in geology or a cognate subject and who have at least two years' relevant postgraduate experience, or not less than six years' relevant experience in geology or a cognate subject. A Fellow with a minimum of five years' relevant postgraduate experience in the practice of geology may apply for chartered status. Successful applicants are entitled to use the designatory postnominal CGeol (Chartered Geologist). Fellows of the Society may use the letters FGS. Other grades of membership are available to members not yet qualifying for Fellowship.

The Society has its own Publishing House based in Bath, UK. It produces the Society's international journals, books and maps, and is the European distributor for publications of the American Association of Petroleum Geologists (AAPG), the Society for Sedimentary Geology (SEPM) and the Geological Society of America (GSA). Members of the Society can buy books at considerable discounts. The Publishing House has an online bookshop (*http://bookshop.geolsoc.org.uk*).

Further information on Society membership may be obtained from the Membership Services Manager, The Geological Society, Burlington House, Piccadilly, London W1V 0JU (e-mail: *enquiries@geolsoc.org.uk*; tel: +44 (0)171 434 9944).

The Society's Web Site can be found at *http://www.geolsoc.org.uk/*. The Society is a Registered Charity, number 210161.

Published by The Geological Society from:
The Geological Society Publishing House
Unit 7, Brassmill Enterprise Centre
Brassmill Lane
Bath BA1 3JN, UK

(*Orders*: Tel. +44 (0)1225 445046
Fax +44 (0)1225 442836)
Online bookshop: *http://bookshop.geolsoc.org.uk*

First published 1999

British Library Cataloguing in Publication Data
A catalogue record for this book is available from the British Library.

ISBN 1-86239-050-9

Typeset by E & M Graphics, Midsomer Norton, Bath, UK.

Printed by The Alden Press, Osney Mead, Oxford, UK.

Distributors

USA
AAPG Bookstore
PO Box 979
Tulsa
OK 74101-0979
USA
Orders: Tel. +1 918 584-2555
Fax +1 918 560-2652
e-mail *bookstore@aapg.org*

Australia
Australian Mineral Foundation Bookshop
63 Conyngham Street
Glenside
South Australia 5065
Australia
Orders: Tel. +61 88 379-0444
Fax +61 88 379-4634
e-mail *bookshop@amf.com.au*

India
Affiliated East–West Press PVT Ltd
G-1/16 Ansari Road, Daryaganj,
New Delhi 110 002
India
Orders: Tel. +91 11 327-9113
Fax +91 11 326-0538

Japan
Kanda Book Trading Co.
Cityhouse Tama 204
Tsurumaki 1-3-10
Tama-shi
Tokyo 206-0034
Japan
Orders: Tel. +81 (0)423 57-7650
Fax +81 (0)423 57-7651

Contents

References to this volume

It is recommended that reference to all or part of this book should be made in one of the following ways:

MARRIOTT, S. B. & ALEXANDER, J. (eds) 1999. *Floodplains: Interdisciplinary Approaches*. Geological Society, London, Special Publications, **163**.

ADAMS, W. M. & PERROW, M. 1999. Scientific and institutional constraints on the restoration of European floodplains. *In*: MARRIOTT, S. B. & ALEXANDER, J. (eds) *Floodplains: Interdisciplinary Approaches*. Geological Society, London, Special Publications, **163**, 89–97.

Preface

Floodplains are major components of fluvial systems wherein physical, chemical and biological processes combine over a range of temporal and spatial scales varying with, and contributing to, environmental change. Floods have a major (defining) impact on floodplains and have significant socio-economic importance. The relatively flat, generally fertile, land with an adjacent water supply has attracted a large proportion of the world's human population to dwell on floodplains at the mercy of the hazards of major flooding, landslides and mudflows.

Floodplains are areas of natural sediment storage and sites for contaminant capture and remobilization. Sediment accumulation over relatively recent geological time has formed substantial natural resources (loose aggregates, peat, gold and diamond placer deposits, shallow groundwater aquifers), that are of major socio-economic importance in many areas of the world. Floodplain accumulation over longer geological time periods has formed much of the world's coal reserves and ancient deposits include many aquifers and hydrocarbon reservoirs. Despite the resulting economic importance of floodplain deposits, their architecture and processes of formation are only more recently becoming better understood, since previously most research concentrated on channel processes.

Over at least the past 3000 years human activity has altered the state of floodplains so that very few, if any, are still in an anthropogenic unimpacted state. With increasing population pressures, floodplains are continuing to change, and the character and implications of these changes are poorly known and often ignored. Much of the development on floodplains and their 'management' has been piecemeal, often without regard for natural processes in the catchment as a whole and in general ignorance of, or disinterest in, the long term effects of planned activity on the system. This situation has arisen partly as a result of differing interests of residents, land owners and local, regional and national administrative bodies. But in addition to this there is a common lack of communication between practitioners in the fields of planning, civil engineering, geomorphology, ecology and sedimentology, and likewise between any of these 'experts' and local population.

The Floodplains '98 meeting held at the University of East Anglia, which led to this book, was convened with the intention of bringing together those at the forefront of research into many aspects of floodplains. Hydrologists, ecologists, environmentalists, geomorphologists, sedimentologists and geologists presented and discussed research addressing problems relating to floodplain processes, ecology and morphology, deposit character and architecture and environmental management. This book includes papers on many of the projects presented at the meeting and additional noted contributions, in an attempt to represent the complex and very broad subject of floodplains in a truly interdisciplinary way.

The preparation of this book relied on the unpaid assistance of a large number of people. We would particularly like to thank the following who reviewed papers: N. E. M. Asselman, M. D. Blum, R. R. Boar, J. S. Bridge, G. Brierley, A. Brookes, R. Bryant, P. A. Carling, S. Crooks, D. L. Dent, M. H. Dinnin, P. F. Friend, B. Gomez, A. M. Gurnell, A. J. Hartley, G. L. Heritage, K. M. Hiscock, J. A. Howell, M. J. Kraus, M. R. Leeder, A. A. Lovett, M. G. Macklin, A. E. Mather, S. J. McLelland, G. J. Nichols, C. P. North, A. V. Panin, M. Provensal, J. E. Rae, A. J. Russell, R. H. J. Sellin, F. D. Shields Jr. , R. L. Slingerland, C. Spencer, M. S. Stoker, M. Street, T. J. Stuart, J. C. A. Taylor, J. A. Taylor, K. G. Taylor, T. E. Tornqvist, B. Turner, D. E. Walling, W. Woodland, M. van der Perk, V. P. Wright, Y. Zhao.

Jan Alexander and Sue B. Marriott

Introduction

JAN ALEXANDER[1] & SUSAN B. MARRIOTT[2]
[1]*School of Environmental Sciences, University of East Anglia, Norwich, NR4 7TJ, UK (e-mail: j.alexander@uea.ac.uk)*
[2]*School of Geography and Environmental Management, University of the West of England, Coldharbour Lane, Bristol, BS16 1QY, UK (e-mail: Susan.Marriott@uwe.ac.uk)*

Floodplains are of major socio-economic and ecological importance, ranging as they do from intensely inhabited and industrialized areas, through high-productivity agricultural land to sites of extraordinary biodiversity and biological productivity that have suffered little management or other human intervention such as some of the flooded forests of the Amazon Basin. Natural floodplains vary in character depending on their climatic setting, catchment size and character and, as a consequence, discharge character and sediment load. Biological communities are sensitive to these variations and the major floodplains of the world may be dominated by plant communities with very different evolutionary histories. On more local scales, there may be much closer ecological, if not taxonomic similarities. Floodplain character has changed through geological time because of the evolution of land plants and animals, and changing atmospheric chemistry, global climate and sea level.

Over the relatively recent past (*c.* 50 ka) human activity has brought about rapid change through, for example, forest clearance, water use and channel engineering. This book examines both natural features of floodplains whilst taking into account the human impacts on them. This demands a multi-disciplinary approach and documents the evolution of recent research.

Floodplain deposits reflect the diversity of mechanisms by which sediment is transported and deposited. These include transfer from the channel during overbank flow, by slope wash from terraces and valley sides on distal parts of a floodplain and by aeolian processes. Apart from colluvial deposits at the edges of a floodplain, most of the material deposited is generally fine-grained – clay/silt to fine sand. Floodplains are sinks for this fine-grained material and account for most of the transport loss as sediment moves through the system. Mechanisms for the transfer of fine-grained sediment from a channel to its floodplain have been studied extensively, both in numerical and computer simulation models and by field experiment. The latter is less well documented because of the difficulties inherent in collecting data during flood events. This information, together with the results of sediment budget studies, has also helped in the study of transport and storage of contaminants such as heavy metals, pesticides and fertilizers. These contaminants tend to be associated with particular sediment grain sizes and can, thus, be stored in floodplains for long periods. Eventually, these may be mobilized and have serious biological consequences.

The diverse aspects of floodplains, both in terms of subject area and geographic location, have been studied by various research communities, often with little mutual communication. Some floodplains and particular processes, such as channel bank accretion, have been studied intensively, sometimes with duplication of research effort; while other areas and processes have been largely ignored. Temperate–climate floodplains such as that of the Rhine-Meuse (**Asselman; Schoor *et al.***) are far better documented and understood than either their high- or low-latitude counterparts. But even in well-studied cases such as the Rhine, the response of the floodplains and their biological communities to management or environmental change are difficult to predict.

For millennia, floodplains have been favoured sites for human habitation, because of the combination of water supply, fertile soil, navigable waterways and flat terrain for building and communication. Many of the world's most densely populated areas are on floodplains: yet other floodplains remain sparsely populated. Floodplains are managed in many ways and for many purposes; some have been managed for long periods and their

From: MARRIOTT, S. B. & ALEXANDER, J. (eds) 1999. *Floodplains: Interdisciplinary Approaches*. Geological Society, London, Special Publications, **163**, 1–13. 1-86239-050-9/99/$15.00

natural character has been obscured or destroyed. Floodplain management takes many forms including construction of various flood defence systems, engineering for navigation or water-powered mills, farmland irrigation, fertilization and drainage, seasonal vegetation burning, and wetland management for wildlife. Additionally, indirect management results from engineering works on floodplains, which affect floodwater paths. Mismanagement can result from poor understanding, education, government procedure or conflicting interests. There is increasing realization in some countries that 'hard engineering' to control river and floodplain processes has its limits and now engineers are increasingly looking to reinstate 'natural' systems where appropriate.

Current research into floodplains addresses a very broad range of physical, chemical, biological, ecological, economic and social problems using very differing techniques. This book consists of contributions on many different aspects of a very broad subject area in an attempt to increase interdisciplinary study. Here, we introduce the definitions and importance of floodplains and give an overview referring to the topics covered and the authors represented in this volume.

Floodplains have been variously defined by geomorphologists and hydrologists, and may mean different things to ecologists, engineers and economists. An individual's concept of a floodplain seems to depend on their training (discipline) and perspective, particularly in respect of geographical location and time-scale considered.

This volume includes studies on areas with a wide variety of characteristics under the name of floodplains, ranging from small areas of temperate farmland in Britain (e.g. **Cotton *et al*.**; **Dinnin & Brayshay**; **Nicholas & McLelland**; **O'Donoghue**) to vast areas of tropical Australia and Bangladesh (**Alexander *et al*.**; **Hasan *et al*.**) and includes marine-influenced wetlands (**Crooks**). The studies are of modern floodplains and deposits of various geological ages back to the Ordovician (**Liu**; **Wright**; **McCarthey & Plint**).

What is a floodplain?

A floodplain is a functional part of a fluvial system. Its form is the product of a large number of interrelated processes that change over time in response to external factors. These allocyclic factors such as climate change cause variation in, for example, runoff, biological communities, weathering rate and sediment flux. The floodplain, as interpreted by most of the authors in this volume, can be summed up by the broad terms of a definition given by Schmudde (1968)

> … as a topographic category, it is quite flat and lies adjacent to a stream; geomorphologically, it is a landform composed primarily of unconsolidated depositional material derived from sediment being transported by the related stream; hydrologically, it is perhaps best-defined as a landform subject to periodic flooding by the parent stream.

Although this definition is vague enough to be applicable to most situations, it may not be adequate for many purposes, such as delineating the floodplain for administrative decision making. In a similar way wetlands can be defined broadly as

> areas where water table is at or above the land surface for long enough each year to promote the formation of hydric soils and to support the growth of vegetation much of which is emergent (Cowardin *et al.* 1979).

Although in many instances wetlands may be equivalent to floodplains there are a lot of cases where the wetland forms a sub-area of a floodplain and other cases (coastal wetlands) where it is debatable if they have any correspondence.

Flooding defines natural floodplain environments. Floods control the morphology, the ecology, and the sediment distribution of a floodplain. Thus a floodplain may be defined as an area of relatively low relief, adjacent to a stream that floods at least once in a given period. Many works (cf. Nanson & Croke 1992) appear to consider that the floodwater in such a definition should be derived as overflow from the parent channel. Others would include floods resulting from local runoff or intense rainfall (cf. **Alexander *et al*.**), high groundwater (water-table rising above the topographic surface) and, even, storm surge events which periodically introduce marine floodwater into many lowland and coastal areas.

Hydraulic definitions of floodplain area (area inundated by floods of a particular return period) are used widely for channel management, insurance rate calculations and as design criteria for major engineering projects. There is considerable variation in the frequency of inundation used to define a hydraulic floodplain. Wolman & Leopold (1957) suggested that an **active floodplain** is an area subject to annual inundation, but Leopold *et al.* (1964) found that, on average, rivers flood every 1.5 years. Such short return periods define areas along most rivers that are very small in comparison to what is regarded commonly as a floodplain. The areal extent becomes nearer to the general concept of a floodplain (but not the same) if the criterion of flood frequency is amended to inundation intervals of up to 10 years (Schmudde 1968). Most flood engineering programmes and many administrative decisions rely on the concept of a **design flood** of perhaps 100 or 200 year return period (e.g. Philippi

1996) and this defines a floodplain area based on flood risk that is much nearer to the concept of a floodplain held by most geomorphologists, geologists and the general public. The delineation of the floodplain area, then, depends on historical records of inundation and discharge, and empirical models of runoff and flood storage. The position of the margins will change with time as a result of natural autocyclic processes (e.g. channel migration), allocyclic factors (such as climate change to reduced precipitation and runoff) and management (e.g. embankment construction, damming and landuse change).

The hydrological definition or delineation of a floodplain does not consider the processes that formed the landform nor the nature of the material that makes up the area. A hydraulic floodplain may be underlain by floodplain deposits (sediment deposited in the floodplain environment) or anything else including, for example, glacial deposits, volcanic material, basement geology or reclaimed land.

Nanson & Croke (1992) proposed the term **genetic floodplain**, which relates a landform to the contemporary climatic and hydrological conditions of the parent stream. They defined a genetic floodplain as a

> largely horizontally-bedded alluvial landform adjacent to a channel, separated from the channel by banks, and built of sediment transported by the present flow-regime.

Although the concept of genetic floodplains is useful, this definition needs improvement to include areas where a lot of the sediment accumulation is by *in situ* organic growth (as in the Norfolk Broads, cf. **Boar *et al.***) or those which have a large wind transported component as in the Kuiseb River, Namibia (Ward & Swart 1997). In addition, many workers would wish to incorporate areas where a significant part of the floodplain is constructed from channel bar and bank material as a result of channel migration (cf. Howard 1992, Bridge *et al.* 1998; **Panin *et al.***). Nanson & Croke's (1992) definition is concerned with contemporaneous conditions, with the provision that environmental change will produce a new genetic floodplain, related to the new conditions (see also **Bravard & Peiry**). Therefore, most modern floodplains that have evolved through Holocene climate and sea level changes (e.g. **Cotton *et al.***; **Crooks**; **Panin *et al.***) are composed of more than one genetic floodplain and are, thus, described as **polyphase floodplains** by Nanson & Croke (1992). This concept is comparable to that of an **alluvial plain**, although the latter may be related to more than one parent stream.

Given that the character and distribution of sediments deposited during each successive genetic floodplain period are likely to be different, the concept of **genetic-floodplain depositional units** may be useful. This concept is similar but not quite equivalent to the suite of facies deposited between successive channel avulsions (rapid changes of channel position, Kraus & Aslan 1993; Smith *et al.* 1989) and may be related to the development of **parasequences** (cf. sequence stratigraphy, Emery & Myers, 1996).

For the purposes of this volume, the **active floodplain** is a relatively flat area adjacent to a stream that is periodically (over a period of 100–200 years) inundated by flood water, at least part of which emanates from the channel. This is, by definition, a **genetic floodplain**. Modern floodplains often include inactive as well as active flood areas with areas that are now infrequently or never inundated as a result of channel migration or avulsion, discharge or channel capacity change, incision (and terracing) or artificial restriction (Fig. 1).

The difficulty in defining a floodplain increases with discharge variability. In areas with erratic discharge (for example in areas of unreliable monsoon rainfall, erratic tropical cyclone patterns or glacial areas that experience jökulhlaups), the definition of a 100–year flood (or 200–year flood) and delineation of inundation area is difficult – due to often short monitoring periods and the extent of channel and overbank change that occurs with successive large discharge events. If a shorter duration return period is used to define the hydraulically active floodplain, then it may be contained within the flood channel and be of little use for prediction of flood risk. In yet more extreme cases where flow is ephemeral, any hydraulic definition of the floodplain becomes problematic, especially in extreme cases where the parent channel may be defined poorly and may change with each discharge event.

The upstream and downstream limits of floodplains

The upstream and downstream limits of floodplains are debatable (Fig. 2). Streams where the bed is the same width as the valley floor (with steep slopes rising from the channel), cannot really be said to have a floodplain or to support riparian vegetation, although they may overtop their banks and cause considerable damage. They do not generally store much sediment, except during periods of channel-bed aggradation. Central to this problem is the criterion for delineating the point when a bank-attached bar becomes part of the floodplain. There is no well-defined threshold between an upland

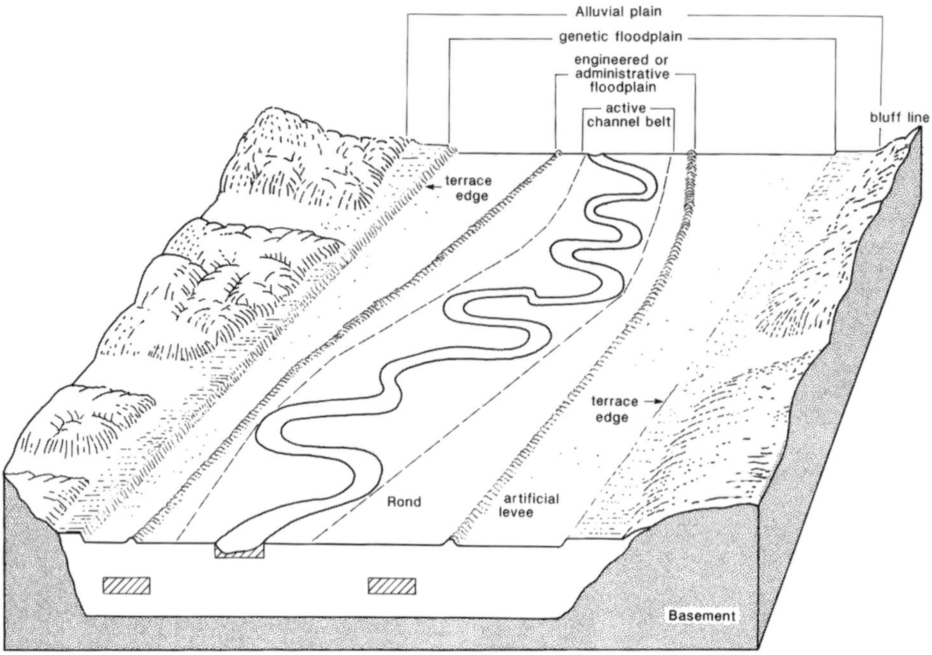

Fig. 1. Diagrammatic representation of floodplain and alluvial plain. The floodplain has been modified by engineering such that the active floodplain is restricted by artificial levees. It is debatable whether in this situation the genetic floodplain is also confined by the levees or includes the protected land area between the levees and terrace edge.

stream that has a floodplain and one that has not. The same valley reach may have a floodplain at some point in geological time and not at others because of changing runoff. For example, during Holocene pluvial periods a valley may have been occupied fully by the channel, while in drier periods the valley may have been occupied by an underfit stream with floodplains yet underlain by genetically unrelated channel deposits. Although such upland floodplains, and also those within canyon or gorge settings, are small in area, discontinuous, persist for relatively short periods, and are insignificant in the geological sedimentary record, they are locally of major importance as building land despite their inherent flood hazard.

The downstream limit of floodplains is defined variously. The downstream limit is relatively easy to define where a stream enters a lake or sea where tidal range is small. In such sites, the boundary may be defined as a shoreline and the downstream part of the floodplain may be part of a delta plain. Even in these low-gradient coastal areas, the boundary may alternatively be defined as the transition to salt flat, sand or mud flat, mangrove swamp or back barrier lagoon.

In settings with a high tidal range, definition of the seaward limit of a floodplain is more problematic. Many people include intertidal and supratidal flats as part of the floodplain system (e.g. **Crooks**), while others do not. In the geological record the limit of marine influence has been suggested as the delineation point as in the distinction between upper and lower delta plain (cf. Fielding 1985). Such delineation on many modern alluvial surfaces may be made on the basis of dominant plant communities or soil conditions, but in other modern sites is difficult and even more difficult in ancient deposits. The extent of marine influence on channel and overbank environments varies through time on both short- and long-time scales. We consider that the floodplain should include all dominantly terrestrial surfaces adjacent to a stream that are periodically inundated with non-marine floodwater. This does not exclude areas that are periodically inundated by marine water and does not require definition of a boundary based on the limit of marine influence.

The distinction between channel and floodplain

Most definitions of floodplains refer to the area adjacent to the main stream and the channel is not

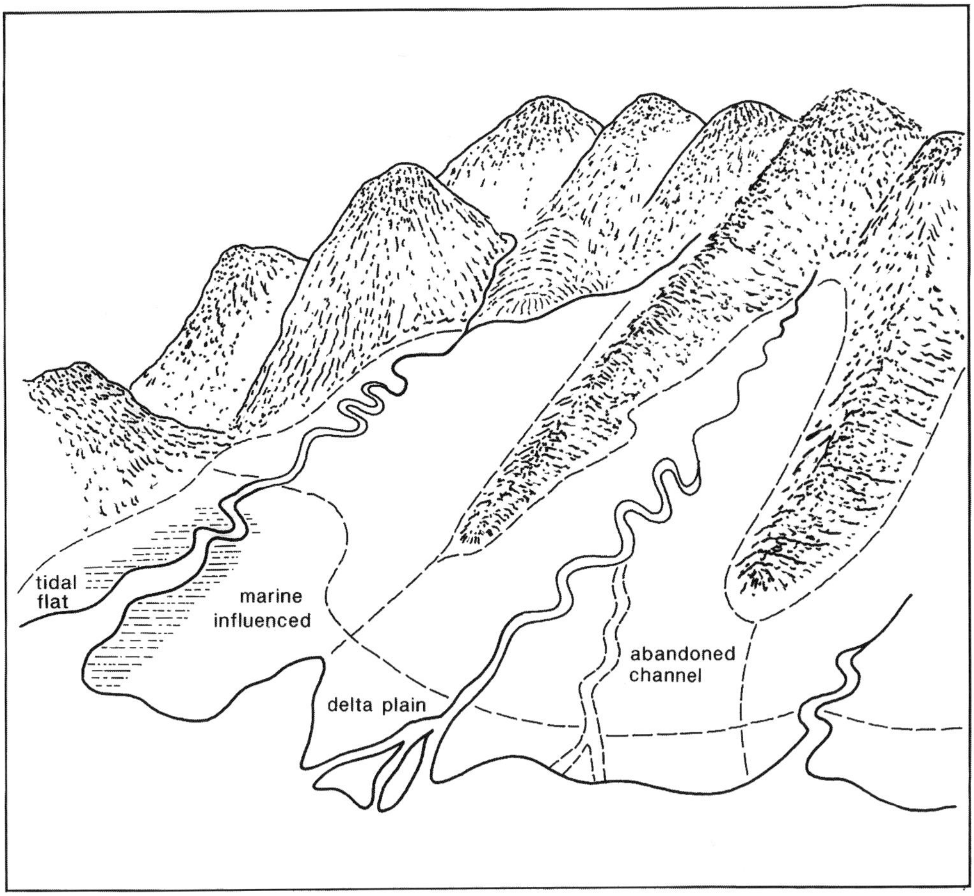

Fig. 2. Diagrammatic representation of floodplain limits. (A) with a high tidal range downstream limit of the floodplain passes into tidal flats, whereas (B) with a small tidal range the downstream limit approximates to the shoreline.

included in the floodplain although tributary and distributary streams may be. In contrast, some regard the floodplain as the whole low-lying area including the channel(s). The latter obviates the problems associated with distinguishing the channel in ephemeral or erratic discharge systems but is less satisfactory for many purposes, such as planning regulation.

If the main channel(s) is regarded as distinct from the floodplain area it is necessary to define the boundary between floodplain and channel and establish criteria for the point where a channel bar becomes part of the floodplain or when splay deposits change to channel deposits on avulsion (cf. Pérez-Arlucea & Smith 1999). It has been demonstrated that avulsion deposits (those sediments deposited during an avulsion rather than between subsequent events) may make up a significant proportion of a floodplain deposit (cf. Kraus & Gwinn 1997).

Aerial photography and other forms of remote sensing demonstrate that large proportions of many floodplains are underlain by channel deposits. This results from situations where

(a) channel-belt width approaches floodplain width (cf. **Bridge**)
(b) avulsion may occur relatively frequently despite a low aggradation rate, or
(c) where channel belts migrate (cf. Alexander *et al.* 1994).

So what determines when a bar surface becomes part of the floodplain? One criterion would be on the basis of the mean shoreline position, which in perennial channels approximates to the vegetation line (the boundary between lower bar and upper bar deposits cf. Bridge *et al.* 1998) or the marginal zone (National Rivers Authority 1992). This is unsatisfactory in many (most?) cases as the criterion for delineating floodplain boundary and totally inadequate in ephemeral systems and those with very variable discharge. Better distinction may be made by defining areas of dominantly riparian and dominantly floodplain plant communities (cf. Fielding *et al.* 1997) or possibly by defining hydraulic bankfull level, for example, on mean annual flood. However defined, channel migration results in channel bars becoming part of the floodplain. This is addressed in part in a case example in the Brahmaputra-Jamuna River by **Hasan *et al.***

Flooding

Flooding represents a major human hazard. Hazards such as drowning or structural damage by currents are fairly obvious, whilst others are more insidious. Flooding also brings many benefits, particularly for ecological variability and, locally, soil fertility. The predominant human desire to reduce or prevent flooding may not be the best management strategy in the long term, and in several areas this practice is being reversed (cf. **Andrews**; **Schoor *et al.***; **Adams & Perrow**).

Many of the world's floodplains support the life cycles of the 60 or so species of the mosquito genus *Anopheles* that transfers the malaria parasite plasmodium between humans. Over two billion people are exposed to malaria and around 250 million suffer from it at any one time. Floodwater may form standing pools that persist for long enough to allow mosquito reproduction and consequent increases in malaria, dengue fever and bilharzia. Many parasites, diseases and pests become more widespread during or following floods (e.g. liver fluke, *leptospira*; causing for example Weil's disease in people). Any infectious pathogen (such as the bacteria *Brucella* and *Salmonella*) that is shed in urine or faeces in high numbers and survives for weeks or months, can be spread in floodwater from one area to another (one farm to another) and deposited on pasture land to infect new herds and flocks.

An insidious hazard of flooding is its effect on pollutant transport and redistribution. Flooding can directly cause pollution incidents by, for example, flooding waste sites or preventing free drainage of sewage. Floodwater may remobilize previously deposited pollutants, an example of this is the redistribution of radiocaesum originating from the 1986 Chernobyl accident (**van der Perk *et al.***). The radiocaesum is readily absorbed by soil particles, particularly illites, and these are seasonally eroded and redeposited on the banks and floodplains downstream that are used for agricultural production.

Near coasts, floodwater may be augmented by storm surge, driving saline floodwater inland. Saline water in channel settings can poison fish, while on farmland it can damage standing crops with effects lasting for some years.

The beneficial aspects of flooding are less obvious to many people, and particularly to those whose dwellings are at risk of flood inundation. Flooding promotes exchange of materials and organisms between habitats and plays a key role in determining the level of biological productivity and diversity (Petts 1996). Flooding is particularly beneficial for soil fertility. One classical illustration of this is in the Nile Valley agricultural land which before the construction of the High Aswan dam in 1964 was supplied with nutrient rich sediment and water with each annual flood. In other areas, such as the Kakadu National Park, Australia, flood distribution influences forest fire risk and fire type.

Flood risk

The risk of flooding may be predicted from empirical data and catchment monitoring. To assess flood risk fully requires full documentation of hydrological characteristics, including records of peak discharges, flood stage and duration, flood-wave velocity, rate of rise to the flood crests, water velocities, sedimentation and degradation patterns in the channel, flood channels and flood basins and the effects of floods on water quality. It is impossible to obtain such data for all points in every catchment, and in many cases there are few or no such records, consequently, hydrological modelling based on available data has to be used.

For floodplain management it is important to understand flood flow behaviour, but it is generally difficult to measure the nature of flow and sediment transport during flood events. This difficulty is compounded where the floodplain covers large areas. Investigation of the nature of flow in a channel bounded on one or both sides by relatively shallow flow on the floodplains (the problem of compound channel flow) has been studied theoretically and using experimental approaches (Sellin *et al.* 1993; James 1985). Most experimental studies have used fixed-bed models to study flooding problems (e.g. Loveless *et al.* 1999) but as bed, bank and floodplain changes are common during flooding, research is needed with mobile-bed experiments (cf. Smith 1998; Ashworth *et al.* 1997). The study of flow patterns in natural floods is more difficult. A small case study of the hydrodynamics of a floodplain recirculation zone is described in this volume by **Nicholas & McLelland**.

Increased population pressure on floodplains results directly in increased risk to human life and increased economic loss in any individual flood event. In many areas, land use change and associated increased rates of surface runoff have increased the risk of flooding because of reduced lag times and more peaked flood hydrographs.

Change in catchment or channel management often leads to changes in flood character and intensity, either intentionally or unintentionally. One example of this is the change in flood size distribution along the lower Burdekin River of NE Australia that resulted from the construction of the Burdekin Falls Dam (**Alexander *et al.***). The long-term response of the Burdekin system to the changed flood distribution is unknown and may have serious connotations for channel stability. Another, now classic example is the High Aswan Dam that was built in 1964 for water storage, predominately (90%) to supply year-round irrigation. The dam catches all of the annual flood such that the Nile downstream of the dam is now effectively a canal (Said 1993).

The nature of floodplains changes where the flood's size distribution and character change. For example there is natural variation in the area of inland wetlands along the Nile River system that results from changing inflows due to upstream changes in rainfall distribution; notable area reductions occurred during the 1960s (Conway in press; Conway & Hulme 1993; Howell *et al.* 1988). Similar changes can be caused by managed changes in discharge patterns.

Many fluvial systems are being managed to reduce flood risk (both predictability and scale). This is the aim of the floodplain restoration along the Cosumnes River, California, USA (**Andrews**). In this example, re-establishing cottonwood forest and riparian oak woodland is reversing the changes caused by earlier land clearance. The author has used an understanding of floodplain ecology to improve management. Similarly, the establishment of alluvial forest species in European floodplains is underpinned by information on, for example, the timing of seed dispersal, germination ecology and the responses of tree seedlings to water level fluctuations (van Splunder *et al.* 1995, 1996).

Water resources on floodplains

Floodplain water bodies (rivers, tributary and distributary streams, lakes and ponds) are major water resources for domestic, agricultural and industrial use. Water is often extracted from floodplain deposits (alluvial aquifers) and also intercepted (precipitation) before entering the fluvial system.

Interception of runoff and extraction from rivers locally make major impacts on channel and floodplain behaviour downstream. A striking example of this is increased water usage from rivers flowing into the Kruger National Park in South Africa (Heritage *et al.* 1997). For example, the increasing extraction from the Letaba River has resulted in that river becoming ephemeral in the Park and in Mozambique. Predicted increased demand in the adjacent Sabie catchment suggests that rivers in this catchment will change in a similar way, with consequent changes to sediment movement and ecology (Heritage *et al.* 1997). The problems of water usage from rivers are increasingly complicated where rivers flow across administrative boundaries when water extraction and damming in one country affects others. This is illustrated by Conway's (in press) discussion of water usage along the Nile.

Floodplain sediments form important aquifers in many areas. Given anticipated human population growth and threats in many parts of the world water resources in floodplain sediments are of great importance both as primary water sources and as water storage sites. One example of this is the groundwater storage in the North China Plain (**Zhao *et al.***), and another is the proposed use of aquifers in palaeochannels of the Ken River in India to alleviate acute water shortages (Gautam 1994).

Floodplain waterways are important for fisheries, navigation and recreation. There is economic interest in maintaining navigation routes and also considerable environmental impact of such traffic on bank erosion, ecology and so channel behaviour.

Sedimentation and floodplain deposits

Floodplains are natural sites of sediment storage and also act temporarily, semi-permanently or possibly permanently as pollution traps. Sediment accumulates in floodplains by: channel bank accretion (point bar growth and translation, braid bar bank attachment, lateral bar expansion), sediment fall out from overbank flow, authigenic mineral precipitation (associated with flood and groundwater evaporation), organic growth, aeolian dust and sand input, and more rarely, volcanic or marine sedimentary input. Of these, channel bank accretion has been documented best because it is often relatively rapid, dynamic and studied relatively easily: it is easier to measure a change in channel position than a vertical change in field elevation. This predominance of channel studies may also partly relate to the greater ease in observation and also in teaching a visually dynamic subject. Vertical accretion has been measured in a variety of ways and has become more achievable with modern geochemical/isotopic and geodetic methods (cf. **Asselman; Walling**). Channel avulsion is demonstrably important in floodplain construction, but because of its relatively infrequent occurrence it has only very recently become the topic of significant research (e.g. Pérez-Arlucea & Smith, 1999).

Bank accretion deposits

There have been numerous studies of channel bar growth and bank accretion resulting in floodplain construction (cf. Bridge 1993; Bridge *et al.* 1995, 1998 and others) and studies of floodplains composed of ancient bar deposits (cf. Alexander *et al.* 1994; **Panin *et al.***; and others). This results in a belt of dominantly channel deposits (channel-belt deposits) and the ratio between the channel-belt width and floodplain width varies greatly (cf. **Bridge**). Processes of channel migration and bar and bank accretion lead to deposition of relatively coarse grade sediment above a basal erosion surface. The channel sediment body is commonly much larger (particularly wider) than the channel responsible for its formation, although, in most cases, the top of the coarse sediment body will not represent the level of the top of the channel banks. Rates of channel migration have been estimated from historical records (e.g. Lecce 1997; Mack & Leeder 1998), radiocarbon dating (e.g. Bridge *et al.* 1995) and dendrochronology (e.g. Everitte 1968; Kalicki & Krapiec 1995; Burckhardt & Todd 1998) but the complete three-dimensional depositional history of any bar (let alone channel belt) has not been recorded.

In-channel processes are largely omitted from this volume except where relevant directly to floodplain interaction. This does not suggest that in-channel processes are unimportant but rather is an attempt to redress an imbalance with abundant documentation of channel processes and products. Howard's (1992) model of meander migration and floodplain aggradation suggests that meander-belt width is generally very slow to increase beyond about two or three times the size of the largest meander (which itself is controlled largely by discharge; Schumm 1968) and so, in the absence of channel-belt migration or repeat avulsion, wide floodplains must be composed largely of overbank deposits.

Channel migration and bank accretion are important in maintaining ecological habitat variability. Bank accretion forms sites for primary vegetation successions, which unlike clearance sites do not have residual vegetation or soil seed banks and so primary plant successions develop. Primary successions of this type are a major mode of natural forest regeneration in many areas, such as the tropical lowlands of the Amazon (Kalliola *et al.* 1992) and bottomland areas of the SE Coastal Plain of USA (Shankman, 1993). In such settings pioneers of longer-term ecological successions colonize the newly emerged bar surfaces and later stages occur on older surfaces, ensuring habitat spatial heterogeneity. Channelization (stablizing channels) prevents large-scale natural disturbance and thus may cause more homogenous forests with less ecological habitat variation.

Overbank or flood basin deposition

Overbank deposits are those sediments deposited or formed on or beyond the river bank (outside the active channel environment) normally resulting in vertical accretion of the floodplain surfaces. Of predominant importance in most cases is sediment deposited from floodwater, although other sediments, such as peat or wind blown dust, may predominate in some sites.

Overbank, flood-derived sediment is generally finer grained than the contemporary channel thalweg deposits but may be coarse-grained sand (cf. Burdekin flood deposits, **Alexander *et al.***) and locally, gravel (cf. Shaw & Kellerhals 1977). Sediment distribution (size, composition, and thickness) resulting from individual events is generally poorly documented. However, Marriott (1992, 1996) has recently compared grain-size distributions derived from James' (1985) model for overbank sedimentation with the distribution obtained from sediment collected following the 1990 flooding on the River Severn (UK) floodplain and found similar patterns.

Because of the difficulties in examining sedimentation in or deposits of individual flood events, many workers have attempted to measure or model sediment aggradation over longer time periods (years to thousands of years (**Bottril *et al.***). Most of these approaches require dating and this has been achieved by various methods including palynological studies (e.g. pollen and spore stratigraphy, **Dinnin & Brayshay**), archaeological material, radiocarbon dating (e.g. **Panin *et al.***), tephra chronology and use of other radionuclides from, for example, resulting from nuclear weapon testing and the 1986 Chernobyl accident (cf. **van der Perk *et al.***; **Walling**).

This rapidly increasing but still sparse data set on floodplain aggradation generally tends to confirm that the rate of aggradation varies with surface age, elevation and distance from the main (supplying) channel (cf. Pizzuto 1987; Walling *et al.* 1992; Mertes 1994).

Abandoned channels

Channels abandoned by avulsion, reach or meander cut off, river capture or plan-form change make up a significant proportion of many floodplains, for example see the abandoned channels of the River Eden terraces, UK (**Cotton *et al.***) or the lower Burdekin River (**Alexander *et al.***) (Fig. 3). Before channel abandonment, in-channel processes build sediment bodies by active bar and bank migration and bed aggradation, but the abandonment of the channel inevitably leaves a topographic depression – processes of bank and bar accretion and bed aggradation cannot lead to complete filling of a river channel. Decreasing channel discharge will lead to sedimentation within the channel, but such discharge reduction will be accompanied by declining sediment load capacity and later stages of channel fill, therefore, will be finer grained. It would be unusual for a stream channel to fill completely with sand during gradual discharge reduction. Many channel abandonment processes are relatively rapid, for example, resulting from abrupt avulsion of a stream to a new course. This will leave the abandoned reach with either reduced flow (misfit stream), standing water or dry depressions.

Abandoned channel sediment bodies consist of active-channel, syn-abandonment and post-abandonment deposits in varying proportions. The deposits can be classified on the basis of channel-fill mechanism into active-channel deposits (bar and bank accretion and bed aggradation); abandonment fill associated with waning flow (including products of bank collapse); post-abandonment misfit channel deposits; alluvial overbank deposits (sediment brought in by flood water); organic deposits (such as peat); lacustrine deposits; and other deposits including

(i) marine sediments following transgression,
(ii) aeolian dust or sand,
(iii) glacial sediments, or
(vi) volcanic ash or lava.

Although active channel processes are responsible for the development of extensive coarse alluvial sediment bodies, the channel depression is usually filled by post abandonment processes and such deposits may be preserved more commonly in the geological record.

Abandoned channel deposits are often of major socio-economic importance for a variety of reasons. Coarse grained sediment often forms significant shallow aquifers that are exploited

Fig. 3. Photograph of an abandoned channel of the Lower Burdekin River at Kalamia sugar mill near Ayr (**Alexander *et al.***). The abandoned channel is about 500 m across.

easily for both water production and storage (cf. **Zhao *et al.***). Young abandoned channel reaches may have pronounced topography and standing water which, depending on its situation, may be beneficial (for example, water supply or recreation) or may be detrimental (for example, as mosquito breeding areas).

Abandoned channel deposits are exploited for aggregates (as in the Thames Valley, UK). Such bulk extraction may create large expanses of water such as the Cotswold Water Parks in the Thames Valley, UK, or be used for landfill sites. Differential entrainment, transport and selective deposition in alluvial channels causes local concentration of heavy minerals and consequently economic accumulations of, for example, diamonds and gold, are often found in palaeochannels (e.g. Jones 1991; Mazzucchelli 1996).

Floodplains as records of environmental change

The wide variety of deposits and features preserved in many floodplains represent a record of environmental change. The ecology and sedimentology of floodplains are environmentally sensitive, so that the sediment and fossil record reflect the sequence of past change. These records are being studied increasingly to elucidate both natural and anthropogenic change (cf. **Dinnin & Brayshay**).

The morphology of abandoned channels and character of their deposits can be used to interpret changing hydrological conditions (cf. **Panin *et al.***) which may relate to changing catchment runoff.

Socio-economic importance of floodplain deposits

Floodplain deposits are important in various ways especially as substrate for agricultural production, inherent value for mineral and water extraction, surface for construction, urban and industrial development and storage of contaminants. Floodplains provide much more than simply land for agriculture. Much of the history of human development has been influenced by floodplain processes and the wealth of natural resources found in floodplain deposits.

Floodplain deposits form soil of enormous spacial variability. This causes problems resulting from variation in drainage and nutrient supply and consequently, crop growth over the area of a single field. Another example is the quality of the crop grown on differing floodplain deposits (e.g. the quality of thatching reed, **Boar *et al.***). A more insidious example of the importance of deposit character is that the extent to which radiocaesum derived from the Chernobyl 1986 accident is taken into the food chain depends on the geochemical characteristics of the soils (**van der Perk**). Another is the sporadic and difficult-to-predict occurrence of acid sulphate soils and consequent hazards to aquatic life where some sediments are sulphidic (Dent 1986).

Organic-rich floodplain deposits have been exploited extensively for peat. Such exploitation may radically change floodplain form and channel behaviour. For example, digging and dredging in Norfolk and Suffolk removed about 25.5 million m^3 of peat between about 900–1350 AD and flooding of the workings created over 1.057×10^7 m^3 of interconnected waterways with a mean depth of 2.4 m forming a complex of shallow lakes known as the Norfolk Broads (Smith 1960; Lambert *et al.* 1965; George 1992). Much larger areas have been exploited in the Low Countries (Pons 1992). In coastal areas the grade of agricultural land is influenced by the nature of sediment deposited. Sandy areas like the Wash (UK), and large parts of the Netherlands have some of Europe's highest quality agricultural land. Where mud dominates the quality is much poorer.

Floodplain deposits in the rock record

Alluvial deposits are volumetrically important in the rock record and have considerable local economic significance as hydrocarbon and water reservoirs, and as hosts for mineral deposits such as gold, diamonds and coal. In addition, they preserve evidence of changing terrestrial conditions (colonization of the land by plants, evolution, climate and sea level change).

In areas of good exposure, fluvial channel and overbank deposits can be examined together. Overbank deposits tend to be finer grained than in-channel deposits because, during normal conditions, in-channel flow velocities are greater than overbank velocities. Despite this, coarser grained sediments may be found within floodplain deposits and fine-grained sediment often fills channels (particularly abandonment, or post-abandonment deposits). Channel deposits normally rest on a basal erosion surface and often contain complex hierarchies of sedimentary structures between bounding surfaces which can be interpreted as the products of growth and amalgamation of bars (e.g. Allen 1983; Alexander *et al.* 1994; Bridge *et al.* 1995, 1998). In most cases it is practical to divide alluvial deposits into channel-belt and non-channel-belt deposits and assume that the bulk of the former will be relatively coarse-grained and the latter relatively fine-grained. This conceptual division is particularly prevalent in the hydrocarbon industry where exploration for and production of oil and gas require some understanding of coarse deposit (sand and gravel) distribution.

This distinction between channel-belt and overbank deposits facilitates modelling of deposit accumulation. Such alluvial architectural simulation is illustrated by **Bridge**. It allows consideration of allocyclic factors such as tectonic tilting or sea level change. Studies of alluvial architecture have largely concentrated on the distribution of channel-belt deposits (notably hydrocarbon reservoir-quality sandstones) and largely ignored other facies, even though in most fluvial settings active channels occupy a small percentage of floodplain area.

Although the coarse-grained sediment of channel belt deposits are the major reservoir rocks in alluvial deposits, the sand- and gravel-grade sediments in overbank deposits should not be ignored. Although they contribute little to the net reservoir volume, they may increase hydraulic connection between channel-belt bodies (either during production or possibly during hydrocarbon migration).

Studying subsurface alluvial deposits is considerably more difficult than either examining exposures or studying modern fluvial processes. In the context of oil exploration/production there are often problems in correlating fluvial sandstone bodies between boreholes and in predicting the location of channel-belt deposits from boreholes penetrating overbank deposits. Recent advances in the study of overbank deposits, promise to be useful for understanding sedimentary facies distribution and, also, recognizing palaeogeographic and environmental change, e.g. in the interpretation of palaeoclimate from the micromorphology of palaeosols (cf. **McCarthy & Plint**).

Conclusion

Floodplains are a big and complex subject. We now know enough about them to know that we know very little. In view of their socio-economic and ecological importance it is imperative that we continue to undertake research into the character of floodplains so that they can be managed and developed in a sustainable manner and the deposits can be exploited to best advantage.

Writing this introduction, organizing the floodplains conference, and editing this volume have all been thought provoking. Discussions with many people have helped to stimulate our ideas. Thank you to all participants at every stage. Particular thanks to Rosie Cullington and Phil Judge for their help in preparation of this paper. This paper has been greatly improved by constructive comments on earlier versions by R. B. Boar, D. Conway, S. Crooks, D. L. Dent and M. S. Stoker.

References

ALEXANDER, J., BRIDGE, J. S., GAWTHORPE, R. L., LEEDER, M. R. & COLLIER, R. E. L. 1994. Holocene meander belt evolution in an extensional basin, SW Montana, USA. *Journal of Sedimentary Research*, **64B**, 542–559.

ALLEN, J. R. L. 1983. Studies in fluviatile sedimentation: bars, bar-complexes and sandstone sheets (low-sinuosity braided streams) in the Brownstones (L. Devonian), Welsh Borders. *Sedimentary Geology*, **33**, 237–293.

ASHWORTH, P., BEST, J. PEAKALL & LORSONG, J. 1997. Aggradation controls on braided alluvial architecture: Ashburton River, Canterbury Plains, New Zealand. *6th International conference on Fluvial Sedimentology abstract volume*, International Association of Sedimentologists.

BRIDGE, J. S. 1993. The interaction between channel geometry, water flow, sediment transport and deposition in braided rivers. *In*: BEST, J. L. & BRISTOW, C. S. (eds) *Braided Rivers*, Geological Society, London, Special Publications, **75**, 13–71.

——, COLLIER, R. E. L. & ALEXANDER, J. 1998. Large-scale structure of Calamus River deposits (braid bars, point bars and channel fills) revealed using ground-penetrating radar. *Sedimentology*, **45**, 977-986.

——, ALEXANDER, J., COLLIER, R. E. L., GAWTHORPE, R. L. & JARVIS, J. 1995. Point-bar deposits documented in 3-D using ground-penetrating radar and coring, and their genetic interpretation, South Esk River, Scotland. *Sedimentology*, **42**, 839–852.

BURCKHARDT, J. C. & TODD, B. L. 1998. Riparian forest effect on lateral stream channel migration in the glacial till plains. *Journal of the American Water Resources Association*, **34**, 179–184.

CONWAY, D. *in press*. Water Resources Development on the Upper Blue Nile: Some Environmental and Hydropolitical Considerations. *In:* The proceedings of *Ethiopia and Eritrea since 1991: Development Experience and Prospects*, Natural Resources Institute.

—— & HULME, M. 1993. Recent fluctuations in precipitation and runoff over the Nile subbasins and their impacts on Main Nile discharge. *Climatic Change*, **25**, 127–151.

COWARDIN, L. M., CARTER, V., GOLET, F. C. & LA ROE, E. T. 1979. *Classification of wetlands and deepwater habitats in the United States*. US Department of Interior Fish and Wildlife Service Report FWS/OBS–79/31.

DENT, D. L. 1986. *Acid sulphate soils: a baseline for research and development*. International Institute for Reclamation and Improvement, Publication, **39**, Wageningen, Netherlands, 1–203.

EMERY, D. & MYERS, K. J. 1996. *Sequence Stratigraphy*, Blackwell Science, London.

EVERITTE, B. L. 1968. Use of the cottonwood in an investigation of the recent history of a floodplain. *American Journal of Science*, **266**, 417–439.

FIELDING, C. R. 1985. Coal depositional models and the distinction between alluvial and delta plain environments. *Sedimentary Geology*, **42**, 41–48.

——, ALEXANDER, J. & NEWMAN-SUTHERLAND, E. 1997. Preservation *in situ* vegetation in fluvial channel deposits – data from the modern Burdekin River of north Queensland, Australia. *Palaeogeography, Palaeoclimatology, Palaeoecology*, **135**, 123–144.

GAUTAM 1994. Paleochannels of Ken River as potential groundwater reservoirs and geomorphic evolution of adjoining area, Banda and Hamirpur Districts, UP. *Indian-Minerals*, **48**, 181–184.

GEORGE, M. 1992. *The Land Use, Ecology and Conservation of Broadland*, Package Publishing Limited, Chichester.

HERITAGE, G., BIRKHEAD, A. & VAN COLLER, A. 1997. *Catchment influences on the geomorphology and biota of the Sabie and Letaba rivers in the Kruger National Park.* Field guide 3, 6th International Conference on Fluvial Sedimentology, University of Capetown International Association of Sedimentologists.

HOWARD, A. D. 1992. Modelling channel migration and floodplain development in meandering streams. *In*: CARLING, P. A. & PETTS, G. E. (eds) *Lowland Floodplain Rivers*, John Wiley and Sons, Chichester, 1–42.

HOWELL, P. P., LOCK, M. & COBB, S. 1988. *The Jonglei Canal: Impact and Opportunity*. Cambridge University Press, Cambridge.

JAMES, C. S. 1985. Sediment transfer to overbank sections. *Journal of Hydraulic Research*, **23**, 435–452.

JONES, R. 1991. *River diversions at Aredor Mine, Guinea, West Africa.* Institution of Mining and Metallurgy – Section A. **100**, A115–A120.

KALICKI, T. & KRAPIEC, M. 1995. Problems of dating alluvium using buried subfossil tree trunks – lessons from the black-oaks of the Vistula Valley central Europe. *Holocene*, **5**, 243–250.

KALLIOLA, R., SALO, J., PUHAKKA, M. & RAJASILTA, M. 1992. New site formation and colonizing vegetation in primary succession on the western Amazon floodplains. *Journal of Ecology*, **79,** 877–901.

KRAUS, M. J. & ASLAN, A. 1993. Eocene hydromorphic paleosols: significance for interpreting ancient floodplain processes. *Journal of Sedimentary Petrology*, **63**, 453–463.

—— & GWINN, B. 1997. Facies and facies architecture of paleogene deposits, Willwood Formation, Bighorn Basin, Wyoming, USA. *Sedimentary Geology*, **114**, 33–54.

LAMBERT, J. M., JENNINGS, J. N. & SMITH, C. T. 1965. The origin of the Broads. *In*: ELLIS, E. A. (ed.) *The Broads*, Collins, London, 37–65.

LECCE, S. A. 1997. Spatial patterns of historical overbank sedimentation and floodplain evolution: Blue River, Wisconsin. *Geomorphology*, **18**, 265–277.

LEOPOLD, L. B., WOLMAN, M. G. & MILLER, J. P. 1964. *Fluvial Processes in Geomorphology*, W. H. Freeman, San Francisco.

LOVELESS, J. H., SELLIN, R. J. H., BRYANT, T. WORMLEATON, P. R. CATMUR, S. & HEY, R. D. 1999. *Experiments with meandering mobile bed channels having overbank flow.* IAHR Symposium on river coastal and esturine morphodynamic, Conference volume.

MACK, G. H. & LEEDER, M. R. 1998. Channel shifting of the Rio Grande, Southern Rio Grande rift: implications for alluvial stratigraphic models. *Sedimentary Geology*, **177**, 207–219.

MARRIOTT, S. B. 1992. Textural analysis and modelling of a flood deposit, River Severn, UK. *Earth Surface Processes and Landforms*, **17**, 687–697.

—— 1996. Analysis and modelling of overbank deposits. *In*: ANDERSON, M. G., WALLING, D. E. & BATES, P. D. (eds) *Floodplain Processes.* John Wiley, Chichester, 63–94.

MAZZUCCHELLI, R. H. 1996. The application of soil geochemistry to gold exploration in the Black Flag area, Yilgarn Block, Western Australia. *Journal of Geochemical Exploration*, **57/1–3,** 175–185.

MERTES, L. A. K. 1994. Rates of floodplain sedimentation on the central Amazon River. *Geology*, **22**, 171–174.

NANSON, G. C. & CROKE, J. C. 1992. A genetic classification of floodplains. *Geomorphology*, **4**, 459–486.

NATIONAL RIVERS AUTHORITY 1992. *River Corridor Surveys: Methods and Procedures.* Conservation Technical Handbook No. 1, NRA Bristol.

PONS, L. J. 1992. Holocene peat formation in the lower parts of the Netherlands. *In*: VERHOEVEN, J. T. A. (ed.) *Fens and bogs in the Netherlands*, Kluwer, Germany, 7–79.

PÉREZ-ARLUCEA, M. & SMITH, N. D. 1999. Depositional patterns following the 1870s avulsion of the Saskatchewan River (Cumberland Marshes, Saskatchewan, Canada). *Journal of Sedimentary Research*, **69**, 62–73.

PETTS, G. E. 1996. Sustaining the ecological integrity of large floodplain rivers. *In*: ANDERSON, M. G., WALLING, D. E. & BATES, P. D. (eds) *Floodplain Processes,* John Wiley, Chichester, 540–551.

PHILIPPI, N. S. 1996. *Floodplain Management – Ecologic and Economic Perspectives.* Academic Press, California.

PIZZUTO, J. E. 1987. Sediment diffusion during overbank flows. *Sedimentology*, **34**, 301–317.

SAID, R. 1993. *The River Nile. Geology, Hydrology and Utilization*. Elsevier Science, Oxford.

SCHMUDDE, T. H. 1968. Floodplains. *In*: FAIRBRIDGE, R. W. (ed.) *The Encyclopedia of Geomorphology*. Reinhold Book Corporation, New York, 359–362.

SCHUMM, S. A. 1968. Speculations concerning palaeo-hydraulic controls of terrestrial sedimentation. *Geological Society of America Bulletin*, **79**, 1573–1588.

SELLIN, R. H. J. ENVINE, D. A. & WILLETTS, B. B. 1993. Behaviour of meandering two stage channels, *Journal of Water and Maritime Engineering*, **101**, 99–111.

SHANKMAN, D. 1993. Channel Migration and vegetation patterns in the southeastern coastal plain. *Conservation Biology*, **7**, 176–183.

SHAW, J. & KELLERHALS, R. 1977. Palaeohydraulic interpetation of antidune bedforms with application to antidunes in gravel. *Journal of Sedimentary Petrology*, **47**, 257–266.

SMITH, C. E. 1998. Modelling in high sinuosity meanders in a small flume. *Geomorphology*, **25**, 19–30.

SMITH, C. T. 1960. Historical evidence. *In*. MURRAY, J. (ed.) *The Making of the Broads*, Royal Geographical Society, London, Research Series, **3**, 63–111.

SMITH, N. D., CROSS, T. A. DUFFICY, J. P. & CLOUGH, S. R. 1989. Anatomy of an avulsion, *Sedimentology*, **36**, 1–23.

VAN SPLUNDER, I., COOPS, H., VOESENEK, L. A. C. J. &

BLOM, C. W. P. W. 1995. Establishment of alluvial forest species in floodplains: the role of dispersal timing, germination characteristics and water-level fluctuations. *Acta Botanica Neerlandica*, **44**, 269–278.

——, VOESENEK, L. A. C. J., COOPS, H., DE VRIES, X. J. A. & BLOM, C. W. P. W. 1996. Morphological responses of seedlings of four species of Salicaceae to drought. *Canadian Journal of Botany*, **74**, 1988–1995.

WALLING, D. E., QUINE, T. A. & HE, Q. 1992. Investigation of contemporary rates of floodplain sedimentation. *In*: CARLING, P. A. & PETTS, G. E. (eds) *Lowland Floodplain Rivers*, John Wiley and Sons, Chichester, 165–184.

WARD, J. D. & SWART, R. 1997. *Flash flood fluvial systems of the Central Namib Desert. Field Excursion Guidebook 10*, 6th International Conference on Fluvial Sedimentology, University of Capetown, International Association of Sedimentologists.

WOLMAN, M. G. & LEOPOLD, L. B. 1957. River flood plains: some observations on their formation. *US Geological Survey Professional Paper*, **282-C**, 87–109.

Hydrodynamics of a floodplain recirculation zone investigated by field monitoring and numerical simulation

A. P. NICHOLAS & S. J. McLELLAND

Department of Geography, Exeter University, Exeter EX4 4RJ, UK
(e-mail: a.p.nicholas@exeter.ac.uk)

Abstract: Results are presented from a combined field monitoring and numerical modelling study of flow dynamics within a floodplain recirculation zone. An array of four acoustic Doppler velocimeters (ADVs) was employed to monitor 3D flow velocities within a backwater site on the River Culm, Devon, UK. Patterns of measured mean velocity and turbulent kinetic energy are compared with the results of a numerical simulation of 3D flow carried out using the computational fluid dynamics software package FLUENT. Simulation results illustrate the presence of flow recirculation within the backwater zone, which is consistent with lateral flow convergence monitored in the field along the upstream portion of the channel–backwater interface. Both the simulation results and ADV data highlight the existence of a free-shear layer at the mixing interface between these flows, which is characterized by high levels of turbulent kinetic energy. However, simulated turbulent kinetic energy levels are significantly lower than those monitored in the field. In addition, ADV data illustrate deviations from isotropic turbulence that may not be simulated using a two-equation k–ε turbulence model. These results provide quantitative evidence of the hydraulic mechanisms responsible for promoting high rates of suspended sediment deposition in such backwater recirculation zones, and highlight several modelling issues that require further attention.

It is well known that lowland floodplains may act as sinks for suspended sediment deposited during periods of overbank flow (see Walling *et al.* 1996) and that floodplain sedimentation rates may spatially be highly variable (Kesel *et al.* 1974; Asselman & Middelkoop 1995). Very high rates of deposition typically occur in backwater sedimentation zones and former channel cut-offs (Lewis & Lewin 1983; Nicholas & Walling 1997*a*). Levees, drainage channels and floodplain depressions are also commonly sites of rapid sedimentation, whereas elevated floodplain regions and those areas located farther from the main channel tend to be characterized by low to negligible rates of deposition (Gretener & Strömquist 1987; Walling *et al.* 1992; Gomez *et al.* 1997).

The field studies cited above have done much to highlight the variability in floodplain deposition rates and to illustrate the significance of the relationships between complex floodplain topography, flow hydraulics and sediment deposition (see Lewin & Hughes 1980; Simm 1995; Nicholas & Walling 1998). However, efforts to provide an improved understanding of the processes involved in overbank sedimentation have been largely restricted to the use of physical and numerical models. Research conducted in laboratory flumes has identified the existence of a free-shear layer and momentum transfer across the interface between fast-moving channel water and slower floodplain flow (e.g. Shiono & Knight 1991; Sellin & Willetts 1996). The results of such experiments have been used to test the accuracy of numerical simulations of compound channel flow (e.g. Naot *et al.* 1993; Younis 1996; Sofialidis & Prinos 1998). Numerical models of suspended sediment transport driven by turbulent mixing at the channel–floodplain interface have also been constructed for simple compound channels (e.g. James 1985; Pizzuto 1987). Yet relatively few attempts have been made to model flow and sediment transport processes in topographically complex floodplain environments (see Nicholas & Walling 1997*b*, 1998).

The paucity of research conducted on natural floodplains, involving either monitoring or modelling of flood hydraulics and associated suspended sediment dynamics, may reflect three main factors: (1) the infrequent nature of flooding and logistical problems inherent in working during flood conditions; (2) a lack of appropriate field instrumentation, which has hindered accurate measurement of 3D floodplain flows; (3) in the case of numerical modelling, the complexity of topographic boundary conditions on natural

From: MARRIOTT, S. B. & ALEXANDER, J. (eds) 1999. *Floodplains: Interdisciplinary Approaches*. Geological Society, London, Special Publications, **163**, 15–26. 1-86239-050-9/99/$15.00

floodplains and, as a consequence of factors (1) and (2), the lack of high-resolution field datasets required for model operation and validation.

Here we present some results from a study of 3D flow hydraulics across the interface between the main channel and a backwater sedimentation zone on the River Culm, Devon. Patterns of mean and turbulent flow are investigated through a combination of field monitoring, using an array of four acoustic Doppler velocimeters (ADVs), and numerical modelling, using the computational fluid dynamics (CFD) software package FLUENT.

ADV system description

The ADV is a relatively new 3D flow monitoring instrument that has been deployed in a number of recent field and laboratory studies (e.g. Kraus *et al.* 1994; Nikora & Goring 1998; Sukhodolov *et al.* 1998). The ADV works by transmitting short acoustic pulses at a frequency of 10 MHz from a transmitter at the centre of the probe head (Fig. 1a). As each pulse propagates through the fluid, acoustic energy is reflected by air bubbles and small suspended particles. Three receivers are arranged around the transmitter and focus on a sampling volume *c.* 0.05 m below the probe head. These receivers detect acoustic signals reflected back from the sampling volume, which have a Doppler-shifted phase as a result of the velocity difference between the scatterers and the probe. The ADV monitors radial velocities along the bisector of the transmitted and reflected beam axes (Fig. 1c). These are converted into a Cartesian coordinate system by a transformation matrix determined empirically for each probe by the manufacturer.

The system employed in the current study incorporates a non-standard ADV configuration with the probe head mounted on a 0.007 m diameter, 0.15 m long stem that is attached by a 0.8 m flexible cable to the signal conditioning module (Fig. 1b). This module contains analogue electronics that detect the weak acoustic signals from the probe. The signal-conditioning module is connected by a 15 m cable to the ADV processor, which is mounted on a single circuit card installed in a portable personal computer. This card controls the ADV system and performs digital signal processing computations to estimate the three velocity components and record the signal properties at frequencies of up to 100 Hz.

Field data collection

Data were monitored at the interface between the main channel and a backwater sedimentation zone at a site on the River Culm, Devon, UK (Fig. 2a). Hydraulic data were monitored at this site during a period of gradually declining stage following a near-bankfull flow event (stage fell by *c.* 0.04 m during the measurement period). The ADVs were deployed using a custom-built aluminium wading rod, which was supported by a tripod. The wading rod was located upstream of the tripod and attached to it by a 1 m long steel rod. This arrangement provided maximum stability while ensuring that the tripod did not interfere with the flow. Four targets mounted on the wading rod were surveyed at each measurement location to allow the data to be located and reoriented into the chosen frame of reference (e.g. Lane *et al.* 1998). A total of 112 measurements was obtained at 14 vertical profiles along a 13.5 m transect, each containing 4–16 measurements depending upon flow depth (Fig. 2b). At each measurement position ADV data were collected for *c.* 4 min (24 576 data points) at a sampling rate of 100 Hz and using a sampling volume height of 9 mm.

Concurrent with measuring 3D flow velocities the ADV also outputs the signal amplitude, signal to noise ratio (Ω) and the correlation (r^2) between paired radial velocities measured by each receiver. The latter variables provide indices of data quality (Nortek 1997). During processing of velocity time-series, data points were rejected if an individual measurement exceeded the limits $r^2 < 0.6$ or $\Omega < 15$ dB, or if the average of all three measurements exceeded the limits $r^2 < 0.7$ or $\Omega < 25$ dB. Data were replaced using linear interpolation from adjacent values (see Lane *et al.* 1998). Digital filtering during post-processing of the time-series was used to eliminate the variance contribution from white noise at frequencies greater than 25 Hz. This was achieved using a low-pass, frequency domain, Chebyshev type-I filter that has a uniform response up to the cut-off frequency.

Numerical simulation

Numerical simulations were carried out using the FLUENT CFD package. This software has been used previously in modelling studies of flow hydraulics in single-thread rivers (e.g. Hodskinson 1996; Hodskinson & Ferguson 1998), braided streams (e.g. Nicholas & Sambrook Smith, 1999) and estuaries (e.g. Hardisty *et al.* 1996). FLUENT may be used to solve the Reynolds-averaged form of the Navier–Stokes equations within a 3D domain. The equations of mass and momentum conservation in the *i*th direction solved by FLUENT for steady-state conditions may be written in Cartesian tensor form as

$$\frac{\partial}{\partial x_i}(u_i) = 0 \qquad (1)$$

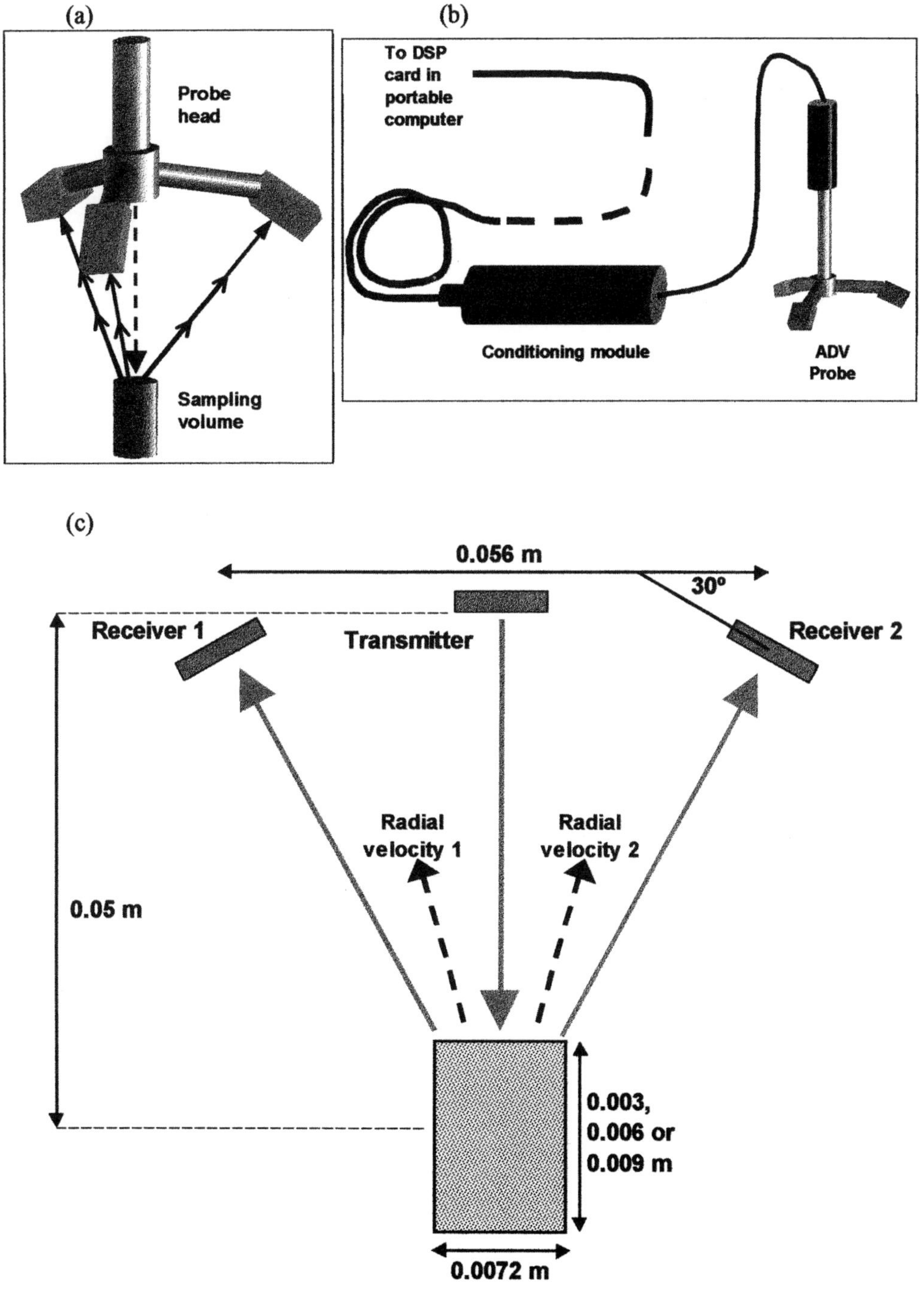

Fig. 1. (**a**) ADV probe head geometry and the position of the sampling volume. Dashed line shows transmitted beam path and continuous lines show reflected beam paths. (**b**) Schematic diagram showing ADV probe and signal conditioning module. (**c**) Two-dimensional view of the transmitter and two receiver elements showing the path of the transmitted beam and back-scattered signals and variable size of the sampling volume.

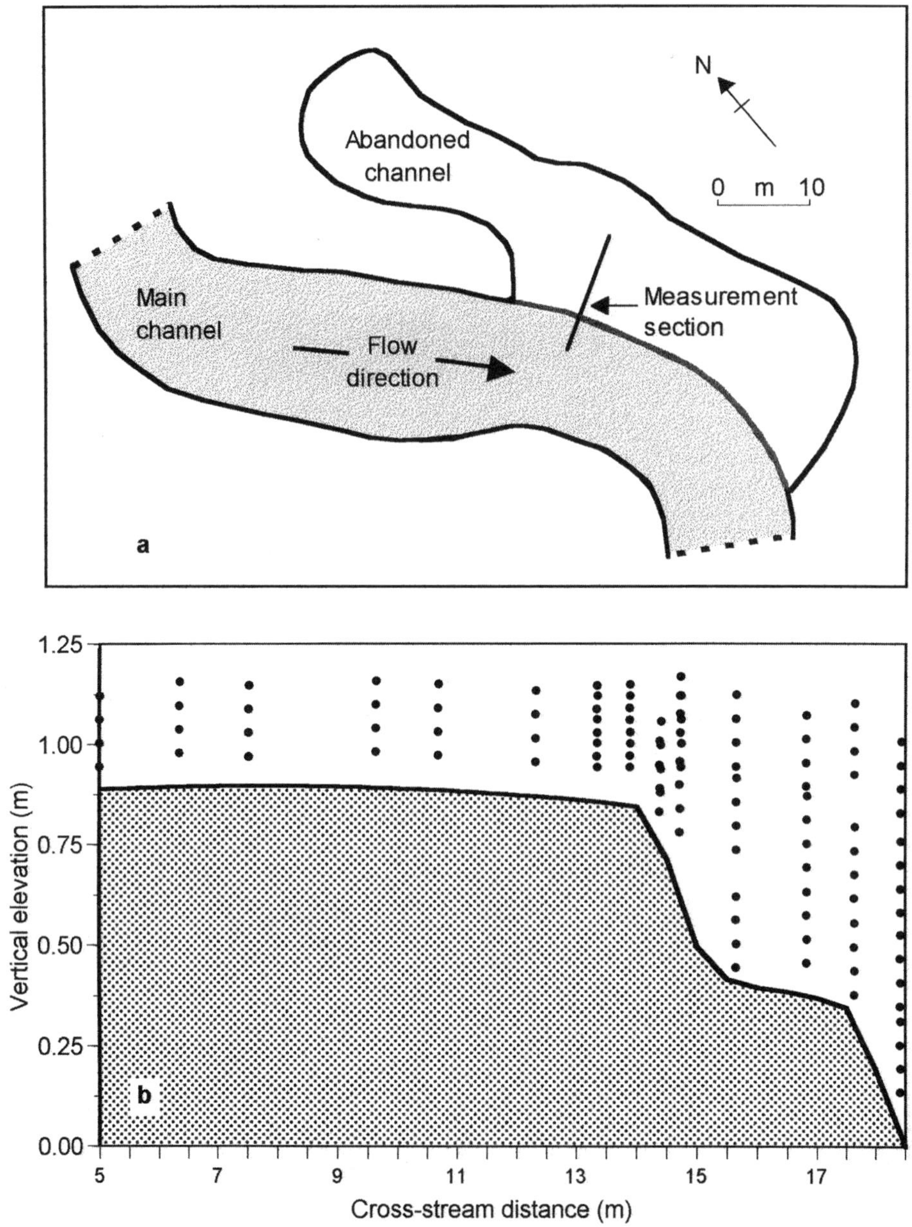

Fig. 2. (**a**) Schematic diagram showing the planform shape of the study reach and location of the measurement cross-section. (**b**) Distribution of data collection points across the measurement section.

$$\frac{\partial}{\partial x_j}(\rho u_i u_j) = \frac{\partial}{\partial x_j}\left[\mu\left(\frac{\partial u_i}{\partial x_j} + \frac{\partial u_j}{\partial x_i}\right) - \left(\frac{2}{3}\mu\frac{\partial u_l}{\partial x_j}\right)\right] - \frac{\partial p}{\partial x_i} + \frac{\partial}{\partial x_j}\left(-\rho\overline{u_i' u_j'}\right) \qquad (2)$$

where ρ is density, p is the static pressure, u_i is the velocity in the x_i direction, μ is the molecular viscosity and the final term in equation (2) represents the Reynolds stresses resulting from the decomposition of instantaneous velocities into their mean and fluctuating components. FLUENT models the Reynolds stresses using a standard k–ε model, RNG k–ε model (where k is the turbulent kinetic energy and ε is the energy dissipation rate) or full Reynolds stress model. The full Reynolds stress model allows a more detailed treatment of turbulence but involves a significant increase in computational effort. In the current study the RNG k–ε model was used because it is better suited than the standard k–ε model to situations involving flow separation and the existence of a free-shear layer. The RNG model employs the following transport equations for k and ε:

$$\frac{\partial}{\partial x_i}(u_i k) = \frac{\partial}{\partial x_i}\left(\alpha \, \nu_t \frac{\partial k}{\partial x_i}\right) + \nu_t S^2 - \varepsilon \qquad (3)$$

$$\frac{\partial}{\partial x_i}(u_i \varepsilon) = \frac{\partial}{\partial x_i}\left(\alpha \, \nu_t \frac{\partial \varepsilon}{\partial x_i}\right) + C_{1\varepsilon}\frac{\varepsilon}{k}\nu_t S^2 + C_{2\varepsilon}\frac{\varepsilon^2}{k}\varepsilon \qquad (4)$$

where $C_{1\varepsilon} = 1.42$, $C_{2\varepsilon} = 1.68$ (see Rodi 1980), S is the modulus of the mean rate of strain tensor S_{ij}, α is the inverse Prandtl number for turbulent transport, v_t is the eddy viscosity and R is a rate-of-strain term. Further details on the development and application of k–ε turbulence closures have been given by Launder & Spalding (1974), Rodi (1980) and Yakhot & Orszag (1986).

The model results reported herein were obtained using the fixed-lid version of FLUENT. In this approach the water surface must be specified as a boundary condition and is not free to adjust as part of the solution. A mesh composed of *c.* 90 000 hexahedral cells was constructed using high-resolution survey data, obtained with an electronic theodolite and electronic distance measurement (EDM) unit, to specify the geometry of the channel bed, floodplain surface and the water surface. The discharge at the entrance to the study reach must be specified as a boundary condition and this was determined using the established Environment Agency stage–discharge relationship for the River Culm at Rewe. Distributed velocity information at the boundary was generated by assuming that the flow conformed to the law of the wall. To overcome any uncertainty in this assumption the inlet boundary was located well upstream of the recirculation zone under investigation.

At cells adjacent to the channel bed FLUENT calculates hydraulic variables using a semi-logarithmic wall-function of the form

$$u \, \frac{\rho C_\mu^{1/4} k^{1/2}}{\tau} = \frac{}{\kappa} ln\left[\frac{\rho C_\mu^{1/4} k^{1/2} E y}{\mu}\right] - \Delta B \qquad (5)$$

where u and k are the flow velocity and turbulent kinetic energy at height y above the bed, κ is von Kármán's constant, E is an empirical coefficient that takes a value of 9.79 (Hodskinson 1996; Fluent, Inc. 1998), C_μ is a turbulence model coefficient that takes a value of 0.0845 (Yakhot & Orszag 1986) and ΔB is a roughness factor that is determined as a function of the roughness length scale k_s using the equations of Cebeci & Bradshaw (1977). For hydrodynamically rough boundaries ($k_s^+ > 90$ where $k_s^+ = k_s \rho u_* / \mu$ and u_* is the shear velocity) such as those in the current study ΔB is given by

$$\Delta B = (1/\kappa)\ln(1 + C_{ks} k_s^+)$$

where C_{ks} is a roughness constant that takes a value of 0.5 (Fluent, Inc. 1998). The roughness length scale (k_s) employed in the model wall-function at cells adjacent to the channel bed and floodplain surface must be specified as a boundary condition. However, considerable uncertainty surrounds the selection of an appropriate value for k_s. Bridge & Dominic (1984) recommends $k_s = 0.5D$ for uniform size sediments, whereas studies conducted for gravel mixtures and incorporating both grain and form roughness have suggested $k_s = 3.5D_{84} = 6.8D_{50}$ (Bray 1982). Treatment of roughness effects for vegetated surfaces also vary widely. Masterman & Thorne (1992) calculated an effective roughness height based upon the vegetation stiffness index of Kouwen (1988); however, this variable is intended for use in mean velocity calculations. In contrast, Naot *et al.* (1996) proposed an alternative wall-function to model vegetation effects. Such an approach is beyond the scope of the current version of FLUENT. Clearly, considerable uncertainty surrounds the specification of roughness values, particularly in the case of vegetated surfaces, and further research is required into this issue. In the current study a uniform k_s value of 20 mm (equal to the D_{50} of the channel-bed sediment as recommended by Fluent, Inc. (1998)) was applied throughout the study reach. This should be considered as a first approximation for k_s. However, it should be noted that the dominant features of both the monitored and modelled flow field described below are associated with the mixing interface between the main channel and backwater flow regions, rather than with turbulence production at the channel boundaries. In addition, spatial variations in boundary roughness contributed by roughness elements coarser than the model mesh are defined by the detailed topographic survey. Consequently, the effects of spatial uncertainty in the specification of k_s values are considered to be minimal.

Results

Figure 3 shows the simulated planform pattern of mean horizontal velocity within the study reach at the water surface (contours indicate the velocity

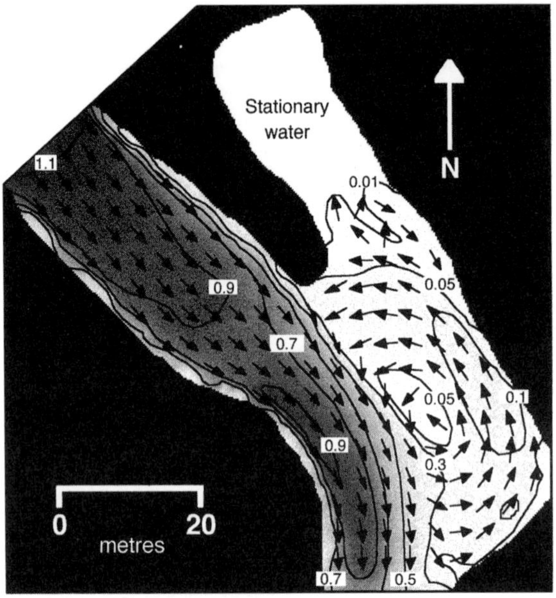

Fig. 3. Pattern of simulated horizontal mean velocity (m s^{-1}) within the study reach at the water surface. Contours indicate velocity magnitude, vectors indicate flow direction.

magnitude and vectors show the flow direction). Maximum downstream velocities of 1.05–1.1 m s^{-1} occur in the centre of the channel at the entrance to the study reach. Heading downstream this high-velocity fluid decelerates to *c.* 0.9–0.95 m s^{-1} as the flow depth increases and moves across to the right-hand bank on the opposite side of the channel to the cut-off. Flow within the backwater is characterized by recirculation of water, which enters the cut-off at its downstream margin, turns upstream, flowing fastest at the back of the cut-off (>0.1 m s^{-1}), and then returns to the main channel at the upstream end of the backwater. Velocity magnitudes within the cut-off are generally <0.1 m s^{-1} and decline to zero at the centre of the circulating flow and at the upstream limit of the backwater, where a former channel segment is occupied by standing water. This pattern of flow recirculation is consistent with the results of flume studies of open-channel dead zones (e.g. Kimura & Hosoda 1997).

Figure 4a and b shows the monitored and modelled patterns of mean 3D velocity across the measurement section marked in Fig. 2a. It should be noted that the main channel extends for a further 8 m to the right of this section and has a maximum depth of *c.* 1.5 m. Data have been reoriented to yield one component perpendicular to the section (termed streamwise) and two components within the plane of the section (termed lateral and vertical). Both the field data and model results show similar patterns of streamwise velocity, with maximum values of >0.5 m s^{-1} occurring at the edge of the main channel and a steep lateral velocity gradient between the channel and backwater zone. Streamwise velocities decline into the cut-off, where a region of weak negative (i.e. upstream) flow occurs. In general, both modelled and monitored streamwise velocities exhibit similar patterns; however, the region of upstream flow within the backwater is stronger and more extensive for the model output than that measured in the field. In addition, vertical gradients of streamwise velocity in the deeper flow are greater for the monitored velocity data than for the simulation results, possibly suggesting that the bed roughness length scale employed in the model has been underestimated.

Both the modelled and monitored lateral velocities illustrate a pattern of flow convergence at the interface between the main channel and the backwater zone, which is consistent with the pattern of flow recirculation identified in Fig. 3. However, significant differences between simulated and measured velocities are evident. First, left to right flow in the backwater is considerably weaker in the field (<0.03 m s^{-1}) than in the simulation results (0.03–0.06 m s^{-1}). Second, the location of the convergence differs slightly between measurements and model results. In the former, the zone of flow convergence coincides

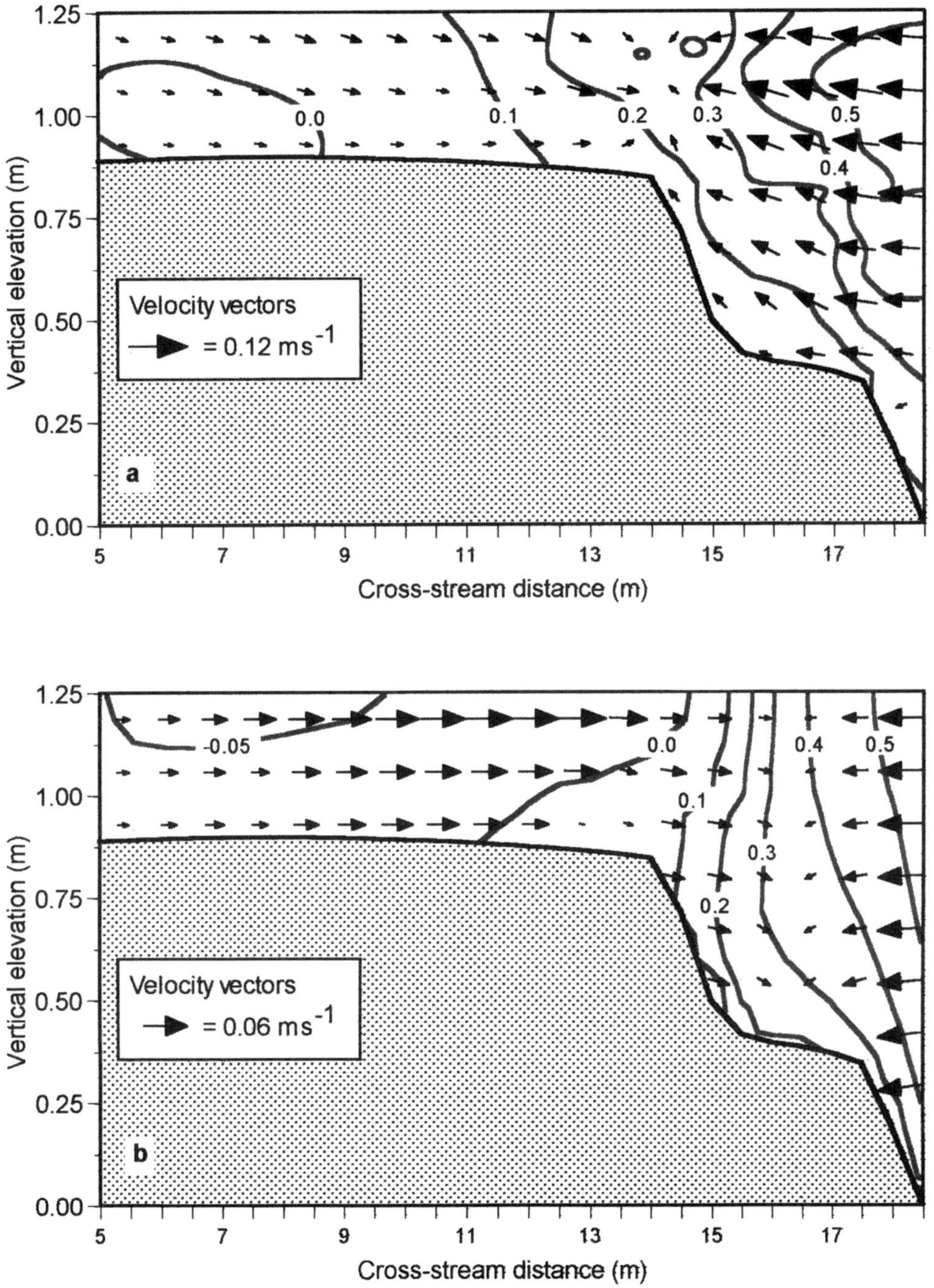

Fig. 4. Pattern of 3D mean velocity (m s^{-1}) at the measurement section: (**a**) monitored; (**b**) simulated. Contours indicate streamwise velocities, vectors indicate lateral and vertical velocities.

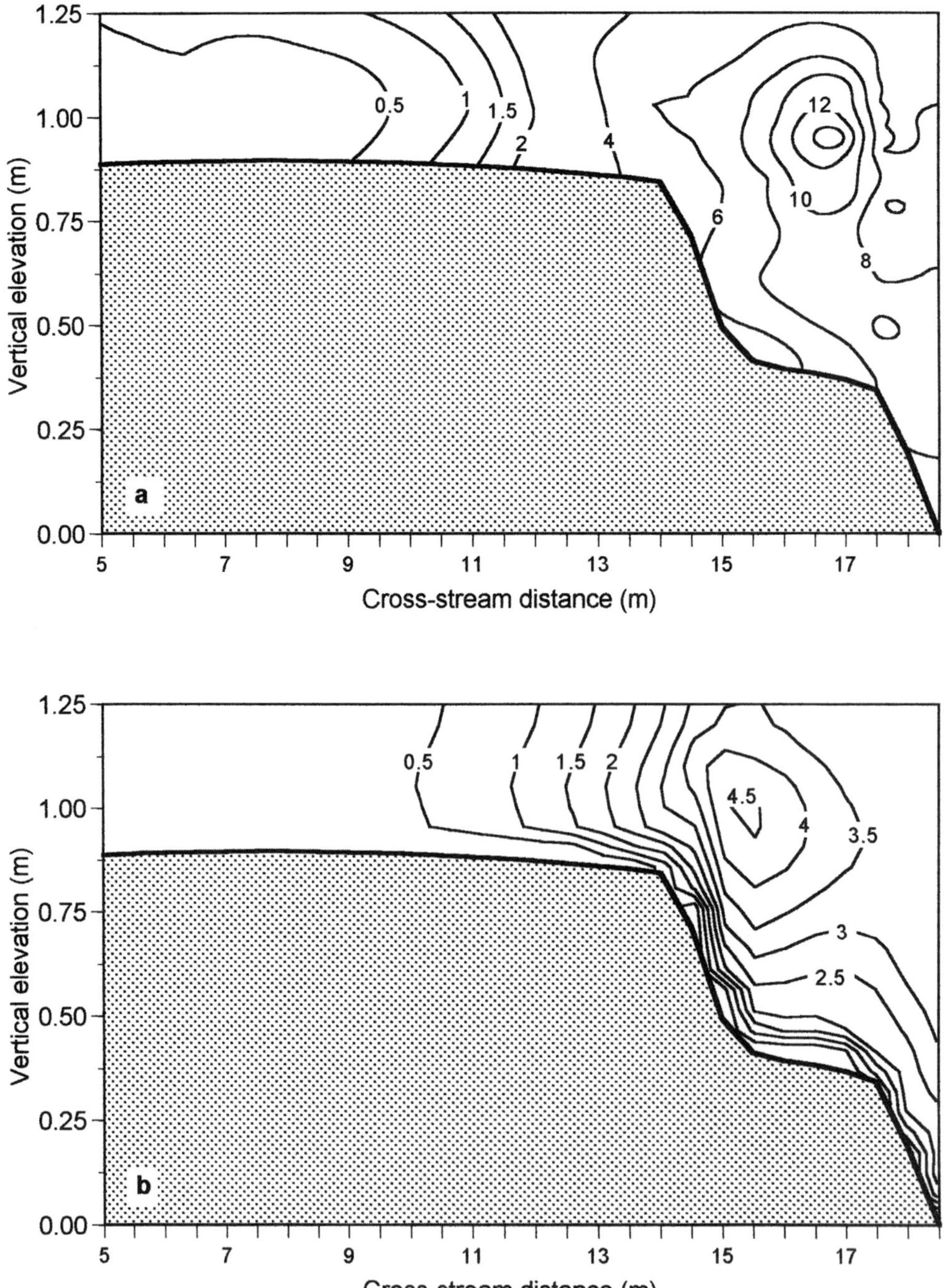

Fig. 5. Pattern of turbulent kinetic energy ($m^2 s^{-2} \times 10^{-3}$) at the measurement section: (**a**) monitored; (**b**) simulated.

with the rapid increase in depth at the left hand margin of the backwater zone. In contrast, the zone of convergence is displaced by 2 m towards the main channel in the model output.

Figure 5a and b shows the monitored and modelled patterns of turbulent kinetic energy across the measurement section. Turbulent kinetic energy (k) is defined as

$$k = 0.5\left(\overline{u'^2} + \overline{v'^2} + \overline{w'^2}\right) \qquad (7)$$

where $\overline{u'^2}$, $\overline{v'^2}$, $\overline{w'^2}$ are the mean square values of the velocity fluctuations in the streamwise, lateral and vertical directions respectively. In the field situation a core of high kinetic energy (0.010–0.015 $m^2 s^{-2}$) occurs at the interface between the shallow backwater flow and the deeper main channel flow. The results of the simulation also exhibit a core of high kinetic energy at the juncture between the two flow regions, although it is displaced slightly further to the left of the core shown in Fig. 5(a). In addition, it should be noted that the maximum simulated kinetic energy levels are 0.004–0.005 $m^2 s^{-2}$, which is 2–3 times lower than those monitored in the field. Lower than observed kinetic energy levels near the bed may suggest that the value of the roughness length scale employed in the numerical model is too low. However, the core of high kinetic energy is located well above the bed and is associated with the shear layer rather than with bed-generated turbulence.

Discussion

The monitored and modelled flow patterns presented in this study provide quantitative evidence of the hydraulic mechanisms responsible for promoting high rates of suspended sediment deposition in such backwater zones. The simulation results illustrate flow recirculation within the backwater that is consistent with weak upstream velocities and lateral flow convergence monitored in the field. This circulation will drive advective suspended sediment transport into the backwater region. Both the simulation results and ADV data are characterized by a core of high turbulent kinetic energy, which is indicative of the existence of a free-shear layer between the main channel and backwater zone. Such a shear layer at the channel–floodplain interface has formed the basis of much recent research into compound channel hydraulics (e.g. Shiono & Knight 1991; Sellin & Willetts 1996), but has yet to be monitored in detail in natural floodplain environments. Turbulent mixing along the shear layer will drive suspended sediment transport into the cut-off, so enhancing sediment supply to the backwater resulting from flow recirculation.

Figure 6 shows the monitored pattern of the r.m.s. of the vertical velocity at the measurement section. Comparison of these data with the monitored turbulent kinetic energy shown in Fig. 5a highlights the close relationship between vertical

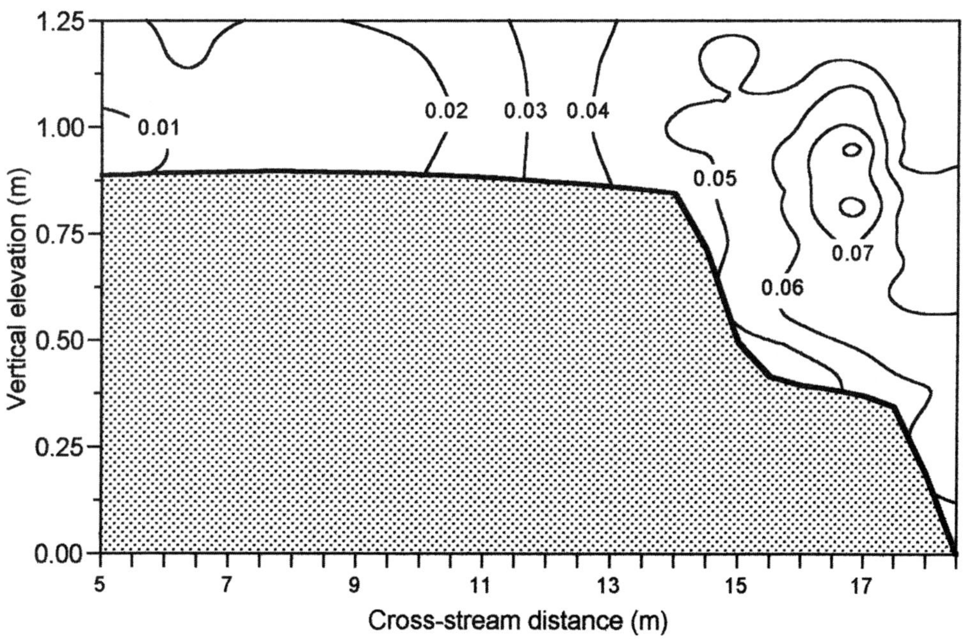

Fig. 6. Pattern of monitored vertical velocity r.m.s. (m s^{-1}) at the measurement section.

turbulence intensities and kinetic energy levels. Despite this, the lateral gradient of the r.m.s. of the vertical velocity is somewhat weaker than that of the kinetic energy, indicating that fluctuations in streamwise and lateral velocities contribute increasingly to turbulent energy levels within the shear layer. In combination with the patterns of monitored mean vertical velocity shown in Fig. 4a the data illustrated in Fig. 7 provide further evidence to suggest that the backwater zone will be characterized by relatively high rates of sedimentation. In the main channel, and on the narrow shelf between the channel and the backwater, the mean flow rises away from the bed and vertical turbulence intensities (as indexed by the vertical r.m.s.) are relatively high. Both these factors will oppose the settling of fine particles and so keep sediment in suspension. In contrast, within the backwater, mean vertical velocities are weak and directed towards the bed. Furthermore, turbulence intensities are almost an order of magnitude lower than within the shear layer. Consequently, sediment transported into the backwater by advective and diffusive mechanisms will settle rapidly, thus promoting high rates of deposition.

The simulation results and the field data obtained using the ADV illustrate similar patterns of 3D flow, with both modelled and monitored hydraulics being characterized by lateral flow convergence and the existence of a free-shear layer at the channel–backwater interface. However, despite the broad consistency between simulated and monitored flow patterns, a number of significant differences are evident between the two. These are largely associated with the location of the zone of flow convergence between the main channel and backwater flow and the magnitude of the core of turbulent kinetic energy that characterizes the shear layer between these two regions. As noted above, inaccurate specification of the boundary roughness length scale k_s may provide a partial explanation for these differences. However, the use of a two-equation k–ε turbulence closure rather than a more sophisticated Reynolds stress model may be more significant. Reynolds stress models solve transport equations for each of the normal and shear stresses resulting from turbulent velocity fluctuations, hence they provide a more complete representation of turbulent processes and allow the quantification of turbulence anisotropy. Figure 7 shows the monitored pattern of turbulence anisotropy between the lateral and vertical flow components ($\overline{v^2} - \overline{w^2}$) at the measurement section. Within the backwater, lateral and vertical normal Reynolds stresses are approximately equal. However, the interface between the channel and backwater flows is characterized by regions of strong positive anisotropy at the bed and water surface, and a region of negative anisotropy that intrudes from the main channel into the shear layer immediately below the core of high kinetic energy. Naot *et al.* (1993) and Younis (1996) have shown that accurate simulation

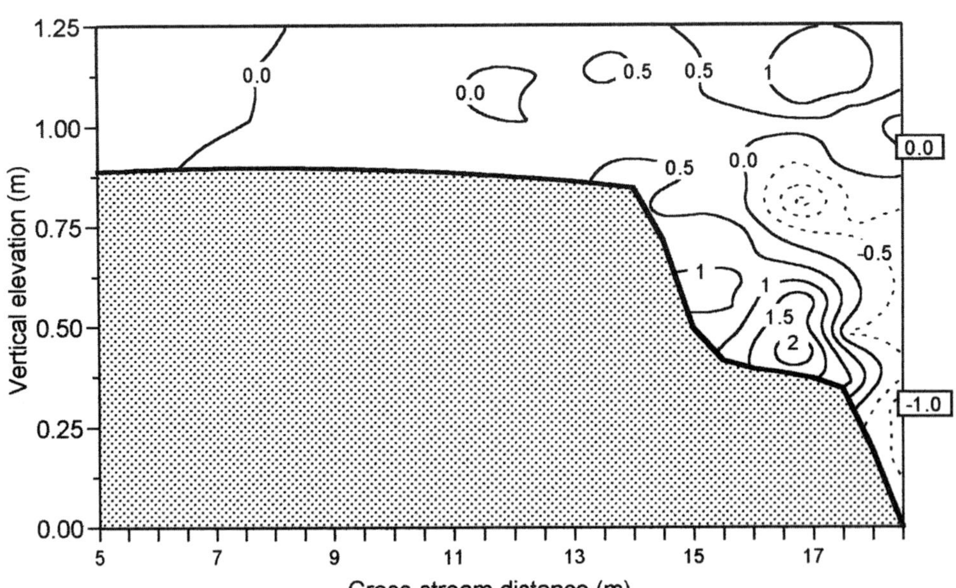

Fig. 7. Pattern of monitored ($\overline{v'^2} - \overline{w'^2}$) turbulence anisotropy ($m^2\ s^{-2} \times 10^{-3}$) at the measurement section.

of anisotropic turbulence in compound channels can be achieved using either algebraic or differential Reynolds stress models. Anisotropic flow mechanisms may strongly influence both the position of the flow convergence and the production of turbulent kinetic energy. Hence, the use of a more sophisticated turbulence closure might provide an improved simulation of shear-layer dynamics at the study site.

Summary

Results have been presented from a combined modelling and monitoring study of 3D flow hydraulics within a floodplain backwater zone. Simulation results illustrate the presence of flow recirculation within the backwater that is consistent with lateral flow convergence monitored in the field along the upstream portion of the channel–backwater interface. Both the simulation results and monitored velocity data highlight the existence of a free-shear layer at the mixing interface between these flows, which is characterized by high levels of turbulent kinetic energy. These results provide quantitative evidence of the hydraulic mechanisms responsible for promoting suspended sediment supply to, and rapid sedimentation within, such backwater recirculation zones.

Future modelling studies of floodplain hydraulics should be carried out to address two important influences on simulation results. First, more information is required on the specification of boundary roughness coefficients for vegetated surfaces and on the sensitivity of model output to roughness parameterization. Second, research should be conducted to identify the potential improvements in model accuracy resulting from the use of a full Reynolds stress model rather than a two-equation k–ε turbulence closure. It is well known that the representation of turbulence anisotropy is essential for accurate simulation of flow hydraulics in simple compound channels. However, in natural floodplain environments that are characterized by complex topography and boundary roughness the importance of adopting a more sophisticated turbulence closure has yet to be evaluated. Such an assessment will only be possible with the aid of appropriate field datasets. In this respect the acoustic Doppler velocimeter represents a valuable new instrument capable of monitoring high-frequency 3D velocities, thus allowing the full decomposition of shear and normal Reynolds stress terms.

This work was funded by NERC Grant GR3/10962 to A.P.N. and Des Walling. The authors are grateful to local landowners for allowing access to the floodplain of the River Culm, Devon.

References

Asselman, N. E. M. & Middlekoop, H. 1995. Floodplain sedimentation: quantities, patterns and processes. *Earth Surface Processes and Landforms*, **20**, 481–499.

Bray, D. I. 1982. Flow resistance in gravel-bed rivers. *In*: Hey, R. D., Bathurst, J. C. & Thorne, C. R. (eds) *Gravel-Bed Rivers*. Wiley, Chichester, 109–133.

Bridge, J. S. & Dominic, D. F. 1984. Bed load grain velocities and sediment transport rates. *Water Resources Research*, **7**, 304–310.

Cebeci, T. & Bradshaw, P. 1977. *Momentum Transfer in Boundary Layers*. Hemisphere, New York.

Fluent, Inc. 1998. *FLUENT/UNS Users Guide*. Fluent, Inc., Lebanon, NH.

Gomez, B., Phillips, J. D., Magilligan, F. J. & James, L. A. 1997. Floodplain sedimentation and sensitivity: summer 1993 flood, Upper Mississippi River Valley. *Earth Surface Processes and Landforms*, **22**, 923–936.

Gretener, B. & Strömquist, L. 1987. Overbank sedimentation rates of fine grained sediments: a study of recent deposition in the Lower River Fryisan. *Geografiska Annaler*, **69A**, 139–146.

Hardisty, J., Rouse, H. L. & Hughes, N. E. 1996. On the origins of large-scale longitudinal flow structures in the Outer Humber Estuary. *In*: Ashworth, P. J., Bennett, S. J., Best, J. L. & McLelland, S. J. (eds) *Coherent Flow Structures in Open Channels*. Wiley, Chichester, 681–704.

Hodskinson, A. 1996. Computational fluid dynamics as a tool for investigating separated flow in river bends. *Earth Surface Processes and Landforms*, **21**, 993–1000.

—— & Ferguson, R. I. 1998. Numerical modelling of separated flow in meander bends: model testing and experimental investigation of geometric controls on the extent of flow separation at the concave bank. *Hydrological Processes*, **12**, 1323–1338.

James, C. S. 1985. Sediment transfer to overbank sections. *Journal of Hydraulic Research*, **23**, 435–452.

Kesel, R. H., Dunne, K. C., McDonald, R. C., Allison, K. R. & Spicer, B. E. 1974. Lateral erosion and overbank deposition on the Mississippi River in Louisiana caused by 1973 flooding. *Geology*, **2**, 461–464.

Kimura, I. & Hosoda, T. 1997. Fundamental properties of flow in open-channels with dead zone. *Journal of Hydraulic Engineering*, **123**, 98–107.

Kouwen, N. 1988. Field estimation of the biomechanical properties of grass. *Journal of Hydraulic Research*, **26**, 559–568.

Kraus, N. C., Lohrmann, A. & Cabrera, R. G. 1994. New acoustic meter for measuring 3D laboratory flows. *Journal of Hydraulic Engineering*, **120**, 406–412.

Lane, S.N., Biron, P.M., Bradbrook, K.F. *et al.* 1998. Three-dimensional measurement of river channel flow processes using acoustic Doppler velocimetry. *Earth Surface Processes and Landforms*, **23**, 1247–1267.

Launder, B. E. & Spalding, D. B. 1974. The numerical computation of turbulent flow. *Computational

Methods in Applied Mechanics and Engineering, **3**, 269–289.

LEWIN, J. & HUGHES, D. 1980. Welsh floodplain studies II: application of a qualitative inundation model. *Journal of Hydrology*, **46**, 35–49.

LEWIS, G. W. & LEWIN, J. 1983. Alluvial cut-offs in Wales and the Borderlands. *In*: COLLINSON, J. D. & LEWIN, J. (eds) *Modern and Ancient Fluvial Systems*. International Association of Sedimentologists Special Publication, **6**, 145–154.

MASTERMAN, R. & THORNE, C. R. 1992. Predicting influence of bank vegetation on channel capacity. *Journal of Hydraulic Engineering*, **118**, 1052–1058.

NAOT, D., NEZU, I. & NAKAGAWA, H. 1993. Hydrodynamic behaviour of compound rectangular open channels. *Journal of Hydraulic Engineering*, **119**, 15–33.

——, —— & —— 1996. Hydrodynamic behaviour of partly vegetated open channels. *Journal of Hydraulic Engineering*, **122**, 625–633.

NICHOLAS, A. P. & SAMBROOK SMITH, G. H. 1999. Numerical simulation of three-dimensional flow hydraulics in a braided channel. *Hydrological Processes*, **13**, 913–929.

—— & WALLING, D. E. 1997*a*. Investigating spatial patterns of medium-term overbank sedimentation on floodplains: a combined numerical modelling and radiocaesium-based approach. *Geomorphology*, **19**, 133–150.

—— & —— 1997*b*. Modelling flood hydraulics and overbank deposition on river floodplains. *Earth Surface Processes and Landforms*, **22**, 59–77.

—— & —— 1998. Numerical modelling of floodplain hydraulics and suspended sediment transport and deposition. *Hydrological Processes*, **12**, 1339–1355.

NIKORA, V. I. & GORING, D. G. 1998. ADV measurements of turbulence: can we improve their interpretation? *Journal of Hydraulic Engineering*, **124**, 630–634.

NORTEK 1997. *ADV Operation Manual*. Nortek AS, Vollen, Norway.

PIZZUTO, J. E. 1987. Sediment diffusion during overbank flows. *Sedimentology*, **34**, 301–317.

RODI, W. 1980. *Turbulence Models and their Application in Hydraulics—a State of the Art Review.* IAHR (International Association of Hydraulic Research), Delft.

SELLIN, R. H. J. & WILLETTS, B. B. 1996. Three-dimensional structures, memory and energy dissipation in meandering compound channel flow. *In*: ANDERSON, M. G., WALLING, D. E. & BATES, P. D. (eds) *Floodplain Processes*. Wiley, Chichester, 255–298.

SHIONO, K. & KNIGHT, D. W. 1991. Turbulent open channel flows with variable depth across the channel. *Journal of Fluid Mechanics*, **222**, 617–646.

SIMM, D. J. 1995. The rates and patterns of overbank deposition on a lowland floodplain. *In*: FOSTER, I. D., GURNELL, A. M. & WEBB, B. W. (eds) *Sediment and Water Quality in River Catchments*. Wiley, Chichester, 247–264.

SOFIALIDIS, D. & PRINOS, P. 1998. Compound open-channel flow modelling with nonlinear low-Reynolds k–ε models. *Journal of Hydraulic Engineering*, **124**, 253–262.

SUKHODOLOV, A., THIELE, M. & BUNGARTZ, H. 1998. Turbulence structure in a river reach with sand bed. *Water Resources Research*, **34**, 1317–1334.

WALLING. D. E., HE, Q. & NICHOLAS, A. P. 1996. Floodplains as suspended sediment sinks. *In*: ANDERSON, M. G., WALLING, D. E. & BATES, P. D. (eds) *Floodplain Processes*. Wiley, Chichester, 399–440.

——, QUINE, T. A. & HE, Q. 1992. Investigating contemporary rates of floodplain sedimentation. *In*: CARLING, P. A. & PETTS, G. E. (eds) *Lowland Floodplain Rivers: Geomorphological Perspectives*. Wiley, Chichester, 165–184.

YAKHOT, V. & ORSZAG, S. A. 1986. Renormalization group analysis of turbulence. I. Basic theory. *Journal of Scientific Computing*, **1**, 1–51.

YOUNIS, B. A. 1996. Progress in turbulence modelling for open-channel flows. *In*: ANDERSON, M. G., WALLING, D. E. & BATES, P. D. (eds) *Floodplain Processes*. Wiley, Chichester, 299–332.

Flood behaviour of the Burdekin River, tropical north Queensland, Australia

JAN ALEXANDER[1], CHRIS R. FIELDING[2] & GEOFF D. POCOCK[3]

[1]*School of Environmental Sciences, University of East Anglia, Norwich NR4 7TJ, UK (e-mail: j.alexander@uea.ac.uk)*

[2]*Department of Earth Sciences, University of Queensland, Qld. 4072, Australia*

[3]*Queensland Department of Natural Resources, Ayr, Qld. 4807 Australia*

Abstract: Strongly seasonal rainfall in north Queensland, Australia, causes extremely variable fluvial discharge, and tropical cyclones commonly produce high-magnitude, short-duration floods. In the Burdekin River catchment (129 500 km^2 of eastern Queensland), rapid runoff and channel floodplain configuration lead to very fast flood-wave speeds (up to 4.34 m s^{-1} calculated). The large catchment and varying tropical cyclone paths result in extreme variations in flood history at different sites. Changes in channel conditions (bed elevation, vegetation, bedforms, sediment surface character) in successive events, combined with complex rainfall patterns and highly unsteady flow, make flood estimates extremely difficult. Because of the short duration of flow events and very rapid rate of change in discharge during individual events, sediment transport rate, bedforms, and bed elevation change continuously. For most of the year the channel is occupied by an underfit stream, but this small perennial stream does not significantly rework the deposits or change channel form before the next major discharge event. Consequently, there is little channel form recovery between major discharge events. Successive events have very variable size and consequently the bedforms and channel morphology are normally out of equilibrium with the flow, and record the result of the most recent major events. High-magnitude floods transport coarse sediment onto the floodplain and the short duration of flooding and rapid runoff results in relatively low preservation of mud in the channel or overbank environment. The Holocene flood deposits include thick, coarse sand and medium sand units and generally thin, fine-grained units.

Conventional views of floodplain inundation and channel behaviour have developed from studies predominantly in northern hemisphere mid-latitudes, and the literature is dominated by studies of small and/or perennial streams and by theoretical studies of equilibrium conditions. In most of these cases channel form can be related to a dominant ('formative') or bankfull discharge, which in many cases approximates to the mean annual flood (Ackers 1992) or discharges with a recurrence interval of about 1–5 years (Williams 1978). In these settings channel form recovers from more extreme events over a few years. However, in small arid catchments channel recovery may take far longer or there may be non-recovery such that the channel form changes progressively with successive flood events (Wolman & Gerson 1978). Likewise in environments with infrequent very high magnitude discharge events, channels may not recover between large events. This paper documents a large, semi-arid to sub-humid catchment where highly variable seasonal discharge results in non-recovery of channel form between major events. Brown (1996) and others have recorded that bankfull capacity of any river is very variable in space and time; this is particularly pertinent to cases such as the Burdekin, where channel form may not recover between successive events. In this example, surface morphology, vegetation and deposits are a record of the most recent event or events.

Flood duration is as important as flood magnitude in determining the degree of channel and floodplain modification (Howard 1996) and bedload transport (Lewin 1989). This is particularly the case in the sub-humid tropical area of north Queensland, where overbank flooding is less frequent than in most natural temperate climate rivers but where individual floods are very large and of short duration. In this setting the duration of individual events is too short for channel–floodplain systems to attain equilibrium conditions

From: MARRIOTT, S. B. & ALEXANDER, J. (eds) 1999. *Floodplains: Interdisciplinary Approaches*. Geological Society, London, Special Publications, **163**, 27–40. 1-86239-050-9/99/$15.00

during a single event, and successive floods are of very variable magnitude.

Because of the low frequency and high variability of flood events in north Queensland rivers, two events with similar discharge may flow over very different bed conditions (vegetation, bedforms and bed and bank topography) and consequently have very different flow properties.

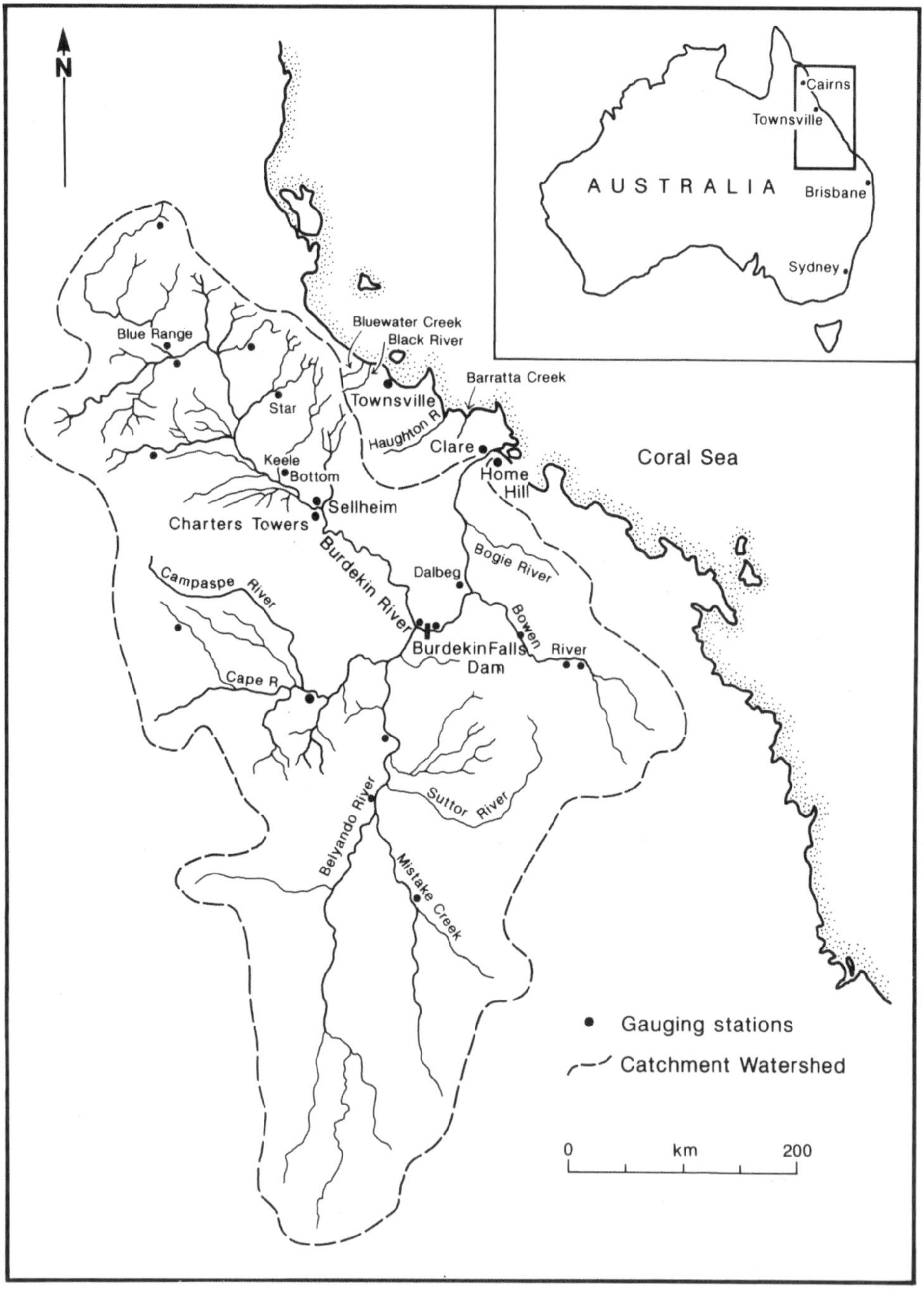

Fig. 1. Map of the Burdekin catchment showing major tributaries, locations of gauging stations and other rivers referred to in this paper.

Fig. 2. (**a**) The Burdekin River at the Inkerman Bridge, July 1998 (dry season). (Note road traffic for scale.) To the left of the Inkerman Bridge are remnants of an earlier bridge damaged by flooding. In the distant background (beyond the sugar mill) is the town of Home Hill. (**b**) The Burdekin Falls Dam during flood. (**c**) Home Hill looking down the main street towards the south end of town in February 1991. (**d**) Flooding near Barratta Creek (treeline in foreground) caused by local intense rainfall associated with Tropical Cyclone Sid. The photograph was taken on 9 January 1998.

As with other catchments with unsteady flow and complex topography and land use, it is difficult or impossible to obtain a unique stage–discharge curve (see Knight & Shiono 1996). Different reaches of the rivers may flood with different recurrence intervals depending on how quickly they adjust to the discharge conditions during any individual flow event, and may have different flood histories as intense rainfall can be very localized and different tributaries may flow at different times.

This paper presents observations from the Burdekin River, North Queensland (Figs 1 and 2), particularly for the 1991 and 1998 flood events, in an attempt to illustrate the problem of understanding and predicting flooding and flood deposition in settings with extreme discharge variability. Some of the Burdekin overbank deposits are described, and their variability and coarse-grained character may be explained by the highly variable discharge behaviour.

The Burdekin River

The Burdekin River drains a large tropical catchment (129 500 km^2) of coastal and inland Queensland, with headwaters in the Great Dividing Range and coastal highlands (Fig. 1). The river ultimately flows across the rich arable land of the Burdekin Delta into the Coral Sea. The coastal area of the catchment is humid (average annual rainfall of 1000–1200 mm) whereas most of the catchment is semi-arid (500–700 mm) and there is significant interannual variability of rainfall throughout the catchment. Tropical cyclone activity between January and April accounts for most of the annual rainfall and causes major flood events. In the summer wet season the rain is frequently very intense, resulting in rapid runoff and a flashy hydrograph (Fig. 3). The Burdekin River's discharge at the Clare gauging station (Fig. 1) varies from long periods of little or no surface flow to peak discharges of more than 35 900 m^3 s^{-1}, with a mean annual maximum discharge of 9784.4 m^3 s^{-1} calculated from 48 years of data.

The short duration of discharge events in the Burdekin system is remarkable, considering the size of the catchment and flood magnitudes. For comparison, the 1993 Mississippi flood peaked at 28 840 m^3 s^{-1} at St Louis and flood conditions lasted about 146 days (Larson 1995), and the

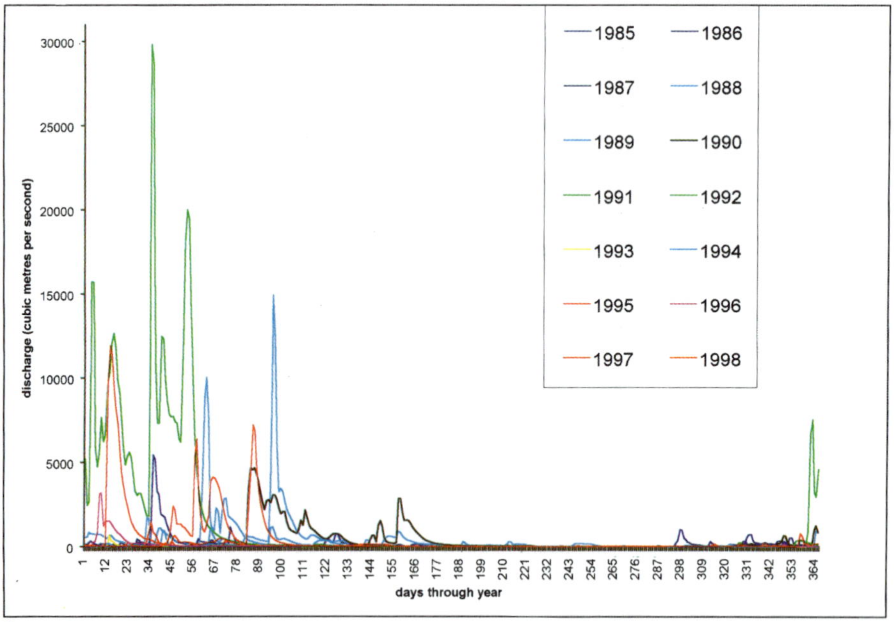

(a)

Fig. 3. (**a**) Superimposed hydrographs for consecutive years from 1985 to 1998 recorded at Clare to demonstrate the strongly seasonal variation in discharge and the flashy nature of the events. (**b**) Graph of the annual maximum series from 1916 to 1998 at Charters Towers, Sellheim, Home Hill and Clare (location shown in Fig. 1). The stations at Home Hill and Charters Towers were not gauged after the mid-1950s, at about which time gauging started at Sellheim and Clare. (**c**) Hydrographs for 1 February–3 March 1991 recorded at six gauging stations, the locations of which are shown in Fig. 1. The data for Burdekin Falls were recorded at longer time intervals than those at the other sites, causing the more angular curve.

1993/1994 Rhine flood reached 11 100 $m^3 s^{-1}$ at Emmerich and flooding continued for many days (Engel *et al.* 1994). In contrast, flood conditions in the Burdekin system rarely last longer than a few days (see below) and although the total runoff is not spectacular, rapid routing leads to enormous flood peaks. The short-duration, high-magnitude discharge events result in repeated entrainment, transport and rapid deposition of very large volumes of sediment within and beyond the channel system (Fielding & Alexander 1996; Alexander & Fielding 1997; Fielding *et al.* 1997, 1999). Several million tonnes of sediment may be transported to the Coral Sea each day of a major flood. For example, Belperio (1979) estimated that 8.4×10^9 kg of sediment were delivered to the Burdekin Delta during 24 h at the peak of the March 1946 flood.

The subaerial part of the Burdekin Delta (*c.* 2112 km^2) has subtle topographic and soil

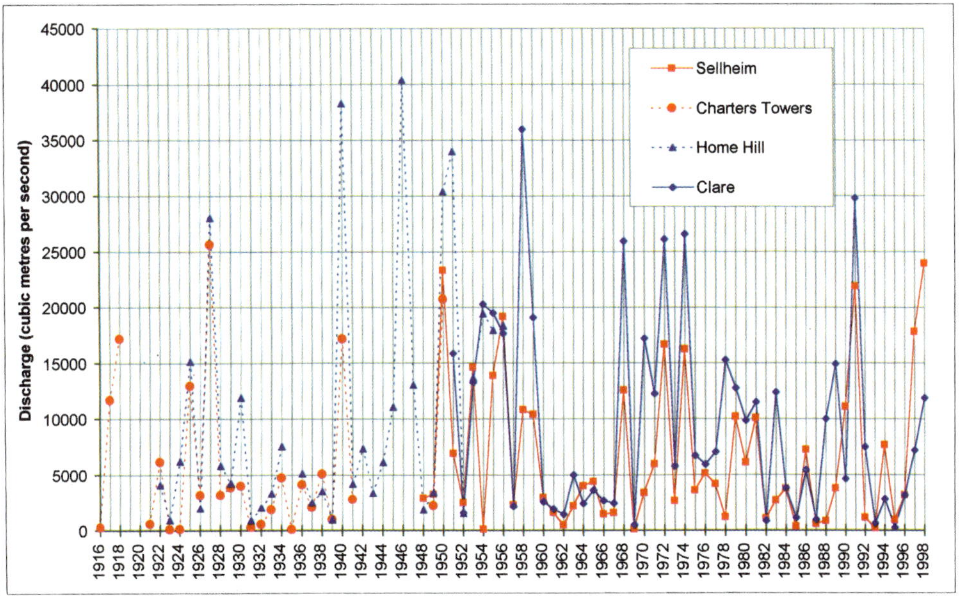

(b)

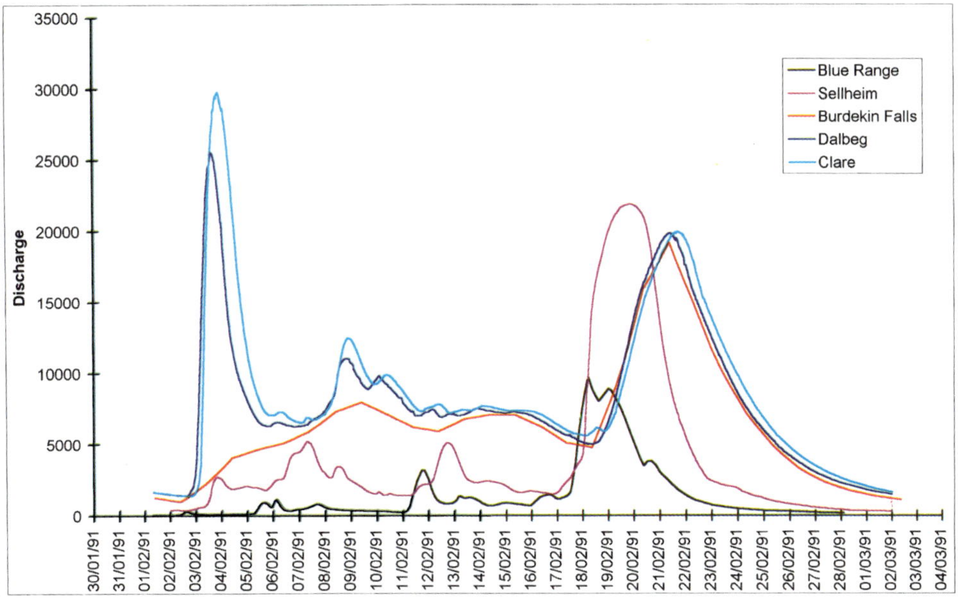

(c)

variations that allow easy mapping of previous courses of the Burdekin River (Holocene palaeochannels; Hopley 1970). Large parts of the delta flood during major discharge events and flow reoccupies some of the palaeochannels.

The Burdekin River's discharge and channel behaviour have been modified significantly by human activity only within the last hundred years or so. The effects of forest clearance on sediment and water discharge in this area are as yet not well known, although areas of recent severe soil erosion have been identified (Isbell & Murtha 1970; McClelland 1996). The channel has been modified by major engineering structures only in recent decades with construction of a few weirs, the Burdekin Falls Dam (Fig. 2b) and bank armouring along some of the lower reaches. The Burdekin Falls Dam was completed in 1987 and has an impoundment capacity of 1.85×10^9 m^3. The dam has affected the discharge behaviour of the lower river, and may have pronounced effects on sediment transport to the Coral Sea.

Discharge characteristics

The annual maximum discharge series (Fig. 3b) demonstrates large interannual variation. Comparisons between upstream sites on the Burdekin River, such as Charters Towers and Sellheim (Fig. 1), and those at more downstream positions, such as Home Hill and Clare, illustrate some significant features of this variability. In most years, maximum discharge at the upstream sites is considerably less than downstream, as a result of contributions from tributaries and direct precipitation on the channel and adjacent floodplain. Some large events at downstream sites had little or no associated event in the upper Burdekin (e.g. 1958, 1988, and the early February event in 1991 discussed further below: Fig. 3c). These represent periods of heavy rainfall in the catchment areas of major tributaries such as the Bogie, Bowen and Belyando rivers (Fig. 1), whereas in the same periods little rain fell on the upper Burdekin catchment. In some years (e.g. 1981: Fig. 3b) the downstream increase is small, which reflects little or no contribution from southern tributaries. In a few flow events (e.g. 1953, 1956, 1964) the natural (pre-construction of the Falls Dam) maximum discharge recorded upstream was a little greater than that downstream as a result of flood-wave attenuation (increasing wavelength but decreasing amplitude as the flood travels downstream: see Knight & Shiono 1996).

Since dam construction, discharge in the upper Burdekin has on occasion (1990, 1994, 1997, 1998) reached much greater magnitudes than below the dam, because of a combination of reservoir filling and flood-wave modification. This behaviour was particularly extreme in January 1998, when one of the highest recorded discharges at Sellheim (flood peak 23 959 $m^3\,s^{-1}$) produced extensive flooding around Charters Towers, but discharge in the lower Burdekin was only moderate, and little inundation occurred.

The January 1998 event resulted from Tropical Cyclone Sid becoming stationary north of Townsville (Fig. 1). Heavy rain began falling in Townsville and the eastern, upper Burdekin catchment on the morning of 10 January 1998, and became extremely heavy around 1800 Eastern Standard Time (EST). The city of Townsville received 237 mm between 1800 and 2100 EST, and by 0900 next day the 24 h total was 549 mm, which the Townsville Weather Bureau reported was 50% higher than the previous 24 h record of 366 mm in 1946. All streams in the immediate area experienced major flooding, with record levels in the Black River and Bluewater Creek (Fig. 1). Peak stage in the Black River reached 9.38 m with a discharge of 2300 $m^3\,s^{-1}$ and the Bluewater Creek reached 9.70 m and peak discharge of 1640 $m^3\,s^{-1}$. Heavy rainfall extended inland into the upper Burdekin catchment and major flooding occurred in the upper Burdekin River system, with only minor flooding in southern tributaries such as the Belyando, Suttor and Bowen rivers (Fig. 1). The Burdekin River at Clare peaked at 12.13 m gauge height, considerably less than the 1958 event, which was 18.39 m at Clare staff gauge (equivalent to 18.00 m at the recorder site). The Haughton River at the Powerline gauging station peaked at 9.72 m gauge height compared with 11.54 m in 1978. Barratta Creek at the Northcote gauging station peaked at 8.69 gauge height compared with 9.52 m in 1991. These January stream levels in the lower Burdekin area were insufficient to cause overbank flow. Some inundation occurred in the area between Haughton River and Barratta Creek as a result of intense rainfall which fell in a decreasing pattern to the south east (Fig. 2d).

Short duration and fast flood wave

Individual discharge events generally are of short duration and travel rapidly down the Burdekin system. Each flood can be regarded as a wave; the hydrograph gives the wave shape and comparisons between records at consecutive gauges give the wave speed. The discharge records for February and March 1991 from six gauging stations at various positions on the Burdekin River illustrate this behaviour of two major events (Fig. 3c).

In January 1991, the catchment remained overcast and rainfall was sufficient to keep soil saturation levels high (K. Klaasen, pers. comm.). In early February, heavy rainfall (but not of extreme intensity) from a monsoon trough over the southern part of the catchment was sufficient to cause major flooding of the Bowen and Bogie rivers and the lower Burdekin River at Dalbeg and Clare (Figs 1 and 3c). The peak discharge recorded at Clare for this event was 29 787 $m^3 s^{-1}$, whereas the Sellheim station recorded a peak of only 2729 $m^3 s^{-1}$ and there was little discharge in the Belyando (Fig. 1). The Dalbeg and Clare records demonstrate the very short duration of this high-magnitude event, which is similar to most major events in this system. The flood wave had a mean speed of 2.77 $m s^{-1}$ over this lower river section (below the dam).

Two weeks later, the monsoonal trough moved over the northern headwaters of the Burdekin, intense rainfall caused inundation and a high-magnitude flood wave travelled down the Burdekin. At Sellheim peak discharge on 19 February reached 21 900 $m^3 s^{-1}$ whereas at Burdekin Falls Dam on 21 February the peak was 19 196 $m^3 s^{-1}$, but the dam was already full. The flood-wave speed between Sellheim and the Burdekin Falls Dam was 1.05 $m s^{-1}$ and it increased to 4.34 $m s^{-1}$ below the dam. Water routing through the dam attenuated the flood peak and contributed to the reduction in the flood height below the dam. The flood level at Clare was calculated to have been reduced by 1.5 m as a result of the presence of the dam (V. Manley, pers. comm.), and this reduction would have been greater had the dam not been full at the time when this event occurred. It is believed that the flooding in the town of Home Hill would have been considerably higher without the dam but that the reduction was less than at Clare.

For comparison with the 1991 events, in January 1972 the Burdekin reached a flood peak of 16 647 $m^3 s^{-1}$ at Sellheim and 26 140 $m^3 s^{-1}$ at Clare and the speed of the flood wave is calculated as 2.7 $m s^{-1}$ over the middle to lower sections of the river. This speed is similar to the early February 1991 flood-wave speed below the dam. The 1972 flood-wave speed in the upper to middle sections of the river is difficult to calculate because of significant variations in hydrograph form between gauging stations, but is estimated as 2.4 – 4 $m s^{-1}$.

The flood-wave speed and change in hydrograph shape are controlled by effects of storage, tributary input and flow resistance. The variation in Burdekin flood-wave speed may be partly explained by the variation in stage. Knight & Shiono (1996) observed that, in British rivers, as in theoretical models, flood-wave speed typically reaches a maximum at about two-thirds bankfull flow and decreases in some stages of overbank flow. This is consistent with the late February 1991 observations in the Burdekin, with relatively slow wave speed (still high compared with most settings) above the dam where flooding was extensive, and very high velocities below the dam where bankfull conditions were not exceeded. Comparison with the wave speeds of both February 1991 events below the dam also supports this theory. Flood-wave speeds reach very high values in the Burdekin system when compared with other rivers and even similar magnitude floods in rivers such as the Mississippi (flood-wave velocities in the 1993 flood <1 $m s^{-1}$, Moody 1993). This may be accounted for by small floodplain storage, relatively low flow resistance and tributary input. The fast flood waves in the Burdekin add considerably to the flood risk and to the problems of reliable flood warning.

Floodplain inundation

Inundation of channel margins and floodplains is common in the Burdekin system. Such inundation predominantly occurs as a result of: (a) intense local rainfall, particularly on flat low-lying areas; (b) major flows in defined streams that cause overtopping of banks and flooding of adjacent land; (c) a combination of the above. An example of inundation resulting from intense rainfall is shown in Fig. 2d. Such inundation persists for short durations (hours to days). Inundation associated with major discharge events is also generally of very short duration, as illustrated by the 1991 and 1998 observations below.

The intense rainfall that deluged Townsville on 10 January 1998 penetrated inland to the upper Burdekin catchment. The Keelbottom Creek gauging station received 565 mm in 24 h on that day with an additional 310 mm over the next 7 days. The peak stage of 8.14 m at 1900 EST on 10 January exceeded the previous highest recorded level of 6.66 m on 15 February 1968 (Fig. 4). The stream is well confined in the vicinity of the gauging station but significant inundation of low-lying land occurred downstream. The peak was flashy and the river would have been confined within its banks for all but a few hours (see Fig. 4). It is expected that similar rainfall would have been received in the Star River catchment (Fig. 1), but unfortunately the river experienced a record flood peak with water flowing a metre and a half deep through the instrumentation shelter at the gauging station, resulting in the loss of further records. The peak height was accurately ascertained from a clear debris line inside the shelter and the missing section of the hydrograph has been derived. The peak stage at the site was gauge height 17.60 m, well above the highest recorded level of 16.28 m in 1997. The period of inundation above the high bank where the recorder is sited was *c.* 7 h (see Fig. 4).

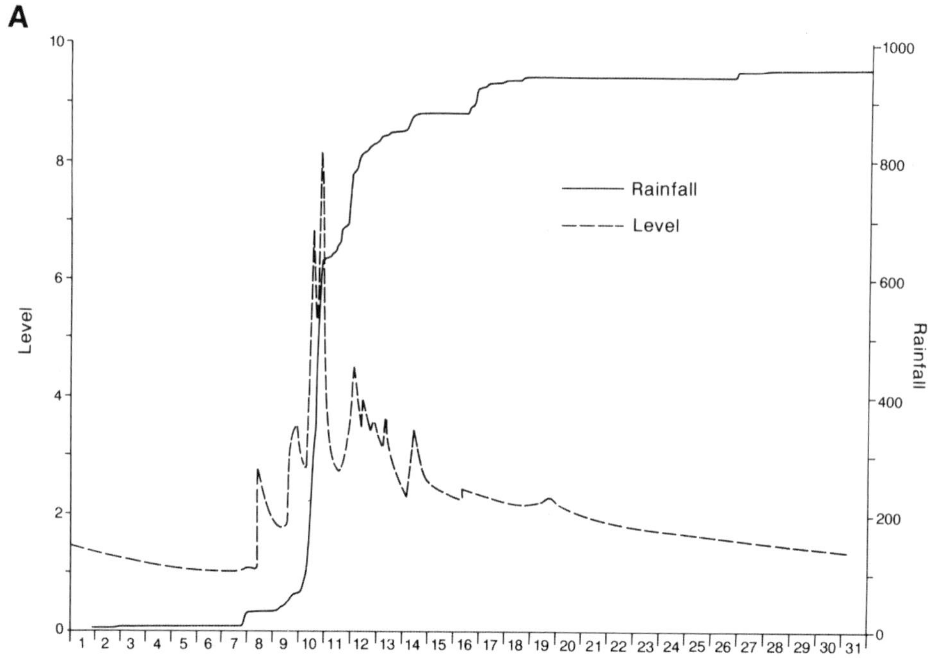

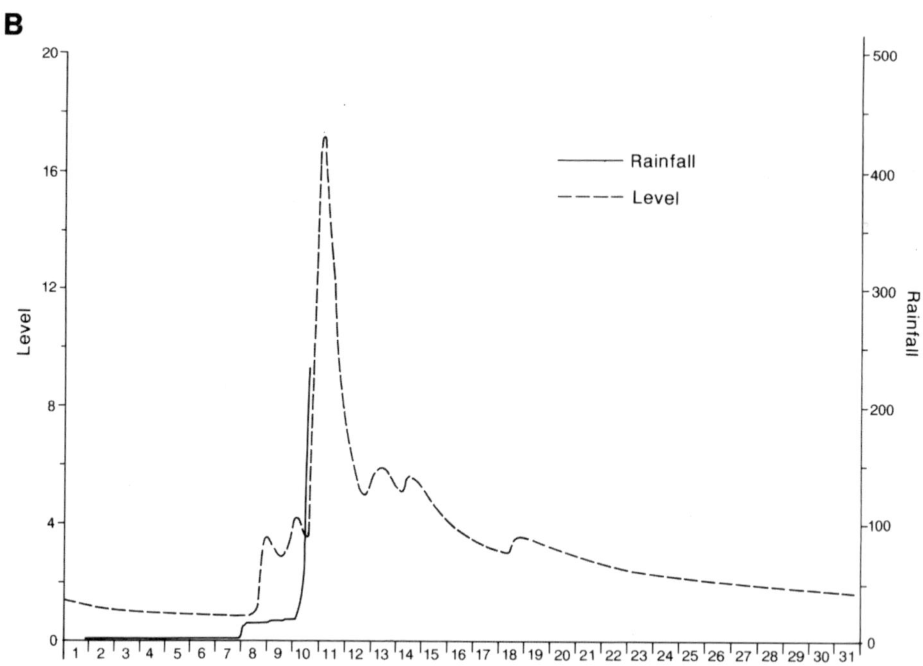

Fig. 4. Cumulative rainfall and water-level data for the January 1998 event. (**a**) Data from Keelbottom gauging station; (**b**) data from Star River gauging station.

The Bowen River flood of early February 1991 (Fig. 3c) discussed above, resulted in widespread inundation of the Home Hill township on 3 February. The Burdekin River, which peaked at gauge height 18.40 m at Clare, overflowed through Kidby Gully and flowed into Home Hill, where water ran 0.6 m deep in the main street (Figs 2c and 5). The water was *c*. 1.5 m deep across the road at the lower south end of town. Water also overflowed through Warren's Gully and into the Saltwater and Yellow Gin creeks (Fig. 5).

The late February 1991 event from the upper Burdekin River was less severe, reaching a peak stage of 15.74 m at Clare. There was no flooding at Home Hill on this occasion but a breakout into Warren's Gully did occur. The floodplain monitoring site adjacent to Warren's Gully recorded a maximum water depth of 0.84 m (elevation level of 15.24 m) during this event. Low-lying country adjacent to the recorder was inundated for a period of 70 h. Unfortunately, the instrument suffered a clock stoppage during the early February event, but it appears that the maximum depth was 1.47 m (elevation level 15.87 m) and the period of inundation would have been considerably longer than 70 h.

During heavy rainfall events in the lower Burdekin, the adjacent floodplain can make significant contributions to river flow through tributaries such as Expedition Pass Creek, Landers Creek, Alligator Creek and Cassidy Creek (these examples are near and just upstream of Clare). However, like many rivers in their lower reaches, the Burdekin has built up levees so that much natural drainage (of water falling on the floodplain) is away from the river, except where definite tributaries exist. During major flooding, there are several places where flow breaks out from the lower Burdekin and some tributaries become distributaries. For example, in 1991, water broke into Plantation Creek from the river, Plantation palaeochannel became a distributary and water rose

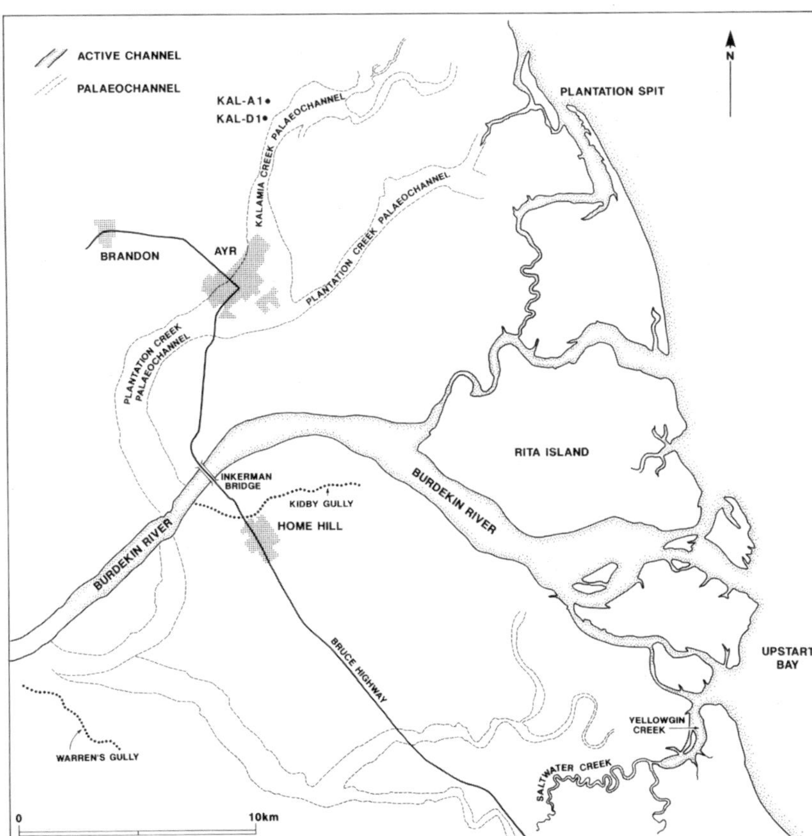

Fig. 5. Map of the area of Ayr and Home Hill to show some of the Holocene palaeochannels, and locations of sites of flooding and boreholes.

into the low-lying parts of Ayr (Fig. 5). Properties and businesses at the southwest end of the main street were affected. Home Hill was more seriously affected with water flowing from the Burdekin River through Kidby Gully, with the business district under water for a period (Fig. 2c). The inundation at Home Hill and a small part of Ayr lasted for about a day.

Bed conditions

Observations of the river bed following successive discharge events show considerable changes in bed elevation, composition and roughness. The magnitude of change varies from year to year and from site to site. Some quantification of the changes in bed conditions has been attempted between Big Bend and the Keelbottom confluence north of Charters Towers (Fig. 6), and qualitative observations have been recorded at many other sites (Fielding & Alexander 1996; Alexander & Fielding 1997; Fielding *et al.* 1999). It is very difficult to observe bed condition changes during these discharge events because of the high suspended sediment concentration, the channel size, water velocity and the mobile character of the bed. When probes are lowered from Inkerman road bridge (Fig. 2a) during major flow events a solid bed is difficult to detect.

Where the banks are not constrained by engineering work (such as vehicle wrecks as rip-rap) or basement rock outcrops, the cross-sectional shape varies following successive discharge events (e.g. Fig. 6). These changes result from variation in sediment storage in the measured reach, and bank movement because of point-bar aggradation and bank erosion. Ground penetrating radar surveys over the Big Bend–Keelbottom confluence site before and after sizeable discharge events demonstrate sediment remobilization of several metres (Fielding *et al.* 1999) and the elevation of a point on the channel floor at Big Bend varies as much as 5 m.

The short duration of flow events and very rapid rate of change of discharge during individual events causes sediment transport rate, bedforms and bed elevation to change continuously. At any one site, bedforms preserved on the exposed dry-season bed vary from year to year; for example, varying from dunes >1.5 m amplitude, to 1 m amplitude antidunes, to near plane beds at Big Bend. The bedform type and size preserved is related to the peak discharge, rate of discharge decline and passage of sediment waves. Although these different bedforms give different bed roughness, the high-magnitude events cause major changes and pre-existing bed conditions may be significant for only a short period at the start of the event. Because of the variable magnitude, short duration, and very rapid discharge change, flow will commonly be out of equilibrium with the underlying bedforms. Because bed levels and

Fig. 6. Cross-section of the Burdekin River at Big Bend environmental park, north of Charters Towers, showing the variation in bed elevations and shape following successive events. The blue lines are water surfaces, other lines are sediment surfaces. In major events such as the later 1991 event the water level rises above the top of this cross-section as drawn.

bedforms vary during individual events, bed elevation changes over 1–10 years or more may not indicate significant change to the risk of local flooding in non-engineered reaches.

The paperbark tree *Melaleuca argentea* lives on the bank and bed, and mature trees with trunks up to 3.20 m circumference have been recorded in the Burdekin near Charters Towers (Fielding *et al.* 1997). Mapping of this arborescent riparian vegetation after successive discharge events shows that there may be significant local stripping of saplings and mature trees during major events, and this would reduce roughness. In contrast, in periods with successive years of small or intermediate flows *Melaleuca argentea* sapling growth would significantly increase bed roughness (Fielding *et al.* 1997).

In the lower river, attempts have been made to stabilize long reaches of river banks to reduce damage to agricultural land. This reduces the channel's degrees of freedom in subsequent discharge events by allowing changes in bed elevation but not width. The protected channel form resulted from relatively recent discharge history and as the system is generally out of equilibrium the long-term consequences of maintaining this channel form for flood risk are unknown. Channel resectioning (widening and/or deepening) to increase the flood-carrying capacity may be a better strategy to reduce flood risk. Channel deepening has recently been proposed for sites near Home Hill, but this would require removal of vast volumes of pebbly sand from the channel floor (which is here about 800 m wide; Fig 2a) and the consequence of such removal is unknown, although, given that the sediment flux is very high, removal would be needed after each event. Such large-scale removal requires economic sale of the removed material and is likely to influence the stability and character of the delta margin. Channel widening might be a more sustainable change but would result in loss of very high value productive agricultural land. The channel could adjust to artificial widening by increasing deposition to narrow itself again, but the newly deposited material would be easily eroded by subsequent large events.

Overbank deposition

Floodplain areas are inundated periodically by local runoff and by riverbank overtopping. In this semi-arid to sub-humid environment both inundation types transport significant volumes of sediment and redeposit it. Depending on the magnitude of the flooding and position relative to channel or channel overtopping point, the sediment calibre varies, ranging locally up to coarse sand and gravel. Direct sampling of flood deposits of individual monitored inundation events has not been undertaken to date, but Recent deposits have been observed in cut banks, trenches, and auger and drill holes.

Channel-bed deposits on the unvegetated or moderately vegetated channel floor consist predominantly of pebbly coarse sand (mean 1 to –0.7ϕ, standard deviation 0.8–1.2ϕ). Bar top and bank sediments are, in contrast, generally medium to fine sand (mean 2.5–3.7ϕ) with mud drapes and thicker mud accumulations in ponded areas. On the upper delta sites, between the active channel and recognized Holocene palaeochannels (Fig. 5), surface grain size varies from coarse sand to mud. The natural deposits of the modern delta floodplain are generally not discernible, because of cultivation of sugar cane and other crops. However, borehole KAL-D1 drilled in August 1998 near Kalamia Sugar Mill, 500 m from the bank of the Kalamia Creek palaeochannel, and an adjacent auger hole (KAL-A1) revealed a stratigraphic succession that may be related to a number of major flood events (Fig. 7). The borehole was drilled to a depth of 12 m from the floodplain surface, approximately level with the top of the palaeochannel bank and about 5 m above the present water level in Kalamia Creek. This borehole encountered a series of sand beds (with minor gravel), interbedded with thinner, grey clay beds. Coarse-grained beds range up to 2.85 m thick (although the thicker ones may record accumulation from more than one event), are sharp bounded and either fine upward or show no vertical grain-size trend. One bed has a gravelly horizon at the base, and dispersed gravel (both extra-formational debris and intraformational mud clasts) is scattered through some beds. The uppermost unit fines upward into silty loam, which forms the cultivated topsoil. Borehole KAL-D1 shows a complex stratigraphic record of (geologically) recent flood events. The sharp-bounded, coarse-grained beds may record the peak of large flows through the Kalamia palaeochannel when the banks were overtopped, allowing coarse sediment to splay outward onto the floodplain.

It is likely that the pattern of variable discharge observed in the modern Burdekin River continues back through Holocene time. This is supported by the similarity of palaeochannel and modern channel size and form and comparable sedimentary facies. During the short-duration, high-magnitude events in the modern Burdekin, flow over the floodplain can be fast and relatively deep (for example >1.4 m south end of Home Hill and Warrens Gully in early February 1991, described above) and this may also have been true of major events when the Kalamia channel was active.

It is generally accepted that during floods, fine-grained sediment may be transported onto

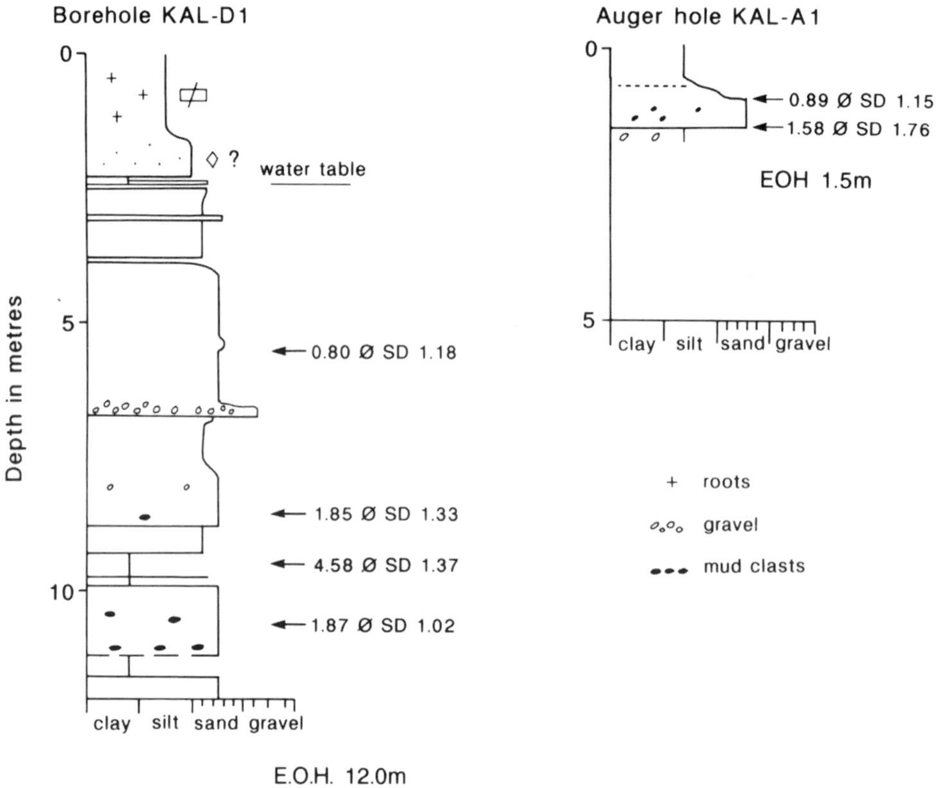

Fig. 7. Sedimentary logs of borehole KAL-D1 and auger hole KAL-A1 near the bank of the Kalamia palaeochannel, which acts as a flood channel in extreme events (see Fig. 5 for locations). The arrows represent sediment samples with mean grain size and standard deviation. The rectangle represents large wood debris and the diamond indicates suspected concretion development.

floodplains as suspended load and deposited as a result of flow deceleration and ponding, whereas coarser-grade material may be transported to near-channel sites as traction load either ramped out of the channel or diverted through levee breaks. In the area of KAL-D1 there is no preserved record of a channel breach (no obvious tributary or distributary channel) and the site is on the outside of a gentle meander; consequently, most of the deposits observed in the borehole must have been deposited by bank overtopping.

The overbank deposits in borehole KAL-D1 are of three types: (a) fine-grained deposits; (b) coarse sand, similar to channel-bed deposits; (c) fine- to medium-grained sand similar to Burdekin River bar and bank top samples.

The fine-grained overbank sediments are easy to explain as sediment deposited from suspension during flooding. These units may record suspension fallout from water emanating from the channel or from runoff as a result of intense local rainfall. Some of these units were probably eroded during subsequent flood events, particularly those transporting coarse sediment over the site. This is confirmed by the amalgamated nature of some beds and the presence of mud clasts in at least two sand beds.

Some of the units in KAL-1D have the same grain-size mode as channel floor deposits but they are a little less well sorted. The poorer sorting may reflect *in situ* weathering and infiltration of fines (both by pedogenic processes) or less efficient hydraulic sorting during transport and deposition at overbank sites. The existence of such coarse-grained material in overbank settings requires either very large floods to drive bedload up over the banks or some sort of channel shift or breach. The former is most likely in the Kalamia site.

Some of the sand beds are fairly well sorted but of a finer mode than channel floor sediments and are more similar to bar top and bank deposits in the modern Burdekin River. The occurrence of this mode on bars and in overbank settings can be explained by the nature of the flood hydrographs, in-channel sediment distribution and sediment transport efficiency. Within-channel discharge events do not mobilize much fine–medium sand from the upper bar and bank tops. The major floods easily erode this material from the catchment, banks and bar tops, and at peak conditions may transport this particle size fraction in suspension (or modified suspension) onto floodplains and beyond the channel mouth.

The succession encountered in KAL-D1 indicates that several flood events occurred during the life of the Kalamia palaeochannel, leading to the gradual aggradation of the floodplain surface to its present level.

Bridge & Leeder (1979) and others studying fluvial architecture have suggested that deposits with low net mud content will result from low aggradation rate and mobile rapid avulsion of channels. The Burdekin, by contrast, is a system with an apparently high percentage of coarse-grade deposits within overbank sediments because of the discharge character.

Conclusions

In many streams in north Queensland, Australia, discharge events exceeding bankfull stage do not occur as frequently as in temperate perennial river systems. Many major discharge events remain in-channel for much of the river's length. Flooding does not occur at all reaches of the channel at the same time or in the same flood, and different parts of the catchment experience different flood histories. Channels do not significantly 'recover' between successive large events.

The flood waves in the Burdekin system are very fast compared with those in rivers in different settings, with calculated speeds up to 4.34 m s^{-1}. Wave speeds vary greatly both between events and between different reaches during one event (ranging from 1.05 to 4.34 m s^{-1} in the late February 1991 event). Because of changes in bed, bank and floodplain conditions flood-wave speeds of events with similar magnitudes may also vary.

Inundation occurs as a result of local intense rainfall and locally as a result of channel overtopping. Both types of flooding are of short duration in this system, even though floodplain water depths can exceed 1.4 m.

Overbank sediments on the delta record deposition of up to gravel-grade sediments, and sand-grade overbank deposition is common. Most of the fine sediment acts as wash load and has low preservation in the overbank deposits examined. Thus, river systems of this type may give rise to more sandy overbank deposits than is common in most fluvial settings.

Thanks are due to the Queensland Department of Natural Resources, Resource Management Group, for provision of data used in this paper. We thank also P. Judge for assistance with some of the diagrams in this paper, and R. Harvey and J. Trueman for field assistance. The paper was improved as a result of constructive comments from M. Blum and R. Slingerland.

References

ACKERS, P. 1992. Gerald Lacey memorial lecture: Canal and river regime in theory and practice: 1929–92. *Proceedings of the Institution of Civil Engineers, Water, Maritime and Energy*, **96**, 167–178.

ALEXANDER, J. & FIELDING, C. R. 1997. Gravel antidunes in the tropical Burdekin River, Queensland, Australia. *Sedimentology*, **44**, 327–337.

BELPERIO, A. P. 1979. The combined use of washload and bed material load rating curves for the calculation of total load: an example from the Burdekin River, Australia. *Catena*, **6**, 317–329.

BRIDGE, J. S. & LEEDER, M. R. 1979. A simulation model of alluvial stratigraphy. *Sedimentology*, **26**, 617-644.

BROWN, A. G. 1996. Floodplain palaeoenvironments. *In*: ANDERSON, M. G., WALLING, D. E. & BATES, P. D. (eds) *Floodplain Processes*. Wiley, Chichester, 95–138.

ENGEL, H., BUSH, N., WILKE, K., KRAHE, P., MENDEL, H. G., GIEBEL, H. & ZIEGER, C. 1994. *The 1993/4 Flood in the Rhine Basin*, Federal Institute of Hydrology, Koblenz, Report **833.**

FIELDING, C. R. & ALEXANDER, J. 1996. Sedimentology of the upper Burdekin river of north Queensland, Australia—an example of a tropical, variable discharge river. *Terra Nova*, **8**, 447–457.

——, —— & NEWMAN-SUTHERLAND, E. 1997. Preservation of *in situ* vegetation in fluvial channel deposits—data from the modern Burdekin River of north Queensland, Australia. *Palaeogeography, Palaeoclimatology, Palaeoecology*, **135**, 123–144.

——, —— & MCDONALD, R. 1999. Sedimentary facies from GPR surveys of the modern, upper Burdekin River of North Queensland, Australia: consequences of extreme discharge fluctuations. *In*: SMITH, N. D. & ROGERS, J. J. (eds) *Current Research in Fluvial Sedimentology*. International Association of Sedimentologists, Special Publication, **28**, 347–362.

HOPLEY, D. 1970. *The geomorphology of the Burdekin delta, North Queensland*. Department of Geography James Cook University, Monograph Series, **1**.

HOWARD, A. D. 1996. Modelling channel evolution and floodplain morphology. *In*: ANDERSON, M. G., WALLING, D. E. & BATES, P. D. (eds) *Floodplain Processes*, Wiley, Chichester, 15–62.

ISBELL, R. F. & MURTHA, G. G. 1970. *Soils, Burdekin—Townsville Region, Queensland*. Resource Surveys, Department of National Development, Canberra.

KNIGHT, D. W. & SHIONO, K. 1996. River channel and floodplain hydraulics. *In*: ANDERSON, M. G., WALLING, D. E. & BATES, P. D. (eds) *Floodplain Processes*, Wiley, Chichester, 139–181.

LARSON, L. W. 1995. The great USA flood of 1993. Paper presented at US–Italy Research Workshop, Water Resources Research and Documentation Center (WARREDOC), Perugia, Italy, 13–17 November 1995.

LEWIN, J. 1989. Floods in fluvial geomorphology. *In*: BEVEN, K. J. & CARLING, P. A. (eds) *Floods: Hydrological, Sedimentological and Geomorphological Implications*. Wiley, Chichester, 265–284.

MCCLELLAND, L. 1996. *Soil and water chemistry and their relationship to land degradation in the upper Burdekin catchment, north Queensland*. Post-graduate Diploma in Science thesis, University of Queensland.

MOODY, J. A. 1993. *Propagation and Composition of the Floodwave on the Upper Mississippi River 1993*. US Geological Survey Circular, **1120-F**, 1–21.

WILLIAMS, G. P. 1978. Bank-full discharge of rivers. *Water Resources Research*, **14**, 1141–1154.

WOLMAN, M. G. & GERSON, R. 1978. Relative scales of time and effectiveness of climate in watershed geomorphology. *Earth Surface Processes*, **3**, 189–208.

Using fallout radionuclides in investigations of contemporary overbank sedimentation on the floodplains of British rivers

D. E. WALLING

Department of Geography, University of Exeter, Exeter EX4 4RJ, UK
(e-mail: geography@exeter.ac.uk)

Abstract: Increased interest in the functioning of river floodplains has generated the need for more information on rates and patterns of contemporary overbank sedimentation. Traditional approaches to documenting rates of overbank sedimentation on floodplains face many practical difficulties, but recent advances in the application of the fallout radionuclides ^{137}Cs and unsupported ^{210}Pb afford greatly increased potential for assembling such information. The potential for using these two radionuclides to investigate contemporary overbank sedimentation is demonstrated by considering examples drawn from recent work undertaken by the author and his co-workers on the floodplains of British rivers. These examples are used to illustrate (1) establishment of recent chronologies for sediment cores collected from river floodplains, (2) documentation of the spatial pattern of overbank sedimentation rates, (3) quantification of conveyance losses associated with overbank sedimentation, and (4) investigation of changing rates of overbank sedimentation over the past 100 years.

River floodplains have attracted increasing attention from geomorphologists, hydrologists and sedimentologists in recent years. This interest reflects a growing awareness of the important role of floodplains as distinctive ecological habitats and key components of the river corridor (see Petts 1996), as buffers between the river and the surrounding land (see Burt & Haycock 1996), and as natural flood control reservoirs (see Penning-Rowsell & Tunstall 1996), as well as a recognition of their potential significance as sinks for river-borne sediment and associated nutrients and contaminants (see Leenaers & Schouten 1989; Marron 1989). This interest in river floodplains has in turn generated the need to develop an improved understanding of their contemporary development and functioning. In considering the geomorphological evolution of river floodplains and their role as sediment sinks, attention is commonly directed either to the coarse channel deposits and the interaction between channel migration and floodplain construction (see Wolman & Leopold 1957), or to the fine-grained overbank deposits that mantle large areas of most floodplains and result in vertical accretion of the floodplain surface. For many lowland river floodplains, particularly those where channelization and river training works limit or prevent channel migration, the overbank deposition of fine-grained sediment will represent the dominant component of floodplain evolution. Information on rates and patterns of contemporary overbank sedimentation must therefore be seen as a central requirement for developing an improved understanding of floodplain development and behaviour.

Traditional approaches to documenting contemporary or recent rates of overbank sedimentation on river floodplains have included the use of sedimentation traps (e.g. Lambert & Walling 1987; Asselman & Middelkoop 1995; Middelkoop & Asselman 1998), post-event surveys of the deposits resulting from individual floods (e.g. Kesel *et al.* 1974; Brown 1983; Marriott 1992), and the identification of datable levels within the overbank deposits (e.g. Lewin & Macklin 1987; Hupp 1988). However, each of these approaches is constrained by practical and logistical difficulties associated with the floodplain environment, including: the infrequent and essentially random occurrence of overbank flood events; the problems of quantifying very small amounts of vertical accretion; lack of datable material; and the temporal variability of sedimentation rates according to the magnitude and duration of individual flood events. Because of this variability, measurements relating to a single event or to a small number of events may not provide representative estimates of longer-term mean annual sedimentation rates required to interpret contemporary floodplain development. Furthermore, the spatial variability of floodplain sedimentation means that it will frequently be necessary to assemble information from a large

From: MARRIOTT, S. B. & ALEXANDER, J. (eds) 1999. *Floodplains: Interdisciplinary Approaches*. Geological Society, London, Special Publications, **163**, 41–59. 1-86239-050-9/99/$15.00

number of points, so as to provide a meaningful representation of the patterns involved. As a result of these limitations, it has generally proved difficult to obtain reliable detailed information on contemporary rates and patterns of overbank deposition.

Recent advances in the application of the fallout radionuclides ^{137}Cs and unsupported ^{210}Pb (see Walling *et al.* 1992; Walling & He 1993, 1997*a*, *b*; He & Walling 1996) to investigations of overbank sedimentation on river floodplains would, however, appear to offer greatly increased potential for establishing the recent chronology of overbank sediments and for estimating associated sedimentation rates, and must been seen as an important development in this field. Use of fallout radionuclides offers many advantages over other techniques for documenting overbank floodplain sedimentation. These advantages include the general applicability of the approach to a wide range of environments, the medium-term time scales involved (i.e. *c.* 35 years for ^{137}Cs and *c.* 100 years for ^{210}Pb), which afford a means of overcoming temporal variability and obtaining representative values of the mean sedimentation rate during the recent past, and the potential for assembling data for a large number of points on a floodplain and therefore for documenting spatial patterns.

Use of fallout radionuclides to document rates and patterns of overbank sedimentation on river floodplains

Detailed explanation of the basis for using ^{137}Cs and unsupported ^{210}Pb to investigate rates of recent overbank sedimentation on river floodplains falls outside the scope of this paper and the reader is referred to Walling & He (1993, 1997*a*), He & Walling (1996) and Walling *et al.* (1996) for further details of the methods involved. In essence, the approach exploits the fact that ^{137}Cs and unsupported ^{210}Pb are rapidly and strongly adsorbed by fine-grained sediment and accumulate within accreting overbank sediment deposits as a result of both direct atmospheric fallout to the floodplain surface and deposition of suspended sediment during overbank flood events. Suspended sediment represents material mobilized by erosion from the upstream catchment and transported by the river and it will contain both ^{137}Cs and unsupported ^{210}Pb originally deposited as fallout on the catchment surface. Both the total inventory, or amount of the radionuclide contained in a floodplain sediment core (Bq m^{-2}), and its vertical distribution will therefore differ from that associated with a core collected from a natural, undisturbed site above the level of flood inundation, that will only have received inputs from direct atmospheric fallout.

Figure 1 presents typical examples of the vertical distribution of ^{137}Cs and unsupported ^{210}Pb in overbank sediment deposits on a river floodplain and in undisturbed soils from an adjacent area above the level of flood inundation, which clearly demonstrate the contrasts described above. ^{137}Cs is an artificial fallout radionuclide with a half-life of 30.17 years, produced by the atmospheric testing of thermonuclear weapons, primarily during the late 1950s and the 1960s, and its depth distribution in the floodplain sediment will reflect the temporal pattern of fallout. The peak ^{137}Cs activity found at *c.* 100–120 mm depth therefore reflects the period of maximum fallout in 1963. Use of this time marker affords one means of estimating the average sedimentation rate over the past 35 years, provided account is taken of the potential for slow downward migration of the ^{137}Cs peak (see Walling & He 1997*a*). Comparison of the total ^{137}Cs inventory of a floodplain core with those of adjacent natural undisturbed soils above the level of inundating flood water permits the excess inventory associated with sediment deposition to be calculated and provides an alternative means of estimating the average sedimentation rate over the period since the onset of significant radiocaesium fallout (i.e. since *c.* 1954). The magnitude of the excess inventory will directly reflect the rate of sedimentation. The grain-size composition of the deposited sediment must, however, be taken into account, as this will influence its ^{137}Cs content and thus the magnitude of the excess inventory (Walling & He 1997*a*). The key advantage of this alternative approach to estimating the sedimentation rate, pioneered by the author and his co-workers, is that only a single measurement of the total inventory of the bulk core is needed. One or more sectioned cores will, however, be required to 'calibrate' the algorithms used to estimate sedimentation rates from values of excess inventory. Because the measurement of ^{137}Cs by gamma spectrometry involves lengthy count times, use of whole-core inventory values to estimate sedimentation rates provides greater scope for assembling information for a large number of points on a floodplain and therefore to investigate spatial patterns of sedimentation.

The depth distributions of unsupported ^{210}Pb presented in Fig. 1 are significantly different from those exhibited by ^{137}Cs, because, although unsupported ^{210}Pb is also a fallout radionuclide, it differs from ^{137}Cs in two important respects. First, it is of natural origin, representing a product of the ^{238}U decay series, with a half-life of 22.26 years. It is derived from the decay of gaseous ^{222}Rn, the daughter of ^{226}Ra, which occurs naturally in soils

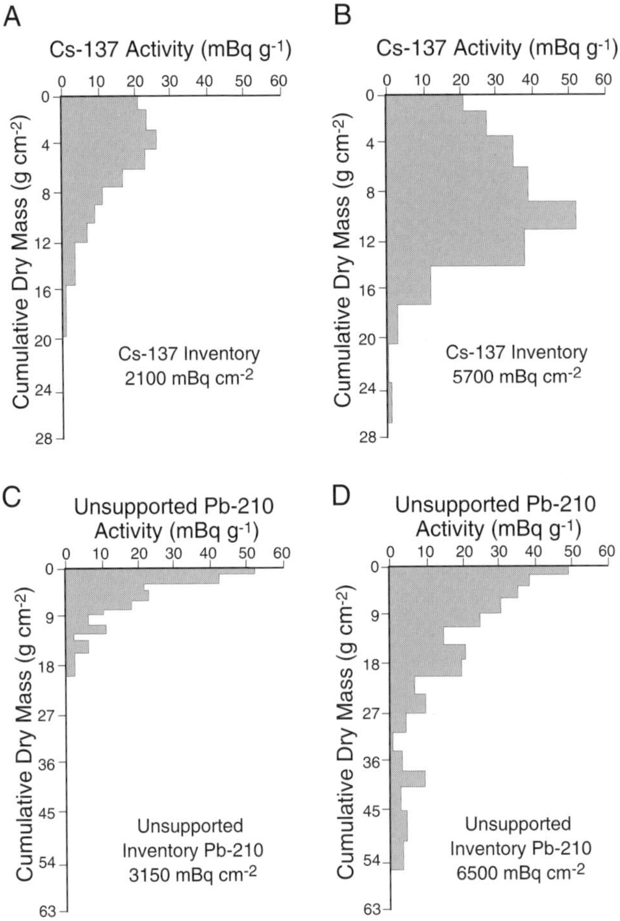

Fig. 1. Representative examples of the vertical distribution and total inventory of ^{137}Cs and unsupported ^{210}Pb in cores collected from undisturbed pasture above the level of inundating flood water (**A**, **C**), and from an adjacent area of the floodplain of the River Culm, Devon, UK (**B**, **D**).

and rocks. Diffusion of a small proportion of the ^{222}Rn from the soil introduces ^{210}Pb into the atmosphere and its subsequent fallout provides an input of this radionuclide that is not in equilibrium with its parent ^{226}Ra. This component is designated unsupported ^{210}Pb, as it cannot be accounted for (or supported by) decay of the *in situ* parent. Second, because of its natural origin, the annual fallout input of unsupported ^{210}Pb may be treated as continuous and essentially constant through time. The continuing input is reflected in the high concentrations found at the surface and the exponential decline of concentration with depth evident for the floodplain core. Because of its continuing input, unsupported ^{210}Pb affords a means of estimating deposition rates over longer periods (i.e. *c.* 100 years or four half-lives). Unsupported ^{210}Pb has been widely used for dating lake sediment cores (e.g. Oldfield & Appleby 1984), but the assumptions underlying the numerical procedures involved are not directly applicable to floodplain sites. However, information on the vertical distribution of unsupported ^{210}Pb in overbank sediments can again be used to derive an age–depth relationship for the sediment core, and thus to date specific levels. He & Walling (1996) have also demonstrated how a single measurement of the total unsupported ^{210}Pb inventory of a sediment core can, as in the case of ^{137}Cs, be used to estimate the average rate of accretion at the point where the core was collected. In this case, the total inventory of unsupported ^{210}Pb associated with a core is similarly apportioned into two components, representing that derived from direct atmospheric fallout and that associated with deposition of sediment mobilized from the upstream catchment. The value for the latter component is then used in conjunction with information on the initial concentration of unsupported ^{210}Pb in deposited sediment to estimate the mean sedimentation rate over the past *c.* 100 years. Use of single whole-core measurements again affords potential for estimating deposition rates for a large number of points on a floodplain and thus for documenting spatial patterns.

Measurements of both ^{137}Cs and unsupported ^{210}Pb can be undertaken by direct gamma spectrometry using an HPGe detector. Provided a suitable Ge detector is available, both measurements can be made simultaneously. In the case of ^{137}Cs, the radionuclide is measured directly at 662 keV, and most standard coaxial detectors are suitable for this purpose. Because of its lower energy (46 keV) and problems of background interference, measurements of unsupported ^{210}Pb require a low-background, low-energy detector. Values of unsupported ^{210}Pb are derived from measurements of the total ^{210}Pb activity by subtracting the ^{226}Ra supported ^{210}Pb activity, and ^{226}Ra must therefore also be determined either directly or by measuring the activity of its short-lived daughter ^{214}Pb (see Joshi 1987).

Examples of the application of fallout radionuclides in investigations of overbank floodplain sedimentation

The potential for using ^{137}Cs and unsupported ^{210}Pb in investigations of recent overbank sedimentation on river floodplains can usefully be demonstrated by considering four examples based on work undertaken by the author and his collaborators on the floodplains of British rivers. These examples aim to illustrate the following: (a) derivation of chronologies for sediment cores from river floodplains; (b) documentation of rates and patterns of overbank sedimentation; (c) quantification of conveyance losses associated with overbank sedimentation; (d) investigation of changing rates of floodplain sedimentation over the past 100 years.

Derivation of chronologies for sediment cores from river floodplains

Increasing concern for non-point or diffuse source pollution in river basins has focused attention on the important role of fine sediment as a carrier of nutrients and contaminants (see Allan 1986; Vaithiyanathan & Correll 1992). Information on changes through time of the nutrient or contaminant content of the suspended sediment transported by a river could provide valuable evidence of the progressive impact of such pollution and the effects of factors such as land-use change, changes in land management, increases in fertilizer and pesticide use, and improvements in effluent treatment. There is, however, a general lack of such records and recourse is frequently made to information from surrogate sources. Lake sediments have been frequently used for this purpose, as they can provide evidence of the nature of the sediment output from the upstream drainage basin over a considerable period of time (e.g. Charlesworth & Foster 1993). One important limitation on the use of lake sediment records is, however, that lakes are frequently only found on smaller tributaries in the headwater areas of drainage basins, whereas it is downstream areas and larger rivers that are often of greatest interest and concern. Overbank floodplain sediments offer considerable potential for providing alternative evidence of the variation of sediment properties through time that relates to downstream areas. To exploit this potential, it is necessary to establish the chronology of sediment cores, so that a record of changes in the properties of transported sediment can be synthesized from information on downcore variations in relevant sediment properties. As part of an investigation of changes in the phosphorus content of suspended sediment transported by British rivers, cores have been collected from the floodplains of several rivers, to document downcore changes in both the total-P content of the sediment and ^{137}Cs activity. These cores were collected using a motorized percussion corer equipped with a 12 mm diameter core tube and subsequently sectioned into 20 or 40 mm increments.

Results obtained for a core from the floodplain of the River Axe at Cownhaye, Devon, are shown in Fig. 2. Figure 2a indicates that the concentration of total-P in the upper portion of the core is relatively high and that there is clear evidence of a downcore reduction in total-P concentration. This situation contrasts with equivalent information obtained for cores collected from adjacent areas of permanent pasture above the level of flood inundation. These cores showed near-surface total-P concentrations that were only about 50% of those associated with the floodplain core, and the total-P concentrations were relatively uniform within the upper 0.30 m of the soil. The increased concentrations found in the floodplain core reflect the deposition of sediment containing high levels of total-P. Although some of the downcore reduction in total-P concentrations found in this core could reflect post-depositional mobilization of P, most is thought to reflect a progressive increase in the P content of deposited sediment through time, in response to increased

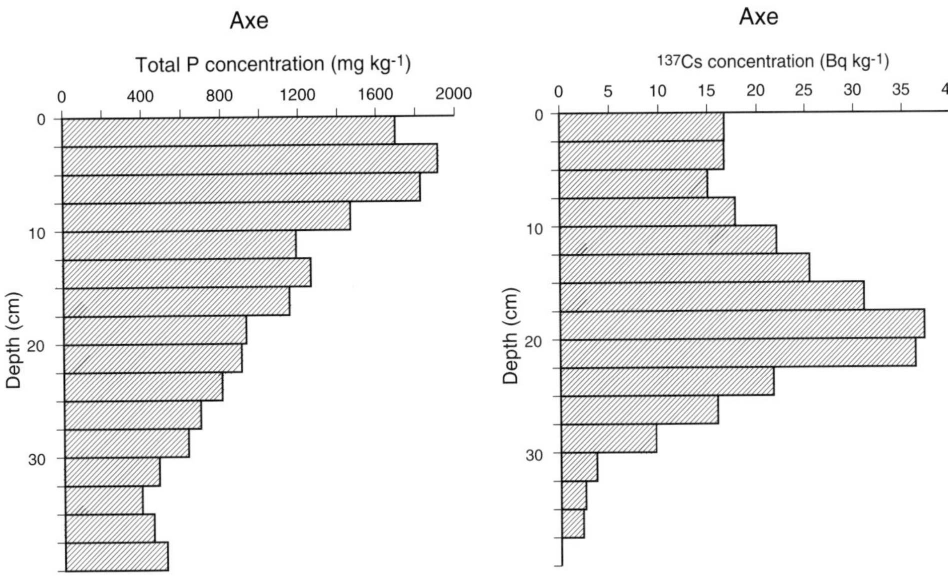

Fig. 2. The vertical distribution of total-P (**a**) and ^{137}Cs (**b**) in a sediment core collected from the floodplain of the River Axe at Cownhaye, Devon, UK.

fertilizer application and, to a lesser extent, increased effluent output from sewage treatment works. Fustec *et al.* (1995) have, for example, reported studies undertaken on floodplain sediments from the River Seine in France, which indicate that *c.* 80% of the total-P content of recently deposited sediment is not readily bio-available and that the relative contribution of this fraction to the total-P content of sediment from a floodplain core shows only a small increase downcore. Such results suggest that post-depositional mobilization is likely to be of limited importance. The position of the 1963 peak in the ^{137}Cs profile for this floodplain core (Fig. 2b) has been used to derive an estimate of the average rate of sedimentation at this site, which is 0.55 g cm^{-2} a^{-1}. This value can in turn be used to locate the approximate depths associated with individual years. If these depths are related to the downcore variation in the total-P concentration of the sediment, it is possible to reconstruct a tentative record of the P content of sediment deposited by the river over the past *c.* 45 years. This record is presented in Fig. 3, along with equivalent results for floodplain cores collected from four other rivers; the River Arun near Billingshurst, Sussex; the River Torridge at Barton Farm, Devon; the River Severn at Atcham, Shropshire; the River Usk near Usk, Monmouthshire.

Clear contrasts are evident between the rivers represented in Fig. 3, and these contrasts may be related to catchment land use. In the case of the Rivers Severn and Usk, which show only limited change in the total-P content of sediment through time, the land use is primarily upland pasture where fertilizer application is limited. In contrast, the catchments of the Rivers Axe and Arun include large areas of arable cultivation and more intensively used pasture, where fertilizer use can be expected to have increased over the past few decades. Here, the total-P content of the deposited sediment shows a substantial increase over the period considered. The catchment of the River Torridge is primarily occupied by permanent pasture, but fertilizer application rates are known to have increased significantly in this area, and a clear increase in the total-P content of deposited sediment can be seen in Fig. 3. By coupling information on the total-P content of suspended sediment with that for average sedimentation rates, it is possible to provide estimates of the rates of total-P input to these floodplain sites. Values of mean annual total-P input for the period since 1963, based on the existing P content of the floodplain core, are listed in Table 1. The values, which range from 3.8 to 9.0 g m^{-2} a^{-1}, are similar to that of 9.0 g m^{-2} a^{-1} reported for the floodplain of the River Seine at Maizières by Fustec *et al.* (1995) and highlight the importance of river floodplains as sinks for fluvially transported P.

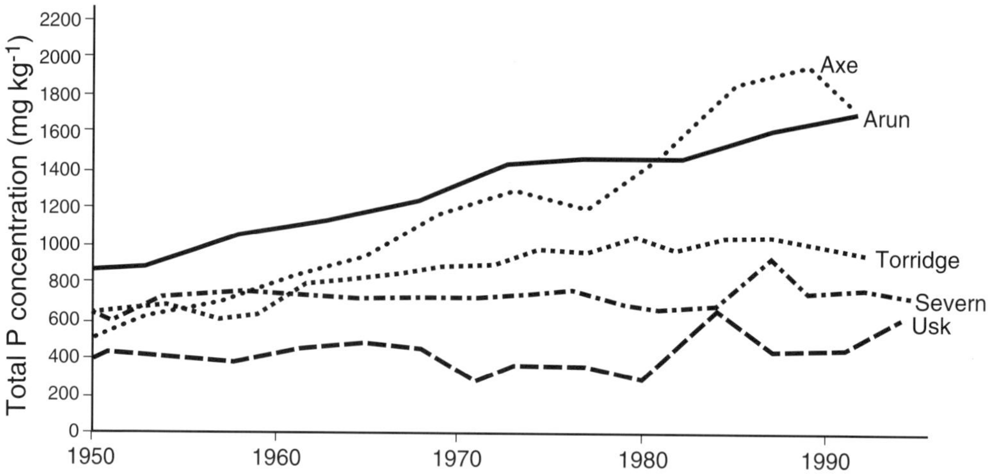

Fig. 3. A tentative reconstruction of variation in the total-P content of suspended sediment deposited on the floodplains of five British rivers over the past 40 years.

Table 1. *Estimates of the mean annual rate of deposition of total-P at the floodplain coring sites on the Rivers Severn, Usk, Torridge, Arun and Axe, since 1963*

Floodplain coring site	Mean annual rate of deposition of total-P (g-P m^{-2} a^{-1})
River Severn near Atcham	9.0
River Usk near Usk	3.8
River Torridge near Great Torrington	6.5
River Axe at Cownhaye	7.6
River Arun near Billingshurst	5.7

Documenting rates and patterns of overbank sedimentation

The results presented above regarding recent changes in the total-P content of overbank sediment deposits are based on estimates of sedimentation rates derived from interpretation of the downcore variation of ^{137}Cs activity in sectioned cores. Approximately 20 separate measurements of ^{137}Cs activity were required to establish the ^{137}Cs depth profile. The use of sectioned cores affords limited scope for detailed investigation of spatial patterns of overbank sedimentation, as the number of ^{137}Cs or unsupported ^{210}Pb analyses involved would preclude obtaining sedimentation rate estimates for more than a very limited number of cores. The use of bulk core measurements (i.e. one measurement per core) pioneered by the author and his co-workers affords a means of overcoming this limitation, by allowing sedimentation rate estimates to be obtained for a large number of sampling points. The potential of this approach can be demonstrated by considering some results from an investigation of the spatial pattern of overbank sedimentation rates on a portion of the floodplain of the River Severn near Buildwas, Shropshire. The study site, shown in Fig. 4, occupies the inside of a meander bend and 124 cores were collected from this area at the intersections of a 25 m $\times$ 25 m grid using a motorized percussion corer. Most of the cores were bulk cores (unsectioned) collected using a 69 mm diameter core tube. A small number of cores were collected using a 120 mm diameter core tube for subsequent sectioning. Coring depths were *c.* 0.70 m, and in all instances the cores included the total depth of sediment containing ^{137}Cs and unsupported ^{210}Pb. Additional cores were collected from undisturbed reference sites above the level of flood inundation to provide estimates of the local fallout inventories for ^{137}Cs and unsupported ^{210}Pb. All bulk cores and sectioned cores were analysed for both ^{137}Cs and unsupported ^{210}Pb, to estimate sedimentation rates over the past 40 years (^{137}Cs) and 100 years (unsupported ^{210}Pb). Measurements of the grain-size composition of sediment associated with individual cores were made using a Malvern Mastersizer laser granulometer, after standard laboratory pre-treatment. Estimates of the mean annual sedimentation rate were derived from the whole core values of excess ^{137}Cs and unsupported ^{210}Pb inventory, respectively, using the procedure described by Walling & He (1997*a*) for ^{137}Cs and the CICCS (Constant Initial Concentration, Constant Sedimentation Rate) model for unsupported ^{210}Pb (see He & Walling, 1996).

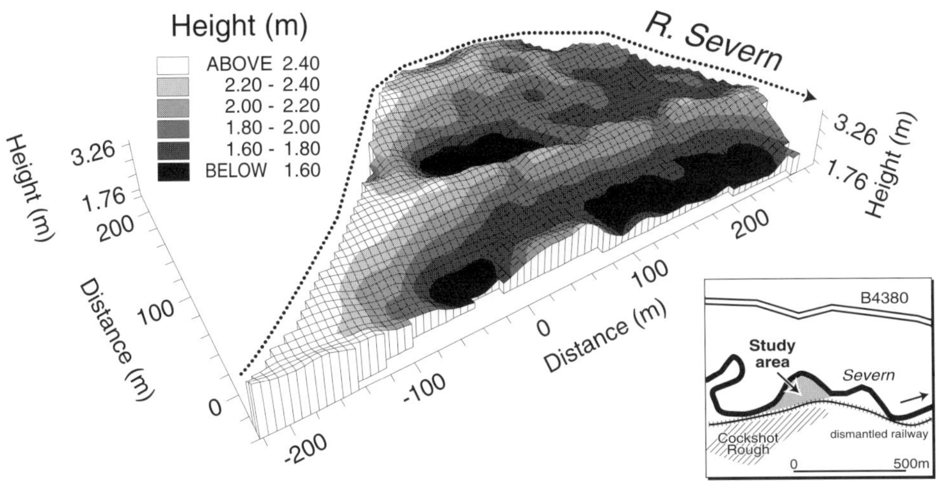

Fig. 4. The location (inset) and topography of the study site bordering the River Severn near Buildwas, Shropshire, UK.

The spatial patterns associated with the mean annual sedimentation rates within the study area over the past 40 and 100 years estimated from the ^{137}Cs and unsupported ^{210}Pb measurements are presented in Fig. 5. There is relatively little difference between either the magnitude or the pattern shown by the results for the two periods, indicating that overbank sedimentation has remained relatively constant at this site over the past 100 years. Sedimentation rates are generally in the range 0.15–0.50 g cm^{-2} a^{-1} and are similar in magnitude to the relatively low sedimentation rates reported by the author and his co-workers for other British rivers (e.g. Walling *et al.* 1996; He & Walling 1997; Walling & He 1998). The detail of the spatial patterns shown in Fig. 5 reflects both the microtopography of the site, with the series of ridges and swales associated with meander development, and a general trend of decreasing sedimentation with increasing distance from the channel and of increased sedimentation in areas occupied by deeper water during periods of flood inundation. The pattern of sedimentation in this area can also be related to the transfer of suspended sediment from the channel onto the floodplain by both diffusion and convection, and is broadly consistent with the models proposed by workers such as James (1985), Pizzuto (1987), Marriott (1992) and Mackey & Bridge (1995).

An attempt has been made to represent the combined effects of distance from the channel and flood-water depth on the observed rates of deposition (R (g cm^{-2} a^{-1})) by fitting the simple deposition function proposed by Walling *et al.* (1996) and Walling & He (1997*b*). In this function, overbank deposition on a floodplain is seen as comprising two components. The first represents the deposition of the coarser fraction across the floodplain at a rate proportional to the flood-water depth and its concentration, which is assumed to decrease exponentially with increasing distance (Y (m)) from the channel. The second reflects the deposition of finer particles, for which the sedimentation rate can be considered to be directly proportional to the flood-water depth (Z (m)), defined as the difference between the maximum floodplain height at the site and the local floodplain height at individual sampling points, because, in the case of fine sediment, concentrations can be assumed to be relatively constant across the floodplain. For a specific point on the floodplain, the concentration of fine sediment particles in the overlying water can be assumed to be relatively uniform in the vertical dimension, but, in the case of coarser particles, concentrations will be higher close to the floodplain surface. These mechanisms may be represented by the following function:

$$R = \alpha Z^{\beta} e^{-Y/\gamma} + \delta Z \qquad (1)$$

The first term on the right-hand side of equation (1) represents the deposition of the coarser particles and the second term the deposition of fine-grained sediment. α and δ reflect the average concentrations of the coarse and fine fractions of the suspended sediment, respectively. The constant γ represents the general decline in the concentration of the coarse fraction with increasing distance from the channel. β serves as a correction factor to take

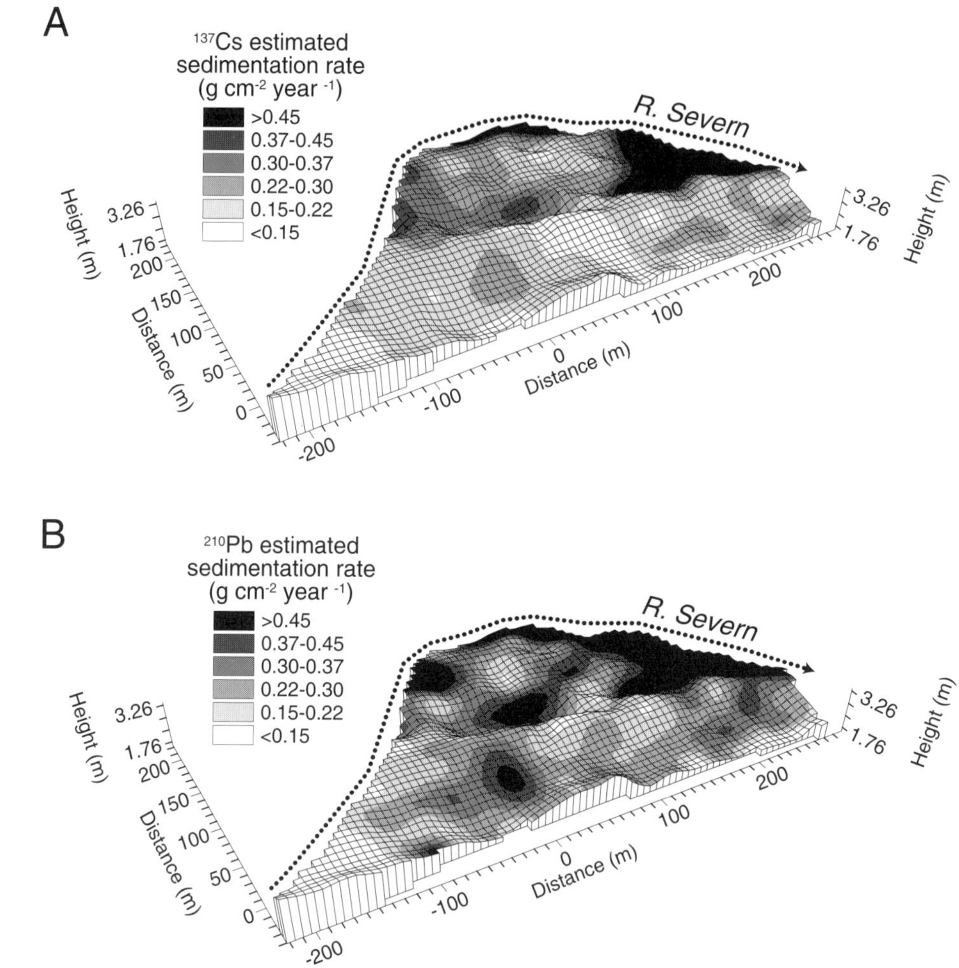

Fig. 5. Deposition rates across the floodplain study site bordering the River Severn near Buildwas, estimated from ^{137}Cs (**A**) and unsupported ^{210}Pb measurements (**B**).

account of the non-uniform vertical distribution of the coarse fraction in the floodwater. Fitting equation (1) to the sedimentation rates estimated using the ^{137}Cs measurements gives

$$R = 0.66Z^{1.9}\, e^{-Y/33} + 0.053Z + 0.14$$

with $R^2 = 0.71$. The sedimentation rates estimated from the unsupported ^{210}Pb measurements give a broadly similar result, namely,

$$R = 0.37Z^{0.8}\, e^{-Y/33} + 0.22Z + 0.03$$

with $R^2 = 0.65$. These results, which are presented graphically in Fig. 6, confirm the importance of both distance from the channel and flood-water depth in accounting for the patterns of sedimentation shown in Fig. 5 and thus the validity of the relationship indicated by equation (1). However, a fuller explanation of the pattern of sedimentation would need to take account of the additional influence of the local microtopography and the local flow patterns on suspended sediment transport and deposition.

The application of a simple function to describe the pattern of deposition in the study area presented above can be seen as demonstrating the potential for using detailed information on sedimentation rates provided by ^{137}Cs and fallout ^{210}Pb measurements to develop, calibrate and validate more complex sedimentation models (see Mackey & Bridge 1995; Nicholas & Walling 1995; Simm *et al.* 1997; Middelkoop & van der Perk 1998). In considering the problems of modelling overbank deposition, Carling & Petts (1992) have stressed the fact that theoretical developments have outstripped empirical validation for reach length scales. Use of ^{137}Cs and unsupported ^{210}Pb measurements may afford one means of rectifying this deficiency and advancing model development.

The potential to assemble detailed information on overbank sedimentation rates for small areas of

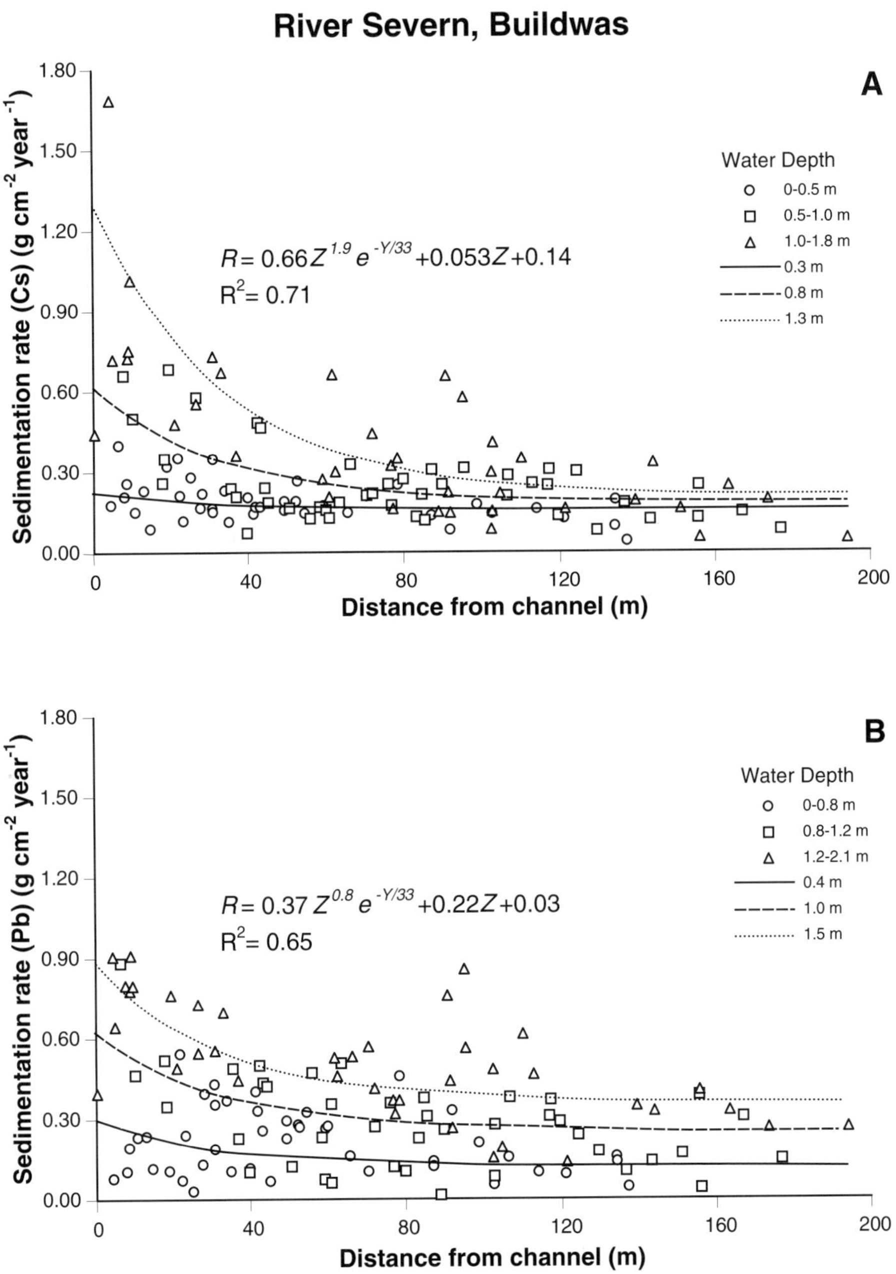

Fig. 6. The interrelationships between sedimentation rate and distance from the channel and water depth established for the floodplain study site bordering the River Severn near Buildwas, based on estimates of sedimentation rates derived from ^{137}Cs (**A**) and unsupported ^{210}Pb (**B**) measurements.

river floodplain also affords scope to examine the interrelationships between sedimentation rates and other aspects of the deposition process. The availability of information on the particle-size composition of surface sediment associated with each of the 124 cores collected from the study area on the floodplain of the River Severn near Buildwas provides, for example, the opportunity to examine the relationship between grain-size composition and sedimentation rate. The grain-size data

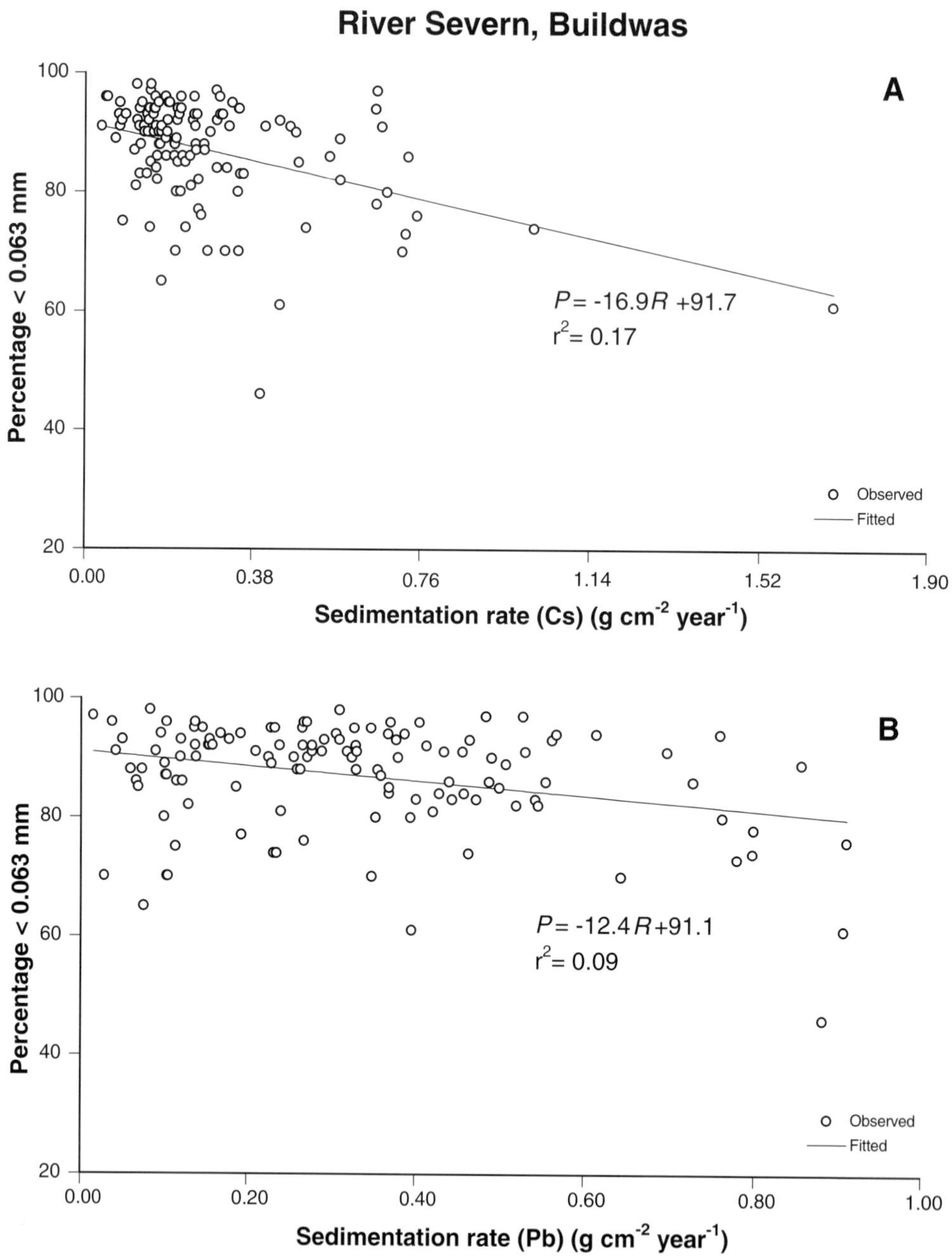

Fig. 7. Relationships between the grain-size composition of overbank floodplain sediment and sedimentation rate established for the floodplain study site bordering the River Severn near Buildwas, based on estimates of sedimentation rate derived from ^{137}Cs (**A**) and unsupported ^{210}Pb (**B**) measurements.

collected for this site show some evidence of the expected inverse relationship between the magnitude of the percentage <0.063 mm fraction and sedimentation rate, but this is not well defined (Fig. 7). The scatter in the relationship is likely to reflect, at least in part, the fact that the suspended sediment is transported and deposited as aggregates or composite particles, rather than as discrete particles. Thus, although the composite particles may respond to control by hydraulic conditions, this control may be masked when relationships involving the primary constituent particles are considered (see Nicholas & Walling 1997).

Quantifying conveyance losses associated with overbank sedimentation

The overbank sedimentation rates reported above for the River Severn, although relatively low, nevertheless confirm the potential significance of such sedimentation as a conveyance loss in the transport of suspended sediment through a river system. This significance was further confirmed by Walling & Quine (1993), who used information on the fate of Chernobyl-derived ^{134}Cs fallout within the 6850 km^2 basin of the River Severn above Upton on Severn, to estimate that sedimentation on the floodplains bordering the main channel system represented a transmission loss equivalent to 23% of the total suspended sediment input to the main channel system during the period 1986–1989. Measurements of ^{137}Cs and unsupported ^{210}Pb can also be exploited at a larger scale to estimate the total mass of sediment deposited on the floodplains bordering the main channels of a river system and thus to estimate the magnitude of the overall conveyance loss. If this loss is related to the suspended sediment output from the river system, it is possible to establish the importance of the conveyance loss relative to the total suspended sediment input to the main channel system (see Walling *et al.* 1998).

Use of this approach to quantifying conveyance losses necessitates scaling-up of sedimentation rate estimates obtained for individual points to the entire floodplain of a river system. Transects across the floodplain, located at representative sites along the main channels, can provide the basis for such extrapolation. If the average sedimentation rate can be established for an individual transect and if these values can be extrapolated to the adjacent floodplain reaches, it is possible to estimate the magnitude of the overall conveyance loss associated with the transfer of fine sediment through the main channel system. The work of Walling *et al.* (1998) on the floodplains bordering the main channel systems of the River Ouse (3315 km^2) and the River Wharfe (818 km^2) in Yorkshire, UK, provides a useful example of the application of this approach. In this study, more than 250 sediment cores were collected from 26 representative transects located along the main channel system of the Yorkshire Ouse and its primary tributaries and the River Wharfe (Fig. 8) for ^{137}Cs analysis. The cores collected from each transect included a single sectioned core taken from a representative location and a series of bulk cores taken from other representative points along the transect. Samples of surface sediment were collected from points immediately adjacent to the cores, for determination of grain-size composition. Reference cores were also collected from a number of sites located along the river system, but above the level of flood inundation, to establish the local ^{137}Cs fallout inventory for individual transects. The value of sedimentation rate obtained for the sectioned core was used to 'calibrate' the algorithm employed to estimate the sedimentation rate for each of the bulk cores based on their respective values of excess inventory and grain-size composition.

The estimates of mean sedimentation rate obtained for the individual transects using the ^{137}Cs measurements ranged from 0.010 to 0.554 g cm^{-2} a^{-1}, and averaged 0.206 g cm^{-2} a^{-1}. Extrapolation of these values to the individual reaches between adjacent transects provided estimates of the mean annual conveyance losses associated with overbank sedimentation on the floodplain bordering the main channel system of the River Ouse and its major tributaries and the River Wharfe (Table 2). By relating these losses to information on the mean annual suspended sediment load of the rivers (Table 2), it was possible to establish the relative importance of floodplain storage to the sediment budget of the main channel system (Table 2, Fig. 9). In the case of the main River Ouse system, floodplain deposition was estimated to account for *c.* 40% of the total mass of suspended sediment delivered to the main channel system over the past 40 years. For the River Wharfe the equivalent value was *c.* 49%.

Investigating changing rates of floodplain sedimentation over the past 100 years

As indicated previously, measurements of both ^{137}Cs and unsupported ^{210}Pb undertaken simultaneously on the same core afford a means of investigating changes in overbank sedimentation rates over the past 100 years, in that estimates of average sedimentation rates relating to the past *c.* 35 years derived from the ^{137}Cs measurements can be compared with equivalent values for the past *c.* 100 years derived from ^{210}Pb measurements. The

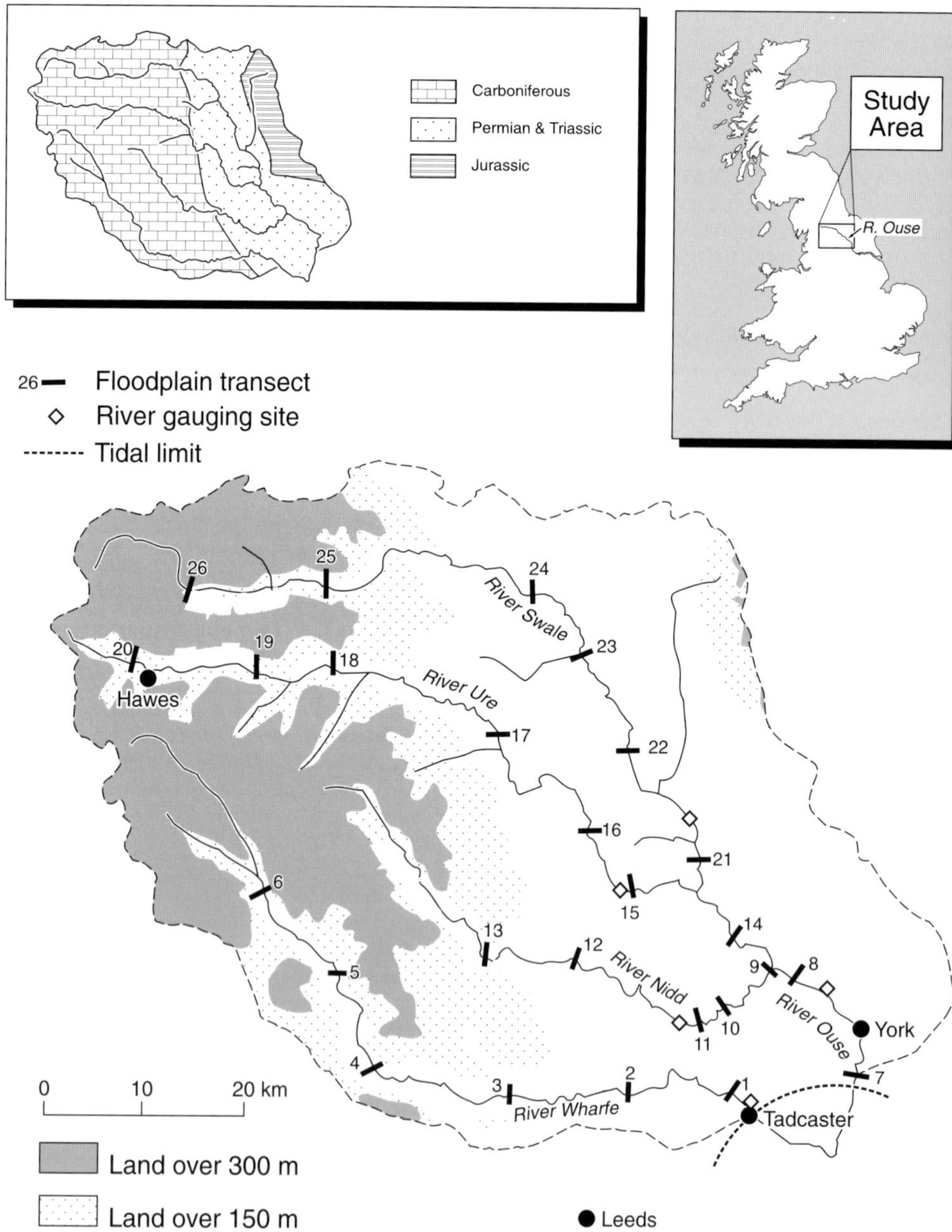

Fig. 8. The catchments of the River Ouse and River Wharfe, Yorkshire, UK, showing the location of the floodplain transects used to estimate sediment deposition rates.

overlap between these two periods necessarily reduces the temporal resolution of the results, but they can, nevertheless, provide evidence as to whether sedimentation rates have increased or decreased or remained essentially stationary in recent years (see Walling & He 1994).

As an example of this approach, Fig. 10 presents ^{137}Cs and unsupported ^{210}Pb depth distributions for representative sites on the floodplains of two British rivers, namely, the River Tone in Somerset and the River Arun in Sussex. In both cases the ^{137}Cs depth distributions show a clearly defined

Table 2. *A comparison of the estimates of total sediment storage on the floodplains bordering the main channel system of the River Ouse and its primary tributaries and the River Wharfe, with the estimated suspended sediment loads for the study rivers (based on Walling* et al. *1998)*

River	Floodplain storage (t a^{-1})	Mean annual suspended sediment load (t a^{-1})	Total sediment delivered to channel (t a^{-1})	Floodplain storage as % of sediment input
Swale	19214	42352	61566	31.2
Nidd	7573	7719	15292	49.5
Ure*	15125	28887	44012	34.3
Ouse*	18733			
Total to Ouse gauging station	49041	75111	124152	39.5
Total to tidal limit	60645			
Wharfe	10325	10816	21141	48.8

* Ure is the River Ure to its confluence with the River Swale, and Ouse refers to the River Ure–Ouse from below this point to the tidal limit.

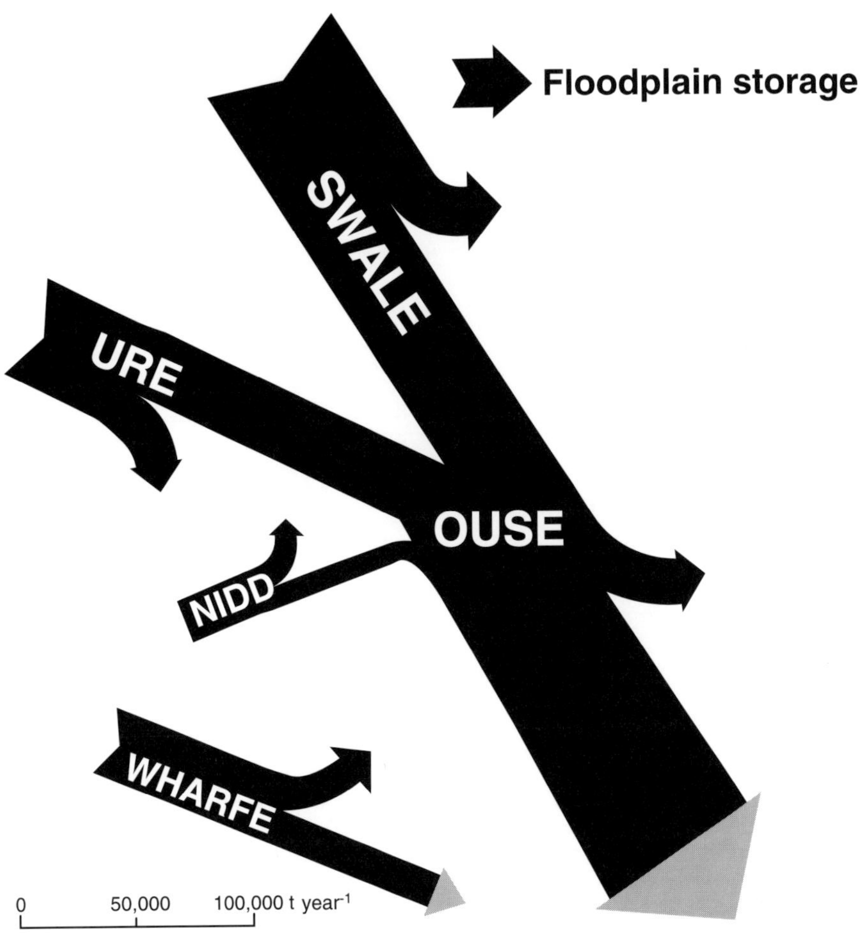

Fig. 9. The role of overbank floodplain sedimentation in the sediment budgets of the main channel systems of the River Ouse and its major tributaries and the River Wharfe, Yorkshire, UK (based on Walling *et al.* 1998).

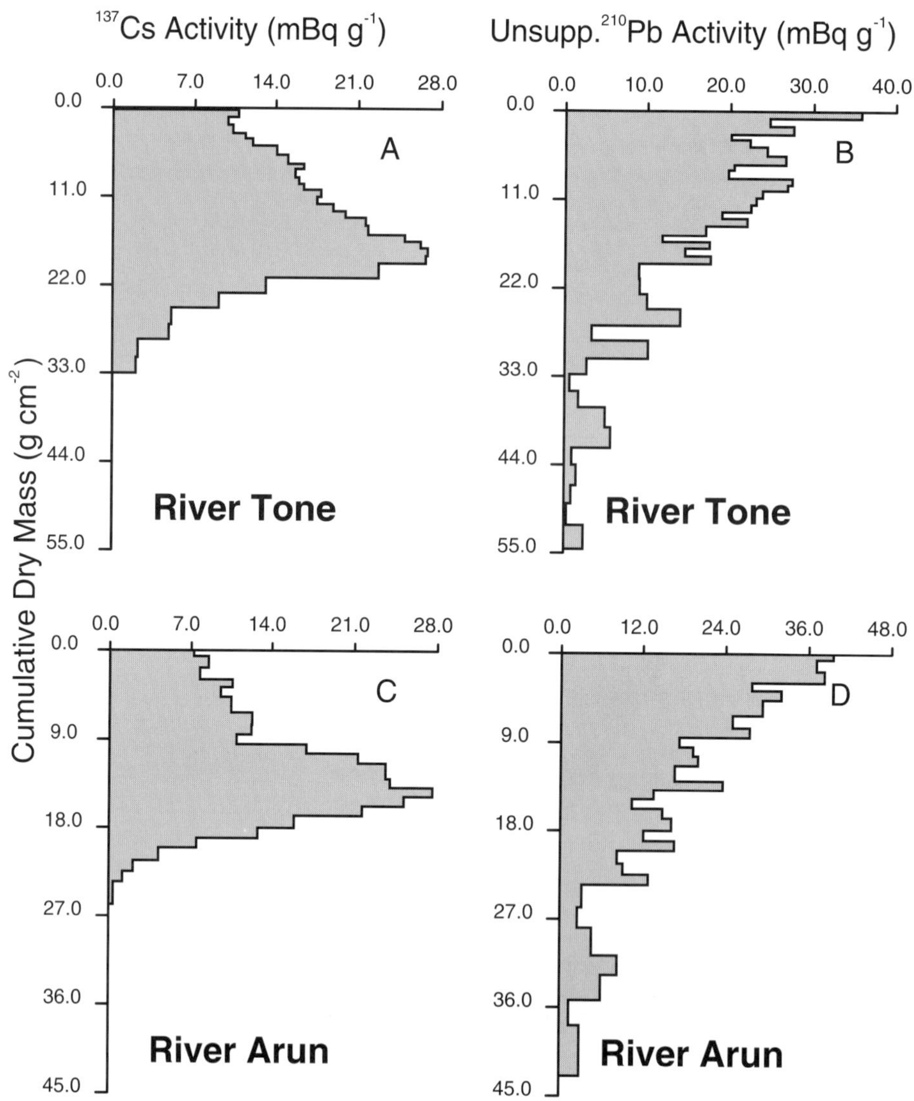

Fig. 10. ^{137}Cs and unsupported ^{210}Pb profiles associated with sediment cores collected from the floodplains of the River Tone near Bradford on Tone, Somerset (**A**, **B**) and the River Arun at Billingshurst, Sussex (**C**, **D**).

subsurface peak, about 180 mm and 130 mm below the surface, respectively, which reflects the level of the floodplain surface at the time of peak fallout in 1963. Estimates of the precise depth of the 1963 surface, and therefore of the average sedimentation rate over the period from 1963 to the time of collection of the cores in 1966, have been derived using the procedures described by Walling & He (1997*a*), which take account of post-depositional redistribution of ^{137}Cs within the sediment profile. The resulting estimates of mean annual sedimentation rate over the past 33 years for these sites are 0.56 and 0.39 g cm^{-2} a^{-1} for the Rivers Tone and Arun, respectively. The unsupported ^{210}Pb depth distributions for the two floodplain sites illustrated in Fig. 10 show the typical exponential decline in concentration with depth, reflecting progressive sediment accumulation and the radioactive decay of ^{210}Pb. The estimates of mean annual sedimentation rate derived for the two cores from the unsupported ^{210}Pb measurements are 0.43 and 0.48 g cm^{-2} a^{-1} for the Rivers Tone and Arun, respectively.

Schematic age–depth relationships for these two sites, based on the estimates of the average sedimentation rates over the past 33 years and 100 years, provided by the ^{137}Cs and unsupported ^{210}Pb data are presented in Fig. 11. This figure indicates that average sedimentation rates have increased towards the present in the case of the site on the River Tone, whereas the core from the River Arun shows evidence of a decreasing sedimentation rate towards the present.

As part of a broader study of rates of overbank sedimentation on the floodplains of British rivers, the approach outlined above has been applied to representative sediment cores collected from the floodplains of 21 rivers (see Walling & He 1999). The locations of the sampling sites are shown in Fig. 12 and listed in Table 3. For each site, the estimate of average sedimentation rate for the past 33 years has been compared with that for the past 100 years, to assess whether the sedimentation rate has increased, decreased or remained essentially constant. The values of average sedimentation rate presented in Table 3 cover the ranges 0.04–1.22 g cm^{-2} a^{-1} and 0.04–1.42 g cm^{-2} a^{-1}, for the ^{137}Cs and unsupported ^{210}Pb measurements, respectively, and are consistent with other data for British rivers presented in this paper. The site-specific nature of the individual values of sedimentation rate preclude detailed consideration of the factors influencing the magnitude of the values listed in Table 3. However, there is some evidence that maximum values are associated with rivers draining catchments with headwaters in upland areas (e.g. Severn, Usk, Yorkshire Ouse and Torridge), whereas rivers draining lowland catchments underlain by more permeable rocks are characterized by lower values (e.g. Dorset Stour, Rother, Arun and Medway).

Significant increases and decreases in deposition rates between the two periods have been identified as those situations where recent, that is, 33 year, sedimentation rates differ by more than ±10% from the 100 year value. Using this criterion, Table 3 indicates that 11 floodplain sites show decreasing sedimentation rates, four sites increasing sedimentation rates and six sites stable sedimentation rates. In all cases, however, both the increases and decreases in sedimentation rate are of limited magnitude and suggest that rates of sedimentation on the floodplains of British rivers have remained relatively constant over the past 100 years, despite significant changes in catchment land use and management practices and evidence for changes in flood magnitude and frequency (see Robson *et al.* 1998) Similar findings are provided by Fig. 5, which compares the patterns of overbank sedimentation on the floodplain of the River Severn at Buildwas derived from ^{137}Cs with those derived from unsupported ^{210}Pb measurements. These show a small decrease in the average sedimentation rate for the study area from 0.33 g cm^{-2} a^{-1} over the past 100 years to 0.28 g cm^{-2} a^{-1} over the past 40 years.

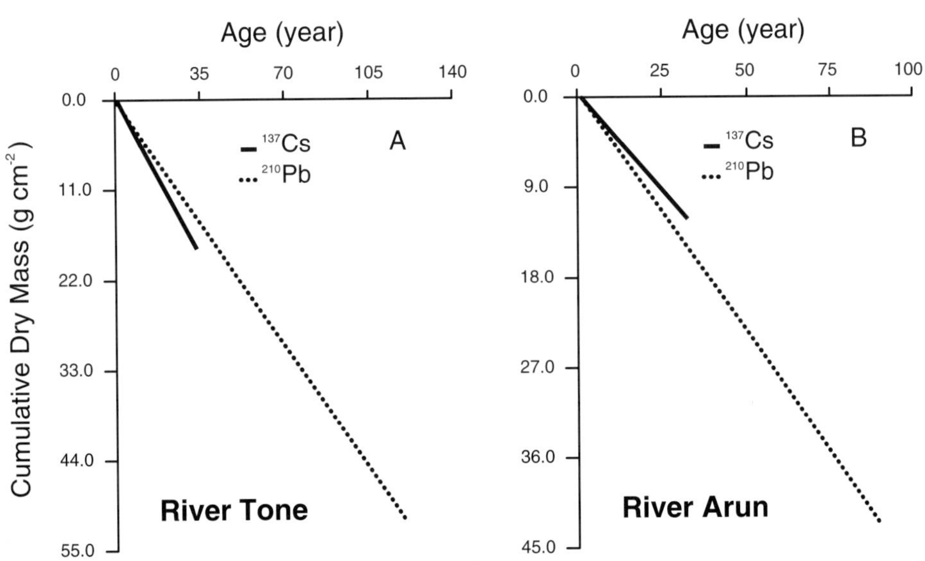

Fig. 11. The age–depth relationships established for the sediment cores illustrated in Fig. 10 using the estimates of mean annual sedimentation rate provided by the ^{137}Cs and unsupported ^{210}Pb measurements.

Where the results presented in Table 3 show evidence of change, small decreases in sedimentation rate predominate. Such decreases could be accounted for by the progressive aggradation of the floodplain surface, resulting in a reduction in the frequency and depth of overbank inundation. However, they could equally reflect reduced magnitude and/or frequency of overbank flooding, changes in flow and ponding patterns as a result of changes in floodplain land use or management, reduced suspended sediment loads and concentrations, and other factors, including local con-

Fig. 12. The location of the floodplain sites listed in Table 3.

Table 3. *Mean annual sedimentation rates estimated for representative sites on 21 British rivers using ^{137}Cs and unsupported ^{210}Pb measurements (based on Walling & He 1999)*

River and location		Sedimentation rate (g cm^{-2} a^{-1})		
		Past 33 years	Past 100 years	Trend*
1.	River Ouse near York	0.95	1.04	stable
2.	River Vyrnwy near Llanymynech	0.21	0.46	decrease
3.	River Severn near Atcham	1.22	1.42	decrease
4.	River Wye near Preston on Wye	0.15	0.28	decrease
5.	River Severn near Tewkesbury	0.86	0.95	stable
6.	Warwickshire Avon near Pershore	0.46	0.66	decrease
7.	River Usk near Usk	0.88	1.01	decrease
8.	Bristol Avon near Langley Burrell	0.39	0.33	increase
9.	River Thames near Dorchester	0.51	0.64	decrease
10.	River Torridge near Great Torrington	0.70	0.93	decrease
11.	River Taw near Barnstaple	0.60	0.65	stable
12.	River Tone near Bradford on Tone	0.56	0.43	increase
13.	River Exe near Stoke Canon	0.45	0.42	stable
14.	River Culm near Silverton	0.35	0.32	stable
15.	River Axe near Colyton	0.51	0.40	increase
16.	Dorset Stour near Spetisbury	0.04	0.04	stable
17.	River Rother near Fittleworth	0.11	0.14	decrease
18.	River Arun near Billingshurst	0.39	0.48	decrease
19.	River Adur near Partridge Green	0.51	0.71	decrease
20.	River Medway near Penshurst	0.15	0.23	decrease
21.	River Start near Slapton	0.51	0.45	increase

* Increases and decreases are identified as cases where recent (33 year) sedimentation rates deviate by more than ±10% from the 100 year value.

ditions. Interestingly, all four of the floodplain sites showing evidence of a small, but significant, increase in sedimentation rate towards the present (i.e. Axe, Tone, Start and Bristol Avon) are located in Southwest England. However, other rivers in this region (i.e. Taw, Torridge, Exe and Culm) are characterized by stable or reduced sedimentation rates. The increases are therefore more likely to reflect changes in land use and land management or other local conditions, rather than climate change, which could be expected to exert a more region-wide influence. The 11 floodplain sites characterized by reduced floodplain sedimentation rates towards the present include rivers draining both upland areas (i.e. Vyrnwy, Severn, Wye, Usk and Torridge) as well as lowland areas (Warwickshire Avon, Thames, Rother, Medway and Arun) and no clear regional differentiation is apparent.

Perspective

This contribution has attempted to demonstrate the potential for using fallout radionuclide measurements in investigations of contemporary overbank sedimentation on river floodplains. All the examples cited relate to lowland reaches of British rivers and it must be accepted that the characteristics of these rivers, which include relatively fine-grained suspended sediment loads, limited channel mobility, and significant areas of overbank deposits with little vertical variation in grain size, make them well suited to this approach. It is likely to be somewhat less successful in upland areas characterized by more mobile channels and where coarser sediments predominate. In some areas of Britain, primarily in the western parts of the country, additional inputs of ^{137}Cs fallout occurred in 1986 as a result of the Chernobyl accident (see Smith & Clark 1989). These additional inputs must be taken into account when interpreting ^{137}Cs depth profiles and, more particularly, inventories. The site bordering the River Severn at Buildwas falls into this category, but Chernobyl fallout was very limited over most areas of lowland Britain represented by the studies reported in this contribution. Where Chernobyl-derived fallout inventories are relatively high, the use of ^{137}Cs measurements to estimate overbank sedimentation rates may be compromised. Unsupported ^{210}Pb measurements should, nevertheless, prove applicable in such areas, and in this context unsupported ^{210}Pb should be seen as a possible alternative to ^{137}Cs, as well as affording opportunities for conjunctive use.

The use of fallout radionuclides must be seen as offering an essentially unique means of establishing the chronology of recent overbank sediment deposits and of providing information on the rates of sedimentation involved. The potential to assemble detailed information on spatial patterns of overbank sedimentation could, in particular, provide a valuable basis for model development and validation. The approach does, nevertheless, have some limitations. These include its relatively low temporal resolution, which means that it is generally only able to provide estimates of mean annual sedimentation rates, rather than values for specific events or periods, and the lengthy count-times required by the associated laboratory measurements.

The work reported in this paper draws heavily on studies undertaken by the author and his co-workers over the past decade. The substantial contributions of Qingping He and P. Owens to these studies, the assistance of J. Grapes, A. Ames and W. Blake with gamma spectrometry measurements and sediment analysis, and the financial support of the UK Natural Environment Research Council through Research Grants GR3/8633 and GST/02/774 are gratefully acknowledged. Thanks are also extended to T. Bacon for producing the figures.

References

ALLAN, R. J. 1986. *The Role of Particulate Matter in the Fate of Contaminants in Aquatic Ecosystems*. Inland Waters Directorate, Environment Canada, Scientific Series, **142**.

ASSELMAN, N. E. M. & MIDDELKOOP, H. 1995. Floodplain sedimentation: quantities, patterns and processes. *Earth Surface Processes and Landforms*, **20**, 481–499.

BROWN, A. G. 1983. An analysis of overbank deposits of a flood at Blandford Forum, Dorset, England. *Revue de Géomorphologie Dynamique*, **32**, 95–99.

BURT, T. P. & HAYCOCK, N. E. 1996. Linking hillslopes to floodplains. *In*: ANDERSON, M. G., WALLING, D. E. & BATES, P. D. (eds) *Floodplain Processes*. Wiley, Chichester, 461–92.

CARLING, P. & PETTS, G. E. (eds) 1992. *Lowland Floodplain Rivers*. Wiley, Chichester.

CHARLESWORTH, S. M. & FOSTER, I. D. L. 1983. Effects of urbanisation on lake sedimentation: the history of two lakes in Coventry, UK—preliminary results. *In*: MCMANUS, J. & DUCK, R. W. (eds) *Geomorphology and Sedimentology of Lakes and Reservoirs*. Wiley, Chichester, 15–29.

FUSTEC, E., ENGUERRAN, F., BONTÉ, P. & FARDEAU, J-C. 1995. Rétention et biodisponibilité du phosphore dans les zones inondables. *In*: *Rapport Piren Seine*. 1995/II, Thème Corridor Fluvial, Laboratoire de Géologie Appliquée, University Pierre et Marie Curie, Paris.

HE, Q. & WALLING, D. E. 1996. Use of fallout Pb-210 measurements to investigate longer-term rates and patterns of overbank sediment deposition on the floodplains of lowland rivers. *Earth Surface Processes and Landforms*, **21**, 141–154.

—— & —— 1997. Rates of overbank deposition on the floodplains of British lowland rivers documented using fallout ^{137}Cs. *Geografiska Annaler*, **78A**, 223–234.

HUPP, C. R. 1988. Plant ecological aspects of flood geomorphology and palaeoflood history. *In*: BAKER, V. R., KOCHEL, R. C. & PATTON, P. C. (eds) *Flood Geomorphology*. Wiley, Chichester, 330–356

JAMES, C. S. 1985. Sediment transfer to overbank sections. *Journal of Hydraulic Research*, **23**, 435–452.

JOSHI, S. R. 1987. Nondestructive determination of lead-210 and radium-226 in sediments by direct photon analysis. *Journal of Radioanalytical and Nuclear Chemistry*, **116**, 169–182.

KESEL, R. H., DUNNE, K. C., MCDONALD, R. C., ALLISON, K. R. & SPICER, B. E. 1974. Lateral erosion and overbank deposition on the Mississippi River in Louisiana caused by 1973 flooding. *Geology*, **2**, 461–464.

LAMBERT, C. P. & WALLING, D. E. 1987. Floodplain sedimentation: a preliminary investigation of contemporary deposition within the lower reaches of the River Culm, Devon, UK. *Geografiska Annaler*, **69A**, 47–59.

LEENAERS, H. & SCHOUTEN, C. J. 1989. Soil erosion and floodplain soil pollution: related problems in the geographical context of a river basin. *In*: *Sediment and the Environment*. (*Proceedings of the Baltimore Symposium*), International Association of Hydrological Sciences Publication, **184**, 75–83.

LEWIN, J. & MACKLIN, M. G. 1987. Metal mining and floodplain sedimentation in Britain. *In*: GARDINER, V. (ed.) *International Geomorphology. Part 1*, Wiley, Chichester, 1009–1027.

MACKEY, S. D. & BRIDGE, J. S. 1995. Three-dimensional model of alluvial stratigraphy: theory and application. *Journal of Sedimentary Research*, **B65**, 7–31.

MARRIOTT, S. B. 1992. Textural analysis and modelling of a flood deposit: River Severn, UK. *Earth Surface Processes and Landforms*, **17**, 687–697.

MARRON, D. C. 1989. The transport of mine tailings as suspended sediment in the Belle Fourche River, west–central Dakota, USA. *In*: *Sediment and the Environment* (Proceedings of the Baltimore Symposium), International Association of Hydrological Sciences Publication, **184**, 19–26.

MIDDELKOOP, H. & ASSELMAN, N. E. M. 1998. Spatial variability of floodplain sedimentation at the event scale in the Rhine–Meuse delta, the Netherlands. *Earth Surface Processes and Landforms*, **23**, 561–573.

—— & VAN DER PERK, M. 1998. Modelling spatial patterns of overbank sedimentation on embanked floodplains. *Geografiska Annaler*, **80A**, 95–109.

NICHOLAS, A. P. & WALLING, D. E. 1995. Modelling contemporary overbank sedimentation on floodplains: some preliminary results. *In*: HICKIN, E. J (ed.) *River Geomorphology*. Wiley, Chichester, 131–153.

—— & —— 1997. The significance of particle aggregation in the overbank deposition of suspended sediment on river floodplains. *Journal of Hydrology*, **186**, 275–293.

OLDFIELD, F. & APPLEBY, P. G. 1984. Empirical testing of ^{210}Pb dating models for lake sediments. *In*: HAWORTH, E. Y. & LUND J. W. G. (eds) *Lake Sediments and Environmental History.* Leicester University Press, Leicester, 93–114.

PENNING-ROWSELL, E. C. & TUNSTALL, S. M. 1996. Risks and resources: defining and managing the floodplain. *In*: ANDERSON, M. G., WALLING, D. E. & BATES, P. D. (eds) *Floodplain Processes.* Wiley, Chichester, 493–533.

PETTS, G. E. 1996. Sustaining the ecological integrity of large floodplain rivers. *In*: ANDERSON, M. G., WALLING, D. E. & BATES, P. D. (eds) *Floodplain Processes.* Wiley, Chichester, 535–551.

PIZZUTO, J. E. 1987. Sediment diffusion during overbank flows. *Sedimentology,* **34**, 301–317.

ROBSON, A. J., JONES, T. K., REED, D. W. & BAYLISS, A. C. 1998. A study of national trend and variation in UK floods. *International Journal of Climatology,* **18**, 165–182.

SIMM, D. J., WALLING, D. E., BATES, P. D. & ANDERSON, M. G. 1997. The potential application of finite element modelling of floodplain inundation to predict patterns of overbank deposition. *Hydrological Sciences Journal*, **42**, 859–875.

SMITH, F. B. & CLARK, M. J. 1989. *The Transport and Deposition of Airborne Debris from the Chernobyl Nuclear Power Plant Accident with Special Emphasis on the Consequences to the United Kingdom.* Meteorological Office Scientific Paper, **42**.

VAITHIYANATHAN, P. & CORRELL, D. L. 1992. The Rhode River watershed: phosphorus distribution and export in forest and agricultural soils. *Journal of Environmental Quality*, **21**, 280–288.

WALLING, D. E. & HE, Q. 1993. Use of caesium-137 as a tracer in a study of rates and patterns of floodplain sedimentation. *In*: *Tracers in Hydrology. (Proceedings of the Yokohama Symposium)*, International Association of Hydrological Sciences Publication, **215**, 319–328.

—— & —— 1994. Rates of overbank sedimentation on the flood plains of several British rivers during the past 100 years. *In*: *Variability in Stream Erosion and Sediment Transport. (Proceedings of the Canberra Symposium)*, International Association of Hydrological Sciences Publication, **224**, 203–210.

—— & —— 1997*a*. Use of fallout ^{137}Cs in investigations of overbank sediment deposition on river floodplains. *Catena,* **29**, 263–282.

—— & —— 1997*b*. Investigating spatial patterns of overbank sedimentation on river floodplains. *Water, Air and Soil Pollution*, **99**, 9–20.

—— & —— 1998. The spatial variability of overbank sedimentation on river floodplains. *Geomorphology*, **24**, 209–223.

—— & —— 1999. Changing rates of overbank sedimentation on the floodplains of British rivers during the past 100 years. *In*: BROWN, A. G. & QUINE, T. A. (eds) *Fluvial Processes and Environmental Change.* Wiley, Chichester, 207–222.

—— & QUINE, T. A. 1993. Using Chernobyl-derived fallout radionuclides to investigate the role of downstream conveyance losses in the suspended sediment budget of the River Severn, United Kingdom. *Physical Geography*, **14**, 239–253.

——, HE, Q. & NICHOLAS, A. P. 1996. Floodplains as suspended sediment sinks. *In*: ANDERSON, M. G., WALLING, D. E. & BATES, P. D. (eds) *Floodplain Processes.* Wiley, Chichester, 399–440.

——, OWENS, P. N. & LEEKS, G. J. L. 1998. The role of channel and floodplain storage in the suspended sediment budget of the River Ouse, Yorkshire, UK. *Geomorphology*, **22**, 225–242.

——, QUINE, T. A. & HE, Q. 1992. Investigating contemporary rates of floodplain sedimentation. *In*: Petts, G. & Carling, P.A. (eds) *Lowland Floodplain Rivers.* Wiley, Chichester, 154–184.

WOLMAN, M. G. & LEOPOLD, L. B. 1957. *River Floodplains: some Observations on their Formation.* US Geological Survey, Professional Papers, **282-C**.

Source and fate of Chernobyl-derived radiocaesium on floodplains in Ukraine

MARCEL VAN DER PERK[1], PETER A. BURROUGH[1], ADRIAN S .C. CULLING[2], GENNADY V. LAPTEV[3], BORIS PRISTER[4], UMBERTO SANSONE[5] & OLEG V. VOITESKHOVITCH[3]

[1]*The Netherlands Centre for Geo-Ecological Research—ICG, Department of Physical Geography, Utrecht University, Utrecht, The Netherlands*
(e-mail: m.vanderperk@geog.uu.nl)

[2]*Institute of Terrestrial Ecology, Merlewood Research Station, Grange-over-Sands, Cumbria, LA11 6JU, UK*

[3]*Ukrainian Hydrometeorological Institute, Kiev, Ukraine*

[4]*Ukrainian Institute of Agricultural Radiology, Kiev, Ukraine*

[5]*Agenzia Nazionale per la Protezione dellAmbiente, Rome, Italy*

Abstract: The accident in the Chernobyl nuclear power plant on 26 April 1986 has resulted in radiocaesium contamination in the Dnieper–Pripyat catchment in Ukraine, Belarus and Russia. As radiocaesium tends to be readily adsorbed by soil particles, by illitic clay minerals in particular, sedimentation and erosion of fine soil particles are the major radiocaesium redistribution processes that have occurred since initial deposition. A part of the eroded material is transported to and through the river channel. The seasonal character of fluvial transport causes deposition of this sediment on the river banks and floodplains. This is particularly relevant if floodplains are used for agriculture. The geochemical characteristics of the floodplain soils determine to what extent radiocaesium is taken up into food chains. In this paper, we try to identify the source and fate of radiocaesium on a floodplain along the Sluč River and its tributaries in the Dubrovitsa district, Ukraine, some 350 km west of the Chernobyl nuclear power plant. In July 1993, a major summer flood occurred, which inundated the floodplain for about 3 weeks. The observed provisional increase of radiocaesium contamination in soil and cow's milk in this area after this flood is discussed in relation to soil type and proximity to the river channel.

The Chernobyl accident on 26 April 1986 has resulted in surface contamination by various long-lived radionuclides over vast areas of Europe. Although most of the contamination was deposited in the direct vicinity of the Chernobyl nuclear power plant, most of the Dnieper and Pripyat catchments became contaminated with radiocaesium (^{137}Cs) deposition >1 Ci km^{-2} (37 kBq m^{-2}) (see Fig. 1). It is well known that caesium tends to be adsorbed effectively by illitic clay minerals (Cremers *et al.* 1998). Thus, the initial deposition of ^{137}Cs has changed with time, partly as a result of physical decay and leaching to deeper horizons and groundwater, but also because of erosion and deposition of fine soil particles (Bonnet 1990; Korobova & Linnik 1993; Walling & Quine 1995; Kudelsky *et al.* 1996; Walling & He 1997). Soil eroded on hill slopes is transported to and through the river. A part of this eroded material leaves the catchment, but most is redistributed within the catchment and deposited on floodplains. Consequently, because of additional inputs of radionuclides adsorbed to deposited sediments, floodplain soils may be particularly vulnerable in terms of environmental pollution, particularly if these are used for agriculture.

In the Sluč catchment near Dubrovitsa, about 350 km west of the Chernobyl nuclear power plant, ^{137}Cs contamination data for soil and cow's milk had been collected for a survey of the transfer of radionuclides to animals (Strand *et al.* 1996). The study site comprised two collective farms in a extremely flat landscape including floodplains of the Sluč River and some of its tributaries (Figs 1 and 2). The mineral soils in the study area vary from soddy podzolic soils on the relatively higher

From: MARRIOTT, S. B. & ALEXANDER, J. (eds) 1999. *Floodplains: Interdisciplinary Approaches*. Geological Society, London, Special Publications, **163**, 61–67. 1-86239-050-9/99/$15.00

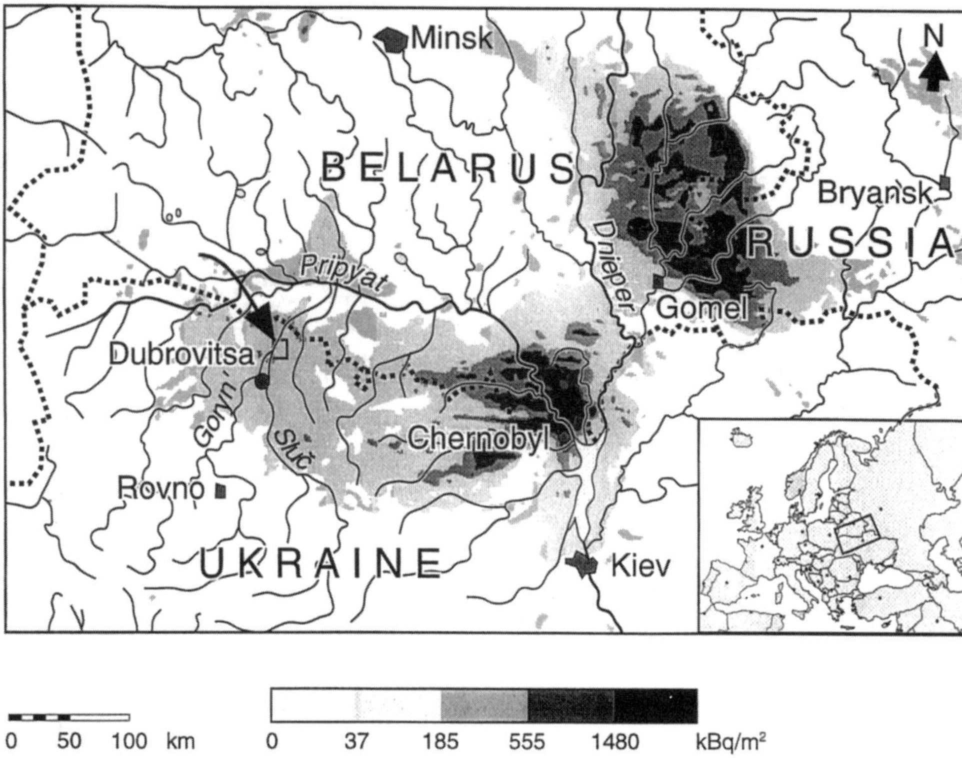

Fig. 1. Location of the study site in the Pripyat–Dnieper catchment.

areas to gley soils in the floodplains, but large parts are covered by peaty soils. The collected data showed that surface contamination by ^{137}Cs and transfer of ^{137}Cs from soil to milk changes in time and previous analysis of these data by Burrough *et al.* (1996) demonstrated that these observed changes were related to flooding.

In this paper we aim to identify the source and fate of ^{137}Cs on the floodplains in the area near Dubrovitsa, Ukraine. This paper discusses the previous observations of contamination of soil and milk by ^{137}Cs in the light of the hydrological, geochemical and ecological processes accounting for changes in floodplain contamination by radionuclides. Moreover, it discusses the role of overbank sedimentation of contaminated sediments and other biogeochemical processes involved in the changes of radioactive contamination.

Available data and data manipulation

The basic data used for this study were obtained from monitoring campaigns by the Institute of Agricultural Radiology, Kiev, Ukraine, and by the Experimental Collaboration Project No. 9 financed by the European Commission (Strand *et al.* 1996). The study site comprises the Chapayev and Kolos collective farms near Dubrovitsa, in the northwest of Ukraine (Fig. 1). The site is part of the Sluč and Goryn' catchments, which drain into the Pripyat River. The available data are briefly summarized below.

Land use, soil type and proximity to flooding

Digital vector maps of land cover, soil, and drainage network and extent of floodplains at an original scale of 1:10 000 were provided in Arc-Info format by the Institute of Agricultural Radiology, Ukraine, and the Institute of Terrestrial Ecology, UK. The maps were gridded to raster maps with the same grid with a cell size of 50 m × 50 m (Burrough *et al.* 1996). For this study, the original very detailed classifications of the soil and land cover maps were simplified to five classes for the land cover map and two classes for the soil map (Fig. 2b and c). Because of their different adsorption characteristics for ^{137}Cs, the soils were divided into organic soils (including meadow swamps and peat bogs) and mineral soils (including soddy podzols and gley soils). As a detailed digital elevation model (DEM) and time series of local water levels were lacking, an indication of flood frequency was obtained by calculating the distance to rivers and/or annually flooded areas (see Fig. 2d).

Seventeen per cent of the study area consists of floodplain that is flooded annually by spring floods, which usually occur during April–May. Occasionally, summer floods inundate the area; for example, during the summer months of 1998 and 1993. About 23% of the floodplains consist of organic peat soils. The annually flooded areas are mainly used as rough unimproved pasture, whereas on the higher areas further away from the river improved pasture and arable land are found.

^{137}Cs surface contamination in 1988, 1993 and 1994

Ukrainian authorities first performed detailed measurements of ^{137}Cs in the topsoil of the study areas in 1988. Bulked samples consisting of five subsamples were cored from the upper 20–30 cm of the soil located within the agricultural fields. Following a summer flood that inundated substantial parts of the study area for about 3 weeks in July 1993, further sampling was carried out and a third set of samples was collected in summer 1994. These latter two sampling campaigns were carried out in the framework of the ECP9 project (Strand *et al.* 1996). Altogether, the ^{137}Cs surface contamination was measured at 72 sites in 1988, 87 sites in

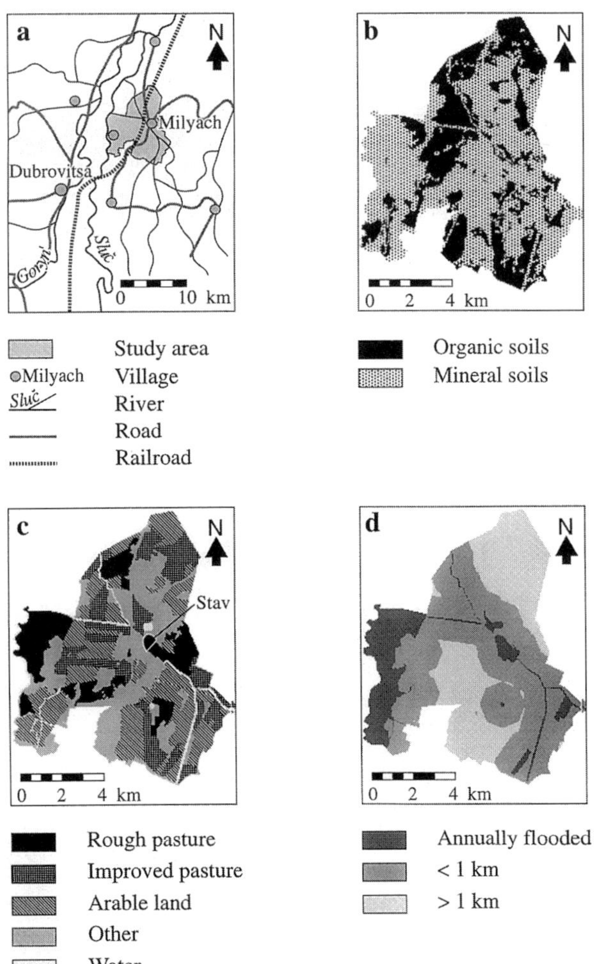

Fig. 2. The Dubrovitsa study site: (**a**) location of the study site in its environs; (**b**) soil types; (**c**) land use; (**d**) flood proximity zones.

Table 1. *Statistics of the transformed data as 1986 equivalents (kBq m^{-2})*

Year	*n*	Min.	Max.	Mean	Median	Mode	CV
1988	72	9.3	406.0	134.4	97.6	93.0	67.1
1993	87	20.6	1267.8	239.4	164.6	107.3	95.2
1994	47	17.3	531.5	135.5	93.6	~80.9	83.0

1993, and 47 sites in 1994. To rule out radioactive decay to allow comparisons between the three datasets, all samples were transformed to equivalent 1986 values using an isotopic ratio for 1986 of $^{137}Cs/^{134}Cs$ = 2:1 (Burrough *et al.* 1996). Because the frequency distribution of the ^{137}Cs contamination values resembled lognormal distributions, the values were also transformed by taking their natural logarithms. Table 1 summarizes the statistics of the three datasets.

Unfortunately, the sampling sites were not the same for the different years, so direct comparison between matched pairs of sites for ^{137}Cs contamination values was not possible. To overcome this problem, the log-transformed data were interpolated in the geographical information system (GIS) to a 50 m × 50 m grid using ordinary point kriging (Deutsch & Journel 1992) and subsequently, back-transformed to obtain median ^{137}Cs contamination values (Burrough *et al.* 1996) (Fig. 3). In the GIS, both the average ^{137}Cs contamination for the different landscape classes based on soil type and proximity to flooded zones (Table 2) and the total ^{137}Cs inventory in relation to flood proximity (Table 3) were examined.

A first inspection of the surface contamination by ^{137}Cs over the three sampled years (Fig. 3, Tables 2 and 3) showed that sites near the river have generally higher contamination values and also greater variability than sites further away from the river. Table 2 also demonstrates that mineral soils in the floodplain contain more ^{137}Cs than organic soils. In 1993, the contamination appeared to be generally higher than in 1988 and 1994 (see Table 1). It is remarkable that in the floodplains, the ^{137}Cs inventory (1986 equivalents) increased between 1988 and 1994, whereas in the rest of the area it remained constant or decreased.

^{137}Cs activity concentrations in cow's milk, 1991–1994

During the period 1991–1995, milk samples were collected from a private herd of about ten cows grazing the Stav pasture near Milyach village (see Fig. 2a). The Stav pasture comprises a part of the peaty floodplain of a tributary of the Sluč River, which is annually flooded. The estimated median surface contamination by ^{137}Cs amounted to 87

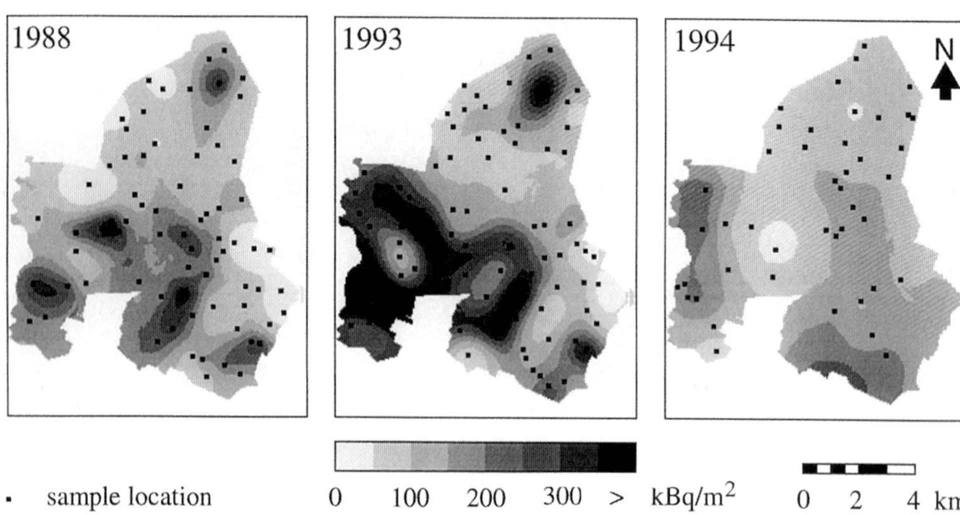

Fig. 3. Median ^{137}Cs contamination values (1986 equivalents) in the study area in 1988, 1993 and 1994 as interpolated from log-transformed point observations.

Table 2. *Median ^{137}Cs contamination values (in kBq m^{-2}) compared with a combination of the incidence and extent of flooding areas that are annually flooded, areas <1 km from water or annually flooded areas, and areas >1 km from water or annually flooded areas and soil type (mineral or organic soil) as calculated from interpolated log-transformed ^{137}Cs contamination values*

	Annually flooded		<1 km		>1 km	
Year	Mineral	Organic	Mineral	Organic	Mineral	Organic
1988	101	90	102	92	121	121
1993	238	199	156	142	174	164
1994	158	137	104	97	83	82

kBq m^{-2} in 1999, 110 kBq m^{-2} in 1993, and 130 kBq m^{-2} in 1994 (1986 equivalents). The pasture has not been improved and has not been influenced by any countermeasure to decrease the transfer of radionuclides from soil to cows. The milk samples collected were analysed for ^{137}Cs activity concentrations (Fig. 4).

The ^{137}Cs activity concentrations in the cows' milk show an exponential decrease since 1991 with an effective half-life of 4.6 years. Such an exponential decrease is a well-known effect of the irreversible ^{137}Cs immobilization by illite, which reduces bioavailability of ^{137}Cs (Livens & Loveland 1988). The only significant departure from this trend is the sudden increase in milk contamination measured in July–August 1993 directly after the summer flood. By the beginning of the 1994 season, ^{137}Cs activity concentrations in milk were consistent with the previous gradual decline. Although slight short-lived increases in ^{137}Cs activity concentrations in milk are noticeable in spring, the annual spring floods do not seem to cause any considerable ^{137}Cs peaks.

Discussion

The following main findings ensue from the preceding spatial analysis of ^{137}Cs contamination of soil and cow's milk in a floodplain area in Ukraine: (1) surface contamination by ^{137}Cs is highest in the floodplains and decreases further away from the river; (2) surface contamination is higher in mineral soils than in organic soils; (3) surface contamination by ^{137}Cs shows a large temporal variation, with the highest values observed in 1993; (4) ^{137}Cs activity concentrations in cow's milk show a considerable temporary peak after the 1993 summer flood, whereas the annual spring floods do not cause an increase in milk contamination.

Because of the affinity of ^{137}Cs with illitic clay minerals, it is likely that the changes in observed patterns of surface contamination by ^{137}Cs are caused by overbank deposition of contaminated sediments. This explains the increase of the ^{137}Cs contamination in the floodplains over the years 1988–1994, whereas in the rest of the area the contamination remained constant or even decreased (Table 3). The fact that the contamination particularly increased on mineral soils supports these findings. The data show that, on average, the soil contamination in 1993 was much higher than in 1988 and 1994. This could be related to the summer flood that inundated the area for 2–3 weeks in July 1993 and apparently caused the enhanced milk contamination of cows grazing a floodplain pasture. Table 3 shows that the total inventory in the study area increased by more than 6 TBq from 1988 to 1993 and that the inventory of the annual flooded floodplains increased by 2.28 TBq in this same period. With typical radiocaesium contamination of the suspended sediment in the rivers of the Pripyat basin of 100–1000 Bq kg^{-1}, this yields a rough estimation of the average floodplain sedimentation rate of 35–350 kg m^{-2} a^{-1}. This is an extremely high value in comparison with

Table 3. *Total ^{137}Cs inventory (in TBq) of the entire study area and compared with proximity of water and annually flooded areas*

Year	Whole area ($76.65\ km^2$)	Annually flooded areas ($13.01\ km^2$)	<1 km ($38.53\ km^2$)	>1 km ($25.11\ km^2$)
1988	9.18	1.52	4.37	3.29
1993	15.78	3.79	7.26	4.73
1994	8.39	2.07	4.06	2.26

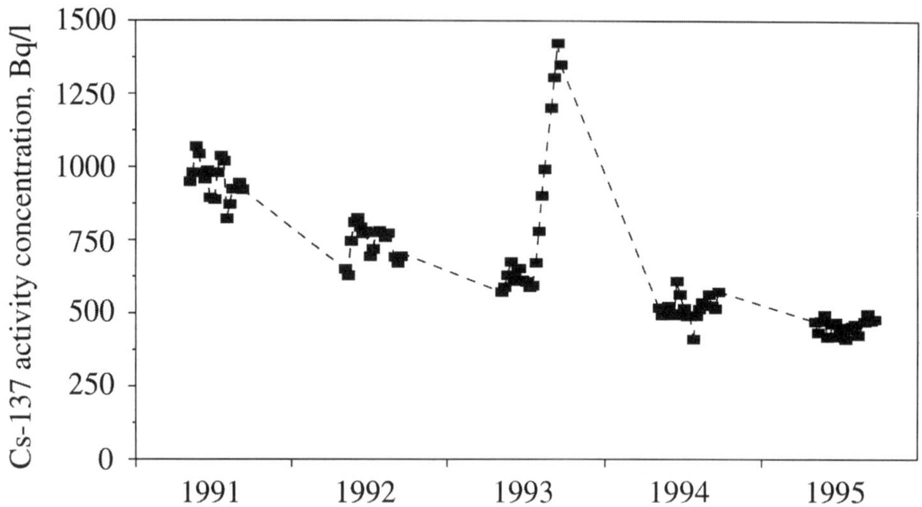

Fig. 4. Observed ^{137}Cs activity concentrations in cows' milk on the Stav pasture for 1991–1995.

sedimentation rates in other floodplain areas (Nicholas & Walling 1997; Walling & He 1997; Middelkoop & van der Perk 1998). In the period 1988–1994, the total ^{137}Cs inventory of the floodplain increased by 0.556 TBq, which yields a more realistic rough estimation of the average sedimentation rate of between 7.2 and 72 kg m^{-2} a^{-1}. This implies that it is very unlikely it was only the 1993 summer flood that caused the high 1993 surface contamination. The fact that in 1994 the surface contamination values were approximately back to the level before 1993 supports this proposition. Figure 3 shows that in the sampled years, the sampling locations were more or less evenly distributed over the study area, which practically rules out the chance that the contamination levels in 1993 were higher because of biased sampling in, for instance, local depressions. At this stage, the high 1993 levels are still unexplained and are still subject to further study

The changes in ^{137}Cs activity concentrations in cow's milk over time showed that the gradual exponential decay was disturbed abruptly by a peak that occurred after the flood. The question remains whether this is due to an increased input of ^{137}Cs by deposition of contaminated sediment in the floodplain or to other flood-related processes that caused an increased ^{137}Cs transfer from soil via grass to cows and cow's milk. The previous analysis of the surface contamination patterns over the different years has shown that overbank sedimentation of contaminated silt has resulted in slightly enhanced surface contamination of the Stav pasture by about 50% over 6 years. However, this enhancement cannot fully explain the short-lived increase in ^{137}Cs activity concentrations in cow's milk of more than 100% following the 1993 summer flood.

There are at least three possible processes explaining the sudden increase of ^{137}Cs transfer to milk during and after flooding: (1) exchange of ^{137}Cs cations at the adsorption complex by ammonium (NH_4^+) during saturation of the peaty soil reducing the redox potential, and consequently increasing the NH_4^+ concentration; several researchers have reported an enhanced uptake of ^{137}Cs by plants in the presence of increased amounts of ammonium (see Livens & Loveland 1988); (2) direct ingestion of ^{137}Cs by cows by drinking the flood water or eating the fresh contaminated sediment that covered the vegetation immediately after flooding; (3) direct uptake of ^{137}Cs from river water by plant roots or submerged leaves.

Both ^{137}Cs exchange by NH_4^+ and direct ingestion of ^{137}Cs do not depend on the time of year. Hence, the fact that floods occurring outside the growing season have only a limited effect on the ^{137}Cs activity concentrations in milk suggests that direct uptake of ^{137}Cs from the river water by the plants plays a major role in the milk contamination after the 1993 summer flood.

Concluding remarks

This study demonstrated the consequence of flooding for ^{137}Cs contamination of soil and milk in the Chernobyl affected catchments of the Pripyat and Dnieper River, which extend over vast flat areas in Ukraine, Belarus, and Russia. Although not all processes controlling the transfer of ^{137}Cs to agricultural products in floodplains are fully understood, it has become evident that flooding entails additional inputs of the ^{137}Cs into the floodplain ecosystems. The enhancement of the ^{137}Cs inventory in the floodplain soils does not necessarily imply increased transfer rates to agricultural products such as cow's milk. Nevertheless, changes in physical and/or geochemical circumstances during flooding may result in substantially, but temporarily, raised ^{137}Cs contamination of agricultural products. During flooding, direct uptake by submerged leaves is thought to be a principal pathway of ^{137}Cs transfer to vegetation.

Further research is needed to unravel the key processes involved in the enhanced ^{137}Cs transfer after flooding, as these processes are not yet fully identified. This research should include detailed studies of changes in the geochemistry of floodplain soils and in the ^{137}Cs uptake by vegetation during and after inundation. In this way, knowledge will be gained about the fluxes of ^{137}Cs through floodplain ecosystems, which is essential to assess the potential risks of the resulting enhanced internal radiation doses to the human population growing their private food products on floodplains.

This research has been carried out in the framework of the EC funded projects 'Restoration of radioactively contaminated ecosystems—RESTORE' (EC Contract FI4P-CT95-0021c) and 'Spatial and temporal radioecological survey systems–STRESS' (EC Contract IC15-CT96-0215). The authors thank collaborators from the former ECP9 project for providing the datasets for this study. B. J. Howard (Institute of Terrestrial Ecology, UK) and G. Voigt (Institut für Strahlenschutz - Germany) are thanked for their comments on an earlier draft of the manuscript.

References

BONNET, P. J. P. 1990. A review of the erosional behaviour of radionuclides in selected drainage basins. *Journal of Environmental Radioactivity*, **11**, 251–266.

BURROUGH, P. A., GILLESPIE, M. K., HOWARD, B. J. *ET AL*. 1996. *Redistribution of Chernobyl ^{137}Cs in Ukraine Wetlands by Flooding*. ITE–ICG, ITE/ICG, Grange-over-Sands–Utrecht.

CREMERS, A., ELSEN, A., DE PRETER, P. M. & MAES, A. 1988. Quantitative analysis of radiocaesium retention in soils. *Nature*, **335**, 247–249.

DEUTSCH, C. V. & JOURNEL, A. 1992. *Geostatistical Software Library and User's Guide*. Oxford University Press, New York.

KOROBOVA, E. M. & LINNIK, V. G. 1993. Geochemical landscape strategy in monitoring the areas contaminated by the Chernobyl radionuclides. *Landscape and Urban Planning*, **27**, 91–96.

KUDELSKY, A. V., SMITH, J. T., OVSIANNIKOVA, S. V. & HILTON, J. 1996. Mobility of Chernobyl-derived ^{137}Cs in a peatbog system within the catchment of the Pripyat River, Belarus. *Science of the Total Environment*, **188**, 101–113.

LIVENS, F. R. & LOVELAND, P. J. 1988. The influence of soil properties on the environmental mobility of caesium in Cumbria. *Soil Use and Management*, **4**, 69–75.

MIDDELKOOP, H. & VAN DER PERK, M. 1998. Modelling spatial patterns of overbank sedimentation on embanked floodplains. *Geografiska Annaler, Series A: Physical Geography*, **80**, 95–109.

NICHOLAS, A. P. & WALLING, D. E. 1997. Modelling flood hydraulics and overbank deposition on river floodplains. *Earth Surface Processes and Landforms*, **22**, 59–77.

STRAND, P. HOWARD, B. J. & AVERIN, V. 1996. *Transfer of Radionuclides to Animals, their Comparative Importance under Different Agricultural Ecosystems and Appropriate Countermeasures*. Experimental Collaboration Project 9, Final Report, **EUR 1653**9.

WALLING, D. E. & HE, Q. 1997. Investigating spatial patterns of overbank sedimentation on river floodplains. *Water, Air, and Soil Pollution*, **99**, 9–20.

—— & QUINE, T. A. 1995. The use of fallout radionuclides in soil erosion investigations. *In*: *Nuclear Techniques in Soil–Plant Studies for Sustainable Agriculture and Environmental Preservation*. International Atomic Energy Agency Publication, **ST1/PUB/947**, 517–619.

Contribution of floodplain sequestration to the sediment budget of the Waipaoa River, New Zealand

BASIL GOMEZ[1], DENNIS N. EDEN[2], D. MURRAY HICKS[3], NOEL A. TRUSTRUM[2], DAVID H. PEACOCK[4] & JANET WILMSHURST[5]

[1]*Geomorphology Laboratory, Indiana State University, Terre Haute, IN 47809, USA (e-mail:bgomez@indstate.edu)*

[2]*Landcare Research Ltd, PO Box 11052, Palmerston North, New Zealand*

[3]*National Institute of Water and Atmospheric Research, PO Box 8602, Christchurch, New Zealand*

[4]*Gisborne District Council, PO Box 747, Gisborne, New Zealand*

[5]*Landcare Research Ltd, PO Box 69, Lincoln, New Zealand*

ABSTRACT: Rapid vertical accretion on the Waipaoa River floodplain is conditioned by the river's high suspended sediment load (30 000–40 000 mg l^{-1} at flood stage). Cumulative sediment accumulation curves derived from three cores suggest an average (post-1850) rate of vertical accretion of *c.* 60 mm a^{-1}, though a 15 year lacuna in flood activity has depressed the post-1948 rate to *c.* 40 mm a^{-1}. Rates of aggradation during floods are several orders of magnitude larger than the time-averaged rate. Within a 44 km long reach, cross-section surveys indicate that 0.2–0.8 m of sediment was deposited between 1979 and 1990. Over this period floodplain storage accounted for 5% of the total suspended sediment load, and 16% of the suspended sediment load transported during events that exceeded bankfull stage. The Waipaoa River floodplain may be representative of floodplains bordering rivers with high suspended sediment loads, produced by rapid, episodic vertical accretion, on which overbank deposition occurs across the entire floodplain, and is complemented by channel aggradation. Such rivers are able to construct high banks. Thus channel capacities are greater and the incidence of overbank flows is less than in rivers where overbank deposition is slow relative to the rate of floodplain destruction by lateral migration. The difference between our time-averaged estimate for sequestration on the Waipaoa River floodplain and comparable estimates for actively meandering rivers, and meandering rivers with low sediment loads, reinforces the notion that there is a link between the sediment transport regime of a river and its sedimentary record. To elucidate this link it is necessary to view vertical accretion in the context of the flood events that generated it, rather than in the context of a time-averaged sediment budget.

Relatively modest changes to a river's flow and sediment transport regimes may have a dramatic effect on the pattern of inundation and the amount of vertical accretion on floodplains (Graf *et al.* 1991; Knox 1993). Climate is considered to have exerted the fundamental control on overbank floods during Holocene time, but its effect on floodplain sedimentation is often eclipsed by hydrological impacts imposed by human activity (Knox 1995). Anthropogenically induced disturbances to land cover enhance runoff and soil erosion, thus engendering episodes of accelerated floodplain sedimentation that may be preserved in floodplain fills (Knox 1987; Marron 1992). Such anthropogenic impacts are well documented, and have been variously attributed to prehistoric, historical, and colonial cultures (Robinson & Lambrick 1984; Macklin 1985; Jacobson & Coleman 1986; Van Andel & Zangger 1990; Brooks & Brierley 1997). In the last (colonial) case, the disturbances that precipitated the changes are a relatively recent (18th–20th century) phenomenon that may still impinge on the contemporary landscape (Kelsey *et al.* 1987; James 1989). Because the circumstances surrounding these changes are embedded in the historical record, they also provide a means of assessing how the behaviour of floodplains and their role as sediment sinks will be affected by modern shifts in land use.

Although attempts have been made to quantify the amount of deposition that occurs at the reach scale, both during annual and rare, high-magnitude

From: Marriott, S. B. & Alexander, J. (eds) 1999. *Floodplains: Interdisciplinary Approaches*. Geological Society, London, Special Publications, **163**, 69–88. 1-86239-050-9/99/$15.00

flood events (Mertes 1994; Gomez *et al.* 1995), most estimates of contemporary deposition rates are location-specific, short-term (i.e. decadal) averages (Simm 1995; Walling *et al.* 1996). Temporal averaging helps obviate bias caused by censoring of the stratigraphic record, the appearance of which may be dominated by events of unusual magnitude or duration. It also constrains the temporal resolution of deposition rate estimates and masks the fact that vertical accretion on floodplains is conditioned by individual events and, as such, is highly discontinuous (Gomez *et al.* 1998). Nevertheless, regardless of whether the data are event based or a product of temporal averaging, the inescapable conclusion is that contemporary rates of overbank deposition are low (see Bridge & Leeder 1979; Ferring 1986; Simm 1995). In this respect, there appears to be a dichotomy between most contemporary estimates of rates of vertical accretion on floodplains and the stratigraphic evidence (Walling *et al.* 1996).

The contemporary estimates reinforce the view that floodplain environments typically experience slow, cumulative vertical aggradation, with fine-grained overbank deposits being formed by episodic overbank flooding of a dominant master channel (Wolman & Leopold 1957). If an average sedimentation rate of only a few centimetres per hundred years is sustained, the flood deposits should be incorporated into the active soil and any internal stratification effaced by pedogenic processes and bioturbation (Kraus & Aslan 1993; Aslan & Autin 1998). In some localities the absence of buried soils and continuity of the near-surface deposits lends support to the argument for the 'continuous' accumulation of overbank sediments over substantial periods in Holocene time (Ritter *et al.* 1973; Putnam 1994). Certainly it is the case that long-term (typically *c.* 10^3 a) average accumulation rates are commonly assumed in simulation models (Bridge & Leeder 1979; Mackey & Bridge 1995). However, overbank sediments preserved in other localities are indicative of episodic sedimentation (e.g. Brakenridge 1984), and the large-scale facies sequences preserved from ancient overbank environments suggest that the basic sedimentation unit on floodplains is the centimetre-to-decimetre-thick depositional sequence generated by individual floods (Bridge 1984). Primary depositional stratification is often preserved in such units, which chronicle individual events characterized by sedimentation rates that are several orders of magnitude higher than the long-term mean rate (Steel & Aasheim 1978; Willis & Behrensmeyer 1994).

In terms of the long-term geomorphological evolution of floodplains, estimates of contemporary deposition rates therefore appear to be unrepresentative both of the stratigraphic evidence for patterns of vertical accretion in ancient overbank deposits that reflect high rates of vertical accretion during individual floods, and of episodes of accelerated floodplain sedimentation in the historical period. Assuming the averaging interval is specified, there may be several reasons for this. If the channel bed is stable, the rate of overbank sedimentation may intuitively be expected to decline as the elevation of the floodplain surface (and thus channel capacity) increases and the frequency of floodplain inundation correspondingly decreases. The storage capacity of floodplains may also be limited by the accommodation space available for sediment accumulation, which may be maximized in transgressive depositional systems and is augmented by tectonic subsidence (Kraus & Bown 1993; Wright & Marriott 1993). However, because the rate of overbank sedimentation declines as the impact of a disturbance to the catchment environment diminishes over time (Trimble & Lund 1982), it may simply be that low contemporary rates of floodplain sedimentation are consistent with catchment conditions that promote low suspended sediment concentrations in the adjacent channel (see Gomez *et al.* 1995; Walling *et al.* 1998). In the United Kingdom, for example, suspended sediment transport rates are typically of the order of 50–100 t km^{-2} a^{-1}, which is low by global standards (Walling & Webb 1996). Thus, even during high-magnitude events, the well-developed floodplains bordering lowland rivers with such low suspended sediment loads probably provide poor analogues for the overbank environment of aggradational river systems in the rock record and historical phases of accelerated floodplain sedimentation. This is not to say that estimates of contemporary deposition rates on such floodplains are of limited relevance to studies of floodplain evolution, rather that they represent one end of a continuum of depositional environments. The other is represented by floodplains bordering lowland rivers that transport high suspended sediment loads, such as the River Amazon and rivers in the coastal region of southeastern Australia (Nanson & Young 1981; Meade 1994; Mertes 1994).

Milliman & Syvitski (1992) have argued that a significant proportion of the sediment load of high-yield rivers draining tectonically active mountain belts is sequestered on floodplains. First-order estimates of the magnitude of floodplain storage in the lower reaches of continental-scale drainage systems lend support to this notion (Allison *et al.* 1998; Dunne *et al.* 1998; Goodbred & Kuehl 1998). However, more detailed information is required to evaluate the contribution that sediment storage on floodplains bordering lowland rivers with high suspended loads makes to a basin's sediment budget, and to determine if the proportionate net loss to the floodplain is comparable with that experienced in meandering rivers with lower sediment loads (see Walling *et al.* 1998). Like Allison *et al.* (1998), we believe that average sedimentation rates, in combination with knowledge of a river's flow and suspended sediment transport regimes, provide a reliable estimate of the role that floodplain storage has in the sediment budget of large river systems. In this paper we document the contribution that contemporary floodplain sequestration makes to the sediment budget of the Waipaoa River, a highly charged, actively aggrading gravel-bed river in the tectonically active East Coast region of New Zealand's North Island.

Study area

The 2200 km^2 Waipaoa River Basin lies on the eastern flanks of the axial Raukumara Range (Fig. 1), adjacent to the boundary of the Australian and Pacific lithospheric plates, and within the zone of active deformation associated with the Hikurangi subduction margin (Moore & Mazengarb 1992). It is underlain by thrusted Cretaceous and Palaeocene mudstone and argillite, Jurassic to early Cretaceous greywacke, and a cover sequence of poorly consolidated Miocene–Pliocene sandstone, siltstone and mudstone (Mazengarb *et al.* 1991). Subduction has induced uplift in the Raukumara Ranges at the head of the Waipaoa River Basin, but slight subsidence is experienced near the river mouth in the Poverty Bay Flats (Mazengarb *et al.* 1991; Brown 1995).

Before early Maori deforestation *c.* 650 years BP, well-drained lowland areas of the East Coast region were covered with dense podocarp–hardwood forest, dominated by *Prumnopitys taxifolia* and *Dacrydium cupressiunum* emergent over a canopy of mixed hardwood trees, which suffered periodic disturbance by fires following volcanic eruptions and lightning strikes (Wilmshurst & McGlone 1996; Wilmshurst 1997; Wilmshurst *et al.* 1997). European colonists continued the clearances on a much larger scale in the nineteenth century; the most recent period of accelerated overbank sedimentation on the Waipaoa River floodplain occurred in the historical period following the

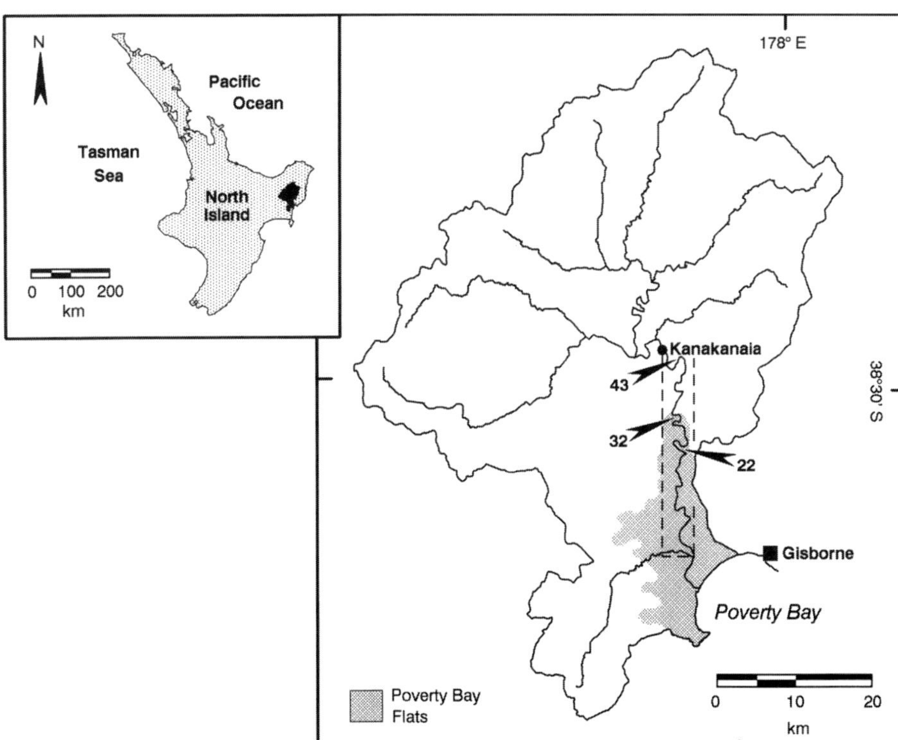

Fig. 1. The Waipaoa River catchment and study reach (outlined by dashed lines). Arrows indicate the locations of the three core sites and numbers the distance in kilometres from the river mouth.

basin's conversion to pasture. Wholesale clearance of the native forest began in the late 1820s (Pullar 1962). By 1880 most of the land on and around the Poverty Bay Flats had been cleared and the headwaters had begun to be converted to pasture. Deforestation of the headwaters was completed by 1920, and today only 2.5% of the basin remains under native forest. Reforestation of selected portions of the headwaters with exotic species, such as *Pinus radiata*, began in 1960 and commercial timber harvesting commenced in 1990.

The combination of weak, highly deformed rocks, the destabilizing effect of deforestation, and a maritime climate that is periodically disturbed by intense cyclonic and more localized storms gave rise to a particularly intense phase of erosion in the historical period (O'Byrne 1967; Gage & Black 1979). Depending on altitude, mean annual precipitation varies from 1000 to 3000 mm (Hessell 1980), and the largest historical storm (March 1988, Cyclone Bola) generated as much as 300–900 mm of rain in a 3 day period. In the headwaters, the crushed and fractured rocks are prone to deep-seated, complex mass movements and support extensive gully complexes (Pearce *et al.* 1981; DeRose *et al.* 1998). Shallow landsliding is the dominant erosion process on the highly erodible Tertiary siltstones and mudstones (Reid & Page, In Press; Trustrum *et al.* 1999). Large quantities of fine sediment are moving out of the gully complexes, and virtually all the sediment contributed to stream channels by the shallow landslides is transported in suspension (DeRose *et al.* 1998; Reid & Page, In Press).

The mainstem gauging site at Kanakanaia is located 48 km upstream from where the river enters Poverty Bay (Fig. 1). The continuous stage record commenced in 1960 and the annual flood record in 1948. A flood history for the period between 1850 and 1948 has been compiled from documentary sources (Todd 1949; Hogan 1997). Suspended sediment gaugings began in 1962 and have continued to be made on an *ad hoc* basis during floods and freshes in conjunction with flow gaugings. The Waipaoa River, at Kanakanaia, has a mean suspended sediment concentration of *c.* 1700 mg l^{-1}, a suspended sediment load of *c.* 10.7×10^6 t a^{-1}, and a specific suspended sediment yield of 6750 t km^2 a^{-1} (Hicks *et al.* In Review). It annually delivers *c.* 15×10^6 t of suspended sediment to the Pacific Ocean. This ranks amongst the highest measured yields for any New Zealand river, and is also very high by global standards (Milliman & Meade 1983; Hicks *et al.* 1996). The bedload yield, estimated using Wilcock's (1998) modified Parker–Einstein formula, amounts to <1% of the suspended load.

From the perspective of suspended sediment transport the moderate flow range is the most important (Hicks *et al.* In Review). Flows <500 m^3 s^{-1} transport 50% of the load, and 83% is transported by flows between the mean flow (34.7 m^3 s^{-1}) and the mean annual flood (1346 m^3 s^{-1}). The most effective flow (360 m^3 s^{-1}) is 0.23 times the mean annual flood and 11 times the mean flow. However, the bankfull discharge at Kanakanaia (2550 m^3 s^{-1}) is seven times greater than the most effective flow. In some rivers the most effective discharge has been equated with the bankfull discharge (Andrews 1980), but this is not necessarily a characteristic of fluvial systems in which overbank deposition dominates over lateral channel migration as a mode of floodplain formation (Nolan *et al.* 1987). This is the case in the lower reaches of the Waipaoa River Basin (Fig. 2a), where the single-thread meandering channel is bordered by a well-developed floodplain (Pullar & Penhale 1970; Brown 1995; Gomez *et al.* 1998). Contemporary rates of vertical accretion on the floodplain are 14–18 mm h^{-1} (Gomez *et al.* 1998), which is at the high end of deposition rates reported for single floods (Gomez *et al.* 1995). The first planimetric survey of the river's lower reaches was accomplished in 1868 and although there have been substantive changes in the configuration of some bends, lateral migration has effected little change to the overall alignment of the channel in the intervening period (Fig. 2b). Gomez *et al.* (1998) argued that the increase in suspended sediment load during the historical period, which promoted the in-channel deposition of fine sediment (see Woodyer *et al.* 1979) and vertical accretion on the floodplain, helped stabilize the planform geometry.

Vertical accretion on the floodplain is complemented by channel aggradation (Fig. 3). For the period 1906–1988, survey data suggest a rate of aggradation in the vicinity of the road bridge at Kanakanaia of *c.* 20 mm a^{-1}. This is similar to the rate that channel cross-section surveys indicate was characteristic of adjoining reaches (30–65 km) in the period 1948–1988. Rates of aggradation in the headwaters were much higher. Localized degradation in the 20 km long reach immediately upriver from the mouth is promoted by diversion cuts that shortened the river by *c.* 8 km, and reinforced by scour during large floods.

Our investigation focuses on the 44 km long reach of the Waipaoa River floodplain located between the mainstem gauging station at Kanakanaia and the confluence of the Waipaoa and Te Arai rivers, 4 km from the coast (Figs 1 and 2). No major tributaries enter the river within the study reach, though the lower 6 km are tidal. The average bed slope is 0.0008, and median grain size of the subsurface bed material is 3.4 mm. Some riparian land has remained under pasture throughout the

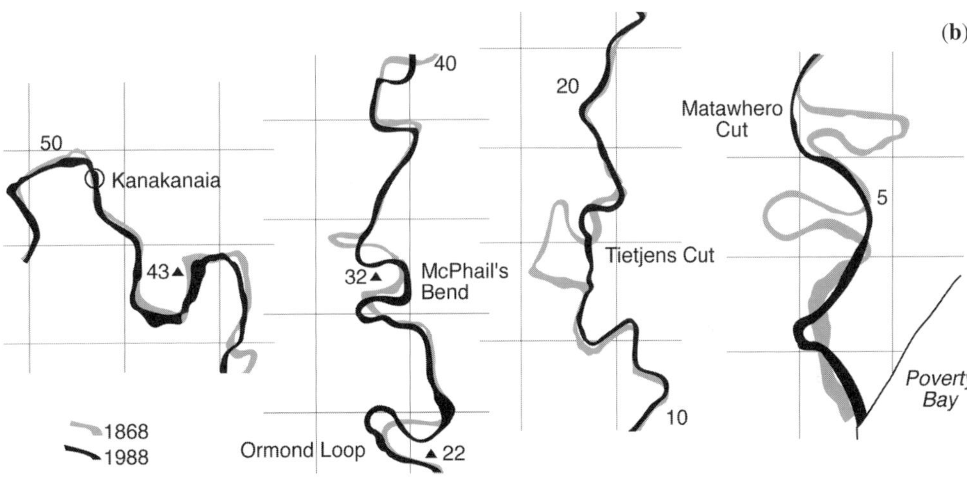

Fig. 2. (**a**) View of the Waipaoa River floodplain and overbank deposition after the March 1988 flood, looking south from *c.* 35 km towards McPhail's Bend. © N. A. Trustrum. (**b**) Channel changes within the study reach in the period from 1868 to 1988. Tietjens (1955) and Matawhero (1953) cuts are engineered diversions. Channel changes in the estuary downstream from the Matawhero Cut are linked to natural shifts in the position of the river mouth. The N–S, E–W grid lines are spaced 2 km apart, numbers denote the distance in kilometres from the river mouth, and the core sites are marked by filled triangles. Data sources: Department of Survey and Land Information—Gisborne Roll Plan 1119 and East Cape Catchment Board Drawing 4257.

historical period (Gomez *et al.* 1998), but most of the floodplain within the study reach has been disturbed by cultivation. Willows have been planted on the banks at many bends, but a close grass cover has been maintained on most riparian land within the study reach since 1948. In addition to the diversion cuts, schemes for flood mitigation and drainage improvement on the Poverty Bay Flats involved the construction of artificial levees. Initially, the standard of levee construction varied, but a continuing programme of rationalization that was initiated in 1953 has improved the flood control scheme to the point where, in the reach downstream from *c.* 38 km, all overbank flow derived from the main channel is confined by a system of levees that connects with bluffs or low terraces bordering the natural floodplain. Upstream from 38 km the pattern of inundation is more variable and colmatage banks enhance aggradation in areas adjacent to the channel. Throughout the lower 18 km of the flood control scheme the distance between the levees is *c.* 325 m. There is no

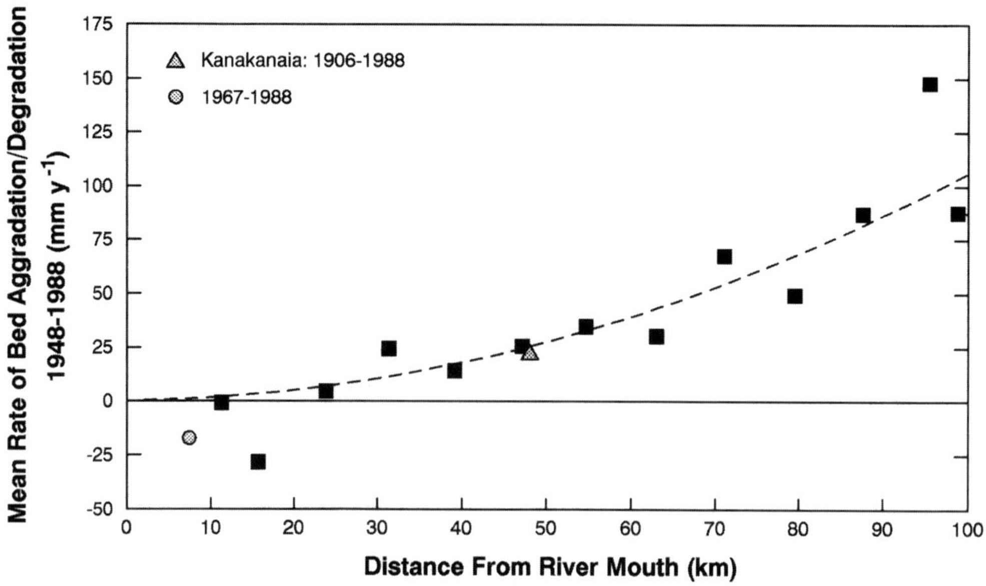

Fig. 3. Mean rate of river-bed aggradation along the Waipaoa River in the period from 1948 to 1988. The regression line (dashed) serves to emphasize the trend.

such consistency in the upper portion of the scheme, where the channel is more sinuous and the width of the active floodplain varies between 300 and 1600 m. Within the 44 km long study reach as a whole, *c.* 15.5 km^2 of the floodplain is still active and has been inundated periodically throughout the historical period. A substantial area of the floodplain was inundated during the record March 1988 flood (Fig. 2a), and the most recent overbank event occurred in March 1996.

Floodplain cores and transects

Cores were obtained from the floodplain at 43 km, 32 km and 22 km (Figs 1 and 2). All the core sites were located on meander bends and are known to have experienced varying degrees of agricultural disturbance. The core from the site at 32 km (McPhail's Bend), which was under pasture throughout the historical period, yielded an exceptionally detailed stratigraphy. Evidence of overbank flooding is preserved in the stratigraphic record by horizontally bedded, fining upwards sequences (typically sand and silt–clay couplets) that are between 0.1 and 1.0 m thick (Fig. 4). Erosional contacts and buried soils are rare, individual flood deposits being differentiated on the basis of textural discontinuities and colour. Sediment density was determined volumetrically. There was no systematic difference in density between sediments at different depths throughout the core. We used magnetic measurements to help establish a stratigraphic correlation between the three cores (Fig. 4). Magnetic susceptibility (χ), the ratio of the induced magnetization of a sample to a magnetic field with which the sample is brought into contact, was determined in the laboratory using a Bartington Instruments MS2 magnetic susceptibility meter.

We previously assembled a composite stratigraphy at the 32 km core site for the period (from 1948 to the present) for which reliable flow data are available (Gomez *et al.* 1998). Here we use the post-1850 flood history compiled from documentary sources (Todd 1949; Hogan 1997) to assign dates to the historical overbank deposits (Fig. 4). All the depositional units in this core can be related to recorded flood events, and the interpretation is constrained by the ^{137}Cs logs and ^{14}C dating, and pollen analysis. The presence of bracken, charcoal and seral vegetation near the base of the core (*c.* 9 m) are indicative of a fire-disturbed, open landscape (Wilmshurst 1997). This, in combination with the ^{14}C date obtained from wood fragments recovered at 9.3 m and the appearance of grass pollen at *c.* 7 m, corroborates our view that the sedimentary record extends back to the beginning of the period of European pastoralism in the Waipaoa River Basin.

In the modern period, starting in 1948, repeat surveys of the channel and floodplain were conducted at fixed locations throughout the area

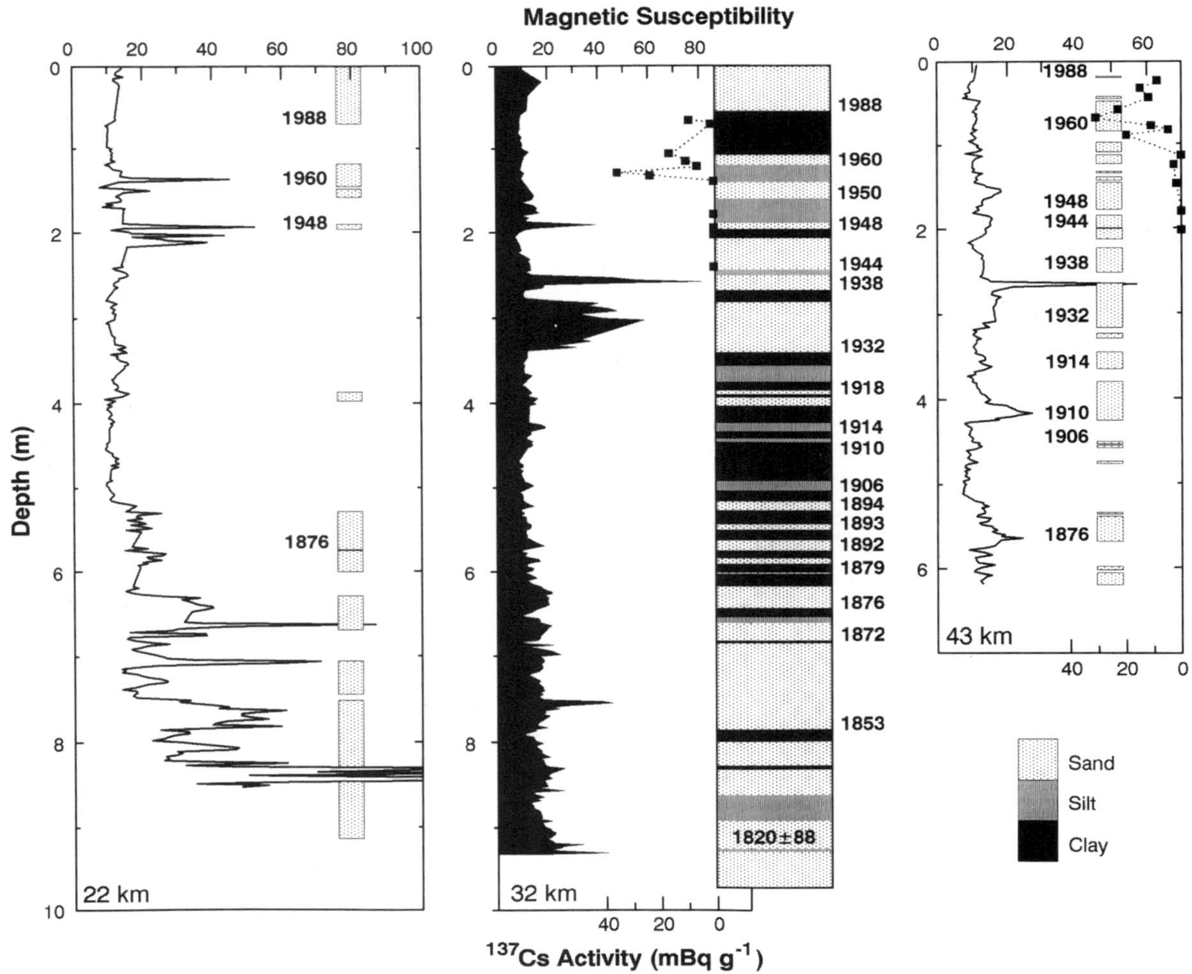

Fig. 4. Probable dates associated with major flood units preserved in the core from 32 km (McPhail's Bend), and their relation to flood units preserved in the cores from 22 and 43 km based on the flood history compiled by Todd (1949) and Hogan (1997) and the magnetic susceptibility logs. ^{137}Cs logs for the cores from 32 and 43 km are shown as dotted lines connecting filled squares. The post-1948 flood history is an abbreviated version of sequence of flood events depicted in the composite stratigraphic record for McPhail's Bend (Gomez *et al.* 1988).

affected by the flood control scheme. The topographic changes that these surveys register provide an accurate indication of the depth and spatial variability of contemporary overbank sedimentation on the floodplain. We used these elevation differences to characterize the mean rate of vertical accretion across the floodplain at different locations within the study reach during the period from 1948 to 1990. The average depth of the flood deposits on each left and/or right bank section of the floodplain was computed separately. These sections sometimes represented up to 90% of the total floodplain width, because the channel centre line rarely coincided with the floodplain axis. Thus the term 'floodplain half-width' (the distance along a transect from the channel centre line to the floodplain boundary) is used to distinguish our measure of floodplain width from the estimates of total valley (or floodplain) width employed by others (see Lecce 1997).

Temporal trends in vertical accretion

Variations in magnetic susceptibility are often related to changes in the overall concentration or relative amount of different ferrimagnetic minerals, and to particle size (Le Borgne 1955; Mullins 1977). They may also be related to primary erosion processes that increase the rate of inorganic influx to a sediment reservoir, and hence to land-use change in a catchment (Thompson & Morton 1979; Turner 1997). In the present case (Fig. 4), major changes in magnetic properties appear to reflect fluctuations in the concentration of magnetic minerals, not variations in grain size. We reason this because, although there was a massive increase in landsliding following the Waipaoa River Basin's conversion to pasture during the period of European settlement, much of the sediment supplied to the Waipaoa River is generated by gully erosion (Hicks *et al.* In review; Trustrum *et al.* 1999). The supply of sediment is maintained because established gullies are activated by small, frequent rainstorms. Extreme events intensify the erosion processes that are responsible for delivering sediment to the channel, and landslides and earthflows generate additional sediment during such events. Thus the net effect of these events is to increase sediment availability. In the historical record extreme events are associated with peaks in susceptibility because landslides in the headwaters generate predominantly minerogenic sediment that naturally enhances the concentration of ferrimagnetic minerals (Turner 1997). The change in the appearance of the susceptibility log for the 22 km core at *c.* 6.25 m probably reflects the adjustment to the sediment supply regime that occurred once the clearances extended into the headwaters, when the amount of sediment generated by gully erosion subsumed that generated by shallow landsliding in all but the largest storms.

The sequence of dated flood deposits can be used to derive a cumulative sediment accumulation curve and time-averaged estimate of the average rate of deposition on the floodplain at 32 km (Fig. 5). However, because floods have a variable duration, which is a small (and typically unknown) fraction of the time between events that is used for averaging, perspectives on average rates of vertical accretion are necessarily tempered by knowledge of the continuum of deposition and event duration. For example, in the case of recent events with a peak discharge <4000 $m^3 s^{-1}$, for which the duration of overbank flooding could be determined from flow records (Gomez *et al.* 1998), there is a well-defined relation between the thickness of individual flood deposits and the duration of overbank flooding (Fig. 6a). Thus we suggest that much of the short-term variability in the cumulative sediment accumulation curve can be accounted for by differences in event duration. Sediment availability does not appear to be a significant factor (Fig. 6b). There is no significant difference (at the 5% level) between rising and falling stage concentrations in the Waipaoa River. Curvature in the suspended sediment concentration–water discharge relation, which tends to level-off at concentrations in the 30 000–40 000 mg l^{-1} range, is a product of the relative rates of supply of water and sediment to the channel during very large storms rather than any inherent limitation on sediment supply (Hicks *et al.* In Review). Indeed, a constant concentration indicates a linear relationship between water and sediment supply. Nevertheless, the rate of vertical accretion is constrained by the relation of particle settling velocity to flow velocity on the floodplain: for much of the March 1988 flood, flow velocities on the floodplain were too high to permit fine particles to be deposited from suspension (Gomez *et al.* 1998). Thus, this protracted event generated a disproportionately thin sequence of overbank sediments (Fig. 6a). Our experience also accords with the observation by Bhowmik & Demissie (1982) that conveyance losses of water to floodplains are greater in low- and intermediate-magnitude floods than in extreme events.

Although vertical accretion on the floodplain is temporally discontinuous, the regression relation for the period from 1853 to 1965 is remarkably coherent ($R^2 = 0.98$) and the slope of the regression line suggests that 63.5 mm a^{-1} is a representative average rate of vertical accretion at the sites of the cores for the entire 112 year period (Fig. 5). The core stratigraphy is not as detailed, but the trend in the cumulative sediment accumulation curve for the

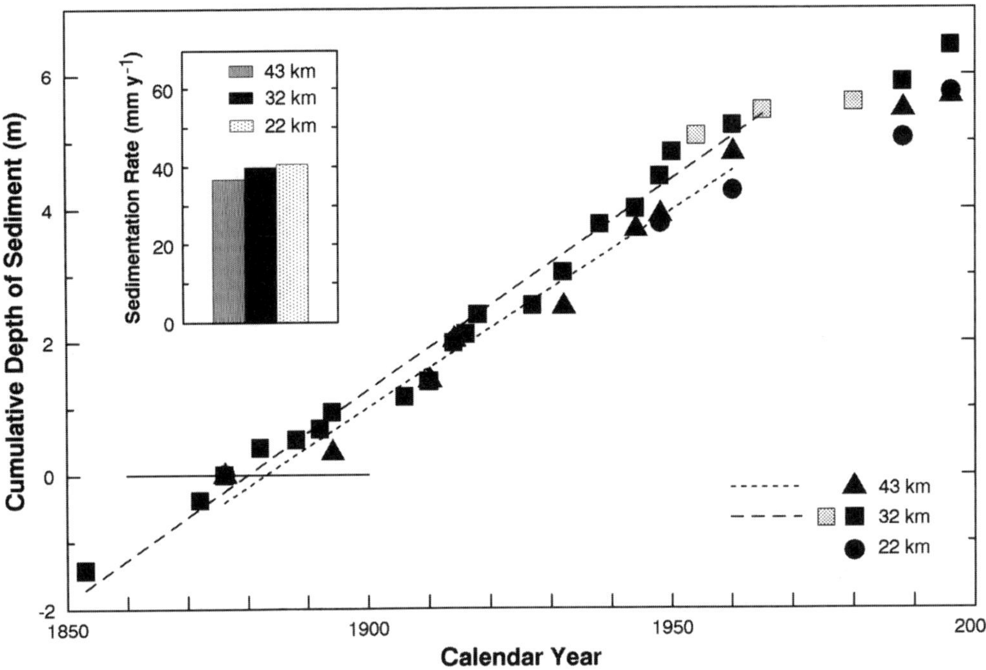

Fig. 5. Sediment accumulation curves (zeroed at 1876) for the 22, 32 and 43 km cores. R^2 values for the regression lines used to define the long-term, mean rate of vertical accretion are 0. 98 and 0. 97 for the 32 and 43 km cores, respectively. The stippled squares indicate data points interpolated from the composite stratigraphic record for McPhail's Bend (Gomez *et al.* 1988). Inset: mean rate of vertical accretion in the modern (post-1948) period.

core from 43 km is similar to that at 32 km, and the slope of a regression relation indicates that for the period from 1876 to 1960 the average rate of vertical accretion on the floodplain at 22 km was 59.3 mm a^{-1}. The stratigraphy for the remaining core is rather more fragmentary, but it does not suggest that the pattern of vertical accretion on the floodplain at 22 km was radically different from that at either of the other two locations. Though the floodplain at all three core sites is affected by artificial levees constructed since the 1960s, their presence does not appear to have affected the long-term rate of vertical accretion at any site. Within the study reach as a whole the contemporary rate of vertical accretion on the floodplain, though variable, is of the same order of magnitude as the rate of aggradation in the channel (Figs 3 and 7). The sedimentation rates are high, even by comparison with the reported range of short-term rates of vertical accretion on floodplains in other river basins which have been subject to disturbance by anthropogenic activity in the historic period (Knox 1987; Marron 1992). They are several orders of magnitude higher than contemporary rates of vertical accretion on floodplains bordering lowland rivers with low sediment loads (He & Walling 1996; Walling *et al.* 1996; Walling & He 1998).

Sediment units deposited by vertical accretion occurring simultaneously across the entire floodplain surface may be characteristic of rivers that transport a high suspended sediment load (see Hereford 1986; Graf *et al.* 1991). The affinity between rates of vertical accretion at the three, widely separated core sites on the Waipaoa River floodplain is emphasized by time-averaged rates for the period from 1948 to 1996 (Fig. 5 inset). It should be noted that the *c.* 20 mm a^{-1} difference between the long-term average rates quoted above and rates of vertical accretion in the modern period (37–41 mm a^{-1}) is attributable to the temporary cessation of vertical accretion during a 15 year lacuna in flood activity. It is not known if the decline in the sedimentation rate since 1965 represents a real departure from the long-term trend, or is a short-term excursion motivated by the absence of overbank floods in the period between 1965 and 1980 (Fig. 5). Less pronounced excursions appear to have occurred twice in the historical period. In the period since 1954 a change in the airflow pattern over New Zealand has modified the

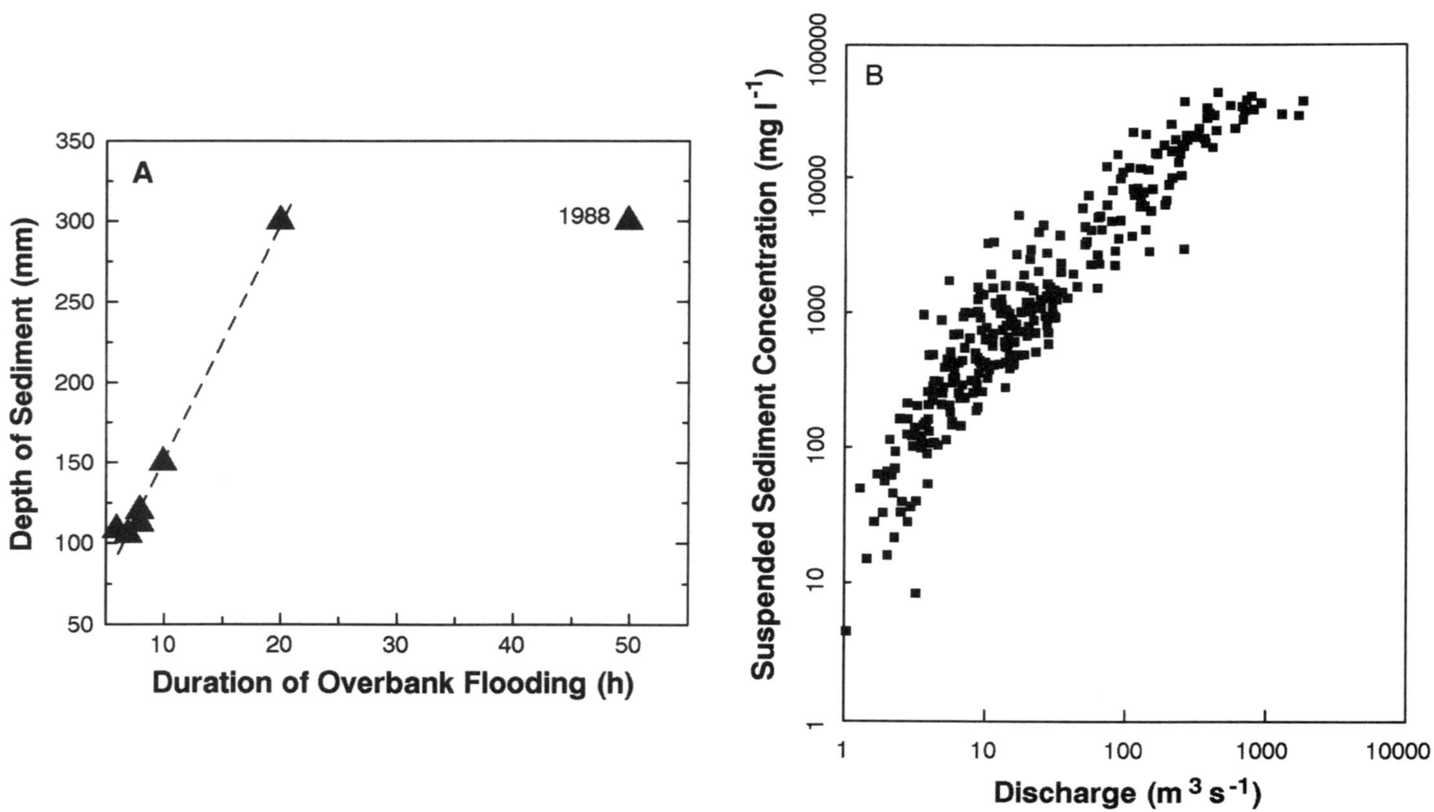

Fig. 6. (**a**) Relation between depth of sediment deposited on the floodplain and event duration. The regression line highlights the trend for events with a peak discharge <4000 $m^3 s^{-1}$. Data source: Gomez *et al.* (1998). (**b**) Log suspended sediment concentration–log water discharge relation for the Waipaoa River obtained from gaugings made at Kanakanaia.

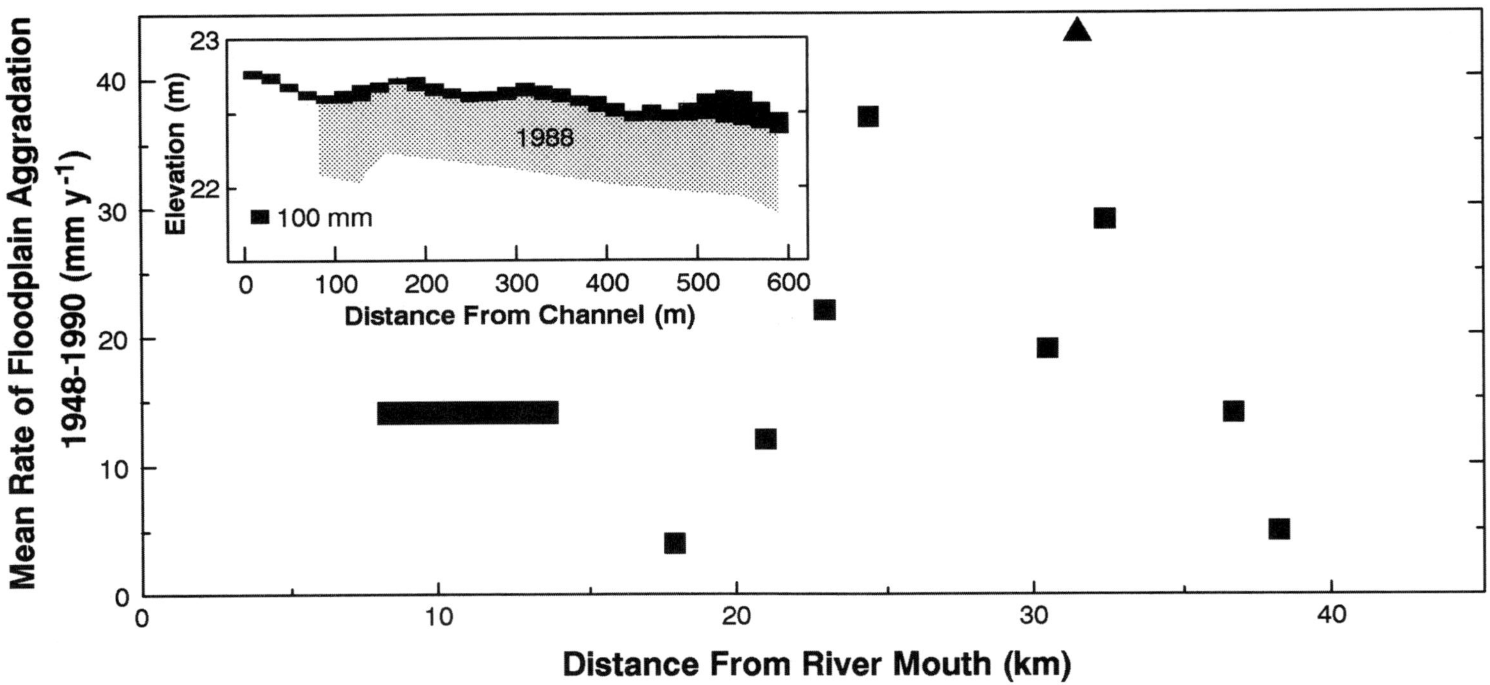

Fig. 7. Mean rate of aggradation on the Waipaoa River floodplain in the period from 1948 to 1990 for transects located within the Gisborne District Council flood control scheme. The filled bar indicates the mean rate for the reach computed from transects spaced *c.* 800 m apart, and the filled triangle the mean rate for McPhail's Bend. Data sources: Peacock (1991) and Gomez *et al.* (1998). Inset: thickness of sediment deposited along the centre line of McPhail's Bend in the March and September floods of 1988 determined by coring, and the March 1996 flood determined by field survey (after Gomez *et al.* 1998, fig. 4). The increase in sedimentation between 500 and 600 m from the channel is associated with inputs from the floodway that bisects the bend.

precipitation regime in the eastern part of the North Island (Hessell 1980). However, the flow record is too short to reveal if the change in precipitation pattern has affected the frequency of overbank flooding in the Waipaoa River Basin, or if reforestation in the headwaters has affected the flood hydrograph.

Spatial variability in the pattern of overbank sedimentation

As in other river basins (see Walling *et al.* 1998), the downstream pattern of sedimentation on the floodplain in the modern period is highly variable (Fig. 7). It is also apparent that contemporary rates of vertical accretion derived from the cores are at the high end of the range of mean rates for individual transects. As there are no variations in floodplain surface roughness or downstream variations of meander belt geometry, we relate spatial variations in the pattern of floodplain sedimentation to variations in floodplain width (see Graf 1983; Magilligan 1985). Narrow zones encourage transport, whereas sediment deposition is promoted in zones where overbank flow is dispersed over a wide area (Lecce 1997).

Relations for the study reach, which are based on the average elevation change along transects that lie perpendicular to the direction of flow, afford a high degree of explanation of the variance in the mean depth of sediment at different locations on the floodplain (Fig. 8). Even though the surveyed transects contain many more control points than are typically obtained by coring, it is unusual to obtain such coherent relations. If the outliers associated with zones of accelerating or decelerating flow are eliminated, for the period from May 1979 to September 1990, 89% of the variation in sediment depth is explained by variations in floodplain half-width. The high suspended sediment load of the Waipaoa River in combination with the subdued topography of the floodplain, which permits the overbank flow to maintain a relatively uniform depth, help efface lateral variations in the depth of the flood deposits at most sites (Fig. 7 inset). Simulation models of floodplain deposition also suggest that high rates of vertical accretion generate a more uniform floodplain surface than do low rates (Howard 1996). In addition, we suspect that (advective) flow across the bankline from the channel (Kiely 1990; Sellin 1995), rather than diffusion (Pizzuto 1987), is the dominant process responsible for transferring sediment onto the floodplain. In these circumstances the thickness of the overbank deposits simply reflects the control that the transport capacity of the flow has on the net accumulation rate. If neither discharge nor slope change appreciably in the downstream direction, transport capacity reflects spatial variations in flood power that are indexed by variations in floodplain width (Magilligan 1992). Much of the residual variability is probably attributable to differences in bank height that give rise to a spatially disjunct pattern of bankfull recurrence intervals (which range between 4 and 35 years on the annual series) within the study reach as a whole. Such variability is characteristic of rivers that have high sediment loads and are bordered by floodplains created by overbank deposition (Pickup & Warner 1976; Nanson & Young 1981; Nolan *et al.* 1987).

Floodplain storage and the sediment budget

To assess the contribution that contemporary floodplain sequestration makes to the sediment budget of the Waipaoa River, we subdivided the study reach into ten segments of varying length (typically 1–5 km, though the lower 13.2 km of the flood control scheme was treated as a single segment). The total floodplain width within each segment was relatively constant (varying by ±25%). We determined the average floodplain half-width for each segment and derived a representative sediment depth from the regression relation (Fig. 8). The volume of sediment stored within each segment was computed as the product of this representative depth and the area of the active floodplain, which was determined from 1:10 000 or 1:50 000 scale drawings. On this basis, in the period from May 1979 to September 1990, *c.* 5×10^6 m^3 of sediment were sequestered within the 44 km long study reach as a whole. On the basis of analyses of the sediments in the core from 32 km, a representative unit weight for the overbank sediments is 1.3 Mg m^{-3}. The conversion from volume to mass yields a figure of *c.* 6.5 Mt of sediment for the 11 year period.

We determined the suspended sediment load for the period covered by the transect resurveys by fitting a suspended sediment concentration–water discharge rating curve to the log-transformed data using a modified version of the locally weighted scatter smoothing (LOWESS) technique (Cleveland 1979). The suspended sediment yield averaged over the period from May 1979 to September 1990 and the distribution of the sediment load by flow were derived by combining the sediment rating with flow record in the manner outlined by Hicks *et al.* (In Review). The total suspended sediment load for the 11 year period was *c.* 146.2 Mt. Floodplain storage thus accounted for 5% of the total suspended sediment load, which equates with an annual loss to storage of 6% of the mean annual suspended sediment load.

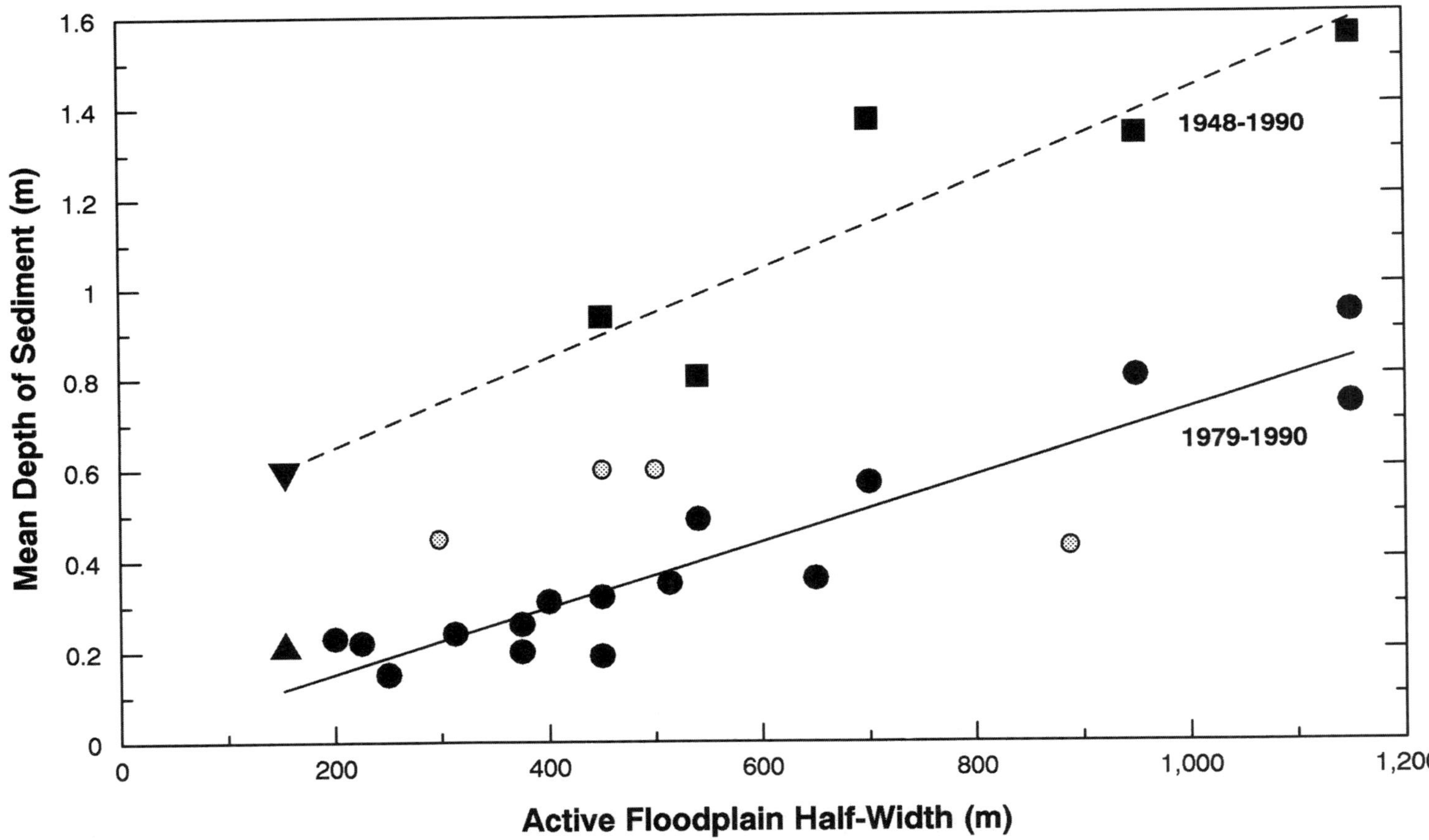

Fig. 8. Relations between the mean depth of sediment on the floodplain and the active floodplain half-width (the distance along a transect from the channel centre line to the floodplain boundary) for the periods from 1948 to 1990 and from 1979 to 1990 determined from surveyed transects perpendicular to the channel that lie within the Gisborne District Council flood control scheme. The filled triangles are average depths for the reach between 8 and 14 km determined from transects spaced *c.* 800 m apart. The outliers (stippled circles) are associated with sedimentation in zones of accelerating or decelerating flow on the floodplain, and were omitted from the regression analysis used to define the relation for the recent, 11 year period ($Y = 0.01 + 0.0007X$, $R^2 = 0.89$). In the case of the relation for the modern (post-1948) period, $R^2 = 0.87$. Data source: Peacock (1991).

In the long term, the amount of suspended sediment transported by a series of events of given magnitude is the product of the effect of a discrete event of that magnitude and the frequency with which it recurs (Wolman & Miller 1960). Events of moderate magnitude and frequency appear more important than large-magnitude, low-frequency flood events in basins where surface wash contributes most of the sediment to the stream channel (Webb & Walling 1984; Hicks 1994). This is also the case in the Waipaoa River Basin, where gully erosion is a significant source of suspended sediment (DeRose *et al.* 1998; Hicks *et al.* In Review). In the 11 year period under consideration, 76% of the suspended sediment was transported by discharges of a lesser magnitude than the bankfull discharge at McPhail's Bend (1800 m^3 s^{-1}), which represents the lower limit for inundation of the study reach as a whole (Fig. 9). In such circumstances it is more realistic to consider the amount of sediment storage in relation to the amount of suspended sediment transported during flood events. Discharges in excess of 1800 m^3 s^{-1} occurred during four floods (including the record March 1988 flood) in the period from May 1979 to September 1990. Floodplain storage accounts for 16% of the total suspended sediment load transported during these events, and 21% of the amount of sediment transported by flows in excess of the bankfull discharge.

We consider our suspended sediment load estimates to be reliable (see Hicks *et al.* In Review), and our estimates of floodplain storage to be temporally representative of contemporary losses to

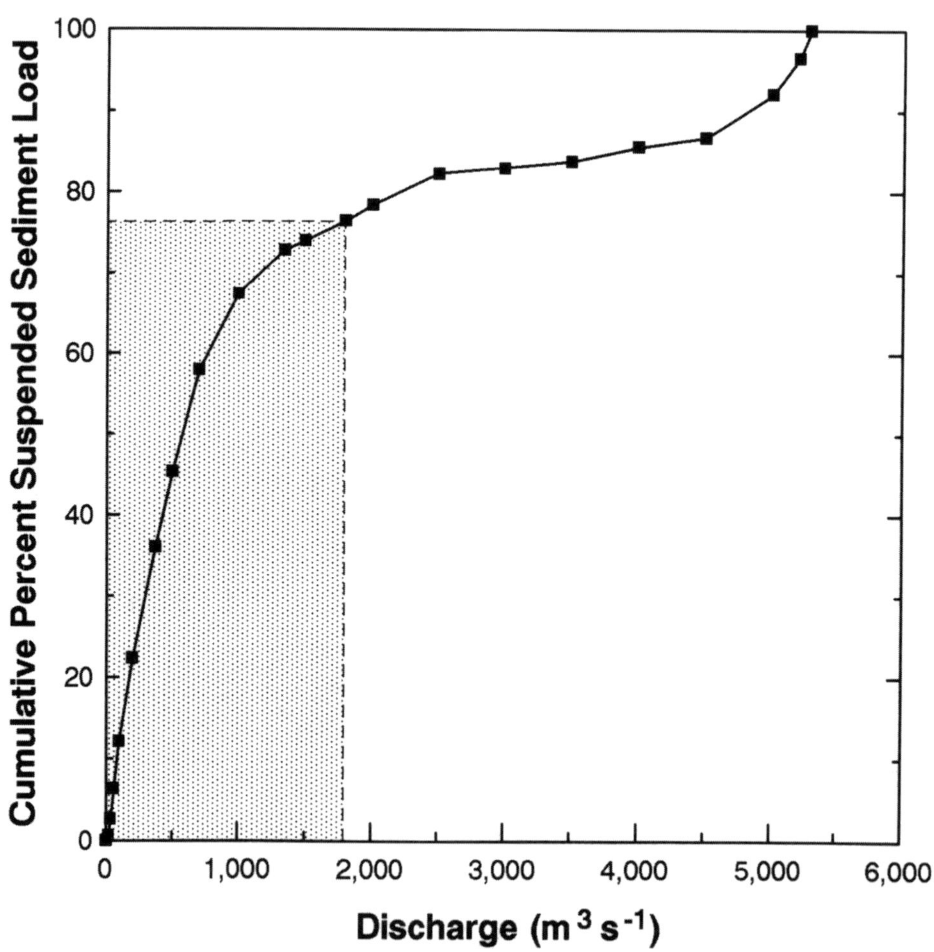

Fig. 9. Cumulative distribution of long-term (May 1979–September 1990) average suspended sediment load of the Waipaoa River at Kanakanaia transported at discharges less than the indicated value. The stippled box delimits the portion of the load transported by flows less than the minimum bankfull discharge for the study reach.

the system in the 44 km long reach downriver from Kanakanaia. The largest errors accrue from extrapolating the average depth of overbank sediment determined from the transect data to an entire segment of the floodplain. We treated the lower 13.2 km of the flood control scheme as a single segment, and the depth estimate from the regression relation is somewhat higher than the average value for the floodplain within the non-tidal section of this reach (Fig. 8), which was derived from transects spaced *c.* 800 m apart, but the error that accrues to the sediment storage estimate is small (±0.25 Mt). However, densely vegetated portions of the floodplain are poorly represented by the transect data and were excluded from the regression analysis, as were sites of enhanced sedimentation associated with zones of diverging flow or recirculation on the floodplain. For this reason we consider that our estimates understate the amount of floodplain storage, though the magnitude of the underestimation is probably small.

Hysteresis effects confound the relation between suspended sediment concentration and discharge in the Rhine–Meuse Delta (Asselman & Middelkoop 1998). However, the estimate by Middelkoop & Asselman (1998) of the transmission loss associated with floodplain storage within a 100 km long reach of the Waal River during the flood of December 1993 (19%) compares favourably with our flood-based estimate (16%), as does the estimate by Goodbred & Kuehl (1998) (15%) of overbank storage during the summer monsoon within a 36 300 km^2 area of the Ganges–Brahmaputra floodplain. However, by comparison with published estimates of time-averaged (annual) transmission losses to the floodplains of low-gradient, meandering alluvial rivers, the loss to storage (6%) within the study reach is low (see Lambert & Walling 1987; Marron 1992; Walling & Quine 1993; Walling *et al.* 1996, 1998).

Perspectives on floodplains as sediment sinks

The hypothesis that floodplain sequestration accounts for a significant proportion of the unmeasured residual component of suspended sediment budgets is predicated on observations made in environments where channel migration helps define the equilibrium relation between the floodplain and the formative (bankfull) discharge, which has a recurrence interval of 1–2 years (Wolman & Leopold 1957). Sediment budgets developed for time periods from 30 to 300 years indicate that floodplain storage (overbank and point-bar sediments) typically accounts for *c.* 30% of the total suspended sediment load in highly disturbed fluvial systems (Marron 1992). Contemporary, time-averaged estimates of transmission losses to floodplains bordering low-gradient, meandering rivers that transport low sediment loads, and upon which rates and patterns of aggradation are controlled by proximity to the channel and floodplain topography (see Middelkoop & Asselman 1998; Walling & He 1998), are broadly in agreement with this figure (Walling *et al.* 1996). However, despite the demonstration by Wolman & Miller (1960) that the product of flow frequency and sediment transport provides a practical means of linking hydrological and sediment transport regimes to channel capacity, to our knowledge only Nolan *et al.* (1987) have previously attempted to relate sedimentation on floodplains to flood effectiveness.

In the Waipaoa River the most effective flow is 0.2–0.1 times the bankfull discharge (Hicks *et al.* In Review), the rate of vertical accretion is *c.* 16 mm h^{-1} (Gomez *et al.* 1998) and, for the period from May 1979 to September 1990, 0.2–0.8 m of vertical accretion was accomplished in 0.08% of the available time. High banks are maintained because the contemporary rate of floodplain construction by vertical accretion is rapid relative to the rate of floodplain destruction by lateral channel migration (typically less than one channel width in the historical period). The return period of bankfull discharge is 4–35 years on the annual series, and infrequent, large floods have a diminished role compared with the cumulative influence of more frequent, lower-magnitude events (Trustrum *et al.* 1999; Hicks *et al.*, In Review). It is estimated that 76% of the suspended sediment load is transported by discharges less than bankfull. In consequence, the proportion of the total suspended sediment load sequestered on the floodplain is low (5%). The difference between our time-averaged estimate for sequestration on the Waipaoa River floodplain and comparable estimates for more active meandering rivers (Marron 1992), and meandering rivers with low sediment loads (see Walling *et al.* 1998), reinforces the notion that there is a link between the sediment transport regime of a river and its sedimentary record (Schumm 1960).

Lateral migration plays an important role in floodplain formation in many meandering river systems. This is the case for the floodplain of the Belle Fourche River, South Dakota, where point-bar sediments account for 40% of the floodplain deposits, and the proportion of sediment in storage (30%) is consistent with that retained on the floodplains of other perennial, meandering rivers (Marron 1992). Actively meandering, alluvial rivers regularly reconstruct their floodplains by lateral migration, a sequence of events that is well illustrated by computer simulations of meandering river and floodplain systems (see Mackey & Bridge 1995; Howard 1996; Gross & Small 1998). The channel and floodplain geometries are such that the most effective discharge is equated with the bankfull discharge, which has a constant 1–2 year recurrence interval (Wolman & Leopold 1957; Andrews 1980). This does not appear to be the case along rivers where overbank deposition is the pre-eminent mode of floodplain formation and some limiting factor inhibits channel migration (Pickup & Warner 1976; Nanson & Young 1981; Nolan *et al.* 1987). The redistribution of boundary shear stress within the cross-section during overbank flow has a profound effect on sediment transport (Ackers 1992), and these rivers are able to construct high banks. Many such rivers have a high sediment load, and although moderate flows are responsible for transporting most sediment the bank height is maintained by rapid vertical accretion during infrequent overbank events (Brakenridge 1984; Nolan *et al.* 1987). Variations in the amount of vertical accretion within a reach promote differences in bank height and in the frequency of flooding. The net result is that channel capacities are greater and the incidence of overbank flows is less than in rivers where overbank deposition is slow relative to the rate of floodplain destruction by lateral migration. It should be noted that if vertical accretion on the floodplain is balanced by aggradation in the channel the frequency of bankfull flow will not progressively decrease with time. However, the equilibrium relation will not be maintained if vertical accretion promotes a progressive increase in bank height and channel capacity, and thus renders the floodplain immune to flooding (Hereford 1984; Nanson 1986). Izumi & Parker (1995) have provided a mechanistic explanation for the tendency of a gravel-bed channel like the Waipaoa River, subject to an increase in the supply of fine sediment load, to evolve into a narrower, deeper river with fine banks and a coarse bed (see Gomez *et al.* 1998). We postulate that the contemporary channel and floodplain geometry are adjusted to the flow(s) doing the most deposition rather than to the flow that accomplishes the most work (see Hey 1975). In short, we believe that the Waipaoa River is representative of that portion of the continuum of floodplain environments where vertical accretion is the dominant mode of floodplain formation. The component of the sediment load accounted for by floodplain storage differs from the published time-averaged estimates because the distribution of the sediment load by flow and the recurrence interval of bankfull discharge are different from those in rivers where lateral accretion affects the process.

Conclusion

As in other areas where overbank sediments have accumulated rapidly (see Knox 1987), the rate of vertical accretion on the Waipaoa River floodplain was conditioned by the river's high suspended sediment load (suspended sediment concentrations in the channel during overbank floods are 30 000–40 000 mg l^{-1}). During the historical (post-1850) period, the average rate of vertical accretion, as determined from three cores, was *c.* 60 mm a^{-1}. A 15 year lacuna in flood activity that occurred between 1965 and 1980 has depressed the rate for the modern (post-1948) period to *c.* 40 mm a^{-1}. For the period from 1948 to 1990, the average rate of vertical accretion within a 44 km long reach, as determined from repeat surveys of transects across the floodplain, ranged between 4 and 37 mm a^{-1} and was complemented by a similar amount of aggradation in the channel. Spatial variations in floodplain half-width (the distance along a transect from the channel centre line to the floodplain boundary) account for 89% of the variation in the depth of sediment present on the floodplain. The unexplained variation is probably attributable to differences in bank height and the consequent variations in bankfull stage. Estimates of the volume of sediment stored on the floodplain were derived by extrapolating the transect data to uniform segments of the floodplain. In the period from May 1979 to September 1990, 6.5 Mt of sediment were stored on the 15.5 km^2 of floodplain within the 44 km long study reach. The total suspended sediment load for the 11 year period was *c.* 146.2 Mt, some 42 Mt of which was transported by floods that generated overbank flow. Floodplain storage thus accounted for 5% of the total suspended sediment load, and 16% of the suspended sediment load transported during events that exceeded bankfull stage. Therefore, although rates of vertical accretion are high, the net loss to the system is comparatively small. This observation has important ramifications for our understanding of how floodplains function as suspended sediment sinks.

Transmission losses to floodplains are often placed into the context of a catchment's time-averaged, suspended sediment budget (see Walling *et al.* 1998), but this may not help elucidate the long-term geomorphological evolution of lowland river floodplains. To do this it is necessary to view vertical accretion in the context of the flood events that generated it, as transmission losses to the floodplain do not occur when the flow is confined within the channel (see Walling *et al.* 1986; Lambert & Walling 1987). This has rarely been attempted in the past and consequently there are deficiencies both in our understanding of the processes that control the architecture of overbank successions and of how floodplain sedimentation influences the delivery of terrigenous sediment to the Earth's oceans. Though lateral accretion influences the morphology of many floodplains, Nanson & Croke (1992) have demonstrated that there is no single mode of floodplain formation. Instead an array of processes is responsible for generating a range of floodplain types (see Willis & Behrensmeyer 1994). The Waipaoa River floodplain may be representative of floodplains bordering rivers with high suspended sediment loads that are produced by rapid, episodic vertical accretion, on which overbank deposition occurs across the entire floodplain and is complemented by aggradation in the channel.

This work was supported by grants from the New Zealand Foundation for Research, Science and Technology (Contract C09612), the National Science Foundation (Grant SBR-9807195) and Indiana State University. The paper is a contribution to Manaaki Whenua–Landcare Research's Waipaoa Catchment Study. T. Pinkney and B. Rosser assisted with the coring, but we are especially indebted to B. Currie and the Gisborne District Council, East Cape Catchment Board and Poverty Bay Catchment Board survey crews, who were responsible for periodically updating the cross-section and transect data in the period since 1948. M. Leeder made a substantive contribution to the review process.

References

ACKERS, P. 1992. Canal and river regime in theory and practice: 1929–92. *Proceedings of the Institute of Civil Engineers*, **96**, 167–178.

ALLISON, M. A., KUEHL, S. A., MARTIN, T. C. & HASSAN, A. 1998. Importance of flood-plain sedimentation for river sediment budgets and terrigenous input to the oceans: insights from the Brahmaputra–Jamuna River. *Geology*, **26**, 175–178.

ANDREWS, E. D. 1980. Effective and bankfull discharges of streams in the Yampa River Basin, Colorado and Wyoming. *Journal of Hydrology*, **46**, 311–330.

ASLAN, A. & AUTIN, W. J. 1998. Holocene flood-plain soil formation in the southern lower Mississippi Valley: implications for interpreting alluvial paleosols. *Geological Society of America Bulletin*, **110**, 433–449.

ASSELMAN, N. E. M. & MIDDELKOOP, H. 1998. Temporal variability of contemporary floodplain sedimentation in the Rhine–Meuse delta, The Netherlands. *Earth Surface Processes and Landforms*, **23**, 595–609.

BHOWMIK, N. G. & DEMISSIE, M. 1982. Carrying capacity of floodplains. *Journal of the Hydraulics Division, American Society of Civil Engineers*, **109**, 1012–1022.

BRAKENRIDGE, G. R. 1984. Alluvial stratigraphy and radiocarbon dating along the Duck River, Tennessee: implications regarding flood-plain origin. *Geological Society of America Bulletin*, **95**, 9–25.

BRIDGE, J. S. 1984. Large-scale facies sequences in alluvial overbank environments. *Journal of Sedimentary Petrology*, **54**, 583–588.

—— & LEEDER, M. R. 1979. A simulation model of alluvial stratigraphy. *Sedimentology*, **26**, 617–644.

BROOKS, A. P. & BRIERLEY, G. J. 1997. Geomorphic responses of lower Bega River to catchment disturbance, 1851–1926. *Geomorphology*, **18**, 291–304.

BROWN, L. J. 1995. Holocene shoreline depositional processes at Poverty Bay, a tectonically active area, northeastern North Island, New Zealand. *Quaternary International*, **26**, 21–33.

CLEVELAND, W. S. 1979. Robust locally weighted regression and smoothing scatterplots. *Journal of the American Statistical Association*, **74**, 829–836.

DEROSE, R. C., GOMEZ, B., MARDEN, M. & TRUSTRUM, N. A. 1998. Gully erosion in Mangatu Forest, New Zealand, estimated from digital elevation models. *Earth Surface Processes and Landforms*, **23**, 1045–1053.

DUNNE, T., MERTES, L. A. K., MEADE, R. H., RICHEY, J. E. & FORSBERG, B. R. 1998. Exchanges of sediment between the flood plain and channel of the Amazon River in Brazil. *Geological Society of America Bulletin*, **110**, 450–467.

FERRING, C. R. 1986. Rates of fluvial sedimentation: implications for archaeological variability. *Geoarchaeology*, **1**, 259–274.

GAGE, M. & BLACK, R. D. 1979. *Slope Stability and Geological Investigations at Mangatu State Forest*. New Zealand Forest Service, Forest Research Institute Technical Paper, **66**.

GOMEZ, B., EDEN, D. N., PEACOCK, D. H. & PINKNEY, E. J. 1998. Floodplain construction by recent, rapid vertical accretion: Waipaoa River, New Zealand. *Earth Surface Processes and Landforms*, **23**, 405–413.

——, MERTES, L. A. K., PHILLIPS, J. D., MAGILLIGAN, F. J. & JAMES, L. A. 1995. Sediment characteristics of an extreme flood: 1993 upper Mississippi River valley. *Geology*, **23**, 963–966.

GOODBRED, S. L. & KUEHL, S. A. 1998. Floodplain processes in the Bengal Basin and the storage of Ganges–Brahmaputra river sediment: an accretion study using ^{137}Cs and ^{210}Pb geochemistry. *Sedimentary Geology*, **121**, 239–258.

GRAF, J. B., WEBB, R. H. & HEREFORD, R. 1991. Relation of sediment load and flood-plain formation to climatic variability, Paria River drainage basin, Utah and Arizona. *Geological Society of America Bulletin*, **103**, 1405–1415.

GRAF, W. L. 1983. Downstream changes in stream power in the Henry Mountains, Utah. *Annals of the Association of American Geographers*, **67**, 323–342.

GROSS, J. L. & SMALL, M. J. 1998. River and floodplain process simulation for subsurface characterization. *Water Resources Research*, **34**, 2365–2376.

HE, Q. & WALLING, D. E. 1996. Rates of overbank sedimentation on the floodplains of British lowland rivers documented using fallout ^{137}Cs. *Geografiska Annaler*, **78A**, 223–234.

HEREFORD, R. 1984. Climate and ephemeral-stream processes: twentieth-century geomorphology and alluvial stratigraphy of the Little Colorado River, Arizona. *Geological Society of America Bulletin*, **95**, 654–668.

—— 1986. Modern alluvial history of the Paria River drainage basin, southern Utah. *Quaternary Research*, **25**, 293–311.

HESSELL, J. W. D. 1980. *The Climate and Weather of the Gisborne Region*. New Zealand Meteorological Service Miscellaneous Publication, **115(8)**.

HEY, R. D. 1975. Design discharge for natural channels. *In*: HEY, R. D. & DAVIES, T. D. (eds) *Science, Technology and Environmental Management*. Saxon House, Farnborough, 73–88.

HICKS, D. M. 1994. Landuse effects on magnitude–frequency characteristics of storm sediment yields: some New Zealand examples. *In*: OLIVE, L. J., LOUGHRAM, R. J. & KESBY, J. A. (eds) *Variability in Steam Erosion and Sediment Transport*, International Association of Hydrological Sciences Publication, **224**, 395–402.

——, GOMEZ, B. & TRUSTRUM, N. A. In Review. Erosion thresholds and suspended sediment yields: Waipaoa River Basin, New Zealand.

——, HILL, J. & SHANKAR, U. 1996. Variation of suspended sediment yields around New Zealand: the relative importance of rainfall and geology. *In*: WALLING, D. E. & WEBB, B. W. (eds) *Erosion and Sedimentation: Global and Regional Perspectives*, International Association of Hydrological Sciences Publication, **236**, 149–156.

HOGAN, A. 1997. *Waipaoa Catchment Study: Flood History and Hydrological Changes*. Gisborne District Council Report, **EWTR97/01**.

HOWARD, A. D. 1996. Modelling channel evolution and floodplain morphology. *In*: ANDERSON, M. G., WALLING, D. E. & BATES, P. D. (eds) *Floodplain Processes*. Wiley, Chichester, 15–62.

IZUMI, N. & PARKER, G. 1995. On fine sand deposition in gravel channels. *Annual Journal of Hydraulic Engineering, Japan Society for Civil Engineers*, **39**, 665–670.

JACOBSON, R. B. & COLEMAN, D. J. 1986. Stratigraphy and recent evolution of Maryland piedmont flood plains. *American Journal of Science*, **286**, 617–637.

JAMES, L. A. 1989. Sustained storage and transport of hydraulic gold mining sediment in the Bear River, California. *Annals of the Association of American Geographers*, **79**, 570–592.

KELSEY, H. M., LAMBERSON, R. & MADEJ, M. A. 1987. Stochastic model for the long-term transport of stored sediment in a river channel. *Water Resources Research*, **23**, 1738–1750.

KIELY, G. 1990. Overbank flow in meandering compound channels—the important mechanisms. *In*: WHITE, W. R. (ed.) *International Conference on River Flood Hydraulics*. Wiley, Chichester, 207–217.

KNOX, J. C. 1987. Historical valley floor sedimentation in the upper Mississippi River valley. *Annals of the Association of American Geographers*, **77**, 224–244.

—— 1993. Large increases in flood magnitude in response to modest changes in climate. *Nature*, **361**, 430–432.

—— 1995. Fluvial systems since 20 000 yrs B. P. *In*: GREGORY, K. J., STARKEL, L. & BAKER, V. R. (eds) *Global Continental Palaeohydrology*. Wiley, Chichester. 87–108.

KRAUS, M. J. & ASLAN, A. 1993. Eocene hydromorphic paleosols: significance for interpreting ancient floodplain processes. *Journal of Sedimentary Petrology*, **63**, 453–463.

—— & BOWN, T. M. 1993. Short-term sediment accumulation rates determined from Eocene alluvial paleosols. *Geology*, **21**, 743–746.

LAMBERT, C. P. & WALLING, D. E. 1987. Floodplain sedimentation: a preliminary investigation of contemporary deposition within the lower reaches of the River Culm, Devon, U. K. *Geografiska Annaler*, **69A**, 393–404.

LE BORGNE, E. 1955. Susceptibilité magnétique anormale du sol superficiel. *Annales de Géophysique*, **11**, 399–419.

LECCE, S. A. 1997. Spatial patterns of historical overbank sedimentation and floodplain evolution, Blue River, Wisconsin. *Geomorphology*, **18**, 265–277.

MACKEY, S. D. & BRIDGE, J. S. 1995. Three-dimensional model of alluvial stratigraphy: theory and application. *Journal of Sedimentary Research*, **B65**, 7–31.

MACKLIN, M. G. 1985. Flood-plain sedimentation in the upper Axe valley, Mendip, England. *Transactions of the Institute of British Geographers*, **NS10**, 235–244.

MAGILLIGAN, F. J. 1985. Historical floodplain sedimentation in the Galena River basin, Wisconsin and Illinois. *Annals of the Association of American Geographers*, **75**, 583–594.

—— 1992. Thresholds and the spatial variability of flood power during extreme floods. *Geomorphology*, **5**, 373–390.

MARRON, D. C. 1992. Floodplain storage of mine tailings in the Belle Fourche River system: a sediment budget approach. *Earth Surface Processes and Landforms*, **17**, 675–685.

MAZENGARB, C., FRANCIS, D. A. & MOORE, P. R. 1991. *Sheet Y16—Tauwhareparae*. 1:50 000 Map and notes. Department of Scientific and Industrial Research, Wellington.

MEADE, R. H. 1994. Suspended sediments of the modern Amazon and Orinoco rivers. *Quaternary International*, **21**, 29–39.

MERTES, L. A. K. 1994. Rates of river flood-plain sedimentation on the central Amazon River. *Geology*, **22**, 171–174.

MIDDELKOOP, H. & ASSELMAN, N. E. M. 1998. Spatial variability of floodplain sedimentation at the event scale in the Rhine–Meuse Delta, The Netherlands. *Earth Surface Processes and Landforms*, **23**, 561–573.

MILLIMAN, J. D. & MEADE, R. H. 1983. World-wide delivery of river sediment to the oceans. *Journal of Geology*, **91,** 1–21.

—— & SYVITSKI, J. P. M. 1992. Geomorphic/tectonic control of sediment discharge to the ocean: the importance of small mountainous rivers. *Journal of Geology*, **100**, 525–544.

MOORE, P. R. & MAZENGARB, C. 1992. Geology and landforms of the Raukumara Peninsula. *In*: SOONS, J. & SELBY, M. J. (eds) *Landforms of New Zealand.* Longman Paul, Auckland, 334–343.

MULLINS, C. E. 1977. Magnetic susceptibility of the soil and its significance in soil science—a review. *Journal of Soil Science*, **28**, 223–246.

NANSON, G. C. 1986. Episodes of vertical accretion and catastrophic stripping: a model of disequilibrium flood-plain development. *Geological Society of America Bulletin*, **97**, 1467–1475.

—— & CROKE, J. C. 1992. A genetic classification of floodplains. *Geomorphology*, **4**, 459–486.

—— & YOUNG, R. W. 1981. Overbank deposition and floodplain formation on small coastal streams of New South Wales. *Zeitschrift für Geomorphologie*, **25**, 332–347.

NOLAN, K. M., LISLE, T. E. & KELSEY, H. M. 1987. Bankfull discharge and sediment transport in northwestern California. *In*: BESCHTA, R. L., BLINN, T., GRANT, G. E., ICE, G. G. & SWANSON, F. J. (eds) *Erosion and Sedimentation in the Pacific Rim.* International Association of Hydrological Sciences Publication, **165**, 439–449.

O'BYRNE, T. N. 1967. A correlation of rock types with soils, topography and erosion in the Gisborne–East Cape region. *New Zealand Journal of Geology and Geophysics*, **10**, 217–231.

PEACOCK, D. H. 1991. *Waipaoa River Flood Control Scheme Aggradation Study*. Gisborne District Council Report, **EW2996**.

PEARCE, A. J., BLACK, R. D. & NELSON, C. S. 1981. Lithologic and weathering influences on slope form and process, eastern Raukumara Range, New Zealand. *In*: DAVIES, T. R. H. & PEARCE, A. J. (eds) *Erosion and Sedimentation Transport in Pacific Rim.* International Association of Hydrological Sciences Publication, **132**, 95–122.

PICKUP, G. & WARNER, R. F. 1976. Effects of hydrologic regime on magnitude and frequency of dominant discharge. *Journal of Hydrology*, **29**, 51–57.

PIZZUTO, J. E. 1987. Sediment diffusion during overbank flows. *Sedimentology*, **34**, 301–317.

PULLAR, W. A. 1962. *Soils and Agriculture of the Gisborne Plains*. New Zealand Soil Bureau Bulletin, **20**.

—— & PENHALE, H. R. 1970. Periods of recent infilling of the Gisborne Plains basin. *New Zealand Journal of Science*, **13**, 410–434.

PUTNAM, D. E. 1994. Vertical accretion of flood deposits and deeply stratified archaeological site formation in central Maine, USA. *Geoarchaeology*, **9**, 467–502.

REID, L. M. & PAGE, M. J. IN Press. Magnitude and frequency of landsliding in a large New Zealand catchment. *Geomorphology*, in press.

RITTER, D. F., KINSEY, W. F. & KAUFMAN, M. E. 1973. Overbank sedimentation in the Delaware River Valley during the last 6000 years. *Science*, **179**, 374–375.

ROBINSON, M. A. & LAMBRICK, G. H. 1984. Holocene alluviation and hydrology in the upper Thames basin. *Nature*, **308**, 809–814.

SCHUMM, S. A. 1960. The effect of sediment type on the shape and stratification of some modern fluvial deposits. *American Journal of Science*, **258**, 177–184.

SELLIN, R. H. J. 1995. Hydraulic performance of a skewed two-stage flood channel. *Journal of Hydraulics Research*, **33**, 43–64.

SIMM, D. J. 1995. The rates and patterns of overbank deposition on a lowland floodplain. *In*: FOSTER, I. D. L., GURNELL, A. M. & WEBB, B. W. (eds) *Sediment and Water Quality in River Catchments.* Wiley, Chichester, 247–264.

STEEL, R. & AASHEIM, S. M. 1978. Alluvial sand deposition in a rapidly subsiding basin (Devonian, Norway). *In*: MIALL, A. D. (ed.) *Fluvial Sedimentology,* Canadian Society of Petroleum Geologists Memoir, **5**, 385–412.

THOMPSON, R. & MORTON, D. J. 1979. Magnetic susceptibility and particle-size distribution in recent sediments of the Loch Lomond drainage basin, Scotland. *Journal of Sedimentary Petrology*, **49**, 801–812.

TODD, A. D. 1949. *Report on Waipaoa River Flood Control: Appendix 1—Newspaper References to Past Floods.* Poverty Bay Catchment Board, Gisbourne.

TRIMBLE, S. W. & LUND, S. W. 1982. *Soil Conservation and Reduction of Erosion and Sedimentation in the Coon Creek Basin, Wisconsin.* US Geological Survey Professional Papers, **1234**.

TRUSTRUM, N. A., GOMEZ, B., REID, L. M., PAGE, M. J. & HICKS, D. M. 1999. Sediment production, storage, and output: the relative role of large magnitude events in steepland catchments. *Zeitschrift für Geomorphologie*, **115**, 71–86.

TURNER, G. M. 1997. Environmental magnetism and magnetic correlation of high resolution lake sediment records from northern Hawke's Bay, New Zealand. *New Zealand Journal of Geology and Geophysics*, **40**, 287–298.

VAN ANDEL, T. H. & ZANGGER, E. 1990. Landscape stability and destabilisation in the prehistory of Greece. *In*: BOTTEMA, S., ENTJES-NIEBORG, G. & VAN ZEIST, W. (eds) *Man's Role in the Shaping of the Eastern Mediterranean Landscape.* Balkema, Rotterdam, 159–182.

WALLING, D. E. & HE, Q. 1998. The spatial variability of overbank sedimentation on river floodplains. *Geomorphology*, **24**, 209–223.

—— & QUINE, T. A. 1993. Using Chernobyl-derived fallout radionuclides to investigate the role of downstream conveyance losses in the suspended sediment budget of the River Severn, United Kingdom. *Physical Geography*, **14**, 239–253.

—— & WEBB, B. W. 1996. Erosion and sediment yield: a global overview. *In*: WALLING, D. E. & WEBB, B. W. (eds) *Erosion and Sediment Yield: Global and Regional Perspectives,* International Association of Hydrological Sciences Publication, **236**, 3–19.

——, BRADLEY, S. B. & LAMBERT, C. P. 1986. Conveyance losses of suspended sediment within a flood plain system. *International Association of Hydrological Sciences Publication*, **159**, 119–131.

——, HE, Q. & NICHOLAS, A. P. 1996. Floodplains as suspended sediment sinks. *In*: ANDERSON, M. G., WALLING, D. E. & BATES, P. D. (eds) *Floodplain Processes*. Wiley, Chichester, 399–440.

——, OWENS, P. N. & LEEKS, G. J. L. 1998. The role of channel and floodplain storage in the suspended sediment budget of the River Ouse, Yorkshire, U. K. *Geomorphology*, **22**, 225–242.

WEBB, B. W. & WALLING, D. E. 1984. Magnitude and frequency characteristics of suspended sediment transport in Devon rivers. *In*: BURT, T. P. & WALLING, D. E. (eds) *Catchment Experiments in Fluvial Geomorphology*. Geo Books, Norwich, 399–415.

WILCOCK, P. R. 1998. Two-fraction model of initial sediment motion in gravel-bed rivers. *Science*, **280**, 410–412.

WILLIS, B. J. & BEHRENSMEYER, A. K. 1994. Architecture of Miocene overbank deposits in northern Pakistan. *Journal of Sedimentary Research*, **B64**, 60–67.

WILMSHURST, J. 1997. The impact of human settlement on vegetation and soil stability in Hawke's Bay, New Zealand. *New Zealand Journal of Botany*, **35**, 97–111.

—— & MCGLONE, M. S. 1996. Forest disturbance in the central North Island, New Zealand, following the 1850 BP Taupo eruption. *The Holocene*, **6**, 399–411.

——, —— & PARTRIDGE, T. R. 1997. A late Holocene history of natural disturbance in lowland podocarp/hardwood forest, Hawke's Bay, New Zealand. *New Zealand Journal of Botany*, **35**, 79–96.

WOLMAN, M. G. & LEOPOLD, L. B. 1957. *River flood plains: some observations on their formation*. US Geological Survey Professional Papers, **282-C**, 87–107.

—— & MILLER, J. P. 1960. Magnitude and frequency of forces in geomorphic processes. *Journal of Geology*, **68**, 54–74.

WOODYER, K. D., TAYLOR, G. & CROOK, K. A. W. 1979. Depositional processes along a low-gradient suspended-load stream: the Barwon River, New South Wales. *Sedimentary Geology*, **22**, 97–120.

WRIGHT, V. P. & MARRIOTT, S. B. 1993. The sequence stratigraphy of fluvial depositional systems: the role of floodplain sediment storage. *Sedimentary Geology*, **86**, 203–210.

Scientific and institutional constraints on the restoration of European floodplains

W. M. ADAMS[1] & M. R. PERROW[2]

[1]*Department of Geography, University of Cambridge, Downing Place, Cambridge CB2 3EN, UK, (e-mail:wa12@hermes.cam.ac.uk)*

[2]*ECON, School of Biological Sciences, University of East Anglia, Norwich NR4 7TS, UK*

Abstract: Most river restoration (or often more appropriately rehabilitation) projects have focused on the river channel, in-stream and channel-edge environments. The desirability of extending restoration to the active floodplain is now widely recognized, because of the critical importance of the floodplain–riparian zone in river structure and function. However, the extension of river restoration to floodplain environments presents the environmental manager with complex scientific and institutional challenges. This paper considers the constraints on floodplain restoration as a result of scientific uncertainty, the ways in which scientific and technical knowledge are applied, the complexity of formal institutions of floodplain governance, and the complexity of informal institutions influencing floodplain environments.

In Western Europe, river management over the past 200 years has chiefly employed channelization, that is, extensive engineering work to deepen, widen, straighten and stabilize river channels. In the UK about 24% of the length of 'main rivers' has been channelized (Brookes *et al.* 1983), 89% of UK rivers are regulated (Petts 1988), and probably about 95% of floodplain area has been modified (ECON & Pond Action 1993). Elsewhere in Europe the impact is slightly less severe, but none the less extensive; for example, in the Morava basin of the Czech Republic 20% of river courses have been engineered for flood defence, and 50% deforested and affected by intensive agriculture (Sterba *et al.* 1997). Throughout Europe, floodplain woodland has effectively disappeared from all but a few fragmented sites (Brown *et al.* 1997; Peterken & Hughes 1998). In addition to these impacts, economic development (through urbanization, manufacturing and agricultural intensification) has contributed a cocktail of pollutants that have further reduced biodiversity and degraded ecosystem function. One response to these impacts has been increasing interest on the part of river managers and conservation planners in river restoration (Petersen *et al.* 1992; Perrow & Wightman 1993; Brookes 1996*a*; Petts 1996, 1998; Petts & Calow 1996). This paper considers the extension of ideas about river restoration to incorporate floodplain environments, and discusses the scientific, technical and institutional constraints on floodplain restoration.

River restoration

Recent interest in river restoration reflects a significant change in river managerial regimes. It also reflects a wider fascination with 'ecological restoration' in the scientific and technical literature and in public debates about approaches to environmental management. The concept of making or inducing positive changes in ecosystems is controversial, both in theory and practice. However, it has caught the attention of conservationists, environmental managers, developers and policymakers. There is a growing literature on the philosophical and scientific dimensions of environmental restoration in general (e.g. Jordan *et al.* 1987; Buckley 1989; Gunn 1991; Hobbs & Norton 1996; Elliot 1997), and on restoration in particular habitats, such as industrial sites (e.g. Bradshaw 1983), as well as rivers (National Research Council 1992).

The history of geological, climatic, biological and chemical processes and forms are unique in any given location, so even in places with minimal human intervention complete restoration to some pristine state is effectively impossible (National Research Council 1992). Indeed, philosophically the very notion of a return to a 'natural' or 'virgin' state through human action is bizarre (Petts 1996). Clearly, however, in human-modified ecosystems, management can be implemented to restore forms and functions identical to, or at least analogous to, those formerly existing.

From: MARRIOTT, S. B. & ALEXANDER, J. (eds) 1999. *Floodplains: Interdisciplinary Approaches*. Geological Society, London, Special Publications, **163**, 89–97. 1-86239-050-9/99/$15.00 ©The Geological Society of London 1999.

Although debate about environmental restoration has advanced very rapidly, it is often based on a very loose terminology. Perrow & Wightman (1993) drew on Cairns (1991) and the National Research Council (1992) to distinguish four distinct categories of action to enhance river environments (Table 1). All four are often commonly given the loose label of 'river restoration', but habitat creation, habitat enhancement and habitat rehabilitation fall short of full restoration. Perrow & Wightman (1993) suggested a logical definition of restoration, as a response to specific processes of degradation. Following Cairns (1991), they defined restoration as 'the complete structural and functional return to a pre-disturbance state' (Table 1). In the context of rivers, 'disturbance' would relate to a specific episode or series of human impacts on riverine systems (Perrow & Wightman 1993). Thus restoration might seek to reverse known episodes of channel widening, deepening, straightening or stabilization, or to remove physical obstructions and chemical pollutants. Following these definitions, the bulk of river restoration projects involves not restoration but something less, an attempt at rehabilitation or enhancement. Although theoretically definable, full restoration is rarely a feasible practical option. Further 'mitigation and enhancement' may be valuable (Brookes 1996*a*, p. 572), but may not necessarily be a step towards the restoration of the whole physical and biological system.

River restoration: the links between channels and floodplains

The active floodplain environment can be distinguished from other low-lying land that is rarely flooded and whose characteristics are not conditioned by the contemporary regime of fluvial processes, and of course from the slopes in topographic sequence above it. The active floodplain might be delimited by the spatial extent of the flood of a given recurrence interval (e.g. the 1 in 100 year flood, or 1 in 50 year flood; Cordes *et al.* 1997). Within this area, fluvial processes of erosion, sedimentation, flooding and channel migration create a set of substrate patches that vary in size, location and longevity, upon which processes of ecological change create a complex of vegetational and faunal assemblages. Floodplain biotic community patterns are closely related to the dynamic interaction of fluvial and terrestrial processes. Floodplain faunal and floral assemblages represent a continuum from terrestrial species tolerating irregular inundation through species requiring regular inundation and desiccation, to aquatic species dependent on persistently wet soil or surface conditions (Malanson 1993; Gurnell 1995, 1997; Hughes 1997).

Experience of river restoration has been growing in northwest Europe (e.g. Brookes 1995, 1996*b*; Brookes & Shields 1996). River restoration can take a wide variety of forms, both structural and non-structural (Table 2). However, experience with environmental restoration beyond the river channel has been more limited (Brookes *et al.* 1996). Most projects in the UK and in much of northwest Europe have been narrowly focused on the river channel and aquatic environments, and have involved intervention in the chemical, physical or biological processes in the channel itself, the lotic ecosystem and its fauna and flora (Brookes *et al.* 1996; Holmes 1998). Few projects have sought to rehabilitate ecosystems beyond the channel in the riparian or floodplain zone.

The floodplain is a critical element in all river restoration, because of the hydrological and geomorphological links between channels and floodplains. Petts (1996) drew attention to the importance of the active hydrological and ecological links between the river channel and flooded land, and suggested that a 'river' should be viewed as including both the active channels and their ecotones.

In Europe, the restoration of floodplain environments has mostly been confined to short stretches of small, low-energy streams (Brookes 1996*b*). Floodplain restoration has rarely been explicitly treated as a distinct element within river restoration, even where experience with channel restoration is extensive, for example, in Germany

Table 1. *Definitions of river environmental improvement*

Form of habitat improvement	Definition
Restoration	'The complete structural and functional return to a pre-disturbance state'
Rehabilitation	'The partial structural and functional return to a pre-disturbance state'
Enhancement	'Any improvement of a structural or functional attribute'
Creation	'The birth of a new ecosystem that previously did not exist at the site'

Source: Perrow & Wightman (1993), drawing on Cairns (1991) and National Research Council (1992).

Table 2. *Techniques of river restoration*

Non-structural techniques
Catchment planning (analysis of surface water, ground water, land use; e.g. EA (Environment Agency) local environmental action plans)
Land use planning (e.g. set-aside, Environmentally Sensitive Areas, countryside stewardship)
Species-centred restoration (e.g. otters, fish)
Restoration of river function through flow manipulation and restoration of the 'flood pulse'

Structural techniques
Channel and in-channel
Reinstatement of natural channels
High-energy streams: sediment supply to enhance recovery
Low-energy streams: physical restructuring replace meanders and sinuosity or riffle–pool structures
Others: multi-stage channels, berms, embankments, bypass and diversion channels in developed streams

Bank modifications
Use of natural material (stumps and logs; willows, emergent vegetation; bankside vegetation)
Artificial alternatives (geotextiles, stone blocks, current deflectors)

In-stream modifications
Substrate reinstatement (cleaning natural gravels; physical reintroduction)
Instream modification using natural materials (debris and boulders; channel vegetation)
Instream modification using artificial materials (fish cover structures; current deflectors; sediment and gravel traps; drop structures)

Riparian zone and floodplain restoration
Reinstatement of natural systems:
- increasing hydrological communication between river and floodplain by raising bed levels, lowering land levels, or removing obstructions
- eliminating and controlling contaminants
- re-establishing native flora and fauna

Livestock control
Creation of alternative systems (buffer strips or zones)
Wetland creation for pollution control
Flood storage areas (to mimic natural floodplain systems and reduce downstream flooding)

Source: adapted from Perrow & Wightman (1993).

and Denmark (Brookes 1996*a*). Even when floodplain restoration is attempted, most projects have been concentrated on only the part of the floodplain immediately adjacent to the channel. Thus, for example, the innovative work of the River Restoration Project on the River Cole near Swindon, UK (Holmes 1998) had as its primary focus the physical re-establishment of former channel form, aquatic chemistry, fauna and flora (Biggs *et al.* 1998; Hoffman *et al.* 1998; Kronvang *et al.* 1998). Riparian and floodplain land provided the context for this channel restoration, but work did not extend to an integrated restoration of the whole riparian corridor. River restoration projects often adopt the 'building block model' (Petersen *et al.* 1992), which starts with the protection of the riparian zone; the recreation of riparian wetlands, forests and ponds form possible later options. More complete floodplain restoration is, however, being attempted on a larger scale within Europe, for example, in the braided channel of the Danube in Austria, between Vienna and the Slovakian border (Tockner & Schiemer 1997). Here works involve reactivation of side-arms by opening channels adjacent to the main stream.

Outside Europe, restoration experiments have been more bold. For example, the Kissimmee River project in Florida, USA, involves the restoration of flooding to 11 000 ha of floodplain (Toth *et al.* 1993), and the US Bureau of Reclamation has passed flood flows down the Colorado River through the Grand Canyon to restore beaches, bars and other habitats (Hecht 1996). In a number of African river basins, artificial floods have been proposed and in some cases released to re-establish ecological function and sustain economic activity in huge downstream floodplains (e.g. Horowitz & Salem-Murdock 1991; Scudder 1991; Acreman & Howard 1996).

There is growing understanding of the significance of the strong hydrological, geochemical and biological links between channel and surrounding land which affect aquatic ecosystems, and with this the importance of the floodplain–river corridor

Table 3. *Functions of natural riparian zones and floodplains*

Increase bank strength, reduce bank erosion, and hence affect width/depth ratio
Provide allochthonous organic matter to the stream
Provide shade (and hence reduce aquatic plant growth)
Provide locations for flood-water storage, and hence reduce flood peaks downstream and increase reach water retention capacity
Provide sites for sediment storage in floods
Reduce sediment inputs from surrounding land
Improve water quality by reducing inputs of nutrients and pollutants from surrounding slopes

Source: adapted from Perrow & Wightman (1993).

within river restoration is being increasingly recognized (Brookes 1996*a*; Brookes & Shields 1996; Holmes 1998). There is wider appreciation by river managers of the 'functions' that floodplains fulfil (Table 3). Rehabilitation and restoration of floodplains can have varying positive effects on river channel attributes and functions (Brookes *et al.* 1996). The separation of floodplains from the hydrological and geomorphological systems of the channel is now seen to be a potentially counter-productive (and sometimes economically costly) aspect of conventional approaches to river control. It is becoming increasingly clear that management for restoration should address the question of the isolation of the floodplain from the channel.

Floodplain restoration

Restoration of floodplain environments may result from management intervention in any of three contexts: the catchment, the river corridor and the channel. In the catchment, it may be promoted by non-structural interventions (e.g. affecting land use), which allow natural recovery of floodplain ecosystems from some anthropogenic perturbation ('recovery enhancement', Perrow & Wightman 1993). In the wider floodplain–river corridor environment outside the confines of the river channel itself, floodplain environments can be positively affected by intervention in the form of habitat creation or improvement. These might include management of the narrow riparian zone, or the creation of specific wetlands (Table 2). River restoration projects focused on the channel can also affect floodplain ecosystems positively, for example, if they involve change in stream flows, channel form or water quality (Table 4).

Floodplain restoration projects include: (1) those projects that explicitly relate to the floodplain (involving direct engagement with the ecology of the terrestrial ecosystems of the floodplain itself); (2) those that focus on the restoration of the structure and function of river channels (and the flow regimes necessary to maintain them) but have wider impacts; (3) those that involve wider intervention in the catchment. The critical element in all aspects of floodplain restoration is the distinction between those actions that directly affect the physical forms or ecosystems of floodplains (e.g. wetland creation or enhancement), and those that affect the processes that give rise to these forms and ecosystems (Hughes & Rood 1999). The former are strictly habitat creation or enhancement, the latter restoration. Full restoration of floodplain ecosystems must involve restoration of the flood pulse (Bayley 1991; Brookes *et al.* 1996).

Floodplain restoration may therefore be defined

Table 4. *Impacts of channel restoration projects on floodplains*

Manipulation of stream discharge (floods and low flows, particularly flood peaks and their distribution in time)
Changes in stream channel or bank form such that patterns of inundation away from the channel are changed
Changes in the spatial patterns of flow within the channel such that channel planform changes and sediments, soils or ground water of the floodplain are affected
Changes in sediment load of the river such that physical interactions between floodplain and channel change and sediments, soils or ground water of the floodplain are affected
Changes in water quality (nitrogen, phosphorus, turbidity, temperature, biocides) such that terrestrial ecosystems outside the channel and levees are affected
Direct or indirect human impacts on channel vegetation (e.g. planting, cutting, grazing) such that ecological patterns or processes outside the channel and levees are affected
Direct or indirect impacts on aquatic or other animals such that ecological patterns or processes outside the channel and levees are affected

as re-establishment of the abiotic and biotic elements of ecosystems within the river environment, including floodplain and riparian land connected to the river, such that appropriate processes of inundation, erosion and deposition are established that can themselves recreate riparian and floodplain ecosystems, and the wildlife and landscape values that are associated with them.

It is clear that river restoration must take account of the links between rivers and their floodplain and riparian zones. The extension of river restoration to include floodplain environments is clearly potentially beneficial from a conservation, amenity and engineering perspective, and consideration of floodplain–channel linkages is obviously an essential element of all river restoration projects. However, floodplain restoration is more complex and problematic than channel restoration. The increasingly sophisticated techniques and growing experience of channel restoration are not sufficient basis for floodplain restoration. Four problems place particular constraints on floodplain restoration: scientific complexity, technical capacity and the formal and informal institutions controlling floodplain management.

Scientific complexity and floodplain restoration

Many of the problems that face would-be floodplain restorers derive from the inherent complexity of floodplain ecosystems, and the limited extent of scientific understanding of the dynamic way they change in space and time. This complexity derives from four sources.

First, floodplains or 'riverine riparian ecosystems' (National Research Council 1992) zones are spatially diverse, both laterally and longitudinally. There is great lateral variability in elevation, sediments, water regime and vegetation. The structure and function of riparian or floodplain environments also varies longitudinally. The National Research Council (1992) has argued that aquatic production is centred on the channel in the upper reaches and in the floodplain further downstream. Riparian zones upstream receive inputs of water, nutrients and sediment from hill slopes and filter them before they reach the river; floodplains are zones of interchange, receiving inputs from the channel, buffering inputs to the channel and generating endogenous processes and products.

Second, there are complex systemic links both laterally and longitudinally within a single river catchment system. There are complex lateral links between floodplains and river channel environments that involve physical processes and pathways (both hydrological and chemical), and biological processes (e.g. plant litter and fish). They are reciprocal, in that under some circumstances floodplain ecosystem processes will have implications for channel ecosystems, whereas in others the reverse will hold.

Third, these lateral links interact with longitudinal interactions between upstream and downstream reaches. Although river restoration managers may have to think in terms of discrete 'sites' or reaches, rivers connect these pieces together in complex ways, so that upstream change (whether natural or anthropogenic) in channel, floodplain or catchment can have significance for the response of downstream floodplains.

Fourth, floodplain ecosystems are complex temporally. There are time lags between changes in different abiotic systems, and between abiotic and biotic systems. These interactions have complex knock-on effects on floodplain systems, such that the evolution of floodplain ecosystems is highly contingent in space and time, and prediction of likely patterns of change is problematic. Furthermore, some characteristic floodplain species have long lifespans (notably floodplain trees). Therefore, where the signature of past events lingers for several generations, this can imply a period of centuries, probably pre-dating direct flow measurements and much documentary evidence on the river. Similarly, present management actions may have effects that resonate well beyond the conventional time horizons used in economic appraisals by project planners (often 30 years).

Technical capacity and floodplain restoration

Scientific uncertainty is perhaps the most obvious constraint on the extension of restoration activities from river channel to floodplain (Brookes & Shields 1996). The inherent complexity of the response of floodplain physical and biological systems to perturbation is an exciting scientific challenge, but it is a severe constraint on technical capacity to make secure predictions of what will happen when river managers intervene in any particular context. There is also a fundamental problem of perception. Many technically trained river managers, as well as members of the public, have little understanding of what a floodplain is and how it functions. Although unfortunate, this is not surprising, as in many countries floodplains no longer exist as geomorphologically active entities. Centuries of river management aimed at controlling the natural flooding patterns of European rivers have tended to erase both the physical terrain and the conceptual category of the floodplain as a dynamic physical entity. For floodplain residents and the engineers who serve them, floods are often perceived as aberrant and unwanted phenomena, threatening valuable real estate.

Technical capacity to restore floodplains is also limited by conventional scientific understanding. Much science is reductionist, and to cope with the complexity of floodplains it needs to be holistic. There needs to be an explicit concern for the interconnectedness of different kinds of processes, and an awareness of their variability in space and time. Scientific understanding also needs to be interdisciplinary, and although the need for such research is widely recognized, it is still hard to obtain funding for it. Lack of effective communication between disciplines (particularly between geomorphology and aquatic and riparian ecology) remains a problem (see Hughes 1997). Multidisciplinary understanding of floodplains needs careful balancing and integration of different skills and perspectives. River engineers too easily dismiss ecologists as 'gardeners' or landscape architects, people to be brought in to re-green a 'restored' physical landscape.

Holistic science needs to feed into holistic planning. This problem is recognized in the context of channel restoration, and strategies have been identified to achieve it; for example, the careful 'building block' approach (Petersen *et al.* 1992). However, river restoration planning still tends to be compartmentalized not holistic. Schemes typically involve physical habitat change, and make the assumption that changes in hydrology, water quality and ecology will all follow. Restoration has therefore tended to be treated as primarily an engineering issue, rather than being based on specific ecological understanding. When restoration extends to the less adequately understood environment of the active floodplain, the lack of a clear understanding of ecosystem change becomes seriously problematic.

Communication between river and floodplain research scientists and those people planning and managing river and floodplain restoration projects tends to be poor. The limited engagement by research scientists in practical floodplain restoration projects may seem surprising, but the lack of post-project appraisal is recognized as a significant problem (Brookes & Shields 1996). Few floodplain restoration projects have been described in the formal academic literature. A Bath Information and Data Services (BIDS) search was carried out on journal volumes published between 1981 and 1998. This yielded only 17 journal papers using one of the following keywords: stream, river, riparian zone, floodplain *and* one of the following: restoration, rehabilitation, enhancement, improvement, rejuvenation. All of these dealt with technical, engineering or physical intervention, but none discussed restoration in a strict sense (six discussed rehabilitation, ten habitat enhancement, one habitat creation); 13 dealt only with the channel, not the floodplain. The papers revealed no common set of environmental attributes monitored. This suggests one of three possible conclusions: that little research to assess restoration impacts is being done; that research is being done, (but not by scientists who publish academic papers, although they may use 'grey' literature); that research is being done but it is not of publishable quality. Whichever conclusion, there appears to be a poor fit between river science and the needs of practical floodplain restoration.

Restoration, organizations and floodplain governance

Floodplain restoration projects have to be integrated into existing systems of land management and land use planning (Brookes *et al.* 1996), and also into existing patterns of responsibility for water distribution and quality. There can be a huge range of organizations whose responsibilities bear on questions of river and floodplain restoration. In England and Wales, for example, statutory bodies (national and local government) include the Environment Agency, the Internal Drainage Boards, the Ministry of Agriculture, Fisheries and Food, the Welsh Office, English Nature, the Countryside Commission and the Countryside Council for Wales (Tunstall 1995). Arrangements in Scotland are different, involving the Scottish Office, District Salmon Fishery Boards, the River Purification Authorities and Scottish Natural Heritage. The array of national, regional and local voluntary organizations is enormous.

This diversity of organizations is both a strength and a weakness. The strength lies in the number of possible avenues for funding and opportunities for collaboration. The weakness lies in the problems of co-ordination between so many different organizations and interests. Widening the focus of restoration to embrace the floodplain has two effects. First, the sphere of interest widens from those organizations with an interest in the aquatic and immediate channel and bank environment to include those of the channel, the active floodplain, other valley floor areas relevant to floodplain environmental function, and lastly to the whole upstream and (to a lesser extent) downstream catchment. Second, the number of organizations with a direct interest in the restoration work increases to include landowners, those with rights to water or other resources in the channel, or on or under surrounding land (navigation, angling, access, water abstraction, minerals), those with statutory responsibilities for, or other interests in, the quantity or chemical quality of water, and those with responsibilities for fauna and flora and landscape quality. Both effects exacerbate the problem of co-ordination significantly.

Floodplain restoration must therefore involve an even more challenging diversity of governmental, voluntary, commercial and private interests than channel restoration. Although collaboration between many of the relevant organizations is well established (Tunstall 1995), these are mostly to tackle familiar problems or issues such as river corridor access, wildlife and landscape conservation, or flood control. Novel ideas, such as allowing flood regimes to re-establish themselves on riparian land, present a considerable institutional challenge. There are ideas that might provide effective strategies for the communication, collaboration and conflict resolution necessary if floodplain restoration is to be accepted; for example, a 'Round Table' process (fairly widely used, for example, in British Columbia, Canada) or the use of Citizens' Juries. Experience with such methods is limited in Europe.

Floodplain restoration and informal institutions

Behind the plethora of governmental and non-governmental organizations with an interest in river restoration, however, lies an even more complex net of informal institutions. The growing field of institutional economics views institutions as much more than organizations, as norms or rules that govern human behaviour, and serve a collective purpose (e.g. Bromley 1989). One approach, following Ostrom (1990), is to distinguish between 'formal' and 'informal' rules. Formal rules would include legislation, and in the UK, Common Law. Law provides the statutory framework of responsibility for water, river channels and floodplain land and resources, and sets limits on who may do what where. Informal rules are much wider. Perhaps the most important informal rules are those of the market (which sets very clear limits on what can be done in floodplains, for example, by determining the price of farmland). Informal rules are also constituted by cultural ideas and traditions, for example, the ideas of river managers as to how much flooding is tolerable, or the ideas of farmers, fishermen or dog-walkers about what a 'natural' river is.

Formal and informal rules bear on different sets of actors. Formal rules bear on riparian owners, those with rights in the channel or surrounding areas, those with responsibilities for planning and land use, water quantity and quality, and environmental conservation. Informal rules bear on these right-holders and statutory bodies, but also on all other stakeholders, including local residents and local and non-local environmentalists. Together they form a web of ideas, interests, mutually agreed restraints, and procedures that somehow have to be negotiated if floodplain restoration is to be successfully implemented.

Concluding remarks

The organizations, formal rules and informal rules that exist in the UK and elsewhere in Western Europe have not been designed to promote river channel restoration, although they can be adapted to make such restoration possible. They are less appropriate to floodplain restoration, and certainly not to the radical idea of restoring ecological function and flood pulses. Government agencies with an interest in rivers are diverse (Tunstall 1995), but in most countries programmes and policies are fragmented (Brookes & Shields 1996). Helpful legislation exists in several European countries, notably Denmark, and there is potential in others, for example, in the Dutch Nature Policy Plan (Government of Netherlands 1990), and in the Local Environmental Action Plans (LEAPs) of the Environment Agency in England and Wales. However, the scientific challenges of predicting the behaviour of flood waters, the economic costs of restoring ecosystem function on floodplains, and the political challenges of devising plans acceptable to all stakeholders are vast.

Floods damage property and sometimes kill people. Professionals engaged in river management, predominantly engineers, have built professional practices on the principle of minimizing the risks and impacts of flooding. New approaches to the management of rivers will demand drastic changes in the informal institutions of river management; in the professions, among statutory and voluntary organizations, and among riparian residents. Through the work of the European Centre for River Restoration in Denmark, the River Restoration Project and its UK successor (the UK River Restoration Centre), parts of the Environment Agency, and organizations such as the National Trust, the Royal Society for the Protection of Birds and the Milton Keynes Parks Trust, experience with river restoration is growing, and support for it is being developed. In time, this work is likely to extend from channel to floodplain restoration. The scientific and technical problems of floodplain restoration are recognized, and to an extent at least are being tackled. The institutional challenge is still ahead.

This paper relates to part of a research contract with the Commission of the European Communities on

'Floodplain biodiversity and restoration: hydrological and geomorphological mechanisms influencing floodplain diversity and their application to the restoration of floodplains' (jointly between Universities of Cambridge, Umeå, Grenoble and Toulouse; Contract ENV4-CY96-0137). The authors are grateful to S. Leigh and F. Hughes for their help with the paper, and to the referees for their comments.

References

ACREMAN, M. & HOWARD, G. 1996. The use of artificial floods for floodplain restoration and management in sub-Saharan Africa. *IUCN* (The International Union for the Conservation of Nature and Natural Resources) *Wetlands Programme Newsletter*, **12**, 20–25.

BAYLEY, P. B. 1991. The flood-pulse advantage and the restoration of river-floodplain systems. *Regulated Rivers: Research and Management*, **6**, 75–86.

BIGGS, J., CORFIELD, A., GRON, P., HANSEN, H. O., WALKER, D., WHITFIELD, M. & WILLIAMS, P. 1998. Restoration of the rivers Brede, Cole and Skerne: a joint Danish and British EU-LIFE demonstration project, V—short-term impacts on the conservation value of aquatic macroinvertebrate and macrophyte assemblages. *Aquatic Conservation Marine and Freshwater Ecosystems*, **8**, 241–255.

BRADSHAW, A. D. 1983. The restoration of mined land. *In*: WARREN, A. & GOLDSMITH, F. B. (eds) *Conservation in Perspective*. Wiley, Chichester, 177–201.

BROMLEY, D. W. 1989. *Economic Interests and Institutions: the Conceptual Foundations of Public Policy*, Blackwell, Oxford.

BROOKES, A. 1995. River channel restoration: theory and practice. *In*: GURNELL, A. & PETTS, G. (eds) *Changing River Channels*. Wiley, Chichester, 369–388.

—— 1996*a*. Floodplain restoration and rehabilitation. *In*: ANDERSON, M. G., WALLING, D. E. & BATES, P. D. (eds) *Floodplain Processes*. Wiley, Chichester, 553–76.

—— 1996*b*. River restoration experience in northern Europe. *In*: BROOKES, A. & SHIELDS, F. D., JR (eds) *River Channel Restoration: Guiding Principles for Sustainable Projects*. Wiley, Chichester, 233–267.

—— & SHIELDS, F. D. R. Jr (eds) 1996. *River Channel Restoration: Guiding Principles for Sustainability*. Wiley, Chichester.

——, BAKER, J. & REDMOND, C. 1996. Floodplain restoration and riparian zone management. *In*: BROOKES, A. & SHIELDS, F. D. R., JR (eds) *River Channel Restoration: Guiding Principles for Sustainability*. Wiley, Chichester, 201–229.

——, GREGORY, K. J. & DAWSON, F. H. 1983. An assessment of river channelisation in England and Wales. *Science of the Total Environment*, **27**, 97–111.

BROWN, A. G., HARPER, D. & PETERKEN, G. F. 1997. European floodplain forests: structure, functioning and management. *Global Ecology and Biodiversity Letters*, **6**, 169–178.

BUCKLEY, G. P. (ed.) 1989. *Biological Habitat Reconstruction*. Belhaven, London.

CAIRNS, J., JR 1991. The status of the theoretical and applied science of restoration ecology. *Environmental Professional*, **13**, 186–194.

CORDES, L. D., HUGHES, F. M. R. & GETTY, M. 1997. Factors affecting the regeneration and distribution of riparian woodlands along a northern prairie river, the Red Deer River, Alberta, Canada. *Journal of Biogeography*, **24**, 675–695.

ECON & POND ACTION 1993. *A Study to Establish the Rationale for a National Rivers Authority (NRA) River Rehabilitation Programme to Further Conservation, Improve Fisheries and Promote Recreation*. Unpublished report.

ELLIOT, R. 1997. *Faking Nature: the Ethics of Environmental Restoration*, Routledge. London.

GOVERNMENT OF NETHERLANDS 1990. *Nature Policy Plan of the Netherlands*. Ministry of Agriculture, Nature Management and Fisheries, The Hague.

GUNN, A. 1991. The restoration of species and natural environments. *Environmental Ethics*, **13**, 291–310.

GURNELL, A. M. 1995. Vegetation along river corridor: hydrogeomorphological interactions. *In*: GURNELL, A. & PETTS, G. (eds) *Changing River Channels*. Wiley, Chichester, 237–260.

—— 1997. The hydrological and geomorphological significance of forested floodplains. *Global Ecology and Biogeography Letters*, **6**, 219–229.

HECHT, J. 1996. Grand Canyon flood a roaring success. *New Scientist* 31 August 1996, 8.

HOBBS, R. J. & NORTON, D. A. 1996. Towards a conceptual framework for restoration ecology. *Restoration Ecology*, **4**, 93–110.

HOFFMANN, C. C., PEDERSEN, M. L., KRONVANG, B. & OVIG, L. 1998. Restoration of the rivers Brede, Cole and Skerne: a joint Danish and British EU-LIFE demonstration project, IV—Implications for nitrate and iron transformation. *Aquatic Conservation Marine and Freshwater Ecosystems*, **8**, 223–230.

HOLMES, N. T. H. 1998. Floodplain restoration, *In*: BAILEY, R. G., JOSÉ, P. V. & SHERWOOD, B. R. (eds) *United Kingdom Floodplains*. Westbury, Otley, 331–348.

HOROWITZ, M. M. & SALEM-MURDOCK, M. 1991. Management of an African floodplain: a contribution to the anthropology of public policy. *Landscape and Urban Planning*, **20**, 215–21.

HUGHES, F. M. R. 1997. Floodplain biogeomorphology. *Progress in Physical Geography*, **21**, 501–529.

—— & ROOD, S.B. 1999. Floodplains, *In*: WARREN, A. & FRENCH, J. R. (eds) *Conservation and the Physical Environment*. Wiley, Chichester, in press.

JORDAN, W. R., III, GILPIN, M. E. & ABER, J. D. (eds) 1987. *Restoration Ecology: a Synthetic Approach to Ecological Research*. Cambridge University Press, Cambridge.

KRONVANG, B., SVENDSEN, L. M., BROOKES, A. ET AL. 1998. Restoration of the rivers Brede, Cole and Skerne: a joint Danish and British EU-LIFE demonstration project, III—Channel morphology, hydrodynamics and transport of sediment and nutrients. *Aquatic Conservation Marine and Freshwater Ecosystems*, **8**, 209–222.

MALANSON, G. P. 1993. *Riparian Landscapes*. Cambridge University Press, Cambridge.

NATIONAL RESEARCH COUNCIL 1992. *Restoration of Aquatic Ecosystems: Science, Technology and Public Policy*. National Academy Press, Washington, DC.

OSTROM, E. 1990. *Governing the Commons: the Evolution of Institutions for Collective Action*. Cambridge University Press, Cambridge.

PERROW, M. R. & WIGHTMAN, A. S. 1993. *River Restoration Project Phase 1. Feasibility Study*. River Restoration Project, Oxford Brookes University, Oxford.

PETERKEN, G. F. & HUGHES, F. M. R. 1998. Limitations and opportunities for restoring floodplain forest in Britain. *In*: BAILEY, R. G., JOSÉ, P. V. & SHERWOOD, B. R. (eds) *United Kingdom Floodplains*. Westbury, Otley, 428–436.

PETERSEN, R. C., PETERSEN, L. B.-M. & LACOURSIÈRE, J. 1992. A building block model for stream restoration. *In*: BOON, P. J., CALOW, P. & PETTS, G. E. (eds) *River Conservation and Management* . Wiley, Chichester, 293–309.

PETTS, G. E. 1988. Regulated rivers in the United Kingdom. *Regulated Rivers: Research and Management*, **2,** 201–220.

—— 1996. Sustaining the ecological integrity of large floodplain rivers. *In*: ANDERSON, M. G., WALLING, D. E. & BATES, P. D. (eds) *Floodplain Processes* Wiley, Chichester, 535–551.

—— 1998. Floodplain rivers and their restoration: a European perspective. *In*: BAILEY, R. G., JOSÉ, P. V. & SHERWOOD, B. R. (eds) *United Kingdom Floodplains*. Westbury, Otley, 29–41.

—— & CALOW, P. 1996. *River Restoration: Selected Extracts from the Rivers Handbook*. Blackwell Scientific, Oxford, 28–41.

SCUDDER, T. 1991. The need and justification for maintaining transboundary flood regimes; the Africa case. *Natural Resources Journal,* **31**, 75–107.

STERBA, O., MEKTOVA, J., KRSKOVA, M., SAMSONOVA, P. & HARPER, D. 1997. Floodplain forests and river restoration. *Global Ecology and Biogeography Letters,* **6,** 331–337.

TOCKNER, K. & SCHIEMER, F. 1997. Ecological aspects of the restoration strategy for a river–floodplain system on the Danube River in Austria. *Global Ecology and Biogeography Letters,* **6,** 321–329.

TOTH, L. A., OBEYESEKERA, J. T. B., PERKINS, W. A. & LOFTIN, M. K. 1993. Flow regulation and restoration of Florida's Kissimmee River. *Regulated Rivers: Research and Management*, **8**, 155–166.

TUNSTALL, S. D. 1995. *Institutional Aspects of River Restoration in the UK*. Middlesex University Flood Hazard Research Centre, R&D Note **265**.

Identification of an ecologically based floodway: the case of the Cosumnes River, California

ELIZABETH S. ANDREWS

Philip Williams & Associates, Ltd, 770 Tamalpais Drive, Suite 401, Corte Madera, CA 94925, USA (e-mail: bandrews@pwa-ltd.com)

Abstract: The opportunities for restoring natural floodplain function along the Cosumnes River in California, USA, were evaluated by integrating traditional methods of engineering floodplain analysis with geomorphological analysis. The specific goal for floodplain restoration was the re-establishment of physical processes that would lead to the creation of cottonwood forest, and ultimately the generation of riparian oak woodland, while preserving or reducing current off-site flood hazards. Long-term trends, current hydraulic conditions, anticipated ecological value, and expected future conditions were evaluated to identify an appropriate floodplain restoration goal for the river.

The Cosumnes River in Sacramento County, California, is the only major Sierra River without significant changes to its hydraulic regime caused by the construction of dams. Adjoining lands are predominantly agricultural, and include the largest remaining stand of riparian valley oak woodland in California's Central Valley. The Nature Conservancy (TNC), which owns property consisting of both valley oak woodland and operating agricultural fields in the lower Cosumnes, contracted with Philip Williams & Associates, Ltd (PWA) to develop a proposal for restoring oak woodland and wetland habitat along the lower 28 km (17.5 miles) of the river.

The establishment of riparian forests begins on mobile gravel bars (Strahan 1984). The processes of deposition and channel migration are responsible for the episodic and frequent disturbance necessary to maintain riparian habitat diversity and quality. The specific mechanism thought to initiate the development of riparian oak woodlands along the Cosumnes is the establishment of cottonwood seedlings through the deposition of both seeds and woody debris on disturbed lands inundated during flood events. As the cottonwoods mature, they provide the shaded setting that favours the development of oak seedlings. Once the cottonwoods have lived out their natural lifespan, they topple and leave the oaks to become the canopy species. At both an accidental levee breach (1985) and an intentional pilot project undertaken by TNC (1995) at the Desmond Ranch, the mere breaching of a levee, thereby allowing disturbance and inundation, had been sufficient to initiate this cycle. Our key restoration goal, therefore, was to evaluate conditions to re-establish the disturbance and flood cycle that would enable this natural progression to recur at a broad scale along the Cosumnes.

Our task was the development of a reasoned floodplain restoration goal for the Cosumnes within the limitations presented by modest funding. The specific goal of the project was to develop a proposal to restore functional floodplains along the Cosumnes River for the purpose of re-establishing physical processes that would lead to the creation of cottonwood forest, and ultimately the generation of riparian oak woodland. Constraints on the execution of this work included the limited amount of available data and the paucity of existing analyses of the flood hydrology, hydraulics and geomorphological trends along the river.

The study was interrupted by the flood of record on the river in January 1997, which damaged approximately 86 homes, caused approximately 24 levee breaks, and provided considerable incentive to consider changes in the management of floods along the river, as was suggested by this study.

This paper describes the key aspects of our analysis, findings, and recommendations, discussed in greater detail in our project report (Vick *et al.* 1997).

Study area

The Cosumnes River flows west out of California's Sierra Nevada Mountains into the heart of the great Central Valley, terminating near the eastern edge of the Sacramento–San Joaquin River Delta (Fig. 1).

From: MARRIOTT, S. B. & ALEXANDER, J. (eds) 1999. *Floodplains: Interdisciplinary Approaches*. Geological Society, London, Special Publications, **163**, 99–110. 1-86239-050-9/99/$15.00

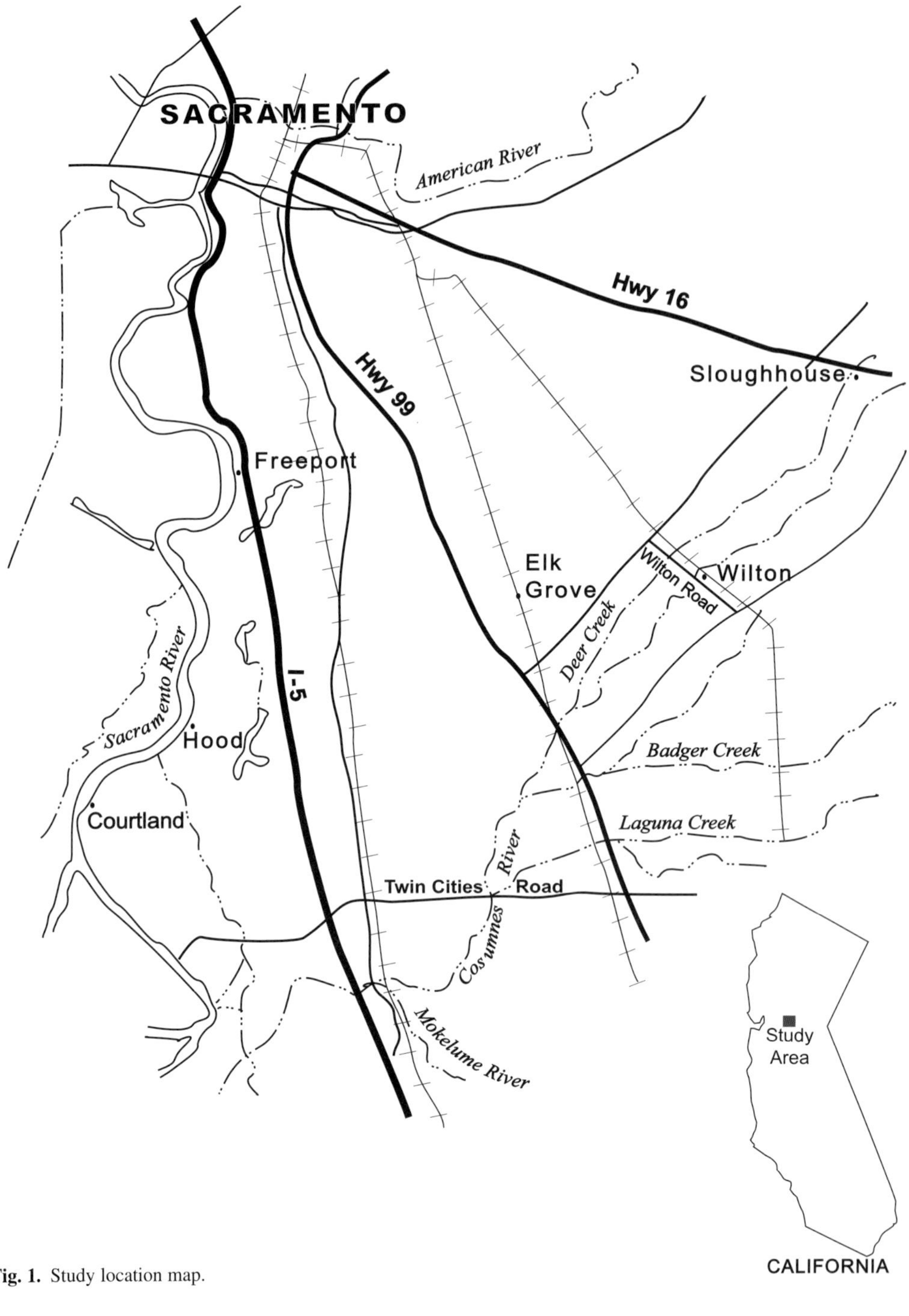

Fig. 1. Study location map.

The river drains a 3276 km^2 catchment ranging in elevation from *c.* 2300 m almost to sea level.

Approximately 20% of the existing land use within the catchment is timber harvest (TNC 1992); grazing, mining, and urban and residential development are also significant land uses within the basin. Approximately 10% of the catchment is in urban or developed land uses (TNC 1992). The lowest 28 km of the river were selected by TNC as the focus for this floodplain restoration study. This lower section of the river travels through lands that are predominantly used for agriculture.

The Cosumnes River corridor has been greatly modified by human-wrought changes. Agricultural settlement of the area began in the 1840s (Peak, pers. comm.), and was associated with the clearing of large stands of riparian forest. Riverside levees were developed in the early 1900s, primarily in the 1930s (Vick *et al.* 1997). These levees are typically low berms, neither engineered nor armoured. Today there are 25.4 km of levees along the right and 5.8 km of levees along the left bank of the lower Cosumnes. Development in the area has also been associated with a significant decline in the groundwater table, on the order of 18 m (60 feet) (Bertoldi *et al.* 1991; Montgomery Watson 1995), and may be responsible for a reduction in or elimination of summer and/or autumn flows. Development in the basin has also affected stream flows in two other ways: (1) direct diversions from the river reduce flows during the irrigation season, generally May–September; (2) catchment changes have affected the timing and magnitude of runoff peaks. However, relative to the hydrological changes that have occurred on other Sierra Rivers as a result of dams and diversions, these changes have left the Cosumnes with a nearly intact hydrological regime during the winter flooding season.

Large tracts of wetlands, backwater sloughs and riparian forest, interspersed by agricultural lands, characterize the lowest reaches of the river. Many public and private land management agencies have purchased lands and easements in these remnant natural areas for conservation purposes.

Methods

We relied on four methods to develop our recommendations: (1) development of hydrological and hydraulic analyses of frequent and rare flood conditions; (2) analysis of geomorphological processes and trends; (3) qualitative weighing of values and costs of floodplain restoration alternatives; (4) qualitative assessment of the long-term trends and their expected effect on floodplain conditions.

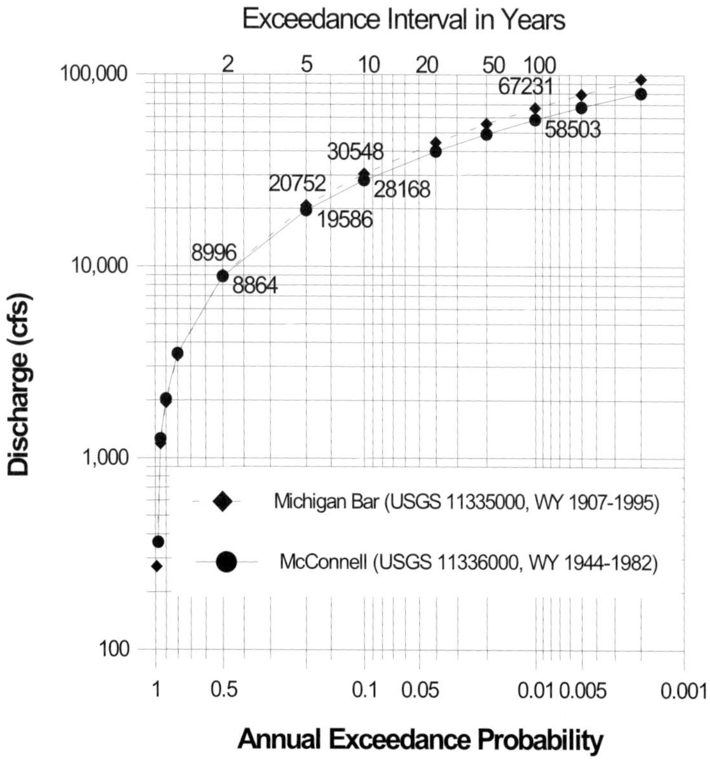

Fig. 2. Cosumnes River flow frequency.

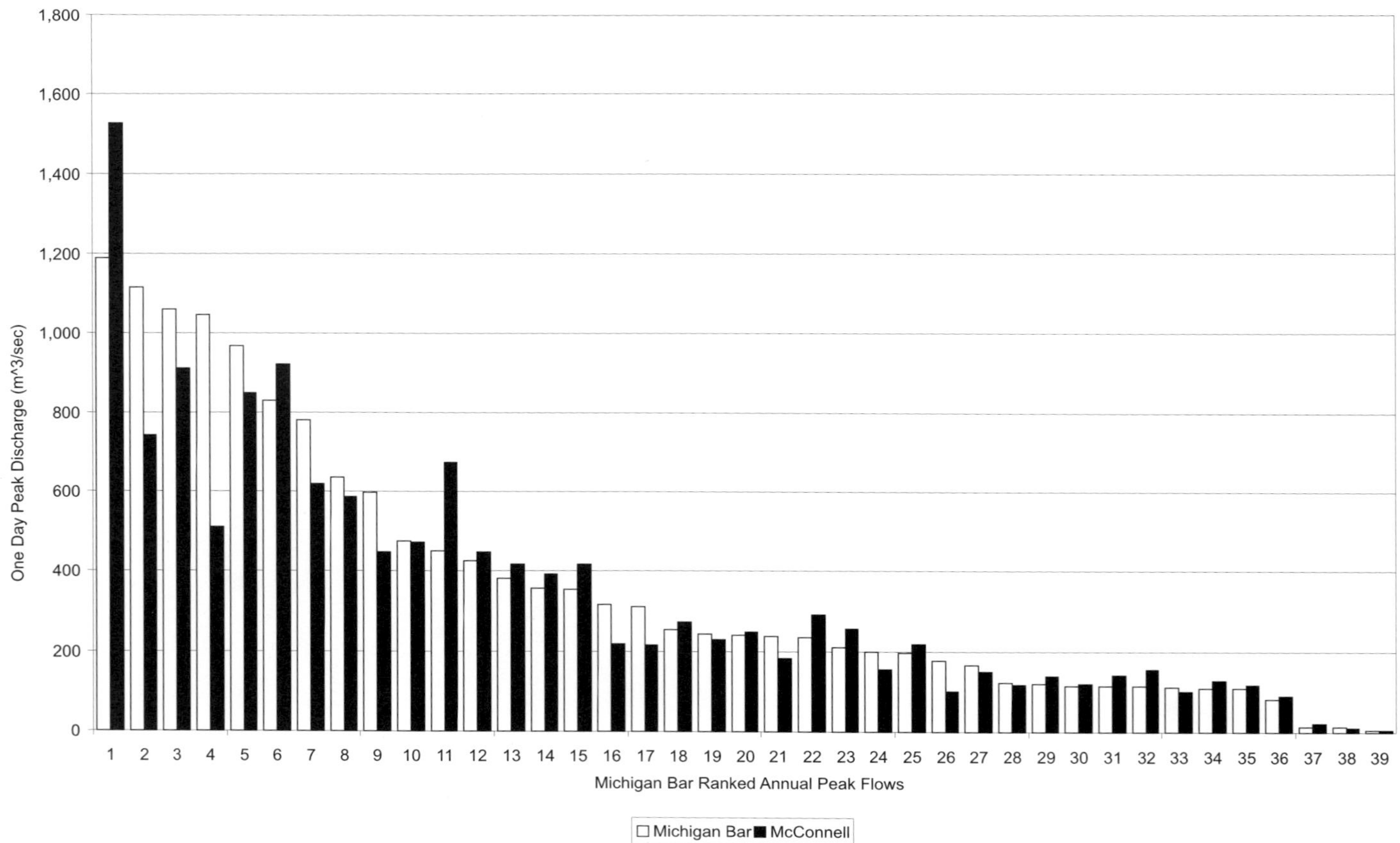

Fig. 3. Comparison of peak flows at the Michigan Bar and McConnell gauges.

Hydrology and hydraulics analyses

Data were available from two stream gauges on the Cosumnes River. One, the McConnell gauge (US Geological Survey (USGS) No. 11336000), was located near Highway 99 (see Fig. 1) in the lower reaches of the river and operated from 1944 to 1982. The Michigan Bar gauge (USGS No. 11335000) is located 40 km (25 miles) upstream and has remained in operation since 1907. Analysis of the data from these gauges was conducted to develop flow frequency curves for the periods of record at each gauge (Fig. 2) and assess the relationship between peak flow levels at each gauge for the same flood events (Fig. 3). Despite the fact that the basin for the McConnell gauge is 35% larger than that of the Michigan Bar gauge, peak flow values for flows of the same recurrence interval, and indeed, for the same flood events, appear to be similar at each gauge. (This fact is likely to be due to the presence of significant floodplain storage and resultant attenuation effects on the flood hydrograph.) Rather than further refine our flood frequency analyses to account for the difference in the periods of record of the two gauges, therefore, we elected to use a single flow rate throughout the modelled reach to represent flood conditions in this approximate study. Assessment of appropriate hydrological assumptions was based both on the flood frequency analyses for the two gauges and on a hydrology study of the North Delta that gave *c.* 2010 $m^3 s^{-1}$ as the 100 year peak flow for the Cosumnes River at the mouth (CDWR (California Department of Water Resources) 1995). Thus, for the purposes of this study, flow rates of 566 $m^3 s^{-1}$ and 1954 $m^3 s^{-1}$ were selected as appropriate estimates of the peak flow for the 5 year and 100 year recurrence interval floods, respectively, throughout the studied reach.

The hydraulic model used was a steady flow, step-backwater model (HEC-RAS) developed by the US Army Corps of Engineers. A prior model of the lowest 5 miles (8.05 km) of the river was extended 12.5 miles (20.12 km) upstream using field-surveyed cross-sections of the channel extended with floodplain topographic data from the USGS 7.5′ topographic maps, which used 5 foot (1.52 m) contour intervals. As-built bridge data were also used where available. Starting water surface elevations were developed using both historical stage data and estimates developed by flood studies for the area downstream of the Cosumnes (CDWR 1994, 1995). With this model, a hydraulic simulation of the approximate 5 year and 100 year flows under existing conditions was developed (see Fig. 4). For the sake of comparison, a companion set of runs showing hydraulic conditions for the same flows but without constraint by levees was also developed.

Geomorphological processes and trends

The analysis of geomorphological processes and trends was developed from the review of historical data and

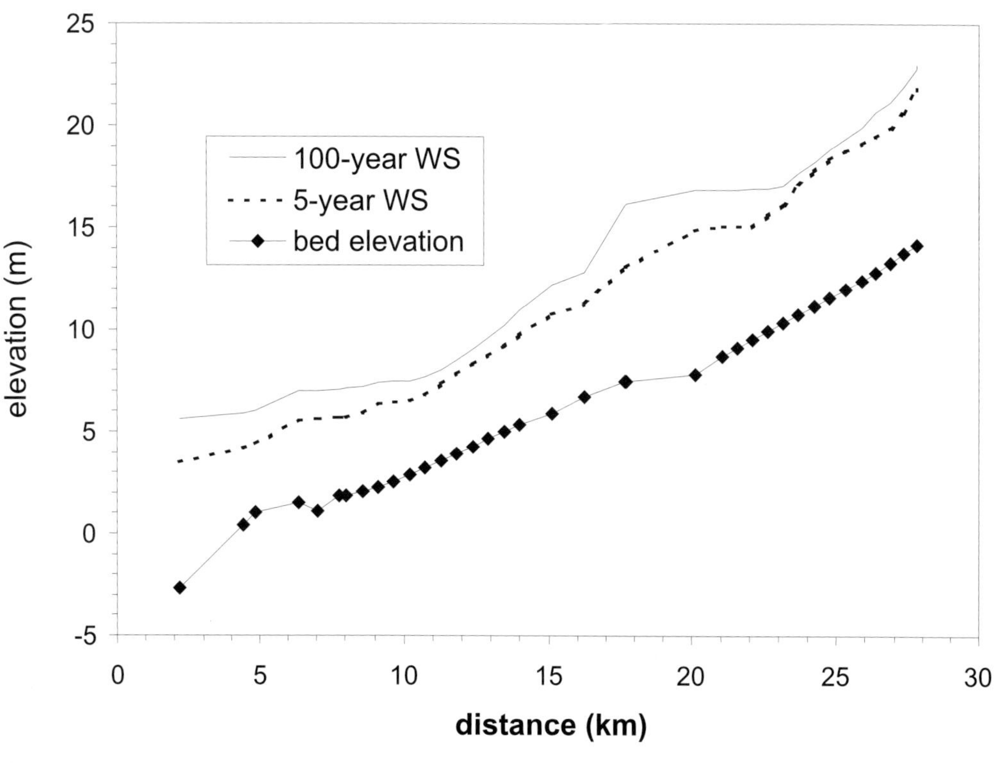

Fig. 4. Existing conditions water surface profile.

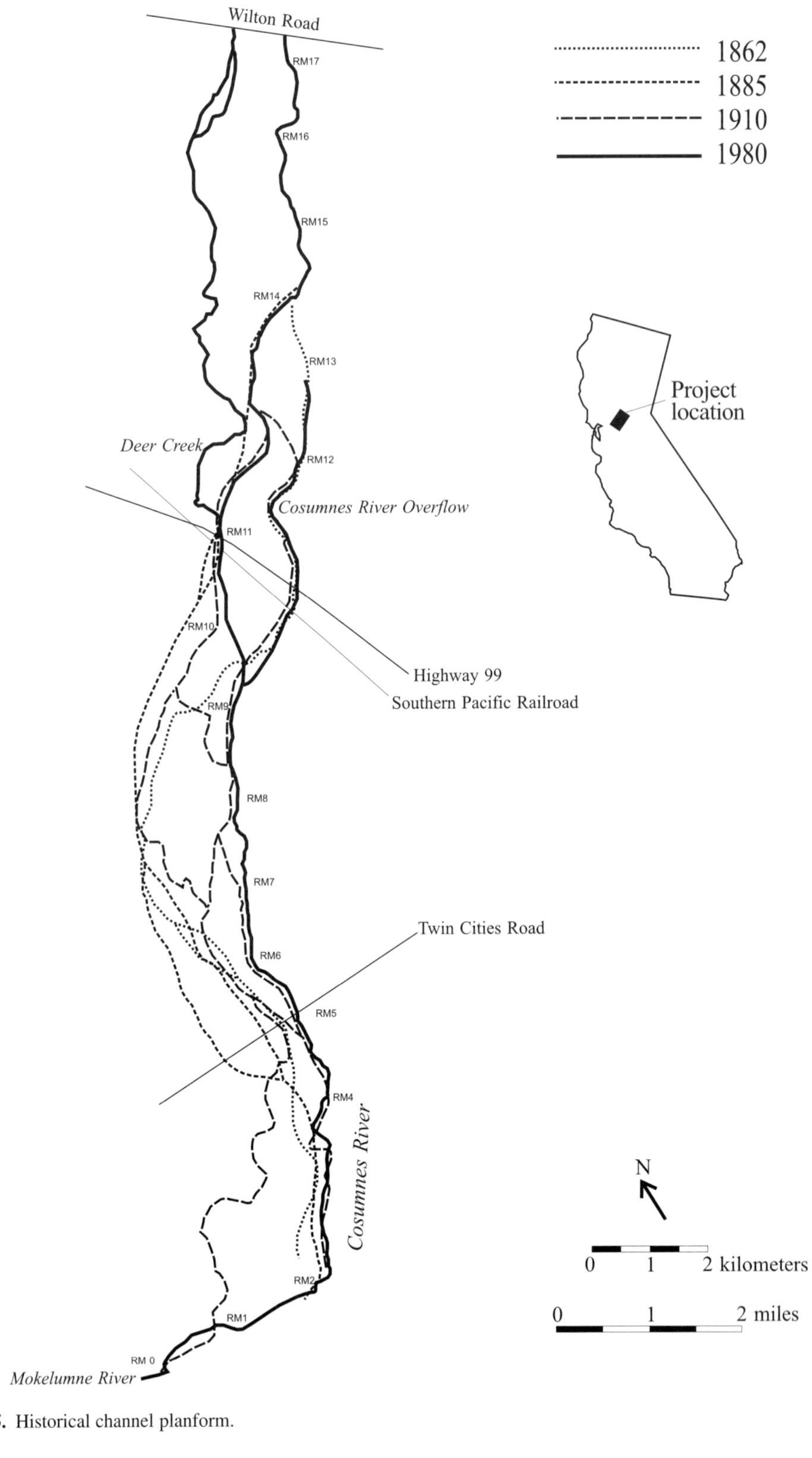

Fig. 5. Historical channel planform.

literature, comparisons of current and historical survey data, analysis of hydraulic results, and field observations. These sources were used to create an approximate description of changes in river channel morphology.

Historical maps and aerial photographs provided snapshot views of dramatic channel planform changes over time (see Fig. 5). The Cosumnes' transition to its present planform is almost certainly due in part to human intervention (Vick *et al.* 1997).

Limited data were available on channel profile and cross-sectional changes. The US Army Corps of Engineers (COE) surveyed a series of cross-sections on the river in 1957. PWA attempted approximately to reoccupy eight of the 1957 COE cross-sections in a 1996 survey. Road management agencies had also surveyed cross-sections at five bridges crossing the lower Cosumnes; PWA resurveyed two of these in 1996 as well.

Although data from the comparison with COE cross-sections are subject to error because of the lack of monuments at the cross-sections, both the COE cross-section comparison and bridge cross-section data are consistent in indicating incision of the channel thalweg in the last 20–40 years. Figure 6 provides the comparison of the 1996 and 1957 surveys at the COE cross-sections. Table 1 provides the comparison of thalweg data from the bridge surveys. Channel incision in recent decades is clearly suggested by these data, with greater levels of channel incision occurring above Highway 99. This hypothesis is consistent with our field observations of channel conditions. Exposure of bridge pier footings, the stranding of bank armour several feet above the thalweg, oversteepened banks, and the lack of unconsolidated alluvial deposits were all observed on the Cosumnes, especially upstream of Highway 99, and are all consistent with a history of channel incision. PWA also reviewed the data derived for hydraulic conditions on the Cosumnes during a frequent (*c.* 5 year) flood, both with and without the current system of levees. These data showed increases in shear stress of 50–100% under the with-levee conditions. Because increased shear stress indicates an increased capability to transport sediment, this analysis suggests that the presence of levees might be one of the causes of channel incision on the river (Vick *et al.* 1997).

Long-term trends

Projections were made of conditions on the Cosumnes River in 100 years under a 'no action' scenario. The following factors were evaluated: degradational trends, rise in sea level, and changes in sediment delivery conditions.

Extrapolating historical trends with an assumption of a power function decline (Simon 1989) beginning in 1930, channel incision in the study reach in the next 100 years is projected to grow from *c.* 0.5–3.0 m since 1957 to as much as 1.0–5.5 m. Actual erosion trends will depend on the nature of bank stabilization actions taken by property

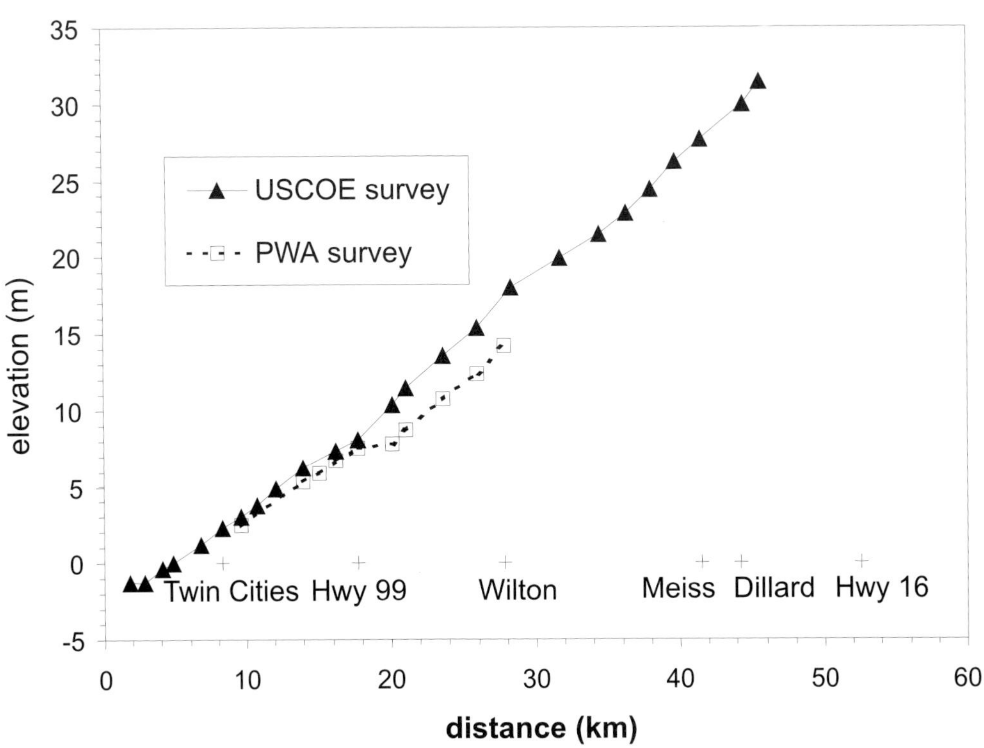

Fig. 6. Cosumnes River thalweg profile.

Table 1. *Change in thalweg elevation at bridges*

Bridge name	First survey thalweg elevation (year) (m NGVD‡)	Last survey thalweg elevation (year) (m NGVD)	Period between surveys (years)	Change in thalweg elevation (m)
Highway 99	8.1 (1957)*	7.5 (1996)†	39	–0.6
Wilton Road	15.3 (1972)	14.2 (1996)†	24	–1.1
Dillard Road	27.9 (1972)	25.0 (1993)	21	–2.9
Highway 16	35.0 (1952)	32.9 (1992)	40	–2.2

Data are from the California Department of Transportation, except as noted.
*From COE survey.
†From PWA survey.
‡National Geodetic Vertical Datum of 1929, approximately equivalent to mean sea level.

owners and the interplay between channel materials and vertical and horizontal erosion. Erosion rates could grow even larger with the additional capacity of the incised channel for both water and sediment transport. Undermining of bridge piers and levees can be expected. Changes in sediment delivery from the catchment are expected to be minor by comparison with the effects of continued channel confinement.

Sea level is expected to rise by less than 0.3 m during the next 100 years (Watson *et al.* 1996). This will result in a slight increase in downstream water surface elevation at the mouth of the Cosumnes, potentially leading to minor

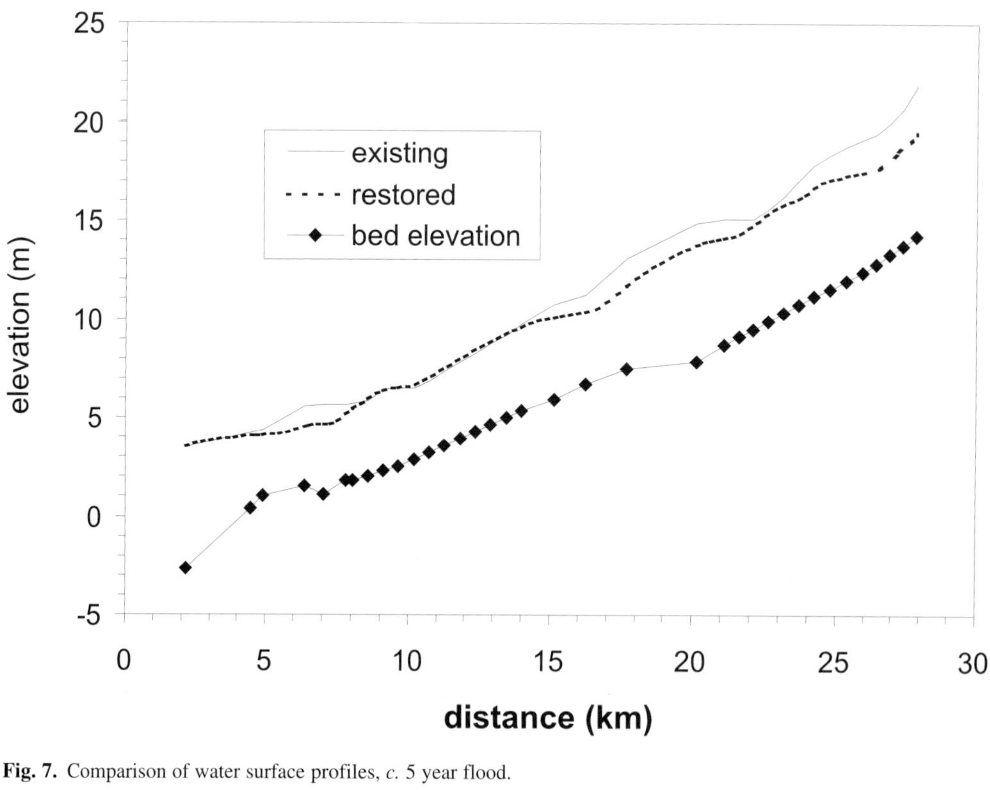

Fig. 7. Comparison of water surface profiles, *c.* 5 year flood.

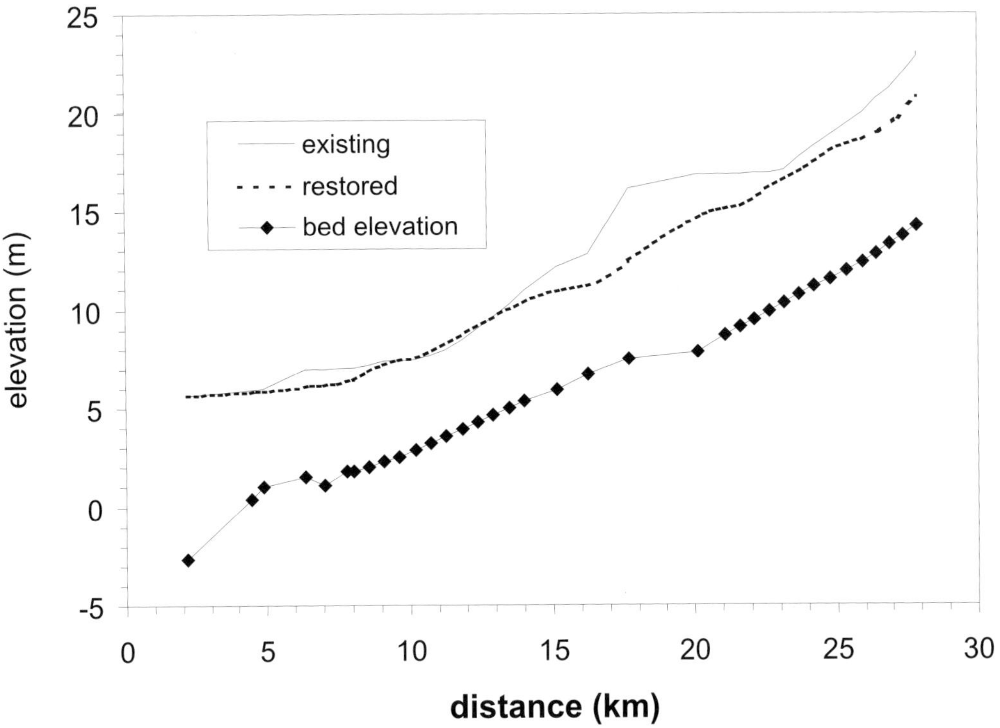

Fig. 8. Comparison of water surface profiles, *c.* 100 year flood.

aggradation in the lower, tidally influenced reach of the river and an increase in the tendency of the floodplain in this region to be inundated. This change is expected to have negligible effect upstream, however.

In summary, the most notable physical changes anticipated on the river in the next 100 years are related to channel incision. This process is expected to result in an increased trend towards bank stabilization by adjacent landowners and loss of riparian vegetation either through placement of revetment or simply through bank erosion, coupled with reduced regeneration of forest as the river becomes further isolated from its floodplain.

Benefits and costs

On the basis of the results of our initial findings, the following objectives were developed as a refinement of the floodplain restoration goal: (1) increase the land area subject to regular flood inundation; (2) arrest and reverse long-term degradational trends; (3) contribute to the stabilization of, or rise in the region's ground-water table; (4) contribute to a flood hazard management strategy that maximizes ecological benefits while minimizing flood hazards.

With these objectives in mind, a qualitative assessment was made of the benefits and costs of pursuing different floodplain restoration options. The range of options extended from no action to restoration of the entire Cosumnes River floodplain. A positive relationship between increasing floodplain area and the following benefits was assumed to exist: (1) reduced flood hazards outside of the study reach: upstream, as result of reduced stage, and downstream, because of flood peak attenuation; (2) increased area of riparian forest and overall river–floodplain ecological value; (3) increased ground-water recharge; (4) decreased shear stresses, leading to decreased channel and levee maintenance costs and decreased channel incision.

The restoration alternative

Given the paucity of data on the relative ecological value of floodplains subject to different flooding frequencies, and the absence of quantitative information on ground-water recharge potential of floodplain inundation, we were unable to quantify fully the expected benefits of floodplain restoration options and to weigh them rigorously against the cost of land purchases in easements or fee title. Instead, we made a subjective assessment of an appropriate balance of these considerations based on the available information, and selected one alternative for specific analysis: restoration of the 5 year floodplain, termed the 'Restoration Alternative'. This alternative includes levee setback as needed to the edge of the 5 year floodplain, as well as promotion of the restoration of bed elevations to 'pre-development' conditions.

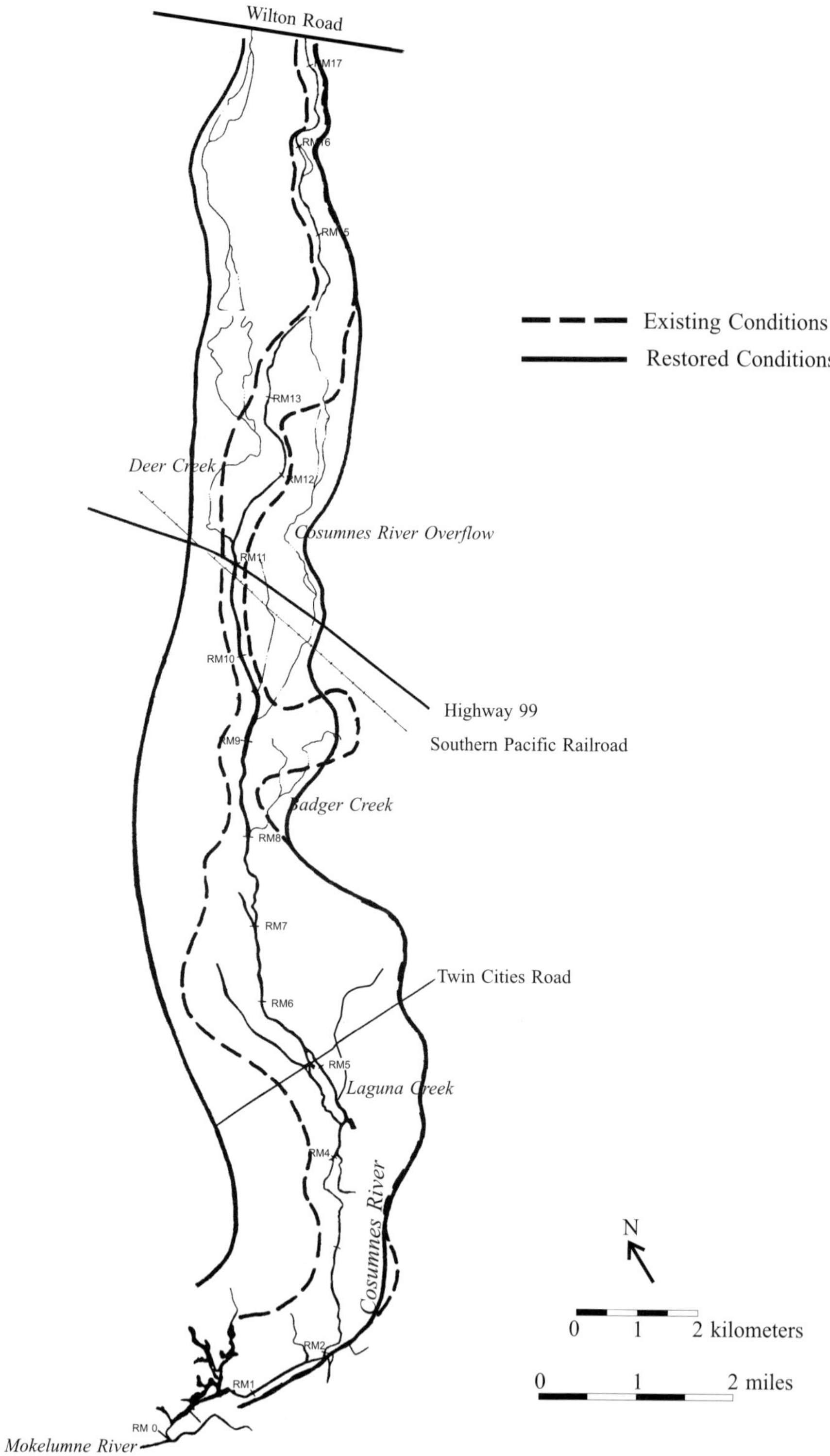

Fig. 9. Comparison of the projected floodplain for the 5 year event, existing and restored conditions.

Evaluation of the restoration alternative

The Restoration Alternative was delineated by performing an hydraulic analysis of the approximate extent of the 5 year floodplain under an levee-free, 'restored bed elevation' condition. The flood management benefits of the Restoration Alternative compared with existing conditions were significant: river stage was significantly reduced beginning *c.* 5 km upstream of Twin Cities Road under both a *c.* 5 year and 100 year events. Stage reductions upstream of this point were as much as 1.3 m in an *c.* 5 year event and as much as 3.5 m in a *c.* 100 year event, as shown in Figs 7 and 8, respectively. Additional benefits from attenuation of the flood peak through floodplain storage were not evaluated as part of this study. Although a greater area would be regularly flooded under the Restoration Alternative, flood stages upstream of this reach, and possibly downstream as well, would be lowered, and lands within the reach but outside of levees would be subject to less hazard in the event of a levee break.

Shear stresses under the Restoration Alternative were reduced by more than 50% from existing conditions under 5 year peak flows, indicating that sediment transport capacity of the Restoration Alternative would also be significantly less. Thus, assuming these frequent floods are approximately the level of channel-forming flows, the tendency of the channel to degrade would be significantly reduced or eliminated. In addition, erosional forces acting on the setback levees would be much less than those acting on the existing riverside levees. An increase in channel migration rates might result from aggradation in the channel bed. However, the distance from the main channel to the setback levees would generally be increased, and the shear stresses exerted by the flowing water would decrease, thereby countering the potential for levee erosion that might otherwise be expected with increased channel migration. The forces acting to undermine the piers of bridges across the river would also be reduced.

The increased frequency and extent of floodplain inundation associated with the Restoration Alternative, as compared with existing conditions (Fig. 9), is expected to improve floodplain habitat and the biotic health of the river corridor. In addition, if channel incision is reversed, the ability of the Cosumnes to actively shift and meander within its floodplain would be increased, enhancing the dynamic conditions critical to floodplain habitats.

Lastly, the increase in frequency and extent of floodplain inundation will increase the amount of groundwater recharge from the river over that experienced under existing conditions.

Conclusion

By examining multiple natural resource management issues on the Cosumnes, we can begin to evaluate the full benefit of a floodplain restoration project. The construction of levees was probably significantly responsible for the degradation of this natural river and floodplain resource even without the construction of major impoundments. Setting the levees back to the limit of the 5 year floodplain could create significant benefits in terms of: reduced upstream flood stage; increased generation of riparian forest and floodplain habitats; reduced or eliminated channel incision; increased groundwater recharge; and reduced maintenance and replacement costs for infrastructure such as levees and bridges.

Failure to address the conditions along the Cosumnes is likely to lead to the further loss of riparian and in-channel vegetation, further degradation of the river channel, and increases in infrastructure installation, replacement and maintenance costs in the future.

The marriage of geomorphological analysis with traditional approaches to flood hazard evaluation allowed consideration of a range of physical processes that define the riverine ecosystem. As interest in restoration and enhancement of river ecosystems grows, it is becoming increasingly apparent that fundamental approaches to the management of river corridors must change. Ecological enhancement may often be more effectively achieved by a fundamental change in the way non-restoration objectives are obtained than by corrective 'restoration' measures that address only the symptoms, not the underlying causes, of ecological degradation. The Cosumnes River study offered an opportunity to rethink the approach to flood management along the river and identify an approach that was supportive of ecological objectives while satisfying fundamental flood management needs.

Funding for this work was provided by The Nature Conservancy. Their support, as well as the time and insights provided by their staff members M. Eaton and R. Reiner, were crucial to our work. Critical research, fieldwork, writing and creative thinking were provided by J. Vick. Discussions with P. B. Williams provided invaluable suggestions for project approach. F. D. Shields, Jr, and A. Brookes provided valuable suggestions that improved the manuscript.

References

BERTOLDI, G. L., JOHNSTON, R. H. & EVENSON, K. D. 1991. *Groundwater in the Central Valley, California—A Summary Report.* US Geological Survey, Professional Papers, **1401-A**.

CDWR (CALIFORNIA DEPARTMENT OF WATER RESOURCES) 1994. *Memorandum Report: Hydrology Report (1), Two-Year Floodplain, North Delta Region.* CDWR Division of Planning, Interim North Delta Program, Sacramento, CA.

CDWR (CALIFORNIA DEPARTMENT OF WATER RESOURCES) 1995. *Memorandum Report: Hydrology Report (2), Low-Frequency Floods in North Delta Region.* CDWR Division of Planning, Interim North Delta Program, Sacramento, CA.

MONTGOMERY WATSON 1995. *Phase II Groundwater Yield*

Analysis: Technical Memorandum No. 2, Impacts Analysis. Prepared for the Sacramento County Water Agency, Sacramento, CA.

SIMON, A. 1989. A model of channel response in disturbed alluvial channels. *Earth Surface Processes and Landforms*, **14**, 11–26.

STRAHAN, J. 1984. Regeneration of riparian forests for the Central Valley. *In*: WARNER, R. E. & HENDRIX, K. M. (eds) *California Riparian Systems: Ecology, Conservation, and Productive Management*. University of California Press, Berkeley, CA.

TNC (THE NATURE CONSERVANCY) 1992. *Cosumnes River Catchment Project Strategic Plan*. The Nature Conservancy of California, San Francisco, CA.

VICK, J., ANDREWS, E. & WILLIAMS, P. B. 1997. *Analysis of Opportunities for Restoring a Natural Flood Regime on the Cosumnes River Floodplain* (PWA Report 1148). Prepared for The Nature Conservancy of California by Philip Williams & Associates, Ltd, San Francisco, CA.

WATSON, R. T., ZINYOWERA, M. C. & MOSS, R. H. (eds) 1996. *Climate change 1995. Impacts, Adoptions, and Mitigation of Climate Change: Scientific–Technical Analyses*. IPCC (Intergovernment Panel on Climate Change). Cambridge University Press, Cambridge.

The use of floodplain sedimentation measurements to evaluate the effects of river restoration works

NATHALIE E. M. ASSELMAN

The Netherlands Centre for Geo-Ecological Research (ICG), Utrecht University, Department of Physical Geography, PO Box 80.115, 3508 TC Utrecht, The Netherlands

Abstract: Over the past century, many rivers throughout Europe have been channelized. The channel of the River Rhine in the Netherlands was straightened, groynes were built, and minor river dykes were constructed to prevent inundation of parts of the embanked floodplain during minor floods. More recently, however, ecological rehabilitation of the rivers Rhine and Meuse and the reduction of flood risk have become major issues for river management in the Netherlands. Various projects for restoration of the embanked floodplains to a more natural state and for the improvement of the discharge capacity of the high-water floodway have been initiated. Proposed measures include the construction of side channels through the floodplain, lowering of the floodplain surface by several decimetres, removal of minor river dykes, and reintroduction of floodplain forests. These measures will alter floodplain sedimentation rates. The aim of this paper is to evaluate the possible effects of such river rehabilitation measures on sediment accumulation using the results of overbank sedimentation measurements. The measurements were carried out during several floods at different floodplain sections along the rivers Waal and Meuse in the Netherlands. Sediment traps made of artificial grass were used to collect the deposited sediment. By comparing amounts and patterns of sediment deposited at floodplain sections characterized by different topography, the possible effects of rehabilitation measures on floodplain sedimentation are assessed. The findings are compared with model predictions. Results indicate that removal of minor dykes and lowering of the floodplain surface will significantly enhance sedimentation, whereas the impact of changes in vegetation cover from grass to grass with clusters of trees will be small.

During the Holocene period, the precursors of the present rivers Rhine and Meuse in the Netherlands were meandering or anastomosing rivers that built up the alluvial ridges and flood basins that now characterize the landscape in the central part of the Netherlands. After the embankment of the rivers in the 12th and 14th century, avulsions no longer occurred, and floodplain sedimentation was limited to a narrow zone between the river dykes and the main channel. Over the past century, the river channels have been straightened, groynes built, and minor river dykes constructed to prevent inundation of parts of the embanked floodplain during minor floods.

These different phases in the morphological evolution of the river are probably characterized by different sedimentation rates. Berendsen (1984) estimated long-term average sedimentation rates in the Holocene flood-basins of about 0.4–1.5 mm a^{-1} using radiocarbon dates of peat samples. Middelkoop (1997) reconstructed overbank sedimentation rates over the past century from heavy metal profiles in floodplain soils. Estimated sedimentation rates varied from 0.18 to 11.55 mm a^{-1} with a mean of 2.78 mm a^{-1}. Contemporary overbank sedimentation rates on the floodplains in the Rhine–Meuse delta were measured during individual flood events by van der Perk *et al.* (1992), Asselman & Middelkoop (1995, 1998) and Middelkoop & Asselman (1998). During a high-magnitude flood with a recurrence interval of about 40 years sedimentation of fine suspended sediment ranged from 0.8 to 7.5 mm, depending on the location within the floodplain section (Middelkoop & Asselman 1998).

Over the past decade, concern about the morphological and ecological condition of the rivers Rhine and Meuse has increased. Initial plans for the rehabilitation of the Rhine distributaries were presented by, among others, De Bruin *et al.* (1987) and Wereld Natuur Fonds (1992). Ecological rehabilitation of the River Rhine has now become a major objective for river management in the Netherlands. Target plans that have been identified

From: MARRIOTT, S. B. & ALEXANDER, J. (eds) 1999. *Floodplains: Interdisciplinary Approaches*. Geological Society, London, Special Publications, **163**, 111–122. 1-86239-050-9/99/$15.00

in national policy plans aim at enhancement of the fluvial dynamics within the floodplain area, and restoration of riverine ecosystems and habitat types, such as river forests, snag wood, side channels, marshes and natural shores (e.g. Cals *et al.* 1998). Simultaneously with the projects for rehabilitation, measures were proposed to improve the discharge capacity of the high-water floodway, so as to reduce the risk of river flooding. Especially after the major floods of December 1993 and January 1995 the reduction of flood risk has become a major issue for river management. Recent policy to reduce the risk of floods and flood damage is to preserve and improve the discharge capacity of the high-water floodway of the rivers Rhine and Meuse (Ministry of Transport, Public Works and Water Management 1997). To fulfil the objectives for both nature conservation and safety, a variety of landscaping measures has been proposed (Fig. 1). These measures include (Silva & Kok 1996): (1) lowering the floodplain surface by 0.5–2 m by excavating the top layer of clay and sand deposits from the floodplain to increase the discharge capacity of the high-water floodway, and to enlarge the extent of wetland ecosystems; (2) removal of minor dykes, to increase the discharge over the floodplain at high discharge and to enhance the exchange of water and organisms between the main channel and the floodplain; (3) reintroducing a wide range of ecosystems, including riverine forests, and reducing the area of agricultural land; (4) removal of obstacles to flow within floodplain areas, to reduce flood-water levels. Together, these measures characterize restoration projects, of which a few have recently been started as pilot projects, and others are at present under consideration. The landscaping measures potentially increase the sedimentation rates and change the dynamics of the floodplain areas. Alhough accelerated morphological development within the floodplain areas may be favourable for many riverine ecosystems, increased sedimentation may reduce the discharge capacity of the high-water floodway over the long term. In addition, there may be ecosystems and species that are vulnerable to increased accumulation of contaminated sediments.

In most rehabilitation plans, the impact of specific measures on floodplain sedimentation has been estimated using model computations (e.g. Van den Brink 1995; Narinesingh 1995). In the present study, however, the possible effects of river rehabilitation measures on floodplain sedimentation rates are evaluated using the results of measurements of contemporary overbank sedimentation. During a high-magnitude flood in December 1993, sedimentation measurements were carried out at 11 floodplain sections along the rivers Waal and Meuse. At two of these floodplain sections measurements were undertaken during a series of floods of different magnitude and duration. In previous studies, the results of these measurements were used to study the relationship between sediment accumulation and different floodplain characteristics such as floodplain elevation and the presence of minor river dykes (Middelkoop & Asselman 1998). The results were also used to estimate the relative importance of floods of different magnitude on overbank sedimentation (Asselman & Middelkoop 1998). In the present paper, the results of the previous studies are applied to assess the effect of rehabilitation works on sediment accumulation. A comparison with model predictions given in the literature is also presented.

Study area

Measurements of overbank sedimentation were carried out at 11 floodplain sections. Seven flood-

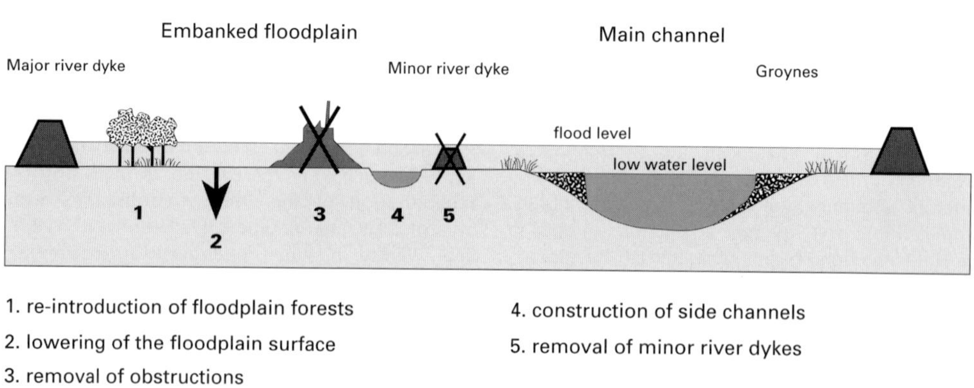

Fig. 1. Measures proposed for floodplain rehabilitation and reduction of flood risk.

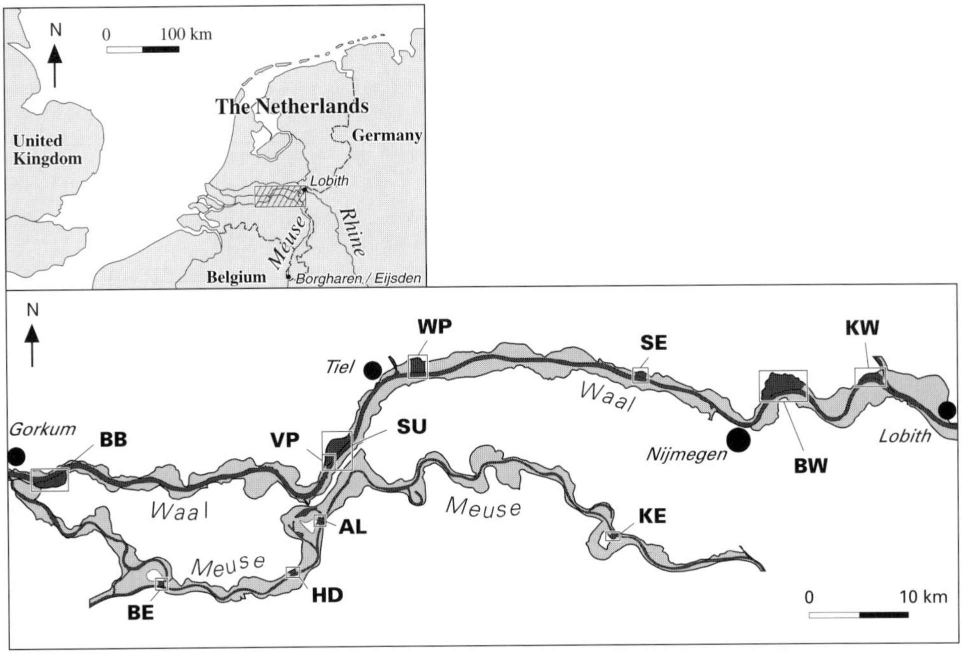

Fig. 2. Location of the investigated floodplain sections (abbreviations are explained in Table 1).

plain sections were located along the River Waal, the largest distributary of the River Rhine in the Netherlands, and four sections along the River Meuse (Fig. 2). The results obtained from five floodplain sections will be discussed in more detail. These sections comprise the Bemmelsche Waard (BW), the Stiftsche Uiterwaard (SU), the Variksche Plaat (VP), the Brakelsche Benedenwaarden (BB) and the floodplain section near Bern (BE). Except for BW and BE these floodplain sections still retain their natural topography of scroll bars and depressions associated with former channels. BE, located along the River Meuse, was formed by a meander cut-off carried out as part of river-bed improvements at the beginning of this century. It is separated from the river channel by a low levee. BW and SU are protected from minor floods by minor river dykes, which are 1–2 m high. SU has an abandoned channel that contains water but is closed off from the main channel. VP lies adjacent to SU, but is not surrounded by a minor river dyke, and has a side channel that still has an open connection with the main channel at its downstream end. Land use within the floodplain sections studied consists of pastureland, with local tree stands. Part of BB, however, is used as arable land. Depending on the presence of minor dykes and differences in elevation, the average number of days per year that the floodplains are inundated and sediment is conveyed onto the floodplain sections ranges from 1 day for BB to 9 days and longer for the lower parts of VP (Table 1).

Water discharge and suspended sediment concentrations

The average Rhine discharge near Lobith (Dutch–German border) is 2200 $m^3 s^{-1}$, of which the Waal transports about two-thirds. Peak discharges usually range between 5000 and 10 000 $m^3 s^{-1}$. Near Lobith, the average suspended sediment concentration is about 30 mg l^{-1}. During periods of high discharge, maximum concentrations vary between 120 and 200 mg l^{-1}. The average discharge of the river Meuse at the Belgian–Dutch border is about 250 $m^3 s^{-1}$. Peak discharge often exceeds 1500 $m^3 s^{-1}$. Average suspended sediment concentrations are 30 mg l^{-1}, whereas maximum concentrations of 150–300 mg l^{-1} occur at high discharge. The suspended sediment load of both rivers mainly consists of clay, silt and fine sand (Rijkswaterstaat 1992).

Sedimentation measurements were carried out during a series of floods of different magnitude and duration. Discharge characteristics of the studied floods are summarized in Table 2.

Table 1. *Characteristics of the investigated floodplain sections*

	Km (km)	Bank	Q_{cr} ($m^3 s^{-1}$)	$H_{cr,L}$ (m)	*n*-avg (days a^{-1})	*n*-1993 (days)
Rhine						
Klompenwaard (KW)	869	L (0.2)	5300	13.5	8.0	23
Bemmelsche Waard (BW)	880	D (1.8)	6450	14.3	3.2	15
Slijk Ewijk (SE)	893	L (0.2)	5720	13.8	6.2	20
Willems Polder (WP)	911	D (1.8)	7110	14.7	2.0	12
Stiftsche Uiterwaard (SU)	921	D (0.6)	6770	14.5	2.6	15
Variksche Plaat (VP)	922	L (0.1)	5860	13.9	5.4	21
Brakelsche Benedenwaarden (BB)	948	L/D (0.5)	8500	15.3	1-2	8
Meuse						
Keent (KE)	178	L (0.2)	1350	–	3.7	8
Alem (AL)	210	X	1500	–	2.2	8
Hoenzadriel (HD)	215	X	900	–	<1	4
Bern (BE)	227	L (0.2)	1600	–	1.4	8

Km, km number in downstream direction along the river. Bank, type of river bank (L, natural levee; D, minor dyke; X, no natural levee or dyke); an indication of the height of the embankment is given in parentheses (m). Q_{cr}, critical discharge for influx of sediment at Lobith (Rhine) and Eysden (Meuse). $H_{cr,L}$, critical water level at Lobith (Rhine) for entire inundation of the floodplain. *n*-avg, number of days per year during which the floodplain is inundated. *n*-1993, number of days of floodplain inundation during the flood of December 1993.

Table 2. *Discharge characteristics of the studied floods*

Flood	Maximum discharge ($m^3 s^{-1}$)	Recurrence time (year)
Rhine		
Jan. 93	5740	<1
Dec. 93	11039	35
Apr. 94	5580	<1
Jan. 95	11815	70
Meuse		
Jan. 93	2320	20
Dec. 93	3120	200

Recurrence times were obtained from Passchier (1993)

Data collection and processing

Sediment traps made of artificial grass with 15 mm long plastic blades, fixed to a pliable (plastic) base, were used to collect sediment deposited during overbank flooding. The traps measured 500 mm × 500 mm and were fixed to the floodplain surface with 100 mm long steel pins. Similar traps were successfully applied by, for example, Mansikkaniemi (1985), Gretener & Strömquist (1987) and Lambert & Walling (1987). They were placed in a semi-regular grid consisting of several transects perpendicular to the main channel and a few clusters. Close to the river channel the sampling interval along the transects was 5 m, but this increased to about 100 m at distances of more than 150 m from the main stream. The spacing between the transects varied from 100 m for small and densely sampled areas to about 250 m for the large floodplain sections. A relatively high sample density was applied in the sections BE and VP, whereas relatively few traps were used at BW, SU and BB. During the high-magnitude flood of December 1993 more than 800 traps were installed. To calculate the total amount of deposited sediment for each floodplain section, and to determine the spatial variability of the sediment deposition, the sediment trap measurements were interpolated using block kriging. More detailed information on the sampling technique has been provided by Asselman & Middelkoop (1995).

Floodplain sedimentation models

The results of the floodplain measurements are compared with results of modelling studies published in the literature. These studies comprise computations performed by Narinesingh (1995), Van den Brink (1995), Middelkoop (1997) and Middelkoop & van der Perk (1998).

Narinesingh's sedimentation model is based on the approach developed by Camp (1946), in which sediment particles settle in a sedimentation basin. For this purpose the floodplain is divided into a number of parallel orientated settling-tubes. Sedimentation in each of the settling tubes is a function of the surface area of the tube, the discharge through the tube, the settling velocities of the sediment particles and the suspended sediment concentration at the inflow point. Processes related to resuspension are neglected.

Middelkoop & van der Perk (1998) developed the geographical information system (GIS)-based SEDIFLUX model, which is based on two-dimensional (2D) patterns of water flow calculated by the hydrodynamic model WAQUA. Sediment accumulation is computed using a sediment balance for 50 m × 50 m raster cells, which is determined by the sediment deposition within a cell, and horizontal sediment fluxes between adjacent cells. Erosion of previously deposited sediment is neglected. Input data for the SEDIFLUX model include raster maps of water level, water depth and flow velocities, suspended sediment concentrations, effective settling velocities and the critical shear stress for deposition.

Van den Brink (1995) applied the WAQUA–DELWAQ model (Delft Hydraulics 1990) to quantify changes in sedimentation in response to rehabilitation measures. The DELWAQ model is also a 2D raster-based sedimentation model, with a multi-layer bed. Sediment transport and deposition are computed under stationary discharge conditions, solving a mass balance equation combined with a convection–diffusion equation. The model also accounts for resuspension.

Table 3. *Sediment accumulation at different floodplain sections during the flood of December 1993*

	Area (ha)	Sedimentation ($kg\ m^{-2}$)	
		Total	<53 μm
Waal			
KW	13.2	6.61	3.98
BW	229.6	3.86	3.74
SE	9.1	2.24	2.20
WP	98.7	2.58	2.50
SU*	105.3	3.57	1.20
VP	28.0	2.60	2.15
BB*	116.1	6.03	1.50
Meuse			
KE	16.4	4.25	1.74
AL	9.0	1.09	1.00
BE	18.1	4.29	2.00

*Without sediment eroded from arable land or sand pit.

Results

The results of the sedimentation measurements carried out at all 11 floodplain sections are summarized in Table 3. Figure 3 shows sediment accumulation at the Waal floodplain sections as a function of the critical water level (H_{cr}) at Lobith. The critical water level is the water level at Lobith that is sufficient to inundate the entire floodplain section. Results for five of the selected floodplain sections are presented in more detail as cross-sections and interpolated raster maps in Figs 4–7. In the following section the effects of different floodplain characteristics on sedimentation rates are considered, and the potential impact of related rehabilitation measures on these sedimentation rates is discussed.

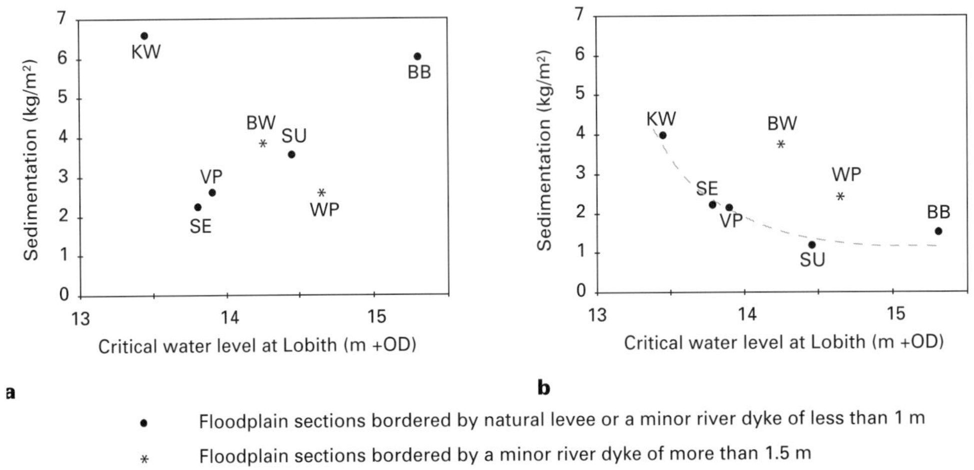

Fig. 3. Sedimentation at different floodplain sections as a function of floodplain elevation: (**a**) total sediment accumulation; (**b**) accumulation of silt and clay.

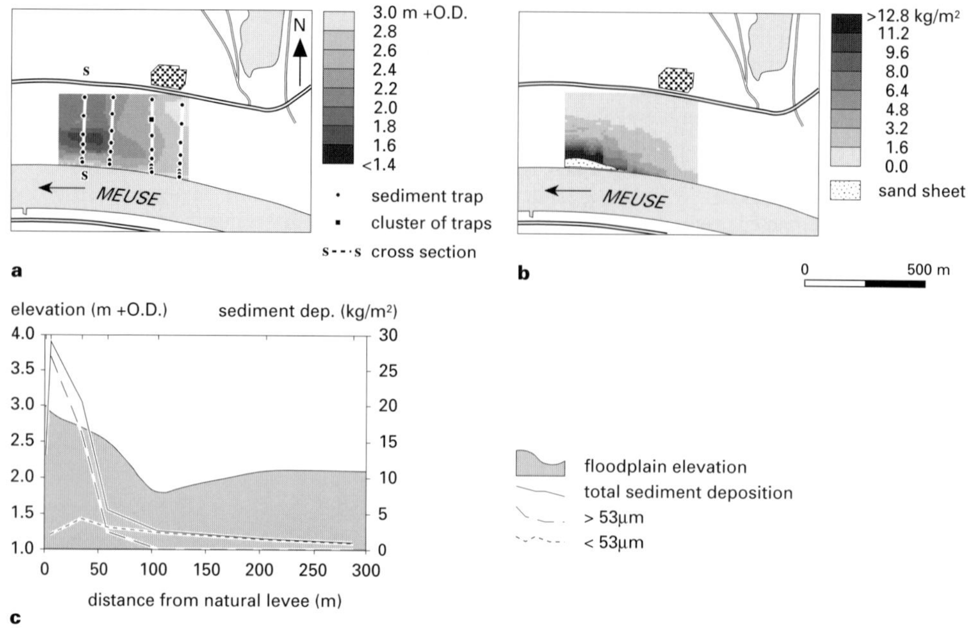

Fig. 4. Bern: (**a**) floodplain elevation; (**b**) interpolated raster map of sedimentation during the flood of December 1993; (**c**) cross-section of floodplain elevation and sedimentation.

Lowering of the floodplain surface

Lowering of the floodplain surface by 0.5–2 m will increase the number of days during which the floodplain section is inundated. Also, the amount of water discharged over the floodplain during a flood will increase. It is therefore expected that a lowering of the floodplain surface will enhance floodplain sedimentation. However, when a minor river dyke prevents flooding of the floodplain during relatively low floods, the annual duration of inundation will remain constant and lowering of the floodplain surface is expected to have a minor impact on the sediment accumulation.

The impact of lowering of the floodplain surface on sediment accumulation can be assessed by comparing sediment accumulation measured at floodplain sections with a different elevation. In Table 1 critical discharges for the complete inundation of the studied floodplain sections are given. At the equivalent stages, the highest parts of the floodplain, the levee, or the minor river dyke are just overtopped by the flood water. Inundation of the lower parts of the floodplain usually starts at lower discharges. The corresponding critical water level at Lobith is also given.

Figure 3a shows that there is no significant relationship between total sediment deposition and floodplain elevation (expressed as critical water level at Lobith). Maximum sedimentation was measured at both the lowest floodplain section KW and at the highest floodplain section BB.

However, when accumulation of bed material load (i.e. sand) is omitted, there is some evidence of a trend (Fig. 3b). On average, less silt and clay accumulated at floodplain sections that were inundated at the highest water levels. Almost twice as much sediment was deposited at KW than at VP and SE, which have a difference in H_{cr} at Lobith of about 0.4 m. For higher floodplain sections the difference in sedimentation seems smaller. However, at floodplain sections that are bordered by a minor river dyke of more than 1.5 m in height, more fine suspended sediment is deposited than at floodplain sections that were inundated at a similar discharge level, but are separated from the main channel by a natural levee or a much lower embankment. Comparison of sediment accumulation at SE, VP and SU with sedimentation at BW and WP suggests that a minor river dyke enhances sedimentation of silt and clay by a factor of two. It can thus be concluded that deposition of silt and clay decreases with floodplain elevation, and that the decrease is greater for lower floodplain sections, but that local differences in trapping efficiency, caused by the presence of minor river dykes, may prevail.

It is important to note that Fig. 3 provides

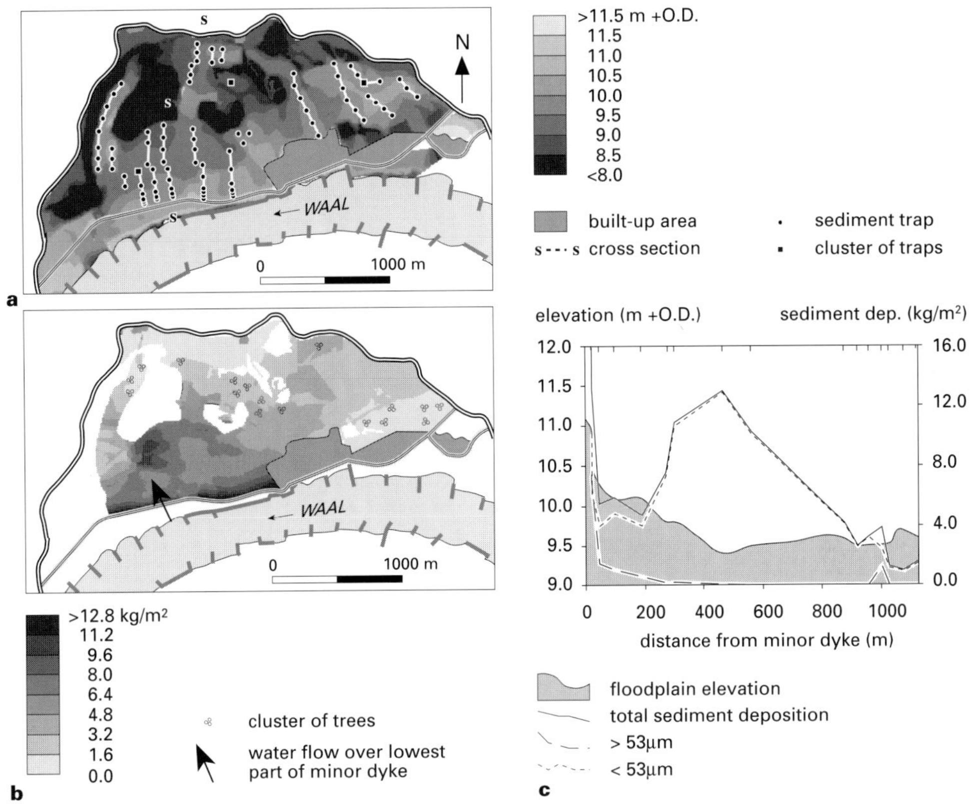

Fig. 5. Bemmelsche Waard: (**a**) floodplain elevation; (**b**) interpolated raster map of sedimentation during the flood of December 1993; (**c**) cross-section of floodplain elevation and sedimentation.

information on differences in sedimentation as a function of floodplain elevation for only a single high-magnitude flood. A different relationship may exist when the results for floods of low magnitude are included, or when the average annual sediment deposition is considered. At VP, sedimentation was measured during a series of four floods (Asselman & Middelkoop 1998). The results for two floods are summarized in Table 4. Although inundation during the high flood lasted ten times longer, and total sediment transport during the entire flood was six times greater, sedimentation was only 2.6 times greater during the high-magnitude flood of December 1993 than during the low-magnitude flood of January 1993 (Table 4). This suggests that the efficiency of low floodplain sections in trapping suspended sediment decreases at very high discharge. Consequently, the amount of sediment deposited during low-magnitude floods is relatively large when compared with the amount of sediment deposited during high-magnitude floods. When the recurrence interval of these low floods is also taken into account, it can thus be expected that, on an annual basis, much more sediment is deposited at low floodplain sections such as VP than at high floodplain sections such as BB or BW, which are inundated only during high floods. Thus, the effect of lowering of the floodplain surface will accentuate the differences in sediment accumulation that were observed after the high-magnitude flood.

Model simulations reported in the literature also indicate that sedimentation will increase when the bed level of the floodplain is lowered. Narinesingh (1995) computed sedimentation as a function of discharge over the floodplain and the efficiency of the floodplain to trap suspended sediment. He concluded that the trapping efficiency of the floodplain decreases when the floodplain surface is lowered. However, the total discharge of water and sediment over the floodplain section increases significantly. According to his model computations, the increase in sediment transport is more significant than the decrease in efficiency, resulting

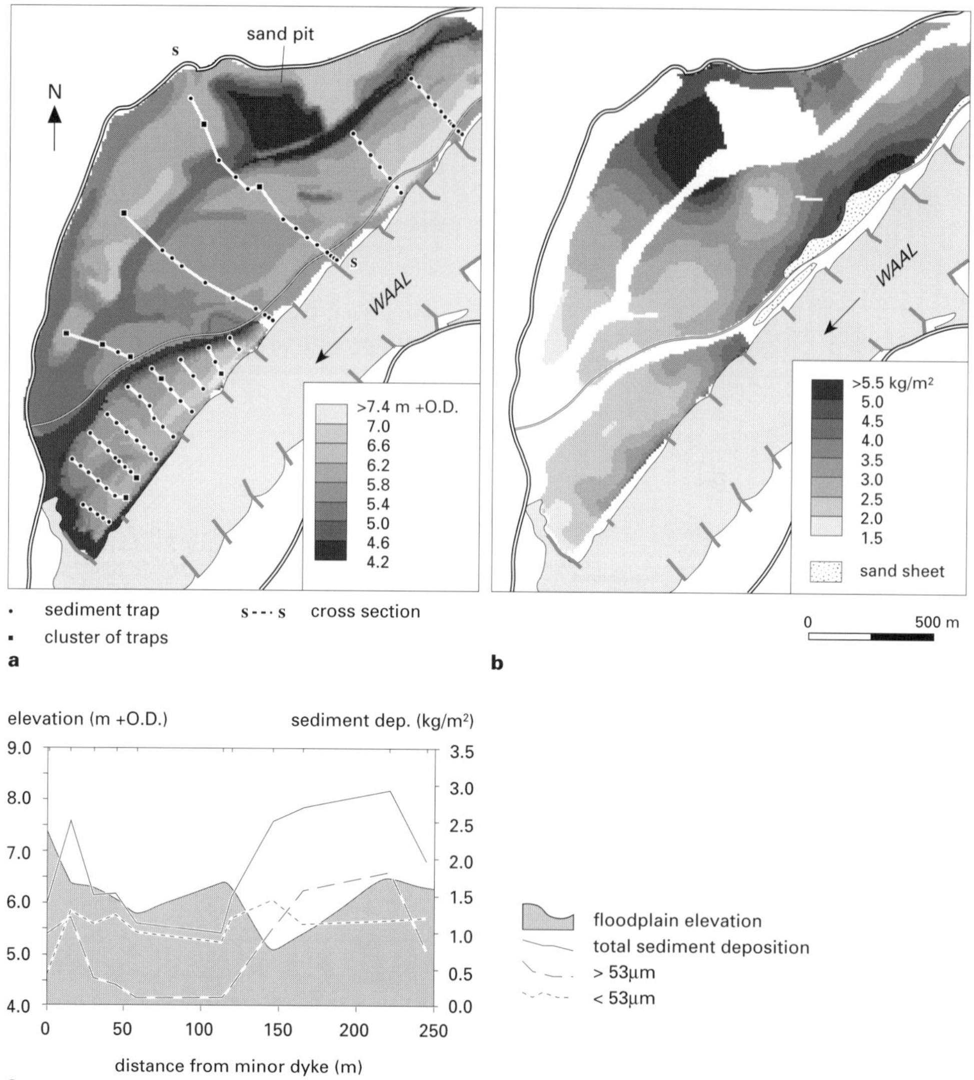

Fig. 6. Stiftsche Uiterwaard: (**a**) floodplain elevation; (**b**) interpolated raster map of sedimentation during the flood of December 1993; (**c**) cross-section of floodplain elevation and sedimentation.

in an increase in sedimentation. According to computations for a hypothetical floodplain section, lowering of the floodplain surface by 1 m may increase the accumulation of sediment by *c.* 60%. These computations do not, however, include an increase in the frequency of inundation and may therefore be underestimates.

Middelkoop (1997) used the SEDIFLUX model to calculate the annual average sediment accumulation at BW, SU and VP (Table 5). Both the measurements and the model results show that during the high-magnitude flood of December 1993–January 1994 most sediment was deposited at BW. Model calculations of average annual sedimentation rates, in which the effect of lower floods is also included, demonstrated that maximum sedimentation rates occur at the lowest floodplain section VP. Hence, lowering of the floodplain surface will enhance sedimentation significantly. It is difficult to quantify the increase, as local factors, such as the presence of minor river dykes, have a strong effect on the sediment accumulation. However, in a different study, Middelkoop (1997) used heavy metal profiles in the floodplain soil to

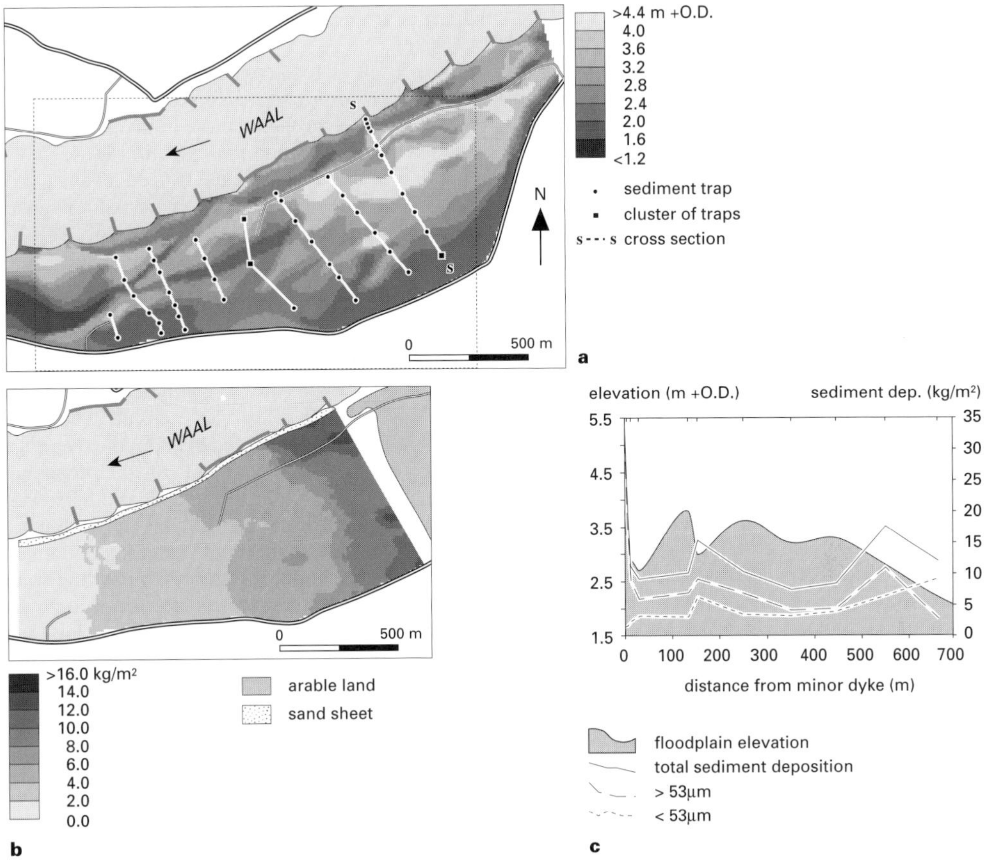

Fig. 7. Brakelsche Benedenwaarden: (**a**) floodplain elevation; (**b**) interpolated raster map of sedimentation during the flood of December 1993; (**c**) cross-section of floodplain elevation and sedimentation.

Table 4. *Sediment transport in the River Rhine and sedimentation at the Variksche Plaat during floods of different magnitude and duration*

	January 1993	December 1993	Dec.–Jan.
Inundation time (days)	2	23	11.5
Sediment transport at Lobith (10^3 tons)	300	1920	6.4
Sedimentation (tons)	303	781	2.6

estimate long-term sedimentation rates. The heavy metal profiles in the floodplain soil at VP suggest average sedimentation rates during the past century of about 10.5 mm a^{-1}. When it is assumed that the discharge and the sediment transport regime of the river have not changed over the past century, comparison of these sedimentation rates with the SEDIFLUX computations indicates that lowering of the floodplain surface by about 1 m may increase sedimentation 4–5 times or even more.

Table 5. *Measured (M) and computed (C) sediment accumulation (kg m^{-2})*

Floodplain section	$H_{cr,L}$ (m)	During flood (kg m^{-2}) Total (M)	<53 µm (M)	Total (C)	Average annual (kg m^{-2}) Total (C)
BW	14.3	3.9	3.8	4.0	1.1
SU	14.5	3.6	1.2	3.6	1.9
VP	13.9	2.6	2.2	3.0	2.7

Removal of minor river dykes

Removal of minor river dykes will affect sedimentation rates in several ways. First, it will prolong the duration of inundation, which will significantly augment sediment deposition. The results presented in Fig. 3 indicate that lowering of the floodplain surface by 0.4 m may double deposition of fine sediment during single floods. Depending on the height of the minor river dyke, its removal may be equivalent to a lowering of the floodplain surface by 0.6–1.5 m or more. Accumulation of fine sediment can thus be expected to increase by more than 100%.

Second, removal of a minor river dyke will decrease the efficiency of the floodplain in trapping suspended sediment, which will result in lower sediment accumulation during high floods. The high trapping efficiency of floodplains enclosed by a minor river dyke can be observed in Fig. 3. The main reason for this high efficiency is that a minor river dyke obstructs the water flow over the floodplain, which results in lower flow velocities and hence in longer residence times and more efficient sedimentation. According to the measurements carried out at the Waal floodplain sections, the presence of a minor river dyke at BW and WP enhanced sedimentation of fine suspended sediment by a factor of two (Fig. 3).

Third, the effect of sediment settling out in ponded water will be eliminated. After recession of the flood, flood water is retained behind the minor dyke, similar to a bathtub effect. At the BW section inundation as a result of ponded water lasted for more than 2 weeks (Middelkoop & Asselman 1998). Assuming that all the sediment in the isolated water body settled out during recession of the flood, this caused an additional sediment accumulation of about 100–200 g m^{-2}, or less than 0.2 mm per flood. The reduction of sediment accumulation because of a decreased bathtub effect will therefore be of minor importance.

Also, removal of minor river dykes will change the spatial pattern of sediment accumulation. At floodplain sections that are bordered by a natural levee, sediment deposition was found to decrease exponentially with distance from the channel. An example of such a floodplain section is BE (Fig. 4). In floodplain sections that are enclosed by a minor river dyke of about 2 m in height, no exponential decrease in sedimentation was found. An example of such a floodplain section is the Bemmelsche Waard (BW, Fig. 5). Maximum sedimentation was measured behind the lowest part of the minor river dyke, i.e. the location where the water starts flowing over the minor river dyke at the onset of inundation (see the location of the arrow in Fig. 5b). Sedimentation generally decreased with distance from the minor dyke, but local differences occurred.

Finally, it is to be expected that the grain-size distribution of the deposited sediment will change as a result of a strong increase in the deposition of sand. The cross-section presented in Fig. 4c shows that large amounts of sand can be deposited at floodplain sections bordered by a natural levee. Locally, sand sheets with a thickness of 100–250 mm were observed (Fig. 4b). The sand must have been eroded from the channel bed and transported by traction over the natural levee, where it was deposited as current velocities decreased. The large total sediment accumulation in the sections KW, KE and BE (Table 3) was mainly the result of the sandbars deposited on the natural levees. In floodplain sections enclosed by a minor river dyke, the percentage sand content of the deposited sediment was low (Fig. 5c) and sandbars were absent. Because of the minor dyke, only the upper part of the water column could enter the floodplain. As little bed material is transported near the water surface, sediment supply and deposition onto the floodplain area was limited to silt- and clay-sized material (Fig. 5c). Sedimentation of sand-sized material is expected to increase after removal of the minor river dyke, as bed material load present in the lower part of the water column can enter the floodplain section.

Similar results were obtained by Narinesingh (1995). His model computations indicate that an increase in the number of days of inundation and an increase in sediment transport over the floodplain outweigh the reduction in sedimentation efficiency, resulting in an overall increase in sediment deposition. No information is given on the composition of the deposited material.

Excavation of side channels and lakes

The construction of a side channel through the floodplain can have many effects on sediment accumulation, depending on the shape of the channel and on the discharge stage. When the side channel is connected to the main channel only at the downstream end, such as around the VP section, sediment can be conveyed into the channel. This sediment may accumulate within the channel and the surrounding low-lying areas. Sedimentation in these areas is large, especially at low but increasing discharge levels (Asselman & Middelkoop 1998).

During large floods, residual channels are reactivated and current velocities in these channels are high. This causes erosion of sediment from the bed. The eroded material is transported through the channel and subsequently deposited on the channel banks. Erosion of sediment in the residual channel at SU resulted in supply of relatively coarse sandy material onto the surrounding floodplain area (Asselman 1997).

During periods of high discharge, erosion also occurs in small ponds such as in the sandpit at the Stiftsche Uiterwaard (SU) (Fig. 6). Much of the eroded material appears to be deposited within an area extending several hundreds of metres downstream of this local source (Fig. 6b). The sand content of this sediment is high (Fig. 6c).

Similar results have been obtained from model predictions for a case study of a side channel in the SU floodplain section (Van den Brink 1995). Sedimentation in the channel is expected to increase at discharges lower than about 6000 $m^3 s^{-1}$ at Lobith. At discharges exceeding 9000 $m^3 s^{-1}$ at Lobith no sedimentation will occur in the side channel, and locally the bed and banks may be subjected to erosion. Narinesingh (1995) concluded that secondary channels and other water bodies enhance the deposition of suspended sediment. However, his model only accounts for sedimentation; erosion and resuspension processes were neglected.

Reintroduction of ecosystems such as riverine forests

Plans for restoration of natural conditions envisage major changes in land use. The area of agricultural land (i.e. productive grassland and arable land) will reduce, and the area of floodplain forest, shrubs, pasture and swamps will increase. Changes in vegetation may lead to changes of the hydraulic roughness of the floodplain. As a result, turbulence in the water may be intensified, and, if rough vegetation extends over large areas, it will reduce the water flow over the floodplain, which will affect sedimentation and erosion.

The investigated areas are at present mainly grass-covered with local tree stands, and are used as pasture. Field surveys after recession of the flood revealed that grass-covered floodplain sections are not susceptible to erosion. Locally some scour may occur where the grass cover is disturbed and bed shear stresses are high. These areas are mainly located on levees or directly behind minor river dykes, and cover less than about 1–10 m^2. Erosion on grass-covered areas can thus be neglected.

Erosion is more pronounced on arable land, where no grass cover is present to protect the underlying floodplain soils. Although it is often believed that cohesive floodplain soils are not susceptible to erosion, the results of this study indicate that significant erosion can occur on floodplains that contain sand as well as silt and clay. Upstream of the Brakelsche Benedenwaarden (BB), a fallow arable field was present. No measurements were carried out in this field. However, sedimentation measured by the traps downstream of the field indicated that erosion occurred on the arable land and that most of the eroded material was deposited in the vicinity of the local source, resulting in a downstream gradient in sediment accumulation (Fig. 7b). As shown in the cross-section of BB, over 50% of the sediment deposited close to the arable land consisted of sand (Fig. 7c). A reduction in the area of arable land will thus result in decreased erosion of the underlying floodplain soil.

The effect of floodplain forests on sediment deposition could not be estimated from the sedimentation measurements, as no real forests are present. However, the Bemmelsche Waard (BW) floodplain section comprised several patches of trees and some lower shrubs (Fig. 5b). As only a small number of traps were placed in these areas, no definite conclusions can be drawn. At first sight, however, tree stands appear to have an insignificant effect on the deposition of suspended sediment.

The modelling results of Narinesingh (1995) suggest that, depending on the forest density, a decrease in sediment accumulation is to be expected. Forests will augment the trapping efficiency, but because of increased resistance, water discharge over the forested floodplain is expected to decrease, which will result in an overall decrease in sediment deposition.

Conclusions

Direct measurements of floodplain sedimentation are valuable for the evaluation of possible effects of floodplain rehabilitation measures on sediment deposition. However, to establish reliable relationships between sedimentation and various floodplain characteristics, measurements of a large number of floodplain sections are required. In this study measurements were carried out at 11 floodplain sections. From the results it is concluded that rehabilitation plans are likely to influence sediment accumulation as follows.

(1) Lowering of the floodplain surface will decrease the efficiency of the floodplain in trapping suspended sediment. However, because of an increase in inundation times, average sedimentation rates will increase significantly.

(2) Removal of minor river dykes will also prolong inundation times. As this effect prevails over the decrease in sedimentation efficiency, average sedimentation rates of fine suspended sediment will increase by a factor of two or more. The increase will be most pronounced for coarser sediment particles (i.e. sand).

(3) Construction of side channels and small lakes enhances sedimentation in this channel at low

discharge. At high discharges, erosion of the lakes and the channel banks will result in an increased supply of sandy material to the surrounding floodplain areas.

(4) Growth of floodplain forests is expected to have a small effect on sediment deposition. The increased trapping efficiency appears to outweigh the decrease in discharge and sediment transport.

The project was funded by the Dutch National Research Programme on Global Air Pollution and Climate Change (NRP). I am grateful to H. Middelkoop, with whom I have carried out the fieldwork and much of the data analysis. He also provided valuable comments and suggestions on a previous version of the manuscript.

References

ASSELMAN, N. E. M. 1997. *Suspended sediment in the river Rhine; the impact of climate change on erosion, transport, and deposition*. PhD Thesis, Utrecht University.

—— & MIDDELKOOP, H. 1995. Floodplain sedimentation: quantities, patterns and processes. *Earth Surface Processes and Landforms*, **20**, 481–499.

—— & —— 1998. Temporal variability of contemporary floodplain sedimentation at the event scale in the Rhine–Meuse delta, the Netherlands. *Earth Surface Processes and Landforms*, **23**, 595–609.

BERENDSEN, H. J. A. 1984. The evolution of the fluvial area in the western part of the Netherlands from 1000 to 1300 AD. *Geologie en Mijnbouw*, **63**, 231–240.

CALS, M. J. R., POSTMA, R., BUIJSE, A. D. & MARTEIJN, E. C. L. 1998. Habitat restoration along the River Rhine in The Netherlands: putting ideas into practice. *Aquatic Conservation: Marine and Freshwater Ecosystems*, **8**, 61–70.

CAMP, T. R. 1946. Sedimentation and the design of settling tanks. *Transactions of the American Society of Civil Engineers*, **111**, 895–936.

DE BRUIN, D., HAMHUIS, D., VAN NIEUWENHUYZE, L., OVERMARS, W., SIJMONS, D. & VERA, F. 1987. *Ooievaar; de toekomst van het rivierengebied*. Stichting Gelderse Milieufederatie, Arnhem.

DELFT HYDRAULICS 1990. *DELWAQ—User's Manual Version 3.0*. Delft Hydraulics, Delft.

GRETENER, B. & STRÖMQUIST, L. 1987. Overbank sedimentation rates of fine grained sediments: a study of the recent deposition in the lower river Fyrisan. *Geografiska Annaler*, **69A**, 139–146.

LAMBERT, C. P. & WALLING, D. E. 1987. Floodplain sedimentation: a preliminary investigation of contemporary deposition within the lower reaches of the River Culm, Devon, UK. *Geografiska Annaler*, **69A**, 393–404.

MANSIKKANIEMI, H. 1985. Sedimentation and water quality in the flood basin of the River Kyronjoki in Finland. *Fennia*, **163**, 155–194.

MIDDELKOOP, H. 1997. *Embanked floodplains in the Netherlands; geomorphological evolution over various time scales*. PhD Thesis, Utrecht University.

—— & ASSELMAN, N. E. M. 1998. Spatial variability of floodplain sedimentation at the event scale in the Rhine–Meuse delta, the Netherlands. *Earth Surface Processes and Landforms*, **23**, 561–573.

—— & VAN DER PERK, M. 1998. Modelling spatial patterns of overbank sedimentation on embanked floodplains. *Geografiska Annaler*, **80A**, 95–109.

MINISTRY OF TRANSPORT, PUBLIC WORKS AND WATER MANAGEMENT 1997. *Fourth National Policy Document*. Ministry of Transport, Public Works and Water Management. The Hague.

NARINESINGH, P. 1995. *Nature restoration and floodplain sedimentation*. MSc thesis, IHE, Delft.

PASSCHIER, R. H. 1993. *Toetsing uitgangspunten rivierdijkversterkingen*. Aanvullend rapport 2: Werklijn Rijn en Maas. WL–RAND, Delft.

RIJKSWATERSTAAT 1992. *Jaarboek Monitoring Rijkswateren 1991*. Rijkswaterstaat, The Hague.

SILVA, W. & KOK, M. 1996. *Integrale verkenning inrichting Rijntakken. Hoofdrapport 'Een weegschaal voor rivierbeheer'*. IVR rapport 1. RIZA–Delft Hydraulics, Arnhem.

VAN DEN BRINK, N. G. M. 1995. *Stroming en sediment gedrag in de Stiftsche Waard voor en na de aanleg van een nevengeul. Resultaten van 2D modelonderzoek van waterbeweging en sedimenthuishouding*. RIZA Nota nr. 95.053. Rijkswaterstaat, Arnhem.

VAN DER PERK, M., ERTSEN, A. C. D. & BLEUTEN, W. 1992. Modellering van slibsedimentatie van zware metalenbelasting in een uiterwaard langs de Waal. *H_2O*, **25**, 233–237 [in Dutch with English summary].

WERELD NATUUR FONDS (WNF) 1992. *Levende rivieren*. WNF, Zeist.

Potential for floodplain rehabilitation based on historical maps and present-day processes along the River Rhine, The Netherlands

MARGRIET M. SCHOOR[1], HENK P. WOLFERT[2], GILBERT J. MAAS[2], HANS MIDDELKOOP[1] & JURRIAAN J. P. LAMBEEK[3]

[1]*Institute for Inland Water Management and Waste Water Treatment RIZA, PO Box 9072, 6800 ED Arnhem, The Netherlands (e-mail: m.schoor@riza.rws.minvenw.nl)*

[2]*The Winand Staring Centre for Integrated Land, Soil and Water Research, PO Box 125, 6700 AC Wageningen, The Netherlands*

[3]*Delft Hydraulics, PO Box 177, 2600 MH Delft, The Netherlands*

Abstract: For the purpose of restoration, a process-based geomorphological and ecological classification of rivers and river reaches has been developed to help identify both past and present-day geomorphological processes along lower branches of the Rhine. Historical maps are used to analyse floodplain development since 1600. The series of maps shows development and erosion of bars and floodplains, and land use reflects both vegetation and geomorphological patterns and indicates that the ecological succession was strongly related to geomorphology. Historical measurements (early 19th century) of the water depth and water levels are used to derive historical river profiles and water duration curves. Present-day hydraulic parameters were calculated which show that they are not related to the geomorphological patterns in floodplains. Changes in the geomorphological activity of the river are strongly related to changes of the width–depth ratio of the main channel. Potentials for bar and levee development are therefore based on differences between the width–depth ratio of the Rhine branches.

Over the past 10 years, ecological rehabilitation of floodplains of the lower River Rhine and Meuse distributaries has become an important issue in river management in the Netherlands. Plans to recover lost habitats, such as floodplain forests, floodplain marshlands, secondary channels and river dunes have been established. Sustainable ecological rehabilitation implies reintroduction of natural processes in regulated rivers and their floodplains, so as to restore a semi-natural geomorphological and ecological setting (e.g. Boon 1993; Silva & Kerkhofs 1994; Van den Brink *et al.* 1996; Pedroli & Postma 1998). The usual effect of this is an increase in the ecological variability of the river area. For the purpose of restoration, a process-based geomorphological and ecological classification of the rivers and river reaches has been used to help identify both past and present geomorphological processes along lower branches of the Rhine.

Most classifications of rivers and floodplains are based on the river planform. Within the river continuum, major distinctions are made between straight, meandering and braided rivers. Anastomosing rivers are classified as multi-channel rivers that consist of single-channel braided, meandering or straight reaches (Schumm 1985; Makaske 1998). Sinuosity and degree of anabranching are often used to classify river planform (Brice 1975; Rust 1987). Many researchers have tried to predict river planform. Leopold & Wolman (1957) showed that channel type is related to discharge and slope. Schumm (1977, 1985) related channel planform to a river's relative bank stability and nature of sediment load. The genetic classification of floodplains by Nanson & Croke (1992) is based on specific stream power and cohesion of floodplain sediment. Rosgen's classification (1994) linked channel type with entrenchment, gradient, width–depth ratio, sinuosity and bed or bank material. Van den Berg (1995) used stream power and bed material size to predict channel planform.

According to the classification of Brice (1975), the Rhine branches are sinuous and meandering rivers with single channels of uniform depth and width. Using the models of Nanson & Croke (1992), all Rhine floodplains can be classified as the same type: laterally stable, single-channel

From: MARRIOTT, S. B. & ALEXANDER, J. (eds) 1999. *Floodplains: Interdisciplinary Approaches*. Geological Society, London, Special Publications, **163**, 123–137. 1-86239-050-9/99/$15.00

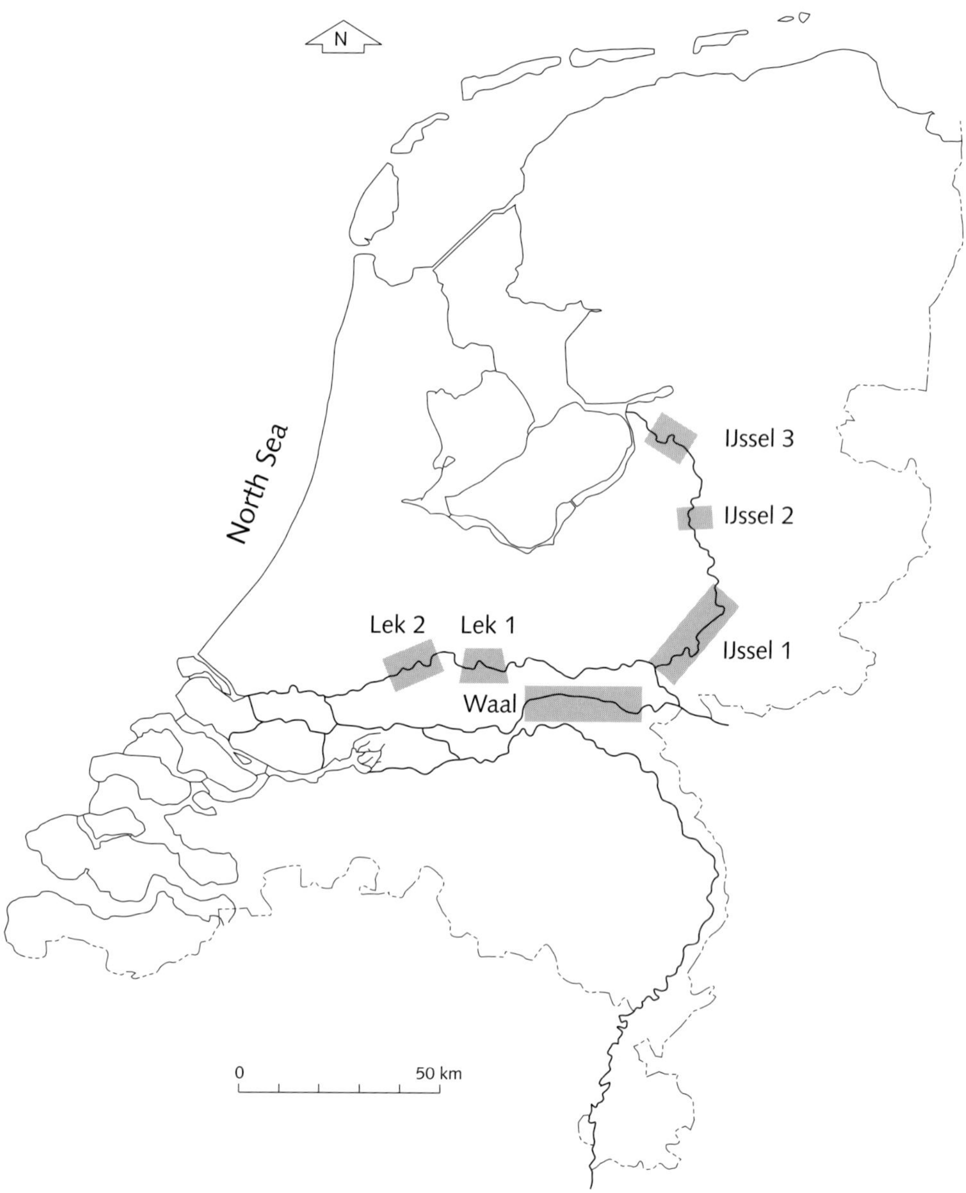

Fig. 1. Map of the Rhine branches in the Netherlands and location of the study areas.

floodplains. However, the geomorphological patterns in the Rhine floodplains vary between river reaches.

From historical maps we know that in former centuries there were branches where width variation, chutes and cut-off loops resulted in anabranching reaches. These features are associated with a less stable channel (Schumm 1977). The ecology of a stable channel is different from that of a less stable channel as, in a less stable channel, bars and shoals occur, which are important habitats for macro-invertebrates, fish and birds (Van de Velde & Van der Brink 1994; Van den Brink *et al.* 1996; Pedroli & Postma 1998). There may also be alternating cliffs and gently sloping banks that give different kinds of bank habitat, leading to high

Table 1. *Historic river maps used to reconstruct river reaches before 1830*

River reach	period				
	before 1650	1650–1700	1700–1750	1750–1800	1800–1830
Waal	Kempinck 1601, 1607*a, b* 1620; Van Geelkercken 1634*a*, 1634*b*, 1639, 1646*a, b*	Van Geelkercken 1651*a, b*, 1654, 1664; Passavant 1688, 1694*a*, 1694*b*	Leempoel 1709, 1723; Couwater 1722*a, b, c*, 1723*a, b*, 1724	Camp 1766; Hendrikman 1776*a, b*, c; Beijerinck 1777, 1778	Beijerinck 1800, 1810
IJssel I	Kempinck 1596*a, b*, 1613; Anonymous 1604; Van Geelkercken 1632, 1633, 1634, 1635, 1637, 1638, 1639, 1640*a, b*, 1641, 1642*a, b*, 1644*a, b*, 1645, 1646	Van Geelkercken 1653*a, b*, 1657, 1660, 1661*a, b*, 1670, undated (2nd half 17th century); Passavant 1676*a, b*, 1696, 1697; De Roy 1681	Van de Heuvel 1714, 1716, 1717, 1719, 1723; De Roy 1738, 1739; Leenen 1740	Leenen, 1753, 1759; Beijerinck 1771, 1776*a, b, c* Van Hooff 1773; Anonymous 1778; Van Heijs 1778; Hottinger 1779	Anonymous 1807*a, b, c, d, e*, 1809; Paulmier 1825
IJssel II		van Geelkercken 1670		van Hooff, 1773	Fijnje 1809*a*
IJssel III	Anonymous 1534; Grooten 1568	Ten Have 1652; Blequin 1664	Van Linden 1712	Leenen 1754; Knollaert 1765, 1768	Fijnje 1809*b*; Beijerinck 1821
Lek I	Schillincks 1620, 1621; Anonymous 1638 ; Van Diepenem 1643	Wiborch 1662; Van Geelkercken 1670		Beijerinck 1778; Engelman 1788; Anonymous 1793	De Liefde 1815
Lek II	Anonymous *c*. 1630, 1632, 1640	Anonymous 1650	Hoofd 1738	Bolstra 1751.	

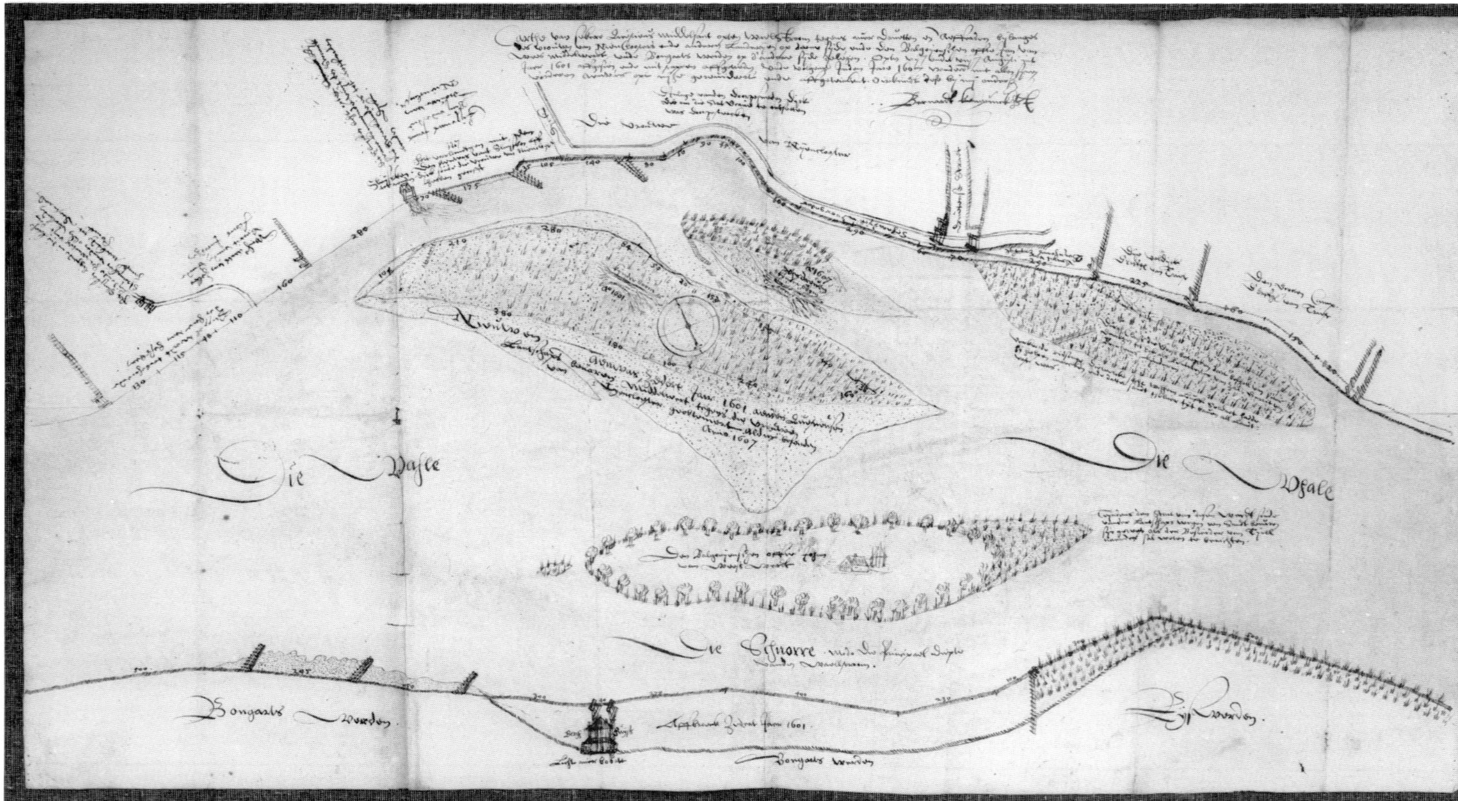

Fig. 2. Map of embanked floodplain along the River Waal near Afferden in 1607 (Kempinck, 1607, Rijksarchief Gelderland, Hof van Gelre en Zutphen, AKV No. 387). There were huge islands in the main channel, which were partly covered by willows and meadow and partly bare sand. Between 1601 and 1607 the surface area of the island in the middle of the channel became much larger.

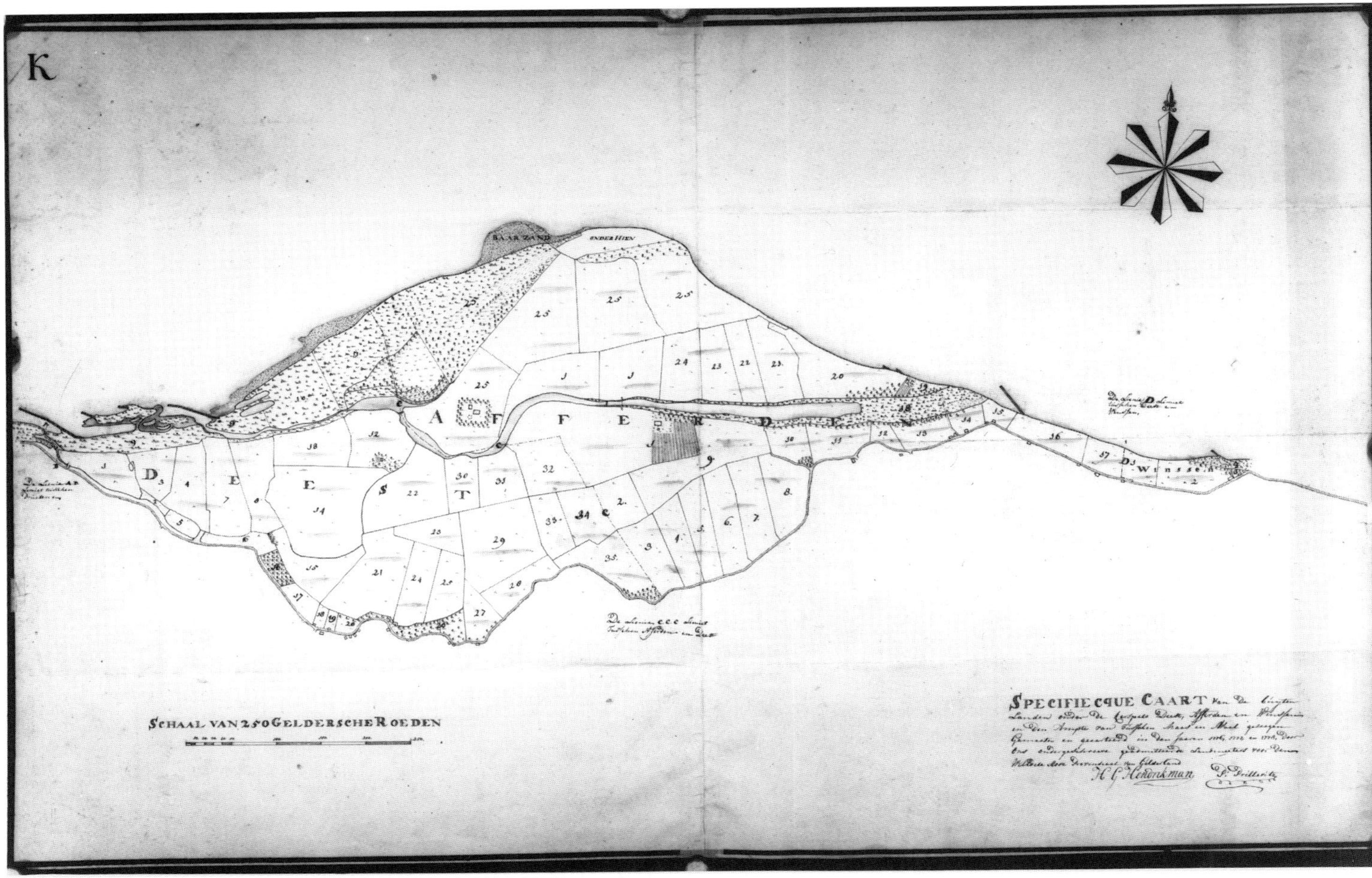

Fig. 3. Map of embanked floodplain along the River Waal near Afferden in 1776 (Hendrikman & Prilleritz 1776–1778, Rijksarchief Gelderland, Archief Verpondingskaarten, krt. No. 171). The youngest part of the floodplain was planted with willows; older parts were used as pasture. In the silted-up abandoned channel grew willows and reed. Numbers refer to detailed information about ownership, land use, area and value of the parcels.

Table 2. *Geomorphological and ecological units*

Mapped geomorphological units (physiotopes) 1:25 000	Mapped ecological units (ecotopes) 1:25 000	Physiotopes and ecotopes for which potentials are given
Main channel	Main channel	Point bar
Point bar	Bare natural river bank	Abandoned channel with open water
Mid-channel bar	Secondary channel	Abandoned channel, silted up
Secondary channel	Abandoned channel	Oxbow lake
Abandoned channel	Oxbow lake	Marshland
Abandoned channel fed with seepage water	Pond	Abandoned channel fed with seepage water
Oxbow lake	Marsh land	
Filled-up channel	Rough land	
Dyke burst pond	Flowerfull floodplain meadow	
Natural levee	Natural grassland	
Floodplain	Pasture land	
Floodplain ridge	Natural floodplain forest	
Floodplain swale	Forest for production	
	Arable land	

species richness (Simons *et al.* 1994). In a less stable channel the geomorphological and riverine processes lead to ecological variety in the river and the floodplains. These features are less likely in a stable, uniform channel.

One aim of this study is to analyse whether or not the Rhine in the Netherlands can be divided into different river reaches each with different opportunities for habitat development. A second aim is to develop a classification method with which the dynamics of a river can be described easily and related to suitable ecological units. This classification has to be applicable not only to present-day rivers, but also for the situation in the 18th and 19th centuries. This allows the linking of historical ecological units to geomorphological processes and present-day processes to ecological potential.

The River Rhine study area

The Rhine River in the Netherlands is a meandering sand-bed lowland river with a low slope (10^{-4}) and an embanked floodplain. The mean annual

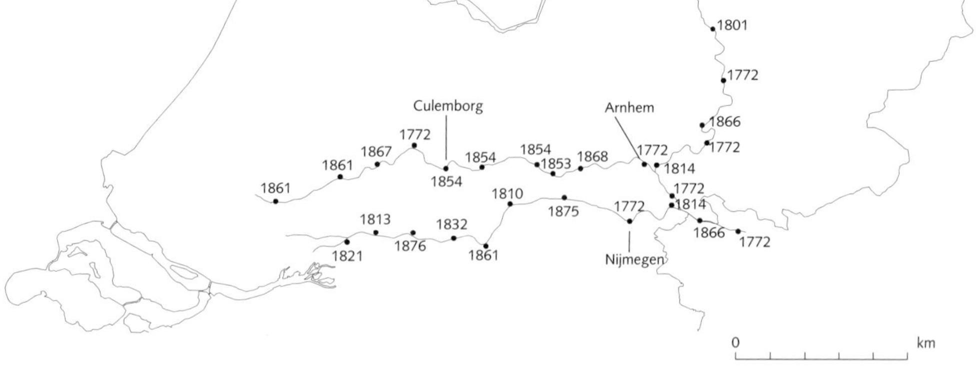

Fig. 4. Gauges along the River Rhine in 1880. The year shows the beginning of the daily measurements of water levels (after Lely 1890).

discharge is 2300 $m^3 s^{-1}$ and peak discharges of 7000 to 12 000 $m^3 s^{-1}$ occur. The lower river Rhine has three major distributaries, the Waal, the IJssel and the Lek (Fig. 1). These can be characterized as follows: the Waal is a wide river, which is relatively straight, the IJssel is a smaller, highly meandering river and the Lek is an intermediate river, today regulated with weirs. Since 1775, 67% of the total discharge flows into the River Waal. Before the 18th century, the discharge distribution was different. In 1600, about 90% of the water flowed into the Waal, which caused the other branches to fill with sand. River regulation started with changes to the discharge distribution and, later, groynes were built all along the banks and the river became narrower and deeper. During the present century, navigation and mining activities became important. Most floodplains are used as pasture land, but there are also nature reserves in the floodplains. Many abandoned channels have an important ecological function as the aquatic plant and helophyte species richness is high, making them important habitats for plankton, invertebrates, amphibians, fishes and birds. For the present study, six river reaches were chosen (Fig. 1), situated both upstream and downstream along the three distributaries of the Rhine.

Method

In this study, two different kinds of information were used. The first was from historical river maps and old records of water levels. These were studied to analyse the eco-morphological conditions in the period between 1600 and 1850. Before river regulation works started, the entire River Rhine, including land use of the floodplains, was mapped between 1830 and 1850. These maps are geometrically correct and are at a scale of 1:10 000. Older maps are less complete and less accurate, but by using time sequences of river maps and comparing them with the geometrically correct maps of 1830 and present geomorphology, the accuracy problem of these old maps could largely be controlled (Middelkoop 1997).

Table 1 shows the old river maps used for this study. The oldest maps originate from the beginning of the 17th century (Fig. 2). Most of these maps were made for judicial cases over disputed ownership of recently formed land, and they display the occurrence and location of sand bars and islands in the main channel. Eighteenth-century maps, made for land administration and tax payment purposes, show detailed information on the land use within the floodplain area (Fig. 3). From the maps of 1750–1800 and 1830–1850, former ecological and geomorphological units were mapped (Table 2). Relationships between land use and geomorphology were derived from these maps. The patterns shown were particularly dependent on the location of abandoned channels and the age of point bars.

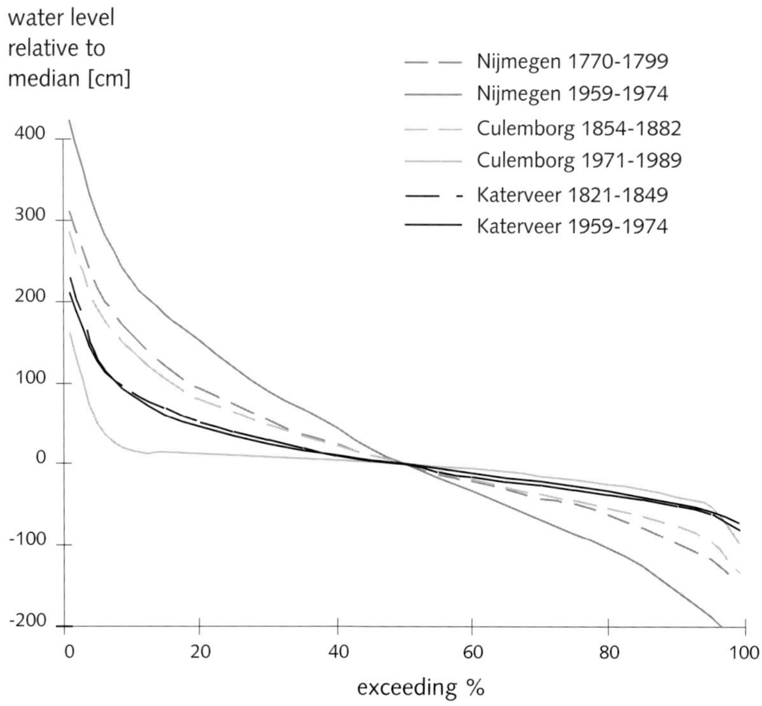

Fig. 5. Water level duration curves.

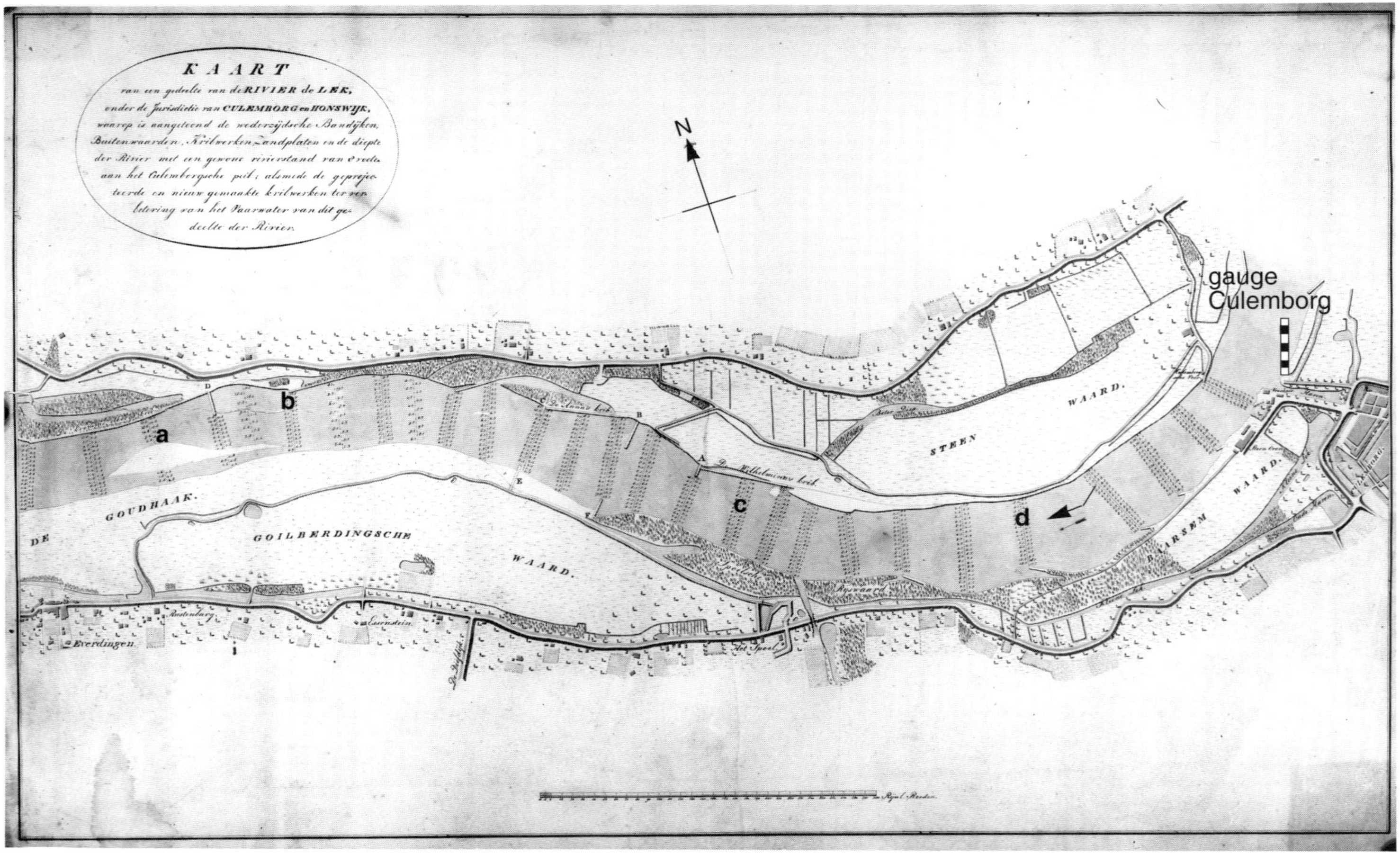
KAART
van een gedeelte van de RIVIER de LEK,
onder de Jurisdictie van CULEMBORG en HONSWIJK,
waarop is aangeteend de wederzijdsche Bandijken,
Buitenwaarden, Kribwerken Zandplaten en de diepte
der Rivier met een gewone rivierstand van 8 voeten
aan het Culemborgsche peil; alsmede de geprojec-
teerde en nieuw gemaakte kribwerken ter ver-
betering van het Vaarwater van dit ge-
deelte der Rivier
N
gauge
Culemborg
a
b
c
d
DE
GOUDHAAK
GOILBERDINGSCHE
WAARD.
STEEN
WAARD.
BARSEM
WAARD.
Everdingen
Rijnl. Roeden

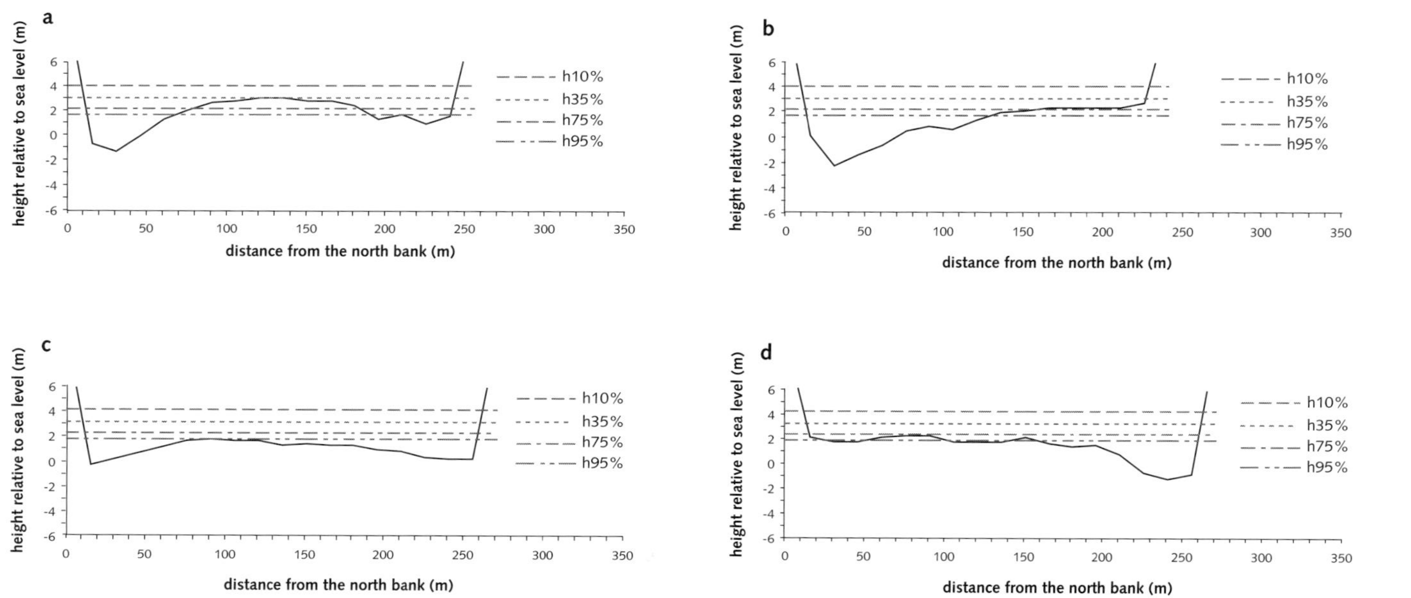

Fig. 6. Map of the River Lek near Culemborg in 1815, with cross-sections of water depth in the main channel. (**a**)–(**d**) are examples of reconstructed cross-sections (After de Liefde 1815, Rijksarchief Gelderland, kaarten Rijkswaterstaat, No. 13a.)

Table 3. *Former linear measurements used on old river maps in the Netherlands (Verhoeff 1983)*

	Name of former linear measurements	Length of a rod (m)	Number of feet in a rod	Length of a foot (cm)	Number of inches in a foot	Length of an inch (cm)
River Waal	Guelders	3.800	12	27.1	10	2.71
River Lek	Lekkendijks	5.030	18	28.0	10	2.80
other rivers	Rhenish with surveyors' inch	3.767	12	31.4	10	3.14
Most commonly used (not only rivers)	Rhenish	3.767	12	31.4	12	2.62

Table 4. *Reconstructed periods*

River reach	Reconstructed periods			
Waal			1780	1830
IJssel I	1596 (partly), 1632 (partly)	1723 (partly)	1750 (partly)	1840
IJssel II				1840
IJssel III	1652 (partly)	1712 (partly)	1770 (partly), 1815 (partly)	1840
Lek I	1630 (partly), 1670		1780 (partly)	1840
Lek II	1640 (partly)		1740	1840

The daily measurements of water levels in the River Rhine started between 1772 and 1876, depending on the gauge (Lely 1890). The mean distance between the gauges was 24 km in 1830 and 11 km in 1880 (Fig. 4). For the gauges at Nijmegen, Arnhem, Culemborg and Katerveer, the daily water levels since the beginning of measurement were available on PC and historical water-level duration curves (30 year period) were calculated (Fig. 5). From these, the water-level percentiles h_{90}, h_{75}, h_{35} and h_{10} were derived, which were assumed to represent the discharge percentiles Q_{90}, Q_{75}, Q_{35} and Q_{10}, respectively. Of all gauges, mean, high and low water levels were documented (Fijnje 1850; Fijnje & Lebret 1852; Van der Kun 1854) and historical slopes between gauges were calculated.

A few maps from the early 19th century indicate the river depth along cross-sections (Fig. 6). Such maps were made for water engineering works such as the building of

Table 5. *Method used to reconstruct river engineering variables in 1800*

Variable	Deducted by
Water level (h)	30 year historical period; the percentiles h_{10}, h_{35}, h_{75} and h_{90} are calculated
Slope (i)	$(h_{\text{gauge 1}} - h_{\text{gauge 2}})$/distance between gauges
Width (w)	Mean width from 20 cross-sections, measured on old map
Depth (d)	Mean depth from 20 cross-sections, read on map, converted to metres (see text)
Width–depth ratio (w/d)	w/d
Hydraulic radius (R)	$wd/(2d + w)$
Discharge (Q)	Discharge and discharge distribution are assumed to be the same as nowadays
Stream velocity (u)	Q/wd
Texture (D_{50})	Assumed to be the same as nowadays
Chezy parameter (C)	$u/\sqrt{Ri}$
Shields parameter (θ)	$u^2/(1.65\, C^2\, D_{50})$
Specific stream power (ω)	$\rho gQi/w$, with $\rho = 1000$ and $g = 9.8$

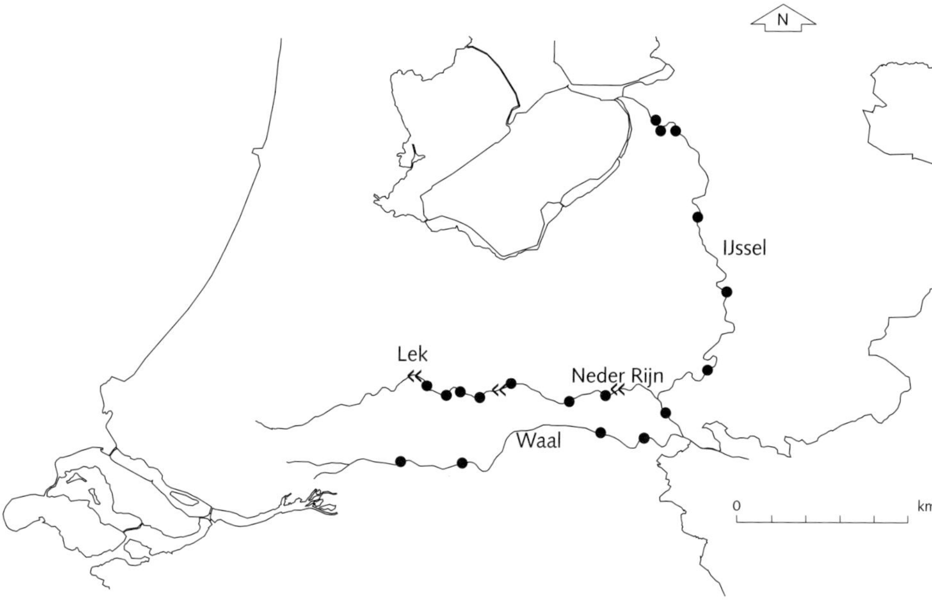

Fig. 7. Location of the weirs (≪) and used present-day cross-sections (●).

new groynes. At that time the water depths were measured with a plumb line and on the map the water level of the nearest gauge at the day of surveying was written. The height of the gauges relative to mean sea level was known (Lely 1890) so that the mean slope between the gauges upstream and downstream from the mapped area could be calculated. By multiplying the slope by the distance between each cross-section and the nearest gauge, it was possible to convert water depths on the map into bottom levels relative to sea level. In the calculations, feet and inches were converted into metres (Table 3). At four locations (Waal, IJssel I and III, Lek I) water depth and river width were reconstructed for about 20 cross-sections each. This was done at different river water levels (h_{90}, $h7_5$, h_{35}, h_{10}), which were converted from the gauge readings using slope and distances between gauge and cross-sections.

Most of the old river maps before 1830, and especially those from the 17th century, show reaches less than 2 km in length, and in many cases only one side of the river is shown. To reconstruct a 10 km reach with floodplains on both sides, different maps of about the same period were combined. Table 4 shows the mapped periods. By comparing sequences of maps from different ages the evolution of the floodplains was reconstructed. For two reaches channel migration was calculated by overlaying composite maps; for the Waal 1780–1830–1990 and for the IJssel I 1750–1840–1990.

With the information from the old cross-sections, several river engineering factors for *c.* 1800 were calculated: slope, width, depth, width–depth ratio, discharge, stream velocity, Chezy and Shields parameters, specific stream power and texture (Table 5). These were calculated for various water levels (h_{90}, h_{75}, h_{35}, h_{10}).

The second approach to classification used a one-dimensional hydraulical model (Barneveld *et al.* 1994) to describe present-day fluvial processes along the various river reaches. This was done by calculating the factors mentioned above for 18 characteristic profiles along the Rhine distributaries (Fig. 7).

Using both types of information, the suitability of river reaches for ecological units were worked out, for example, point bars and different kinds of abandoned channels.

Results and discussion

Figure 6 gives examples of reconstructed cross-sections of the river at the beginning of the 19th century. The development of the River IJssel between 1596 and 1840 is shown in Fig. 8. Comparison of sequences of old river maps of various river reaches demonstrated that most of the variations between reaches and most of the changes since 1600 were related to changing morphodynamics of the river. Morphodynamics depend on the balance between three major factors: flow velocity, the kind of sediment in the river bed and the strength of the banks. To show the changing morphodynamics more clearly we plotted a diagram showing the width–depth ratio and the Shields parameter (Fig. 9). The width–depth ratio describes the balance between the stream power and the strength of the banks, which is determined by geology and bank protection, such as groynes,

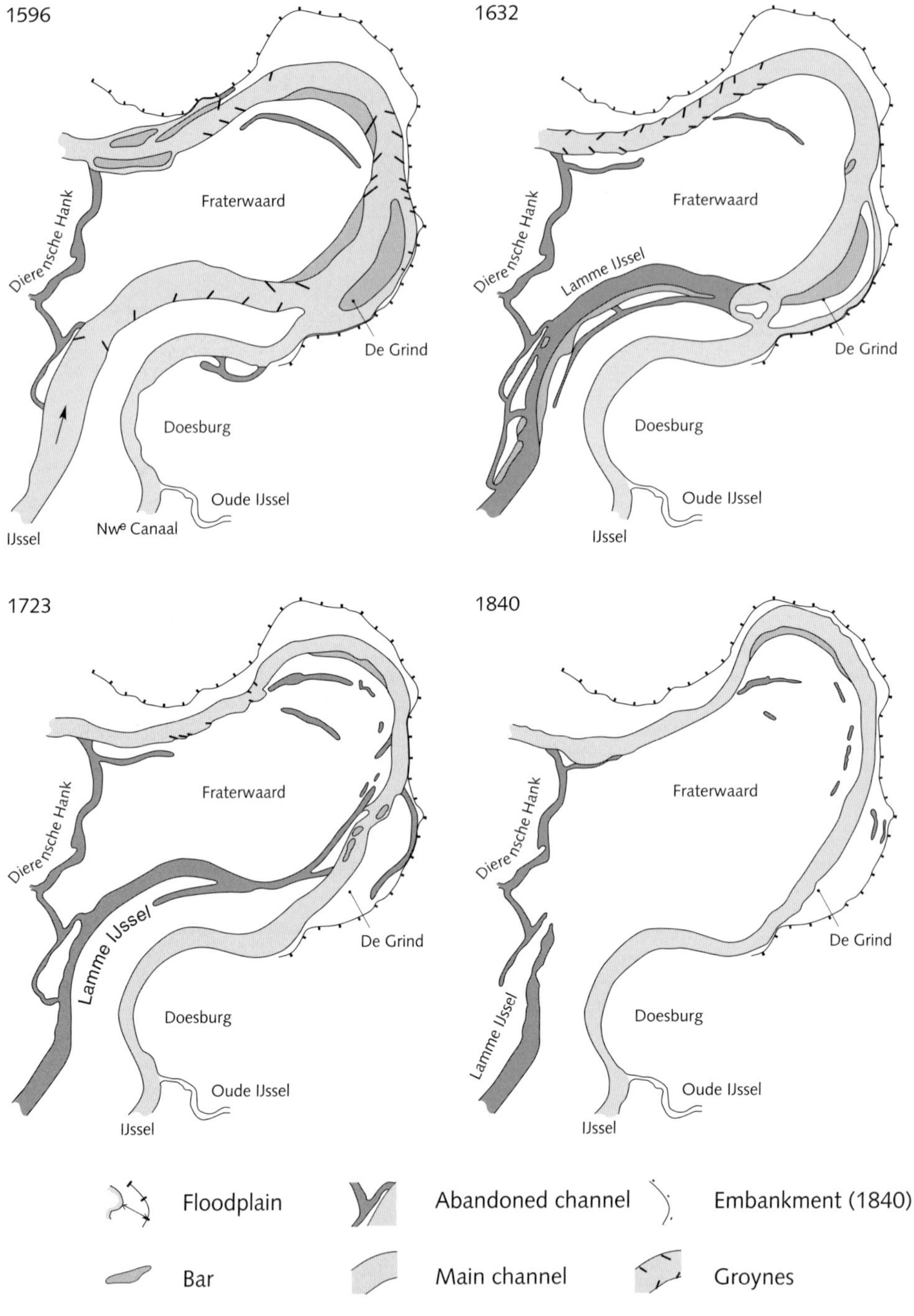

Fig. 8. Development of the River IJssel near Doesburg between 1596 and 1840. By damming the Lamme IJssel, the river was forced to flow by the city of Doesburg. In the Fraterwaard, lateral meandering caused ridge-and-swale topography. The mid-channel bar 'De Grind' evolved on the floodplain. The narrowing of the River IJssel since 1596 should be noted.

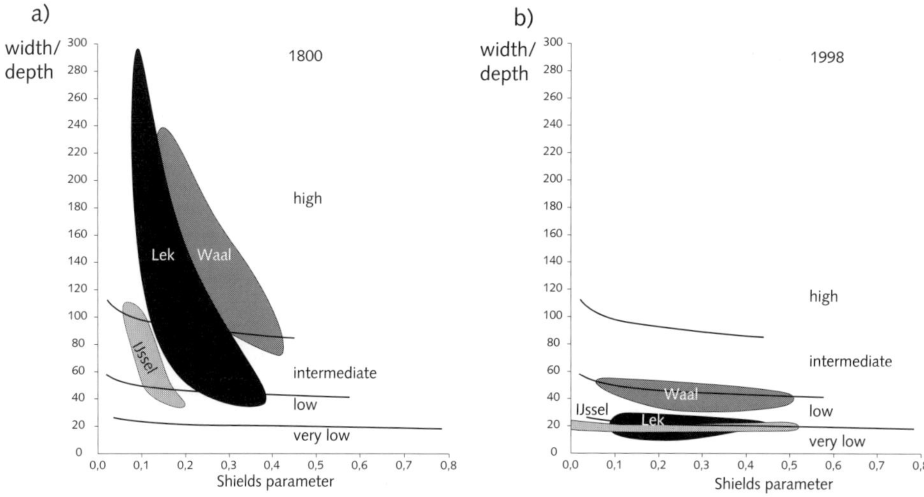

Fig. 9. Changing morphodynamics and classification diagram of the Rhine branches *c.* 1800 (**a**) and in 1998 (**b**).

and is related to sinuosity and the ratio of suspended load to bed load (Schumm, 1985). The Shields parameter describes the balance between the stream and the texture of the river bed. There was no information about the texture of the river bed in former centuries, so it has been assumed that it has not changed. The morphodynamics of a river result in the presence or absence of different kinds of bars; for example, point bars or mid-channel bars. The size, rate of growth and spatial damping of bar development can be estimated from mathematical equations for the movement of water and sand (Struiksma *et al.* 1985; Struiksma & Crosato 1989). The boundaries in the diagram

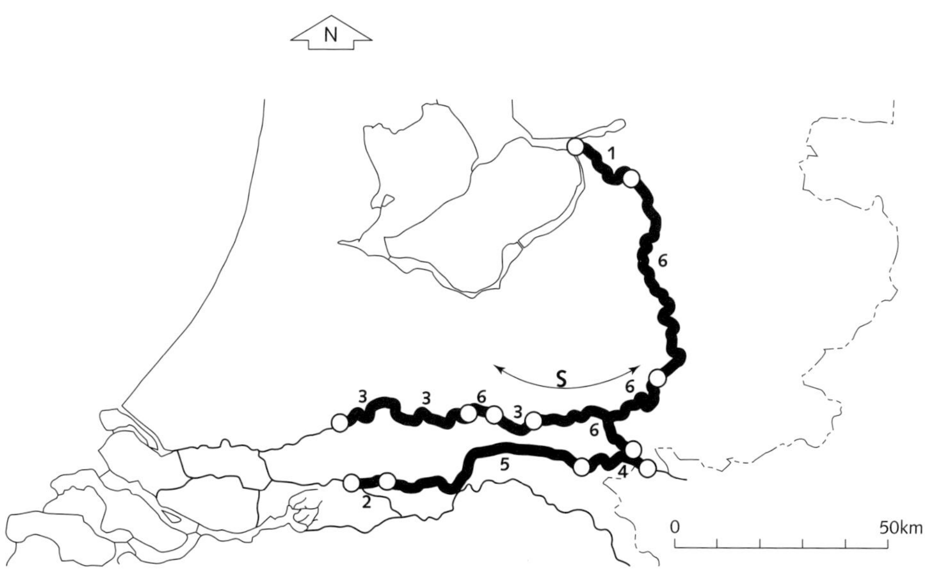

Fig. 10. Suitability of river reaches for ecological land units: 1, high point bars, oxbow lakes; 2, low point bars, silting-up abandoned channels, marshlands; 3, silting-up abandoned channels, marshlands; 4, high point bars, abandoned channels with open water; 5, low point bars, abandoned channels with open water; 6, oxbow lakes; S, abandoned channels fed with seepage water.

(Fig. 9) between high, intermediate, low and very low morphodynamics are derived from those estimates. These boundaries appear to be more strongly related to width–depth ratio than to the Shields parameter.

Figure 9 shows the classification diagram for the Rhine branches *c.*1800 and today. In 1800 mid-channel bars and extensive high point bars were common in the Waal and Lek; the morphodynamics of these distributaries were highly active. In the smaller River IJssel, mid-channel bars were lacking but high point bars were common; the morphodynamics were intermediate. Today it appears that all Rhine branches are characterized as having low or very low morphodynamics. Nevertheless, as shown in Fig. 9b, the River Waal still has possibilities for point bars to develop, although this is not the case for most parts of the rivers Lek and IJssel. During recent floods, the deposition of sand on the banks of the River Waal has been greater than on the banks of the rivers IJssel and Lek (Ten Brinke *et al.* 1998). This can also be explained by the higher width–depth ratio of the River Waal. The River Rhine is a good example of a passive meandering river, as defined by Thorne (1997). It is clear that the present-day river planform and the geomorphological patterns in the floodplains are largely the result of river regulation and former processes. This means that the patterns cannot be connected to present-day processes.

Use of the historical analyses has defined the preconditions for ecological units in floodplains. We focused on point bars, abandoned channels and marshlands. The classification diagram (Fig. 9) shows that islands and mid-channel bars were common in river reaches with a width–depth ratio (w/d) of >100 and that there are no river reaches in the Netherlands where they can develop today. The classification diagram also indicates that high point bars will develop only in river reaches with a width–depth ratio of 40 or more (intermediate morphodynamics). Low point bars can develop in river reaches with a width–depth ratio between 20 and 40 (low morphodynamics). In river reaches with a width–depth ratio <20 (very low morphodynamics) only very low point bars can develop. The boundary values of w/d are higher than those given by Ahmed (1986), who mentioned that a channel will braid when w/d exceeds 60 and it will remain straight when w/d is less than eight. Also, Rosgen (1994) mentioned lower boundaries for w/d at 40 and 12. These differences can be related to the regional, lowland character of the Rhine branches, as Ahmed (1986) and Rosgen (1994) investigated rivers in another spectrum of the river continuum, including gravel bed upland rivers.

Whether an abandoned channel is likely to silt up is dependent on flow velocity and sediment load in the floodplain during floods. The lower the flow velocity, the sooner silting up will take place. Seepage water can give the abandoned channels extra ecological opportunities, as the nutrient concentration of the seepage water is generally lower than that in the river itself. These conditions favour the development of characteristic species (Bornette *et al.* 1994).

Marshlands can develop in areas with low water depth and small water-level fluctuations. This follows from the analysis of water duration curves for the period around 1800. The preconditions for the development of point bars, abandoned channels and marshlands are combined with the actual fluvial processes from hydraulical models of the River Rhine. This resulted in a map indicating the suitability of river reaches for the development of ecological land units (Fig. 10). This map can be used in landscape planning and ecological rehabilitation of floodplains.

Conclusions

The major conclusions of this study are as follows.

(1) The ratio of width to depth is an adequate factor from which the morphodynamics of rivers may be estimated.

(2) Geomorphological patterns in the floodplains of the Dutch Rhine branches are not related to the present-day processes. In the most highly meandering branch, the IJssel, the opportunities for point bars, for instance, are less than in the low sinuosity River Waal.

(3) The fluvial processes are variable along reaches of the Dutch Rhine branches, but less so than they were in previous centuries.

(4) These differences between river reaches can be used to guide the process of floodplain rehabilitation so that optimal use of possibilities of the landscape will increase the ecological variability.

We would like to thank E. Mosselman and N. Struiksma of Delft Hydraulics for the suggestions on the classification diagram. A. Hesselink of Utrecht University was of great help in discovering the evolution of the floodplains of the River Lek.

References

AHMED, M. M. 1986. Meandering and braiding of rivers. *In*: JULIEN, P. Y. (ed.) *Essays on River Mechanics*. Colorado State University, Fort Collins, CO, 1–18.

BARNEVELD, H. J., HUYSKENS, R. B. H., VERMEER, K. & ZEEMAN, M. 1994. *Landscape Planning River Rhine; One Dimensional Model for Water Movement and Morphology of the Rhine Branches, SOBEK*. Verkeer en Waterstaat/Delft Hydraulics/ Geodan, IVR-rapport Nr. 6, Arnhem, The Netherlands. [in Dutch].

BOON, P. J. 1993. Essential elements in the case for river conservation. *In*: BOON, P. J., CALOW, P. & PETTS, G. E. (eds) *River conservation and management*. Wiley, Chichester, 11–33.

BORNETTE, G., AMOROS, C., CASTELLA, C. & BEFFY, J. L. 1994. Succession and fluctuation in the aquatic vegetation of two former Rhône River channels. *Vegetatio*, **110**, 171–184.

BRICE, J. C. 1975. *Air Photo Interpretation of the Form and Behaviour of Alluvial Rivers*. Final report to the US Army Research Office.

FIJNJE, J. G. W. 1850. Tables of highest, lowest and mean water level of the rivers Boven-Rijn, Neder-Rijn and Lek according the daily observations on the gauges in Keulen, Emmerik, Doornenburg (Pannerden), Arnhem, Vianen or Vreeswijk; from 1-1-1772 to 31-12-1849, converted into metres above sea level. *In*: FIJNJE, J. G. W. (ed.) *Register V, Containing the Description of the Gauges and other Marks along the Rivers Nederrijn and Lek from Pannerden to Vreeswijk and Vianen and along the Old Rhine*. Rijkswaterstaat, 's Gravenhage, Appendices III, IV and V [in Dutch].

—— & LEBRET, J. 1852. Tables of highest, lowest and mean water level of the river IJssel according the daily observations on the gauges in Westervoort, Doesburg, Zutphen, Deventer, Katerveer, Kampen and Paardenboer; from 1-1-1772 to 31-12-1850, converted into metres above sea level. *In*: FIJNJE J. G. W & LEBRET, J. (eds) *Register VIII, Containing the Description of the Gauges and other Marks along the River IJssel from Westervoort to the Zuiderzee*. Rijkswaterstaat, 's Gravenhage, Appendices III, IV &V [in Dutch].

LELY, C. 1890. Rivers and river works. *In*: HENKET, N. H., SCHOLLS, CH. M. & TELDERS, J. M. (eds) *Waterbouwkunde*, tweede deel, afdeeling XI. Van Cleef, 's Gravenhage [in Dutch].

LEOPOLD L. B. & WOLMAN, M. G. 1957. *River Channel Patterns; Braided, Meandering and Straight*. US Geological Survey, Professional Papers, **282B**.

MAKASKE, B. 1998. *Anastomosing rivers; forms, processes and sediments*. PhD Thesis, Utrecht University.

MIDDELKOOP, H. 1997. *Embanked floodplains in the Netherlands; geomorphological evolution over various time scales*. PhD Thesis, Utrecht University.

NANSON, G. C. & CROKE, J. C. 1992. A genetic classification of floodplains. *Geomorphology*, **4**, 459–486.

PEDROLI, G. B. M. & POSTMA, R. 1998. Nature rehabilitation in European river ecosystems: three cases. *In*: NIENHUIS, P. H., LEUVEN, R. S. E. W. & RAGAS, A. M. J. (eds) *New Concepts for Sustainable Management of River Basins*, Backhuys Publishers, Leiden, The Netherlands, 67–84.

ROSGEN, D. L. 1994. A classification of natural rivers. *Catena*, **22**, 169–199.

RUST, B. R. 1978. A classification of alluvial channel systems. *In*: MIALL, A. D. (ed.) *Fluvial Sedimentology*. Canadian Society of Petroleum Geologists, Calgary, Memoir, **5**, 187–198.

SCHUMM, S. A. 1977. *The Fluvial System*. Wiley, New York.

—— 1985. Patterns of alluvial rivers. *Annual Review of Earth and Planetary Science*, **13**, 5–27.

SILVA, W. & KERKHOFS, M. J. J. 1994. Ecological recovery of the river Meuse in the Netherlands. *Water Science and Technology*, **29**, 1–8

SIMONS, H. E. J., KOOLEN, J. L. & VERKADE, G. J. (eds) 1994. *Natuurvriendelijke oevers*. CUR, Gouda, rapport 168 [in Dutch].

STRUIKSMA, N. & CROSATO, A. 1989. Analysis of a 2-D bed topography model for rivers. *In*: IKEDA, S. & PARKER, G. (eds) *River Meandering*. American Geophysical Union, Water Resources Monograph, **12**, 153–180.

——, OLESEN, K. W., FLOKSTRA, C. & DE VRIEND, H. J. 1985. Bed deformation in curved alluvial channels. *Journal of Hydraulic Research IAHR*, **23**, 57–79.

TEN BRINKE, W. B. M., SCHOOR, M. M., SORBER, A. M. & BERENDSEN, H. J. A. 1998. Overbank sand deposition in relation to transport volumes during large-magnitude floods in the Dutch sand-bed Rhine river system. *Earth Surface Processes and Landforms*, **23**, 809–824.

THORNE, C. R. 1997. Channel types and morphological classification. *In*: THORNE, C. R., HEY, R. D. & NEWSON, M. D. (eds) *Applied Fluvial Geomorphology for River Engineering and Management*. Wiley, Chichester, 175–222.

VAN DEN BERG, J. H. 1995. Prediction of alluvial channel pattern of perennial rivers. *Geomorphology*, **12**, 259–279.

VAN DEN BRINK, F. W. B., VAN DE VELDE, G., BUIJSE, A. D. & KLINK, A. G. 1996. Biodiversity in the lower Rhine and Meuse river-floodplains: its significance for ecological river management. *Netherlands Journal of Aquatic Ecology*, **30**, 129–149.

VAN DER KUN, L.J.A. 1854. Tables of highest, lowest and mean water level of the rivers Waal and Merwede according the daily observations on the gauges in Hulhuizen, Nijmegen, Tiel, Bommel Gorinchem and Dordrecht from 1-1-1772 to 31-12-1856, converted to metres above sea level. *In*: VAN DER KUN, L. J. A. (ed.) *Registers I, II and III, Containing the Description of the Gauges and other Marks along the Rivers Boven-Rijn, Waal and Merwede from Emmerik to Gorinchem*. Rijkswaterstaat, 's Gravenhage, Appendices III, IV and V [in Dutch].

VAN DER VELDE, G. & VAN DEN BRINK, F. W. B. 1994. Does the Rhine still have characteristics of a river ecosystem? The longitudinal distribution of macroinvertibrates. *In*: VAN DE KRAATS, J. A. (ed.) *Water Science and Technology: rehabilitation of the River Rhine*, **29**(3), 1–8.

VERHOEFF, J. M. 1983. *Old Dutch measures and weights*. Meertens Institute for Research and Documentation of the Dutch Language and Culture, Amsterdam [in Dutch].

Somerset Levels and Moors: buying off the presumptive rights of landholders to manage the land as they see fit

P. J. O'DONOGHUE

Faculty of the Built Environment, University of the West of England, Frenchay Campus, Bristol BS16 1QY, UK

Abstract: Designations as a Site of Special Scientific Interest (SSSI) or as an Environmentally Sensitive Area (ESA) are both intended to contribute to wetland conservation in an important farming area of Somerset. The first is a compulsory scheme and was resisted strongly on its introduction, whereas the second is based on a voluntary regime. Although the relevant SSSIs are within the Somerset Levels and Moors ESA, the two schemes are not alternatives. Both involve restrictions on agricultural practice and on farming operations, and thereby have an effect on capital and revenue values. In many, but not all, cases occupiers or owners are compensated financially. Whether such compensation is adequate from the landholder's point of view, and whether the environmental benefits obtained as a result are equal in value to the cost of the schemes depends on how these values are measured. The future of both schemes is uncertain.

The farming community is often popularly associated with resistance to change, especially in areas where families have long benefited from the use of the same land. Richard Locke of Burnham-on-Sea, a progressive 18th-century farmer, reported graphically how he was 'stoned, bruised and beat by the mob' and his effigy burnt 'by the owners of geese' in reaction to his advocacy of enclosure and land improvement on the Somerset Levels and Moors (Locke 1796). None the less, enclosure and drainage went ahead despite the commoners' protest. The Levels and Moors are now almost entirely privately owned and form the largest area of grazed wetland in western Europe. Consisting mainly of raised bog or alluvium, the area has been subject to increasingly intensive farming in the 20th century, especially during and since the Second World War. Low-lying (mostly at or even below mean sea level) and protected from the sea only by a narrow coastal clay belt, the Levels and Moors (Fig. 1) are increasingly vulnerable to flooding, especially when heavy rainfall coincides with high tides and low barometric pressure. A complicated network of Internal Drainage Boards is responsible for draining agricultural land to pumping stations and to the main watercourses which flow into the Bridgwater Bay area of the Bristol Channel. Ministry of Agriculture, Fisheries and Food payments to support drainage and flood defence amount to about £3 million a year.

Intervention: imposed regulation

Ironically, the protest of the 18th-century commoners was mirrored by another burning of effigies on the Levels, which took place at Stathe in February 1983. The landholders' anger was directed at the Chairman of the Nature Conservancy Council and the Regional Officers of that Council and of the Royal Society for the Protection of Birds (Lowe *et al.* 1986) The protest was targeted at almost the reverse of the land improvements advocated two centuries earlier. Under the Wildlife and Countryside Act 1981 (which updated some relevant provisions of the National Parks and access to the Countryside Act 1949) the Nature Conservancy Council had designated West Sedgemoor as a Site of Special Scientific Interest (SSSI).

SSSI designation involves prohibitions on agricultural operations deemed potentially damaging to the physiographical or geological features of the prescribed area or to the natural or semi-natural fauna or flora, which make the site one of special scientific interest, and any operations that are considered likely to damage these special characteristics. Under the Countryside Act 1968 English Nature (formerly the Nature Conservancy Council) may enter into a management agreement where land is included in a notified SSSI. A management agreement will be offered most

From: MARRIOTT, S. B. & ALEXANDER, J. (eds) 1999. *Floodplains: Interdisciplinary Approaches*. Geological Society, London, Special Publications, **163**, 139–143. 1-86239-050-9/99/$15.00

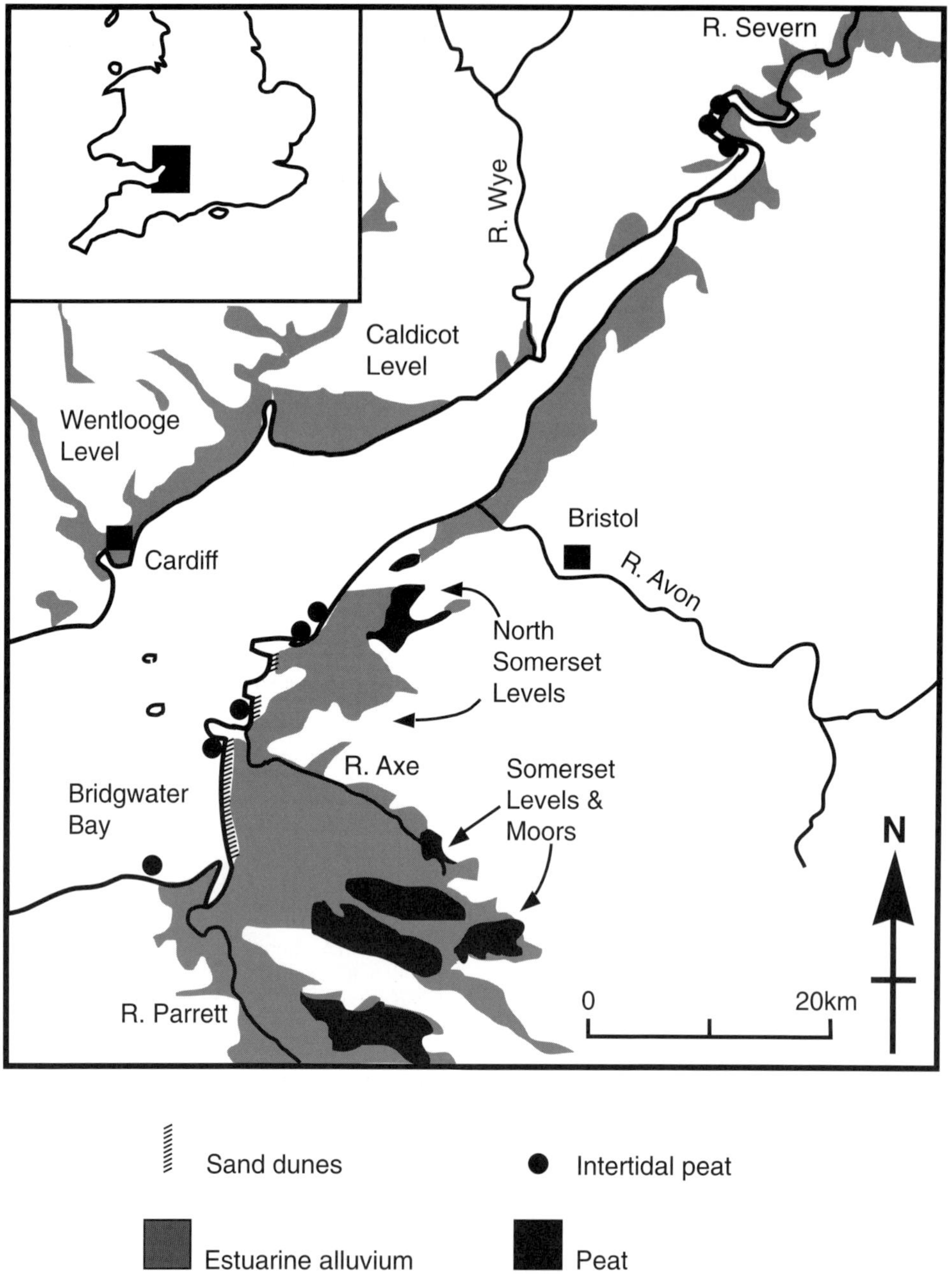

Fig. 1. The Severn Estuary, including the Somerset Levels and Moors (after Bell 1995).

usually when an owner or occupier advises English Nature of an intention to carry out operations notified within the order designating the land an SSSI. Once the notifying order has been made, the landholder is entitled to seek a management agreement from the Council at any time. Although not all SSSIs are covered by management agreements, these are often supported by payments intended to reflect loss of profit arising from restrictions on agricultural use of the land. Management agreements cover some 3–20 years.

In spite of strong reservations by the National

Farmers' Union and the Country Landowners' Association about many of the environmental provisions of the Wildlife and Countryside Act 1981 at the Parliamentary Bill stage, there are now 34 SSSI in the Somerset Levels, comprising 11 376 ha and including some former National Nature Reserves now automatically designated as SSSIs. There is a right of appeal against SSSI designation but a successful outcome is unlikely, and landholders generally remain unhappy at finding their land included in a designated site.

Nationally, around 60% of SSSI are in private ownership and only about 10% of these are the subject of management agreements involving payments to landholders. A major defect of the scheme is the cumbersome process of negotiating individual agreements. The Wildlife Enhancement Scheme was introduced in 1992 to simplify procedures by covering all SSSIs in a locality which share similar conditions and for which payments follow a similar pattern. This scheme has not yet been extended to the Somerset Levels and Moors, where the areas designated as SSSIs follow a patchwork pattern and in some cases are several miles away one from another. However, there are provisions under the Countryside Act 1968 (as amended by the Environment Protection Act 1990) for extending management agreements, and compensatory payments, to agricultural land adjacent to an SSSI but not within the designated area.

Intervention: a voluntary regime

Since the early 1980s countryside management in the agricultural context has come to be based less on regulation and more on a principle by which farmers and landowners voluntarily agree to a more extensive farming pattern and to constraints on production methods in return for compensatory payments. More recently, too, there has been a growing emphasis on positive management rather than on the negative aspect of payment for 'not doing something'.

In an attempt to preserve their wetland character by protecting flora and fauna and by maintaining a higher water table in drier times of the year, without going to the length of an SSSI designation, most of the Somerset Levels and Moors were identified as an Environmentally Sensitive Area in 1987 under the Agriculture Act 1986 (reflecting Article 19 of European Community Regulation 797/85). Section 17 of the Act marks a significant change in direction in countryside management. It lays down that 'in discharging any functions connected with agriculture ... the Minister of Agriculture is under a duty to achieve a reasonable balance between a number of considerations: (1) the promotion and maintenance of a stable and efficient agricultural industry; (2) the economic and social interests of rural areas; (3) the conservation and enhancement of the natural beauty and amenity of the countryside (including its flora and fauna and geological, physiological and archaeological features); and (4) the promotion of the enjoyment of the countryside by the public'.

With revisions effected in 1992, the Somerset Levels and Moors ESA comprises nearly 28 000 ha. Of this, 1027 voluntary agreements cover 15 942 ha, representing almost 58% of the eligible area. The cost of the ESA scheme in the Levels is about £2.5 million per annum. The scheme involves tiered payments to landholders commensurate with the agricultural profits forgone and the environmental benefits arising from the maintenance of raised water tables and from the avoidance of farming operations at critical times (Terry & Case 1994).

Constraints on the voluntary regime

Given that some land within the ESA is also designated as an SSSI with more restrictions than would be the case with an ESA alone, some farmers are reluctant to enter into an ESA agreement on the grounds that this may lead to SSSI designation of the land with the more onerous compliance requirements involved.

A major problem with the Somerset Levels and Moors ESA scheme is that agreements are generally for 10 years and most will expire in 2002. Although agreements are binding on anyone who succeeds to or takes over the land during the life of the agreement, there is no obligation on an owner or occupier to enter into a new agreement thereafter. Although there is scope for a break without penalty at an interval of 5 years, withdrawal from an ESA agreement can be effected with relatively little notice, subject to repayment of the grant already received.

The patchwork pattern already noted in connection with SSSIs is repeated to an even greater extent in the case of land covered by voluntary agreements under the ESA scheme. Although standard terms of all such agreements are laid down in the Statutory Instrument (Government Order) designating an Environmentally Sensitive Area, there is no obligation on an occupier of agricultural land within the area to enter into a management agreement. Further, some land included in an ESA agreement may be within a holding under the same ownership where there are areas not included in the scheme. One effect is that a drained area outside the scheme may be immediately adjacent to land where the water table has to be maintained at an ecologically satisfactory level within the relevant terms of the management agreement. This situation can be aggravated by the withdrawal of different landholders or parcels of land from the scheme or by the inclusion of others not previously covered by it.

Conflicting values

Reconciliation of the use of land for ordinary farming purposes with the other considerations listed in the Agriculture Act 1986 section 17 (above) involves a range of arguments, many of which revolve around economic factors (Turner & Jones 1991), although not the least are political (Winter 1996).

Although a significant number of landholders are themselves enthusiastic in support of environmental protection, farmers look on their land primarily as 'real' property, a tangible and marketable productive asset to be employed for earning profits. Conservationists, on the other hand, attach a different kind of value to the existence of environmental goods. Although clearly it has value, an environmental asset is usually not of itself a source of financial revenue and is difficult or almost impossible to value in money terms as it is not traded in a market which could provide price benchmarks. Conflicts of opinion as to value arise therefore when the agriculturally productive asset and the environmental asset are on the same land; for example, when the habitat of a rare butterfly is threatened by its proximity to intensively grazed grassland. A more complex scenario can arise when, as research in SW England has shown often to be the case, landscapes of high conservation value are found in poorer farming areas from which intensive output is needed to yield adequate income (Battershill & Gilg 1996). How can the question be resolved objectively as to which 'value' should be subordinated to the other?

All economic values are related to the willingness of individuals to pay for benefits or for avoiding costs. But measuring values is more difficult where we wish to attach money values to unpriced assets, that is, to those assets such as environmental benefits for which there is no market and for which therefore there is no scale of prices. Various techniques have been developed (Turner *et al.* 1994) The contingent valuation method revolves primarily around the concept of eliciting from respondents by structured survey their 'wtp' ('how much are you willing to pay for an environmental benefit?') or 'wta' 'how much are you willing to accept in compensation for the loss of a benefit?'). A similar technique, the stated preference method, is based on a survey of respondents' rankings of environmental benefits or losses.

A major drawback of these techniques is the difficulty of verifying or testing responses in the market. A more straightforward approach is the replacement cost method whereby quite simply the cost of replacing a damaged environment is measured. Similarly, the cost of achieving or sustaining an environmental standard or benefit can be estimated. Benefits and costs become harder to identify and quantify when inter-generational considerations come into play.

'Buying off' the landholder

'Compensation' payments to landholders under the SSSI and ESA schemes for ecologically orientated land management are founded on a wtp basis in the sense that Government on behalf of the public has decided on a level of payment for the provision of environmental benefits by landholders. On another view, they are based on a version of the principle of wta, 'how much are you willing to accept' in exchange for refraining from maximizing profits, by agreeing to constraints on land management and farm output. Arguably, this implies that landholders are paid a form of rent for the preservation of environmental benefits on their land. It could be said therefore that we 'lease' these benefits from the property owner or occupier. A lease is an agreement freely entered into between grantor and grantee for a period of time. The analogy with a lease, therefore, may be applicable to management agreements under the ESA scheme to the extent that they are made voluntarily between the Ministry of Agriculture, Fisheries and Food and the landholder.

A complexity lies in the calculation of capital and productive values of land affected by environmental designations. Local land agents consider that land vulnerable to flooding as a result of inclusion in an ESA agreement is worth about £1600–£2500 per hectare whereas other land of the same quality in the same area may be worth upwards of £3000 per hectare. At the same time, it is almost undoubtedly true that grazing land rental values on the Somerset Levels and Moors are depressed by current problems with BSE, but recovery in the beef market or a rise in milk prices could radically alter farmers' perceptions of wta, that is, of what constitutes reasonable payment for abstaining from intensive production. This could mean that 'rent' for the 'lease', or the market price of land carrying environment benefits will undergo substantial change, whereas the objective value of these benefits as 'unpriced assets' remains more or less stable.

Under English law, continuity of environmental protection can be maintained by landowners adopting agreements in return for compensatory payments so that the obligation of ecological conservation passes by covenant with the land to subsequent owners. Covenants may be permanent or for a specific period. Any covenant can be varied or discharged by agreement between both parties who are affected by it (even if not the original parties to the agreement) (Hodge *et al.* 1993; Dwyer & Hodge 1996). Against this is an objection similar to that arising from an ESA agreement, namely, the possibility that burdening the land with such covenants depreciates its capital value to a greater extent than the capitalized return from future compensatory payments. The financial inducement to enter into such a covenant would need to be strong.

Another possibility is for land purchase, perhaps by county Wildlife Trusts. In the longer term this could lead not only to securing the environmental benefit but also to a saving of public money. A grant of £100 000 for land purchase to Essex Wildlife Trust in the early 1990s was reported to be expected to lead to a saving of £600 000 in payments under management agreements (National Audit Office (NAO) 1994).

Conclusions

It is strongly evident that the farming community has proved to be more amenable to the voluntary regime embraced by the ESAs scheme than to the rather more coercive system involved in SSSI designation. On the other hand, the ecological objectives of SSSI designation are more far-reaching than those of the ESA system, hence the inclusion of SSSIs within the ESA. The two regimes are complementary and cannot be regarded as mutually exclusive. The extended life of the ESA scheme may benefit from the fact that, although the concept originated in England (in East Anglia), it represents an extension of European agricultural legislation under the Common Agricultural Policy.

The problem of the continued existence or otherwise of constraints on environmentally damaging wetland management may be overcome by covenants or even by outright purchase by groups engaged in countryside conservation such as wildlife trusts.

References

BATTERSHILL, M. R. J. & GILG, A. W. 1996. Environmentally friendly farming in southwest England: an exploration and analysis. *In*: CURRY, N. & OWEN, S. (eds) *Changing Rural Policy in Britain: Planning, Administration, Agriculture and the Environment*. Countryside and Community Press, Cheltenham.

BELL, M. 1995. Archaeology and nature conservation in the Severn Estuary. *In*: COX, M., STRAKER, V. & TAYLOR, D. (eds) *Wetlands: Archaeology and Nature Conservation*. Proceedings of the International Conference, Wetlands, Archaeology and Nature Conservation, University of Bristol, 11–14 April 1994 HMSO, London.

DWYER, J. C. & HODGE, I. D. 1996. *Countryside in Trust: Land Management by Conservation, Recreation and Amenity Organisation*. Wiley, Chichester.

HODGE, I., CASTLE, R. & DWYER, J. 1993. *Covenants as a Conservation Mechanism*. Department of Land Economy, University of Cambridge, Monograph, **26**.

LOCKE, R. 1796. Historical account of the marsh-land of the County of Somerset. *Letters and Papers, Bath and West of England Society*, **VIII**, 267–283.

LOWE, P., COX, G., MACEWEN, M., O'RIORDAN, T. & WINTER, M. 1986. *Countryside Conflicts: the Politics of Farming, Forestry and Conservation*. Gower, Aldershot.

NAO 1994. *Protecting and Managing Sites of Special Scientific Interest in England*. Report by the Comptroller and Auditor General of the National Audit Office. HMSO, London.

TERRY, A. & CASE, D. 1994. Management of the Somerset Levels and Moors. *Geography Review*, **8**(2), 13–17.

TURNER, K. & JONES, T. 1991. *Wetlands: Market and Intervention Failures*, Earthscan Publications, London.

TURNER, R. K., PEARCE, D. & BATEMAN, I. 1994. *Environmental Economics: an Elementary Introduction*. Harvester Wheatsheaf, London.

WINTER, M. 1996. *Rural Politics: Policies for Agriculture, Forestry and the Environment*. Routledge, London.

Variations in the quality of the thatching reed *Phragmites australis* from wetlands in East Anglia, England

R. R. BOAR[1], J. J. H. KIRBY[2] & D. J. LEEMING[3]

[1]*School of Environmental Sciences, University of East Anglia, Norwich NR4 7TJ, UK*

[2]*Biological Sciences, University of Bath, Claverton Downs, Bath BA2 7AY, UK*

[3]*The Environment Agency, Thames Region, Waltham Cross EN8 8HE, UK*

Abstract: *Phragmites australis* (Cav.) Trin. ex Steud. (common reed) is an emergent grass that grows in a wide belt across the wetlands of North America and Europe. In many European wetlands, dead stems of *P. australis* are harvested annually to supply the roof thatching industry. This study has measured and compared the quality, in terms of mechanical strength and decay rate, of a reed harvest from 12 marshes in the floodplains of the Rivers Bure, Ant, Thurne Waveney and Blythe in East Anglia, UK. The marshes varied from fresh-water peat to marine alluvium and were in floodplains that have been managed traditionally to supply thatching reed. The decay rate of stems varied more within the floodplains than between them. Mechanical properties, however, varied between the different floodplains, with stems of greatest tensile strength from brackish reedbeds not flooded directly by sea water. As neither the mechanical properties, nor the shape and size of reed stems varied with their decay rate, the quality of roof thatch originating from particular growing areas cannot be predicted easily.

The common reed (*Phragmites australis*) is a tall and very productive grass that grows in most natural and semi-natural floodplains in lowland Britain (Haslam 1972), Europe (Van der Putten 1997), Scandinavia (Weisner 1990) and northern America (Shay & Shay 1986). The ecological value of reedbeds may be high because of the role of perennial plants, such as reed, in storage and cycling of nutrients (Howard-Williams 1985), metals (Ye *et al.* 1997), water and particulate materials that might otherwise move quickly from disturbed catchments to open waterways.

Reed-dominated wetland is also valued as landscape and as wildlife habitat, particularly for birds (Bibby & Lund 1982), and reedbed often forms a substantial part of lowland nature reserves. Of the 5000 ha of reedbed that remains in Britain, Painter *et al.* (1995) estimated that around 90% falls under at least one conservation designation. Of this area, *c.* 65% is managed to provide an annual or biannual crop of reed stems to supply the roof thatching industry (Bateman *et al.* 1991).

Management of wetlands to produce thatching reed involves regular wintertime harvesting of standing dead stems and careful maintenance of high water levels and effective drainage during the plants' above-ground growing season (Hawke & José 1996). The effect of this management is to prevent succession of reedbeds to drier, more terrestrial and, in conservation terms, much less desirable communities of trees and shrubs. The maintenance of over half of the wet reedbed in British floodplains depends therefore on its commercial value for thatching reed. This commercial value depends partly upon the quality of the harvest and different growing areas may have very different reputations for reed quality.

The main objective of this study was therefore to measure reed quality in a range of lowland floodplains and to discover also if any mechanical quality or the shape and size of stems might act as a reliable predictor of the resistance of reed stems to decay within thatch. The ultimate contribution of the work is toward maintaining or improving reed quality so that commercial management of habitats continues to maintain reed-dominated wetland. The work forms part of a wider project on the growing conditions of commercial reed (Boar *et al.* 1991) and of the economics of the reed growing and thatching industries in Britain (Bateman *et al.* 1991).

Methods

Five freshwater peat and seven brackish marshes on marine alluvium were chosen to span gradients of both

From: MARRIOTT, S. B. & ALEXANDER, J. (eds) 1999. *Floodplains: Interdisciplinary Approaches*. Geological Society, London, Special Publications, **163**, 145–151. 1-86239-050-9/99/$15.00

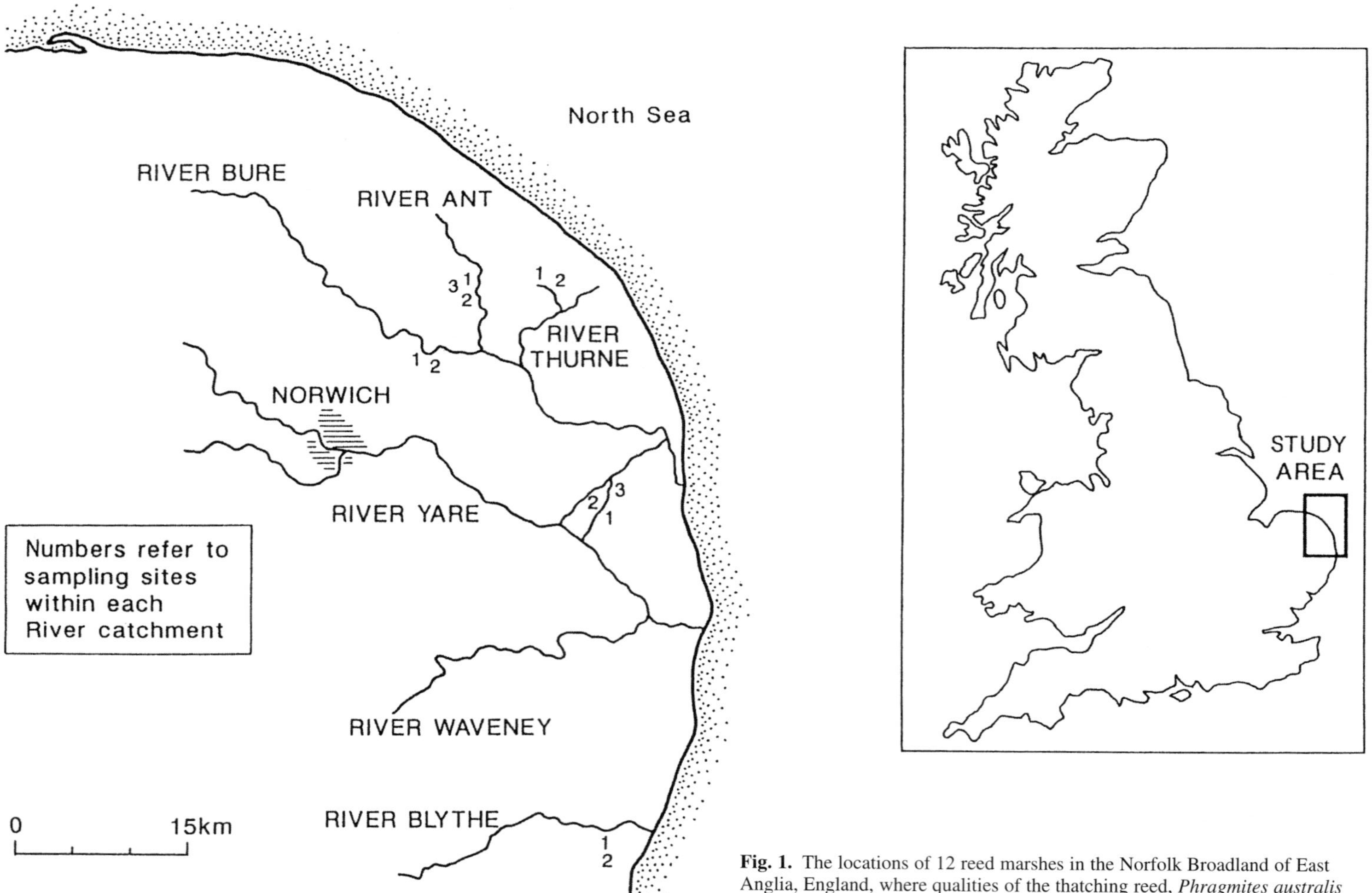

Fig. 1. The locations of 12 reed marshes in the Norfolk Broadland of East Anglia, England, where qualities of the thatching reed, *Phragmites australis* have been measured.

fertility and salinity. The chlorinity of standing water on the marshes varied between about 0.2% and 20% of the chloride ion concentration in sea water, and in January the mean ± SD (n = 3) total phosphorus content of their surface sediment varied between 0.19 ± 0.10 % and 0.63 ± 0.35 %. The 12 marshes all have a history of commercial reed growing and fell within the lower floodplains of the Rivers Bure, Ant, Thurne, Waveney and Blythe (Fig.1). These are five of the largest rivers in the East Anglian region of southeast England. The reedbeds were all dominated by *Phragmites* and varied in area from 0.96 ha to 14.35 ha.

During January 1989, standing dead stems from the 12 marshes were cut at sediment level from ten replicate 0.5 m × 0.5 m quadrats positioned at random in each marsh. In each quadrat, the stems were counted to give values for stem density. The heights, basal diameter and incidence of disease or infection in the stems were recorded. Stems were dried at 60°C for 36 h and weighed to give values for above-ground biomass. Before drying, sub-samples of ten stems from each quadrat sample were prepared for measurement of their ultimate tensile strength (UTS). This involved mounting stem sections cut from the internode closest to 11 cm from the stem base on card with epoxy resin and then equilibration at 65% humidity for 5 days. UTS was measured using an Instrom 1192 tensionometer. The cross-sectional area, width of the stem wall (w) and breadth (a) at the point of fracture were recorded, and UTS was calculated as the force applied at fracture divided by the cross-sectional area of the sample (MN m^{-2}). Strain was calculated as the linear extension of the sample divided by the length under test. Width (w) of the stem wall was also taken as a measure of reed quality. Stem flexibility was measured as Young's modulus, which was calculated as UTS/strain. Details of the methods have been given by Kirby & Rayner (1989).

The rate of decomposition of reed from nine of the same ten quadrats was measured in a laboratory under controlled temperature and water potential. A 400 ± 4 mg macerate of *Phragmites* from a patch of decay taken from a thatched roof was added to tubes containing 2 ± 0.1 g air-dry weight of dead reed stem. Two replicate tubes were incubated at 20°C. A fairly constant water potential of –0.45 M Pa was maintained for 100 days and material was then weighed to the nearest 0.1 g. Stem decomposition was recorded as the loss in dry weight over a 100 day incubation after inoculation with the decomposer organisms.

Results and discussion

Losses in dry weight over the 100 day incubations varied between 3.8% and 11.9% of initial stem weights (Fig. 2a), and mean values per marsh ranged between 6.89 ± 0.98% and 8.48 ± 0.77% loss in weight. The variance around mean decomposition in individual marshes was therefore large and there were no differences between marshes nor between marshes grouped into the five river catchments. This means that the resistance of stems to decay by decomposers of roof thatch from a particular floodplain cannot be predicted. Should the rate at which stems decay depend upon their nutrient contents, then nutrient content would appear to be unrelated to the overall nutrient status of the growing area. Stem nutrient content may instead be related to much finer-scale variations in nutrient availability (Reddy *et al.* 1989) or nutrient storage (Kühl & Kohl 1992; Boar 1996; Kühl *et al.* 1997) that may occur within individual marshes.

The pattern of variation in ultimate tensile strength of stems was different from decay rate in that there were significant differences between marshes although not necessarily in the strength of stems from different floodplains (Fig. 2b). Marshes fell into three groups (Fig. 3): the weakest stems were from a brackish marsh on the River Waveney and from three freshwater marshes flooded by the River Ant; an intermediate group from brackish marshes in the Thurne catchment, two freshwater reedbeds on the River Bure and one brackish site on the Waveney and the strongest stems were from brackish marshes on the Waveney and the River Blythe. The weakest stems were thus from either freshwater or brackish sites, although the strongest were only from brackish marshes. This is plausible because coastal marshes, where sediments tend to contain and retain potassium, are likely to produce strong reed because of the essential role of potassium in the biochemistry of lignin synthesis. Mechanical strength in reed may therefore, to an extent, be predicted from the growing area. However, this may not help thatchers select good quality reed if strong reed does not also resist decay.

Other measurements of reed quality are summarized as mean values in Table 1. Coefficients of variation ((mean/SD) × 100) tended to be around 20% of the mean in all of the growing areas and were similar for the different measures of reed quality. Young's modulus (E), the measure of stem flexibility did not, for example, vary between floodplains nor between individual marshes. Comparison of values of E from the three marshes on the River Ant showed that flexibility can vary (P <0.05) between neighbouring marshes influenced by the same river and underlain by similar sediment. With the exception of ultimate tensile strength, the general pattern in the reed quality was therefore of variations within marshes that were larger than variations between the river catchments.

Relationships between biomass, height, width and density of reed stems are documented well (Haslam 1969; Kvet 1971; Van der Toorn & Mook 1982). Relationships (P < 0.001) between biomass and stem height (R = 0.60) and diameter (R = 0.54), height and diameter (R = 0.85) and (P < 0.001) between stem density, height (R = –0.447) and diameter (R = –0.515) were strong. Associations between the size of the dead stems and their strength were generally weaker than those found in

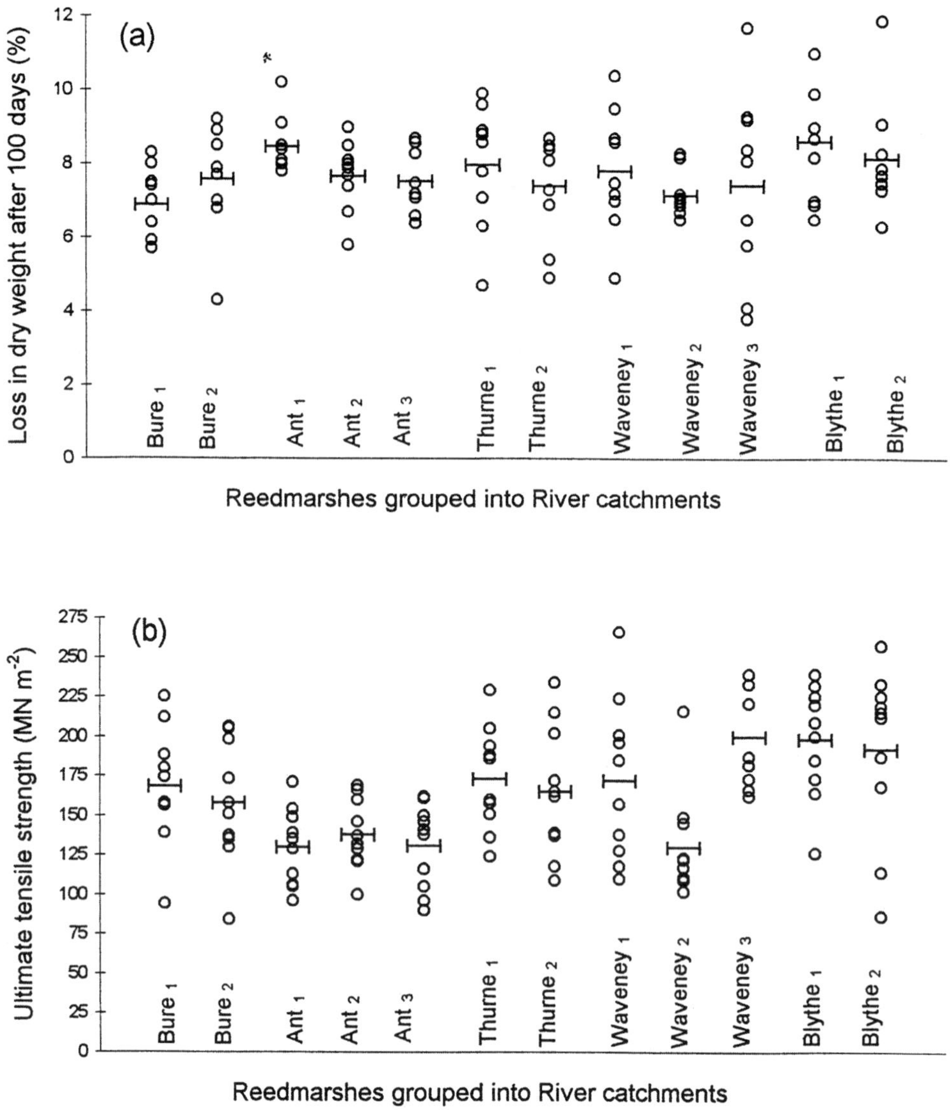

Fig. 2. (**a**) Losses in initial dry weight after laboratory incubation of thatching reed (*Phragmites australis*) from 12 marshes in five East Anglian river catchments. Horizontal bars show the mean weight loss for each marsh (n = 9). (**b**) The ultimate tensile strength of thatching reed (*Phragmites australis*) sampled during January from 12 marshes in five East Anglian river catchments. Horizontal bars show the mean weight loss for each marsh (n = 9).

Dutch stems sampled in August by Coops & Van der Velde (1996). Coops found that the bending stiffness of *P. australis* stems was less in February than in August, and presumably relationships with stem size would also have been weaker for the dead stems.

The rate at which dead stems decomposed was not related to their initial tensile strength (R = –0.04, n = 110) (Fig. 4). An explanation for this may be that although particular ratios of nutrients or major ions may give structural advantages to reed, the role of these ions in an eventual substrate for decomposers may enhance decay. A multiple regression of stem decay against other quality variables (tensile strength, Young's modulus, stem-wall width, internal cavity, stem basal diameter, height and density) accounted for only 2% of the variation in the rate of stem decomposition. None

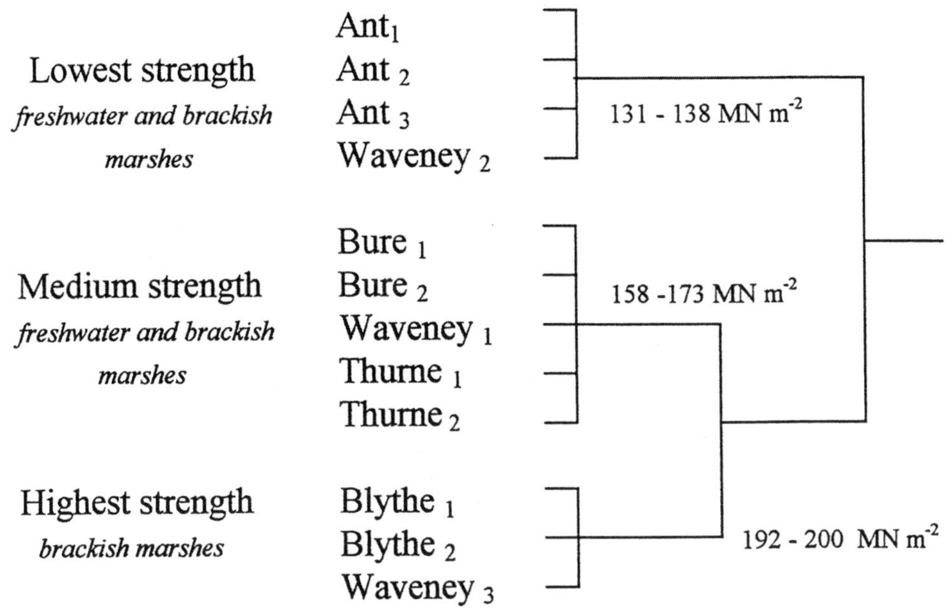

Fig. 3. Commercial reedbeds in East Anglia, UK, clustered according to the ultimate tensile strength (UTS) of samples of stems of the reed *Phragmites australis* (n = 110 stems). Cluster analysis is based upon the average linkage between groups, and the values shown are the ranges of UTS (MN m^{-2}) in the three clusters.

of the t ratios for the predictors exceeded the 95% confidence level.

It is surprising that the rate of stem decay did not vary with any of the other measures of reed quality. This indicates clearly that the rate at which stems are likely to decay on thatched roofs may not be predicted from their strength nor from obvious features of their size or morphology. Further analysis where each of the 12 marshes was dealt with separately revealed that there were no consistent patterns in any of the sites that linked the decay rate of reed with its mechanical properties.

Conclusions

Strongest stems were often, although not invariably, from brackish marshes. The mean chloride ion

Table 1. *Measures of the mean* (n = *10) size, biomass and mechanical properties of thatching reed,* Phragmites australis, *sampled during January 1989 from 12 marshes and five river catchments in East Anglia, UK*

	Decay (%)	Disease (%)	Biomass (g m^{-2})	Density (m^{-2})	Diameter (mm)	Height (cm)	Strength (MN m^{-2})	Width (mm)	Flexibility (E)
$Bure_1$	6.89	2.0	860	302	3.94	172	168	0.48	6021
$Bure_2$	7.58	0.5	924	312	4.10	168	158	0.46	5509
Ant_1	8.48	3.5	910	280	4.34	170	130	0.51	5917
Ant_2	7.67	6.0	734	390	3.51	136	138	0.41	5547
Ant_3	7.54	3.0	812	349	3.69	136	131	0.45	4795
$Thurne_1$	7.99	4.5	959	468	3.53	133	173	0.47	5932
$Thurne_2$	7.41	3.0	744	309	3.85	158	165	0.51	6169
$Thurne_3$	7.82	26.5	488	325	3.08	115	172	0.44	5800
$Waveney_1$	7.16	20.0	390	402	2.61	104	130	0.34	6434
$Waveney_2$	7.43	3.0	892	322	3.84	162	200	0.47	5919
$Blythe_1$	8.62	1.0	745	287	3.90	168	198	0.44	6312
$Blythe_2$	8.16	1.5	684	365	3.67	158	192	0.47	6733

In all cases, variance was less than 20% of the mean value.

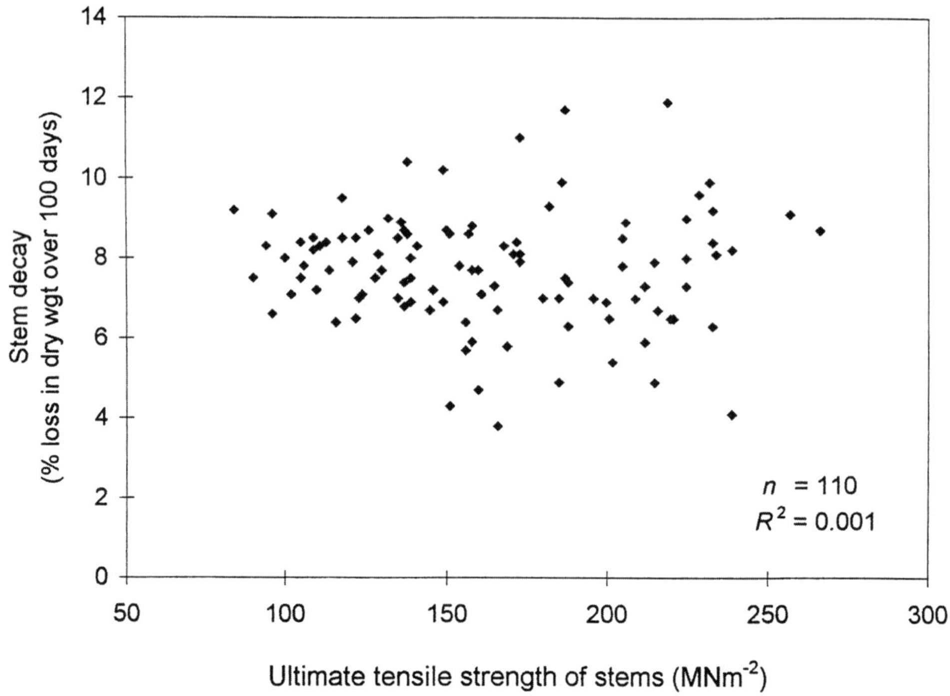

Fig. 4. The relationship between the ultimate tensile strength of thatching reed (*Phragmites australis*) from 12 marshes in East Anglia and its decay rate after laboratory incubation.

concentration in the flood water of such marshes in March varied between 212 ± 88 mg l^{-1} and 830 ± 472 mg l^{-1}, which is about 5% of the concentration in sea water. A connection between estuarine growing areas and mechanical strength is consistent with observations of British reed collected over at least 10 years by Haslam (1995). However, the ultimate tensile strength of dead stems did not vary with their resistance to decay nor with measures of stem size that were independent from tensile strength. The significance of this is that mechanical strength of reed is not a reliable guide to its long-term performance in reed thatch. Those involved in commercial reed growing, buying and thatching have long used reed strength as a major predictor of its overall quality. The results of this work contradict this.

As the rate at which stems decayed did not relate to particular floodplains or areas within them, particular areas need not be taken out of commercial production. This has important and positive implications for conservation in areas where wetland is maintained only by annual cutting and flooding to produce thatching reed. This applies particularly to the lowland, largely agricultural floodplains of Britain and western Europe.

Effort may now be directed toward discovering if reedbed management can improve the resistance of reed to decay and reduce the large variation of quality that we have found within individual marshes.

This work forms part of a larger project funded by the Department of the Environment with subventions from the Broads Authority, English Nature, the Rural Development Commission and the Ministry of Agriculture, Fisheries and Food.

References

BATEMAN, S., TURNER, R. K. & BATEMAN, I. J. 1991. *Socio-economic Impact of Changes in the Quality of Thatching Reed on the Future of the Reed Growing and Thatching Industries and on the Wider Rural Economy.* Rural Development Commission, London.

BIBBY, C. J. & LUND, J. 1982. Conservation of reedbeds and their avifauna in England and Wales. *Biological Conservation*, **23**, 167–186.

BOAR, R. R. 1996. Temporal variations in the nitrogen content of *Phragmites australis* in a shallow fertile lake. *Aquatic Botany*, **55**(3), 171–181.

——, LEEMING, D. J. & MOSS, B. 1991. *The Quality of Thatching Reed and its Impact on the Thatching Industry: the Links between Water and Sediment*

Chemistry and Reed Quality. English Nature, Peterborough.

COOPS, H. & VAN DER VELDE, G. 1996. Effects of waves on helophyte stands: mechanical characteristics of stems of *Phragmites australis* and *Scirpus lacustris*. *Aquatic Botany* **53**(3–4), 175–185.

HASLAM, S. M. 1969. The development of shoots of *Phragmites communis* Trin. *Annals of Botany*, **33**, 695–709.

—— 1972. Biological flora of the British Isles. *Phragmites communis* Trin. *Journal of Ecology*, **60**, 585–610.

—— 1995. *A Discussion of the Strength (Durability) of Thatching Reed* (Phragmites australis) *in Relation to Habitat*. Department of Botany, University of Cambridge, Cambridge.

HAWKE, C. J. & JOSÉ, P. V. 1996. *Reedbed Management for Commercial and Wildlife Interests*. Royal Society for the Protection of Birds, Sandy, Bedfordshire.

HOWARD-WILLIAMS, C. 1985. Cycling and retention of nitrogen and phosphorus in wetlands: a theoretical and applied perspective. *Freshwater Biology*, **15**, 391–431.

KIRBY, J. J. H. & RAYNER, A. D. M. 1989. *Aspects of the Decomposition, Mechanical Strength and Anatomy of the Water Reed* (Phragmites australis) *Used in Thatching*. Report to the Department of the Environment, the Rural Development Commission and the Ministry of Agriculture Fisheries and Food. University of Bath, Bath.

KÜHL, H. & KOHL, J.-G. 1992. Nitrogen accumulation, productivity and stability of reed stands (*Phragmites australis* (Cav.) Trin. ex Steudel) at different lakes and sites of the Lake Districts of Uckermark and Mark Brandenburg (Germany). *Internationale Revue der Gesamten Hydrobiologie*, **77**(1) 85–107.

——, WOIKE, P. & KOHL, J.-G. 1997. Strategies of nitrogen cycling in *Phragmites australis* at two sites differing in nutrient availability. *Internationale Revue der Gesamten Hydrobiologie*, **82**(1), 57–66.

KVET, J. 1971. Growth analysis approach to the production ecology of reedswamp plant communities. *Hydrobiologia*, **12**, 15–40.

PAINTER, M., SMITH, K. & GILBERT, G. 1995. *An Inventory of British Reedbeds 1993*. Conservation Sciences Report, Royal Society for the Protection of Birds, Sandy.

REDDY, K. R., PATRICK, W. H., JR & LINDAU, C. W. 1989. Nitrification–denitrification at the plant root–sediment interface in wetlands. *Limnology and Oceanography*, **34**(6), 1004–1031.

SHAY, J. M. & SHAY, C. T. 1986. Prairie marshes in western Canada, with specific reference to the ecology of five emergent macrophytes. *Canadian Journal of Botany*, **64**, 443–454.

VAN DER PUTTEN, W. H. 1997. Die-back of *Phragmites australis* in European wetlands: an overview of the European Research Programme on Reed Die-Back and Progression. *Aquatic Botany*, **59**, 263–275.

VAN DER TOORN, J. & MOOK, J. H. 1982. The influence of environmental factors and management on stands of *Phragmites australis*. I Effects of burning, frost and insect damage on shoot density and shoot size. *Journal of Applied Ecology*, **19**, 477–499.

WEISNER, S. E. B. 1990. *Emergent vegetation in eutrophic lakes: distributional patterns and ecophysiological constraints*. PhD thesis, University of Lund, Lund, Sweden.

YE, Z. H., BAKER, A. J. M., WONG, M. H. & WILLIS, A. J. 1997. Zinc, lead and cadmium tolerance, uptake and accumulation by the common reed, *Phragmites australis* (Cav.) Trin. ex Steudel. *Annals of Botany*, **80**(3), 363–370.

Island topography mapping for the Brahmaputra–Jamuna River using remote sensing and GIS

A. HASSAN[1], T. C. MARTIN[1] & E. MOSSELMAN[2]

[1]*Environment and GIS Support Project for Water Sector Planning (EGIS), House 49 Road 27, Banani, Dhaka, Bangladesh (e-mail:ahassan@cegisbd.com)*

[2]*Delft Hydraulics, PO Box 177, 2600 MH Delft, The Netherlands*

Abstract: The significance of river morphology and the need to develop reliable morphological prediction methods require better knowledge of river-bed and island topography. This study focuses on techniques for determining the topography of exposed chars (char is defined here as a vegetated island (island-char) or an area of vegetated land within the braid belt that is attached to the floodplain (attached-char)) within the braid belt of the Brahmaputra–Jamuna River, based on land cover and age of the land surface. Traditionally, char topography is assessed from widespread field surveys, which are both costly and time-consuming. In addition, the river is highly dynamic and the utility of field survey is limited beyond 1 year. This study explores the use of remote sensing as a cost-effective and rapid method of acquiring synoptic data on land elevations in the low relief of the Brahmaputra–Jamuna braid belt. The land elevations assessed indirectly by remote sensing are combined with ground surveys for calibration. The results indicate that the rate of char elevation growth is initially high, but decreases as the char matures. It appears that growth ceases after about 7 years, and that char elevation reaches a level that is maintained, provided the char is not subjected to erosion. A multitemporal satellite image database and field survey measurements at corresponding dates are required to develop the method. Extensive survey data are not available, hence verification of the method is limited. Despite this limitation the results give approximate char heights, which can be used for sediment budgeting and other morphological studies.

Bangladesh occupies the lowest part of the Ganges–Brahmaputra basin. One of the major rivers of this basin is the Jamuna (i.e. the Brahmaputra), which is braided with an average width of 11 km and slope 0.076 m km^{-1}. The total length of the Brahmaputra is about 2900 km. The length from its entry into Bangladesh to the confluence with the Ganges is about 240 km (Fig. 1). Along this reach, its slope decreases from 0.085 to 0.065 m km^{-1} and its median bed-material grain size decreases from 0.22 to 0.16 mm. The bed consists mostly of fine sand with a considerable amount of silts. The dominant discharge is 38 000 m^3 s^{-1} and the bankfull discharge is 60 000 m^3 s^{-1}, as described by Flood Action Plan (FAP 1 1993; Thorne *et al.* 1993). The braided Brahmaputra–Jamuna carries an estimated annual average suspended sediment of 402×10^6 tons a^{-1} upstream from its entry into Bangladesh (Goswami 1985) and an estimated 608×10^6 tons a^{-1} after taking in tributaries within Bangladesh (Coleman 1969).

The study of river morphology and morphodynamics is based on physical science and engineering. However, the effects of morphological change are not limited to the physical environment. About one million people reside on the islands, attached-chars and mainland along the Brahmaputra–Jamuna River (ISPAN 1993). Their existence is made precarious by erosion that can destroy land and homesteads, rapid sedimentation that can render land unproductive, and floods that can destroy homes and livestock. Changes in channels, bars and chars driven by the morphodynamic behaviour of the river affect the lives of these hazard-prone people directly and profoundly, but reliable information on their constantly changing environment has long been scarce. The rapid and substantial changes in the morphology of the Brahmaputra–Jamuna River also have an important bearing on the infrastructure of the country, including ferry operations, and construction and maintenance of bridges, embankments and river training structures. Heightened interest in river channel changes has led to the development of morphological prediction methods, but these can only be tested and applied successfully if reliable and up-to-date information on the topography of the river bed is available. Some

From: MARRIOTT, S. B. & ALEXANDER, J. (eds) 1999. *Floodplains: Interdisciplinary Approaches*. Geological Society, London, Special Publications, **163**, 153–161. 1-86239-050-9/99/$15.00

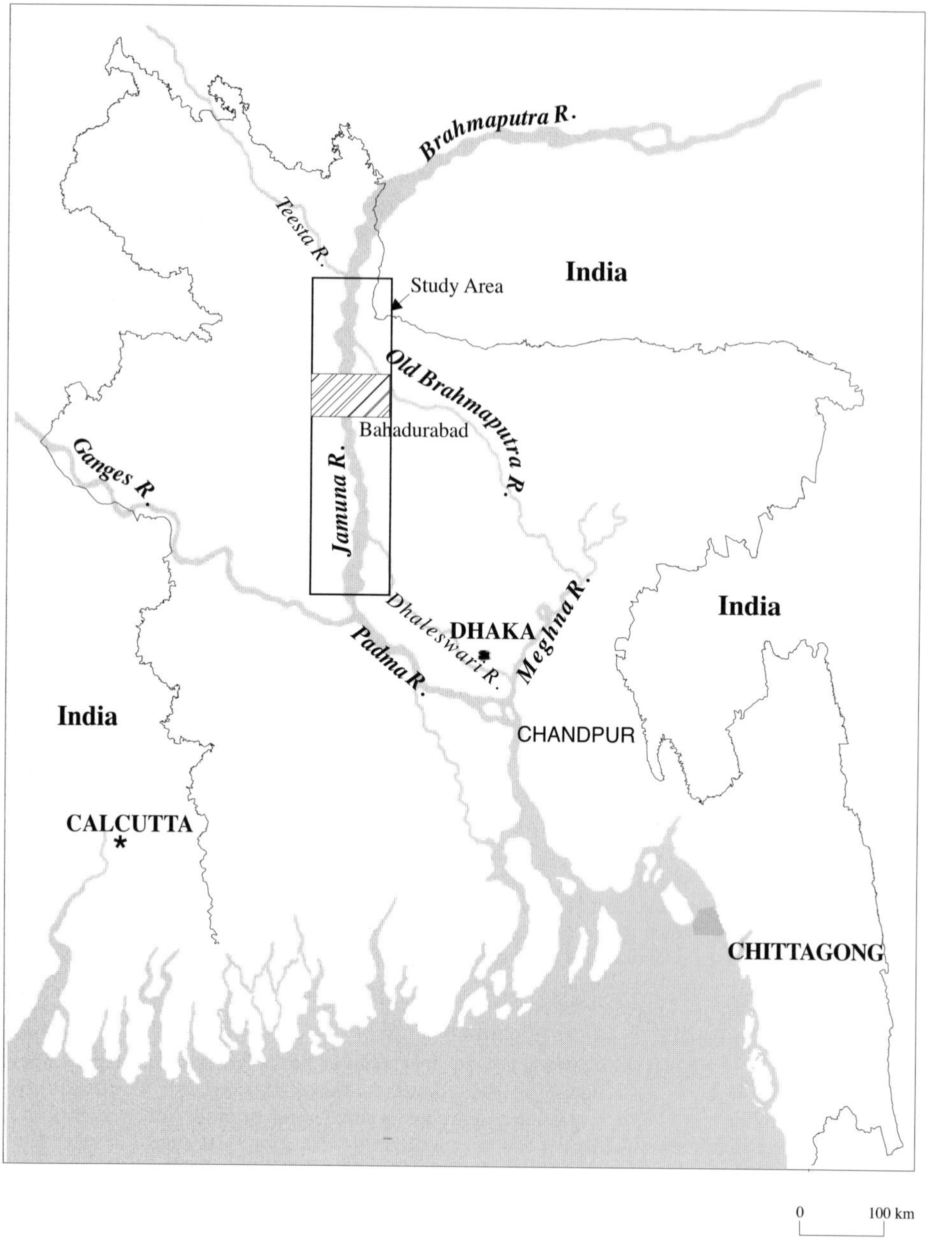

Fig. 1. Area of demonstration application char elevation model.

methods use key reference flows in the regime of the river, including the dominant and bankfull discharges.

The significance of river morphology and the need to develop reliable morphological prediction methods require topographic survey of the river bed at least annually. At present, data are collected exclusively by field surveys, which employ different techniques for the submerged part of the bed and the exposed chars. Techniques used for surveying the submerged bed and analysing the data were studied by EGIS (1996). The focus of this study is on techniques that can be used to determine the topography of exposed chars.

Traditionally, char topography is assessed from widespread field surveys, but this approach is both costly and time-consuming. This is of particular consequence for the Brahmaputra–Jamuna within Bangladesh, with an average width of the braid belt of 11 km, which encompasses about 1700 km^2 of vegetated islands and newly formed bars during the low flow season (EGIS 1997). Regular and comprehensive survey of such a dynamic and enormous system is far beyond the existing capacity of the responsible agencies in Bangladesh. There are numerous examples of modelling of environmental processes using topographic surfaces and a geographical information system (GIS) (Milne & Sear 1997); however, the scale of the river system in Bangladesh and the available resources for field survey justify a more coarse level of inquiry and information content. Using remote sensing and GIS modelling, a rapid assessment of char elevations can be made, which requires only limited ground surveys for calibration. The amount of fieldwork involved is substantially reduced from that required for an exclusively ground-based approach; this suggests that the time and cost of topographical surveys could be optimized through a combination of field surveys and remote sensing.

Methods

Data sources

The Brahmaputra–Jamuna River between the Old Brahmaputra off-take and the confluence with the Ganges was used for the investigation of char elevation in relation to land cover classification and char age. The elevation maps, demonstrating the results of the study, were produced for a 20 km stretch of the river near Bahadurabad (Fig. 1). The following data were used: (1) hard copies of SPOT Satellite images from 1994 at 1:50 000 scale; (2) change maps derived from a series of 13 georeferenced digital Landsat MSS and TM Satellite images from the period 1973–1996; (3) water-level readings and stage–discharge relationships from the hydrometric stations at Chilmari, Bahadurabad, Sirajganj and Mathura (RSP 1996); (4) Bangladesh Water Development Board (BWDB) cross-sections at 34 locations for 1978, 1980, 1987 and 1994; (5) 22 additional cross-sections for 1994. Each raw satellite image was resampled, using the nearest neighbour algorithm, and transformed into a file referenced to the Bangladesh Transverse Mercator (BTM) projection described by ISPAN (1992).

Data processing and analysis

Land cover classification. The digital satellite images were classified using image processing techniques to assign land cover classes to areas with similar spectral characteristics. Stretching and scaling of the range of digital values was done for each of eight Landsat MSS images, using the 1976 image as reference for histogram matching. This modification of the data resulted in images with similar spectral characteristics, which simplified the classification and interpretation of the historic images. A series of tests was carried out using statistical clustering to derive sets of spectral characteristics (signatures) that were used to classify the images.

Each of five Landsat TM images was classified independently using an iterative classification procedure. An unsupervised classification algorithm was used to derive signature statistics, which were examined by an image processing analyst, modified as appropriate and used with a maximum likelihood classifier. Results were examined and acceptable classes were assigned to land cover categories. Image pixels corresponding to inadequate classes were digitally extracted, resubmitted to the unsupervised clustering routine, and reclassified. Four broad land cover classes were assigned to each of the 13 Landsat image pairs in the time series: water, sand, cultivated land and vegetated land (Fig. 2a).

The accuracy of the satellite image classification was considered and, although it was impossible to assess the historical data, which date back to 1973, a field programme was carried out to assess the 1992 data. The 1992 field effort, described by ISPAN (1993), involved several visits to the Jamuna River where fluvial processes, land cover and agronomic practices were observed and documented. The entire course of the river within Bangladesh was covered by 245 sites. An overall accuracy of 88% was found for three broad classes: water, sand and land. Morphological units were interpreted from the land cover classes as follows: from water, low flow anabranch channel; from sand, braid bar or recently deposited sediment on char; from land, island-char or attached-char.

Determination of char age. Previous studies (e.g. Bristow 1987; Thorne *et al.* 1993) have established that the elevations of active sand bars are lower than those of established chars. This is also evident from the fact that the area of sand visible in a low-flow satellite image is sensitive to the water level on the date of image, whereas the visible area of established char land is insensitive. The change from 'sand' or 'land' to water at a given pixel between two consecutive images (which could result from either char destruction or vertical accretion), is always taken to represent char destruction. Similarly, change from 'sand' to 'land' (which could result from either the conversion of a bar to a char through sedimentation, or the re-establishment of vegetation or cultivation on the sediment-covered surface of an existing char), is always taken to represent the creation of a new char area. Therefore, areas classified as sand were excluded from the procedure relating elevation to char age.

The land cover maps created from the satellite images for 1973–1996 were used to determine the age of island and attached-char areas within the respective year's banklines. The maps were created with a GIS to determine the age of the land from the most recent date at which it was classified as being either water or sand. These char age maps were constructed for 1994 (Fig. 2b) and 1996.

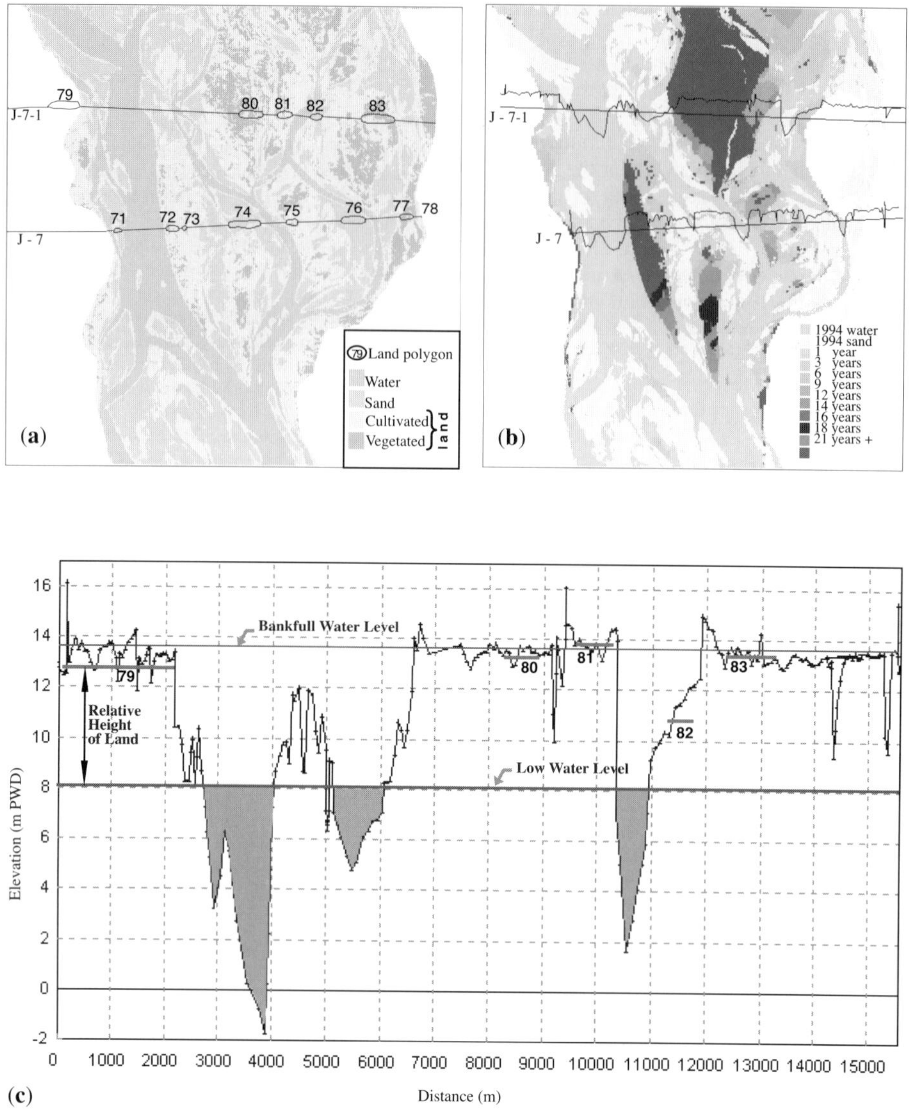

Fig. 2. Procedure for determining relative elevations of sample areas. (**a**) Land cover in 1994; (**b**) char age in 1994; (**c**) cross-section J-7-1 (BWDB) of Jamuna River in 1994.

Georeferencing of cross-sections. Georeferencing of cross-sections was achieved by finding the location of closest correspondence to channels visible in the relevant satellite images. SPOT images were used to georeference cross-sections for 1994 and Landsat images for 1978, 1980 and 1987. In this procedure, a horizontal line was first drawn on each cross-section at the level of the local water surface on the day of the satellite overpass and image composition. The horizontal line, representing the water level at each cross-section location, was estimated by interpolation using measured gauging station water-level data for the respective image overpass date. The points where the horizontal line (i.e. approximate water surface) intersected the cross-section corresponded to the edge of the water in the satellite image. The position of the cross-section was adjusted until a best match was achieved. The georeferencing procedure has been described in detail by EGIS (1996) RSP (1996), and is illustrated in Fig. 2a.

Assignment of relative elevations to sample areas. Representative sample areas were selected on chars along the georeferenced cross-sections. Each sample area was assigned a relative elevation defined as the difference between the average local elevation in the cross-section within the sample area and the low-water level on the day of image acquisition. The procedure is illustrated in Fig. 2c.

Correlation analysis and char elevation mapping. The relationship between relative char elevation and char age was assessed for each sample polygon through various mathematical functions. The success of the various relationships was evaluated by plotting predicted. observed elevations in 'goodness-of-fit' diagrams. A tolerance of ±1 m was used to define an acceptable prediction band. The most suitable functions were then used to construct predicted char elevation maps.

Mapping erosion and deposition. One of the advantages of this method for estimating char elevations is that standard GIS functionality allows rapid calculation of the amount of erosion and deposition involved in changes of char topography. To demonstrate this capability, maps of erosion and deposition were constructed by subtracting earlier char elevation maps from later ones.

Mapping the river planform at different flow stages. Landsat TM images were available for dry season, low-flow conditions, but cloud cover severely limits the availability of images during the high-flow season. Therefore, it is difficult to establish the planform appearance of the river at geomorphologically significant flows such as dominant discharge or bankfull flow from satellite images. The capability to map char elevations could provide a basis for mapping the river planform at any chosen discharge. To demonstrate this, water-level profiles for different flow stages were determined by interpolation from the stage–discharge relationships at the hydrometric stations and relative char elevations were converted to the Public Works Department (PWD) datum. River planform maps were then constructed using the PWD elevations for both water level and chars, and different colours were assigned to chars above and below the water surface (see Fig. 6, below).

Results and discussion

Eighty-seven sample polygons were used in the spatial analysis for statistical values and correlation. Relations between char elevation and age are listed in Table 1 and shown in Fig. 3. The results and relations confirm as hypothesized that sandbars are lower than chars and older chars are higher than younger ones.

Figure 4a and c shows two char elevation functions. The step function in Fig. 4a is based on Table 1, whereas the hyperbolic function in Fig. 4c is based on a curve fitting procedure. Both functions indicate that the rate of elevation growth is initially high but that the rate of growth decreases as the char matures. It appears that growth ceases after a few years, so that char elevation reaches an equilibrium level that is maintained, provided that the char is not subjected to erosion. The corresponding 'goodness-of-fit' diagrams are shown in Fig. 4b and d, and the resulting char elevation maps are presented in Fig. 5a and b.

A limitation of the step function is that it produces an elevation map with only a few discrete levels, whereas the curvilinear hyperbolic function shows a more continuously varying topography. In the step function, three discrete levels were assigned to the land cover type from Table 1 (Figs 4a and 5a). The hyperbolic model used 3.5 m elevation for sand, whereas for vegetated and cultivated land the model used a hyperbolic function for different land ages (Figs 4c and 5b). The result shows that the methods differ mainly in younger chars, those less than 5 years old.

Using the above step and hyperbolic function methods, elevation maps were constructed for 1994 and 1996, for which years cross-sections were available. Using GIS techniques, 1994 and 1996 char elevation maps were overlaid, yielding a map of erosion and sediment deposition for both methods. An example is considered for an area near Bahadurabad. In this analysis, the two methods yielded significantly different volumes of erosion and deposition of the char areas (Table 2). The main reason for this difference is the development of new chars in the period between 1994 and 1996. As the step function only differentiates between chars younger and older than 3 years, these new chars were not recognized by this method. However, the hyperbolic function method detected all recently developed chars and could estimate the corresponding differences.

Table 1. *Relation between elevation (87 samples), land cover class and char age*

Land cover and age class	*n*	Mean relative height (m)	SD relative height (m)
Sand			
age <1 year	30	3.46	0.95
Vegetated + cultivated land			
age 1–3 years	28	4.25	1.18
age >3 years	29	5.38	0.69

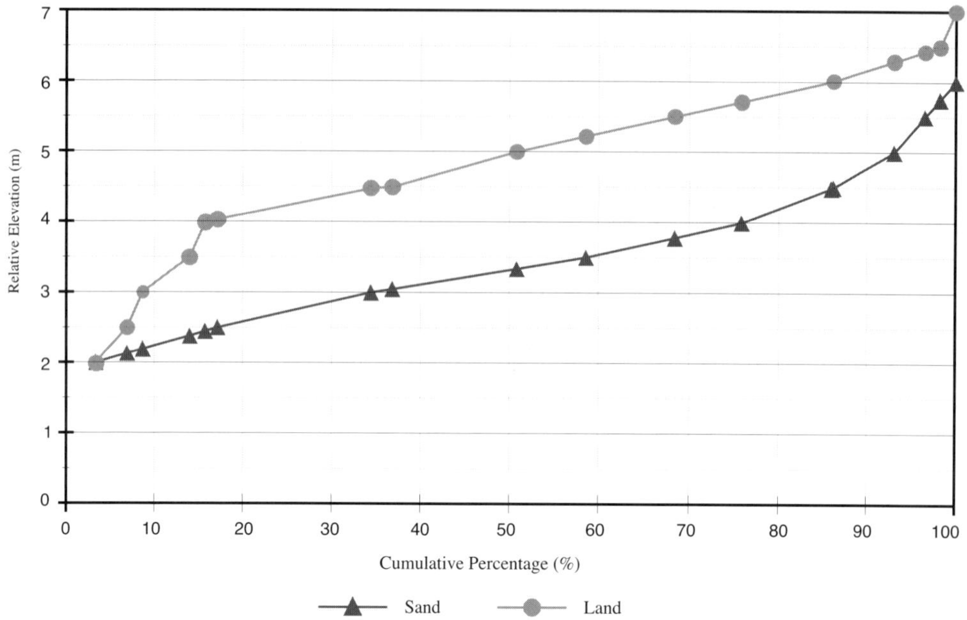

Fig. 3. Elevation distribution for sand and land cover class.

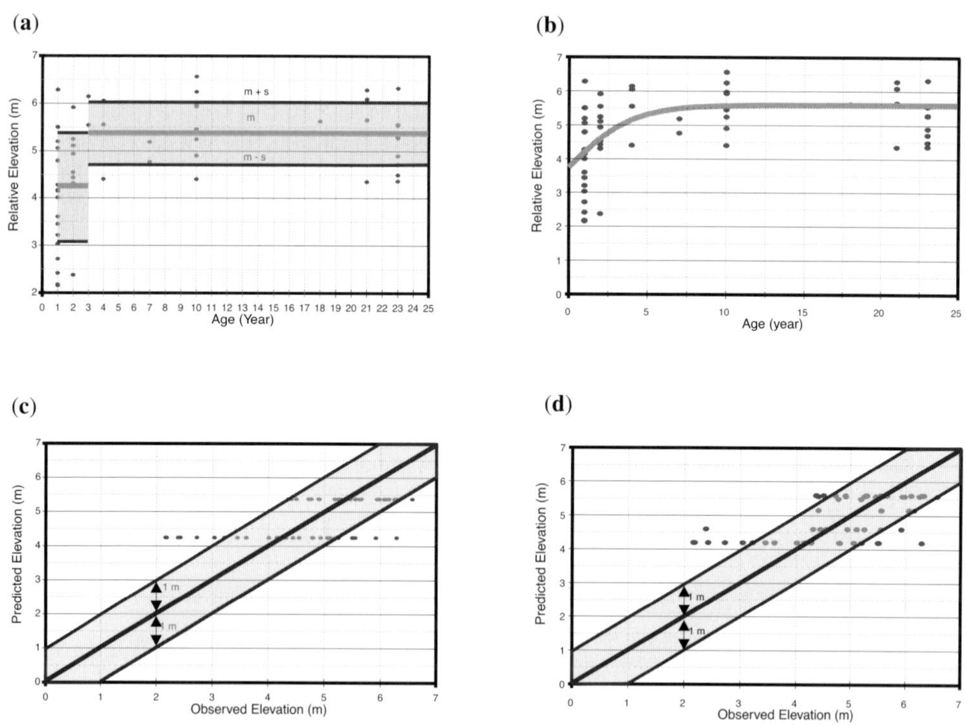

Fig. 4. Char elevation functions. (**a**) Relation between elevation and age with step function; (**b**) relation between elevation and age with hyperbolic function; (**c**) goodness of fit for step function; (**d**) goodness of fit for hyperbolic function.

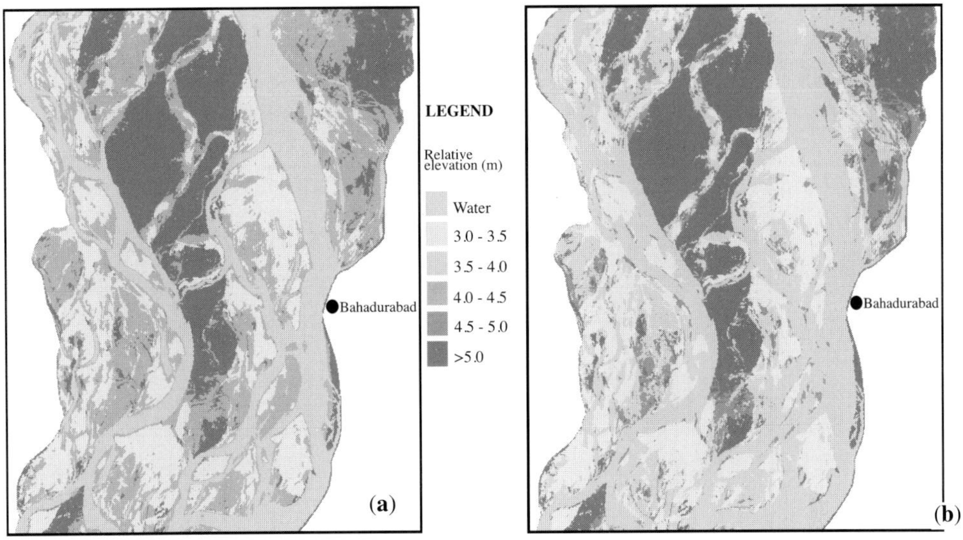

Fig. 5. Relative char elevations near Bahadurabad in 1994. (**a**) Step function; (**b**) hyperbolic function

The char elevation model using the hyperbolic function was applied to a major portion of the river to represent topography and planform. As shown in Fig. 6, low-flow channels, larger anabranches, cross-char channels and island-chars can all be identified at various flow rates. The river planform represented various discharges, at Bahadurabad, as follows: 7000 $m^3 s^{-1}$ to represent low flow and 25 000 and 38 000 $m^3 s^{-1}$ to represent FAP 1's estimate of barfull (barfull discharge is defined as the water discharge at an elevation that submerges the bars in the river) and dominant discharges, respectively (FAP 1 1993). The lower limit of bankfull discharge, 45 000 $m^3 s^{-1}$, as reported by the River Survey Project (RSP 1996) was also represented.

As shown in Fig. 6b, at barfull discharge (25 000 $m^3 s^{-1}$) the water levels were 3.5–4 m higher than levels at low flow and all the sand bars and younger chars (age 1–3 years) were submerged. At these discharges, the map shows a planform of 3–5 anabranches. At dominant discharge (38 000 $m^3 s^{-1}$) flow channels combined to form two to four major anabraches, as expected by FAP 1 (FAP 1 1993; Thorne *et al.* 1993). At 45 000 $m^3 s^{-1}$, the water level is 4.5–5.5 m above the dry season water level and all but the oldest chars are inundated.

Table 2. *Estimates of mass volume of char erosion and deposition using different relationships for the period 1994–1996*

Function	Erosion (Mm^3)	Deposition (Mm^3)
Step	32	19
Hyperbolic	24	29

Conclusion and recommendation

The results indicate that the rate of elevation growth is initially high in young chars but that the rate of growth decreases as the char matures. It appears that growth ceases after a few years, so that the char elevation reaches an equilibrium level that is maintained, provided that the char is not subjected to erosion.

Despite this limitation in the resolution of the models, the maps they predict give a realistic impression of the topography of the river and significant morphological features. Low-flow channels, larger anabranches, cross-char channels and island-chars can all be identified. The potential exists to perform a closed sediment balance for this reach by merging of char elevation maps with bathymetric maps for the low-flow channels.

The hyperbolic function model shows better results than the step function model but it needs more field survey information to increase the accuracy and improve the calibration coefficient. Also, this study considers only 1994 data; for better accuracy, more years of field survey data are needed.

Remote sensing data are very useful for river morphology studies, especially when access to the project area is very difficult, the area is large and

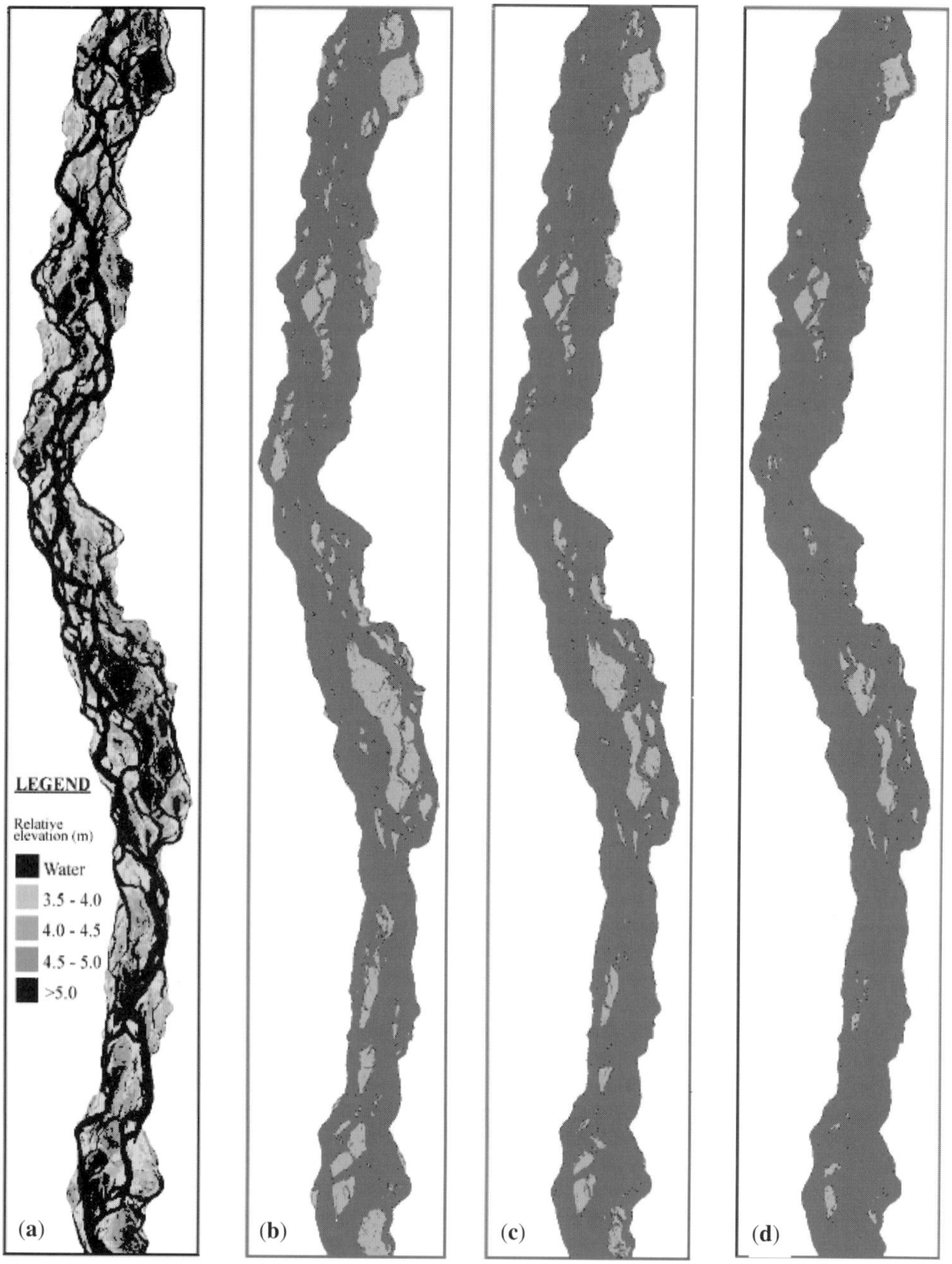

Fig. 6. River planforms at different flow discharge in 1994. (**a**) Planform at a discharge of 7000 $m^3 s^{-1}$; (**b**) planform at a discharge of 25 000 $m^3 s^{-1}$; (**c**) planform at a discharge of 38 000 $m^3 s^{-1}$; (**d**) planform at a discharge of 45 000 $m^3 s^{-1}$.

topographic changes occur frequently. Image processing and GIS tools are very efficient for this type of study.

The research was undertaken by the authors while working on the Environment and GIS Support Project for Water Sector Planning (EGIS). EGIS is a project under the Ministry of Water Resources, Government of Bangladesh, sponsored by the Royal Netherlands Government. Sincere thanks are due to all the team members of EGIS. We gratefully acknowledge C. R. Thorne of University of Nottingham for expertise in fluvial geomorphology and I. Huque of EGIS for remote sensing and image processing.

References

BRISTOW, C. S. 1987. Brahmaputra River: channel migration and deposition. *In*: ETHRIDGE, F. G., FLORES, R. M. & HARVEY, M. D. (eds) *Recent Fluvial Sedimentology*. Society of Economic Paleontologists and Mineralogists, Special Publications, **39**, 63–74.

COLEMAN, J. M. 1969. Brahmaputra River: channel processes and sedimentation. *Sedimentary Geology*, **3**(2–3), 129–239.

EGIS 1996. *Spatial Representation and Analysis of Hydraulic and Morphological Data.* Environmental and GIS Support Project for Water Sector Planning (EGIS), Dhaka.

—— 1997. *Morphological Dynamics of the Brahmaputra–Jamuna River.* Environmental and GIS Support Project for Water Sector Planning (EGIS), Dhaka.

ELAHI, K., AHMED, K. S. & MAFIZUDDIN, M. 1991. *Riverbank Erosion, Flood and Population Displacement in Bangladesh.* Jahangirnagar University, Savar, Dhaka.

FAP 1 1993. *River Training Studies of the Brahmaputra River; Annex 2: Morphology—Draft Final report.* Flood Plan Coordination Organization (FPCO), Dhaka.

GOSWAMI, D. C. 1985. Brahmaputra River, Assam, India: physiography, basin denudation and channel aggradation. *Water Resources Research*, **21**(7), 959–978.

ISPAN 1992. *Irrigation Support Project for Asia and the Near East (ISPAN).* Flood Plan Coordination Organization (FPCO), Dhaka, Technical Notes Series **1**.

—— 1993. *The Dynamic Physical Environment of Riverine Charlands: Brahmaputra–Jamuna, Irrigation Support Project for Asia and the Near East (ISPAN).* Flood Plan Coordination Organization (FPCO), Dhaka.

MILNE, J. A. & SEAR, D. A. 1997. Modelling river channel topography using GIS. *International Journal of Geographic Information Science,* **11**(5), 499–519.

RSP 1996. *Bedform and Bar Dynamics in the Main Rivers of Bangladesh.* Water Resources Planning Organization, Dhaka, Special Report **9**.

THORNE, C. R., RUSSELL, A. P. G. & ALAM, M. K. 1993. Planform pattern and channel evolution of the Brahmaputra River, Bangladesh. *In*: BEST, J. L. & BRISTOW, C. S. (eds) *Braided Rivers*. Geological Society, London, Special Publications, **75**, 257–276.

Biotic response to late Holocene floodplain evolution in the River Irthing catchment, Cumbria

J. A. COTTON[1], G. L. HERITAGE[2], A. R. G. LARGE[1] & D. G. PASSMORE[1]

[1]*Department of Geography, Daysh Building, University of Newcastle upon Tyne NE1 7RU, UK (e-mail: Jacqueline.Cotton@newcastle.ac.uk)*

[2]*Department of Geography, Peel Building, University of Salford, Manchester M5 4WP, UK*

Abstract: This paper presents some results of a geomorphological and palaeoecological evaluation of late Holocene floodplain evolution at a site in the middle reaches of the River Irthing, a tributary of the River Eden, near Brampton, Cumbria, UK. Geomorphological mapping and survey have identified a minimum of seven Holocene alluvial terraces in this reach, each with well-defined palaeochannels developed on their surfaces. Sediment coring and radiocarbon dating of selected channel fills indicate that a large proportion of the Holocene valley floor has been deposited during and after the 3rd millennium BC. Detailed sedimentological and macrofossil analyses undertaken on late Holocene channel fill sediments reveal spatial and temporal variability in patterns of local vegetation development within palaeochannel environments. These reflect local changes in hydrology, the character and rates of sedimentation, and levels of disturbance. Results are interpreted in the context of regional records of Holocene climate and land-use change, as well as existing models of floodplain vegetation succession. The advantages of taking a holistic approach to palaeoenvironmental reconstruction and the utilization of macrofossils are discussed.

Temperate river floodplains exhibit a diverse range of habitats created by channel mobility and flood pulses, which causes the development of a varied set of geomorphological units (Bravard & Gilvear 1996; Marriott 1998) and vegetation communities (Amoros *et al.* 1996). Over the past few decades, research into floodplain characteristics and dynamics has highlighted their importance with respect to biological diversity, biological productivity and nutrient storage (Large & Petts 1994; Burt 1993). However, as a result of frequently intense activity throughout prehistoric and historical times present-day floodplains rarely exhibit the potentially complex mosaic of ecosystem units. Throughout Western Europe from the onset of deforestation and agriculture in prehistoric times through to recent river damming, flood alleviation measures and wetland drainage, floodplain systems have been severely altered (Petts *et al.* 1989; de Waal *et al.* 1998). The continued rise in the intensity of human influence has peaked in the past 50 years with river regulation considerably affecting floodplain environments (Petts *et al.* 1992).

In unregulated, dynamic river systems the mosaic of floodplain landforms is reflected by vegetation communities characteristic of the type of geomorphological unit colonized (Gurnell 1995). Such diversity is enhanced by patterns of ecological succession (as affected by the connectivity between the river and its floodplain), which over time can have the effect of significantly increasing habitat diversity (Prach *et al.* 1997; Bornette *et al.* 1998; Large & Prach 1998). Terrestrialization processes and changes in trophic status influence succession pathways, and flooding and fluctuations in moisture levels can have the effect of reducing species dominance and allowing coexistence of different vegetation communities (Pautou & Decamps 1985; Bravard *et al.* 1986).

The conservation potential of these environments is such that enhancing our knowledge of floodplain vegetation dynamics is central to sensitive management of the riverine landscape. It has recently been acknowledged that more information regarding floodplain processes, particularly relating to the development of floodplain landforms and habitats over various time scales, is required to enhance understanding of these complex systems (Bornette & Heiher, 1994; Gurnell 1995; Armitage 1996; Large & Petts 1996). A particularly important task here is elucidating the nature of system response to changes in natural and anthropogenic disturbance through the Holocene period

From: MARRIOTT, S. B. & ALEXANDER, J. (eds) 1999. *Floodplains: Interdisciplinary Approaches*. Geological Society, London, Special Publications, **163**, 163–178. 1-86239-050-9/99/$15.00 ©The Geological Society of London 1999.

(Brown 1997). Such information can aid in the evaluation of both the present 'status' of habitats (*sensu* O'Keeffe 1997) and implications of past disturbance, and is also of value in assisting in the monitoring of restored sites.

Investigations into the structure and function of the fluvial system, and the effects of natural and anthropogenic disturbance on these attributes, have adopted approaches that have often differed in their time scales of study. Many studies have addressed contemporary floodplain ecology over longer time scales. Research, in particular, into long-term trends in both vegetal and faunal succession (Large & Prach 1998), may be beneficial for increased understanding of floodplain systems (Armitage 1996). Much ecological work in the past, however, has simply concentrated on the spatial rather than temporal nature of habitat change (Bravard *et al.* 1986).

Spatial and temporal floodplain evolution

Holocene fluvial activity in upland rivers in northwest Europe has typically given rise to incised and terraced valley floors (Harvey 1985; Macklin & Lewin 1993), with this fluvial activity recorded in a range of distinct morphological features. Fluvial terraces, and their associated sedimentary sequences and palaeochannels, reflect the type of channel planform and movement, the characteristics of the hydrological regime and the fluctuations in system stability. When not erased by human activity palaeochannels may be evident as terrace surface landforms (Macklin & Lewin 1993; Passmore & Macklin 1997), and therefore can be mapped three-dimensionally to allow reconstruction of fluvial activity over time. Such development can be evaluated across wide temporal scales, and, despite the fact that only a 'single snapshot' of the system can be made by contemporary research, it has been widely acknowledged that increased understanding of floodplain systems can come from investigations of past floodplain development (Large & Petts, 1996; Brown 1997). However, the actual time scales applied significantly influence whether the system is considered stable or unstable. Schumm (1979) recognized that over short time periods channel change could be seen to fluctuate around an equilibrium point, whereas over longer periods of time progressive and more dramatic change in a system becomes evident. Therefore, extending the time scale of research reveals the longer-term general directions of environmental change and responses to variations in external controls on both the meso- and macro-scale.

Aims of the research

Although there have been a large number of studies addressing spatial and temporal change in Holocene river floodplain complexes, the majority have focused on single aspects of change. Recent research has increasingly emphasized the importance of geomorphological, hydrological and ecological interactions in defining floodplain development (Petts *et al.* 1992; Gurnell 1995; Hupp 1996) and have emphasized the fact that these processes all require consideration when defining the nature of responses to environmental change. This preliminary study, carried out on the River Irthing, in the Eden catchment, Cumbria, addresses this holistic emphasis in that it combines geomorphological, hydrological and palaeoecological studies, with the aim of reconstructing biotic response to forcing factors over time.

The primary aim of the investigation was to reconstruct biotic response to environmental change in contrasting geomorphological settings, so as to evaluate the nature and impacts of allogenic inputs, autogenic changes and disturbance regimes. The objectives therefore were to evaluate specific components of the floodplain system at the reach scale, including: vegetation development and succession over time; channel-bed morphology before abandonment; post-abandonment channel infilling.

Vegetation change was reconstructed using plant macrofossils preserved in the organic-rich channel fills. Hitherto, little work has been undertaken on macrofossils in alluvial sediment, despite their potential for reconstructing *in situ* vegetation change in a range of environments (see Birks 1993). The work therefore had added value in that it developed a methodological approach to the investigation of these complex environmental systems. From the macrofossil data, biotic changes could be modelled in relation to abiotic factors associated with fluvial geomorphology.

Unlike its counterpart to the east, the Tyne River basin, the Holocene alluvial histories of the Eden catchment have yet to be investigated. It was felt, however, that probable synchronicity between the fluvial activity in the northwest and the Tyne basin in the northeast, as described by Passmore & Macklin (1997), would assist in elucidating the nature and causes of environmental change for this section of the River Irthing.

Sites

The study reach at Dovecote Farm, *c.* 3 km to the north of Brampton, encompasses 2 km^2 of the Holocene valley floor of the River Irthing, a tributary of the Eden in northern Cumbria (Fig. 1).

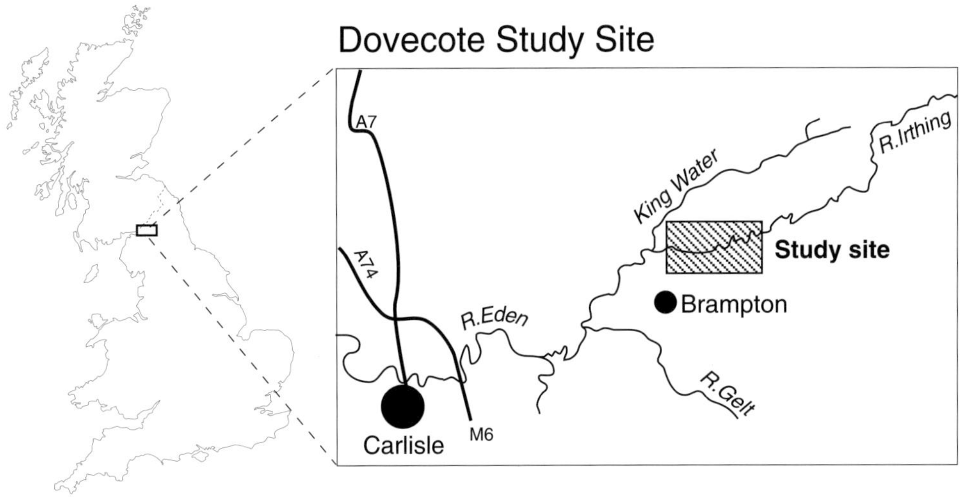

Fig. 1. Study site location.

Lying within the middle reaches of the River Irthing, the valley at this site drains an area of 190 km^2. The underlying geology consists of Triassic sandstone, although the catchment geology is dominated by Carboniferous Limestone and sandstone. A layer of glacial gravel with laminated clays overlies the bedrock.

Situated 2 km to the northwest are Bolton Fell Moss and Walton Moss, which have yielded detailed palaeoclimate and vegetation records (Barber *et al.* 1994). Just 500 m north of the site is the route of Hadrian's Wall. Roman presence within the area has been well documented, though the environmental implications of this have yet to be fully established (Higham 1986; McCarthy 1995).

Methods

Detailed geomorphological mapping has delimited discrete fluvial terraces and the planform morphology of their associated palaeochannels within the study reach. Mapping was undertaken using Ordnance Survey 1:10 000 maps and enlarged aerial photographs as base maps. Terrace elevations were established through a combination of surveyed cross profiles and spot heights.

On the basis of reconnaissance sediment coring, two palaeochannels with organic-rich channel fills were chosen for detailed analyses of palaeohydrology, stratigraphy and palaeoecology. Selected channels were of different age and exhibited contrasting planform sinuosity. It was anticipated that this would allow comparison of differing fluvial regimes before cut-off as well as subsequent patterns of anthropogenic influence and ecological succession. Radiocarbon dates were acquired from peaty sediments extracted from the gravel base of each channel, which provided a more precise date of channel cut-off.

Palaeohydrological methods

Detailed palaeochannel bed morphology reconstructions were carried out on the two channels investigated. Each channel was cored and logged down to the gravel base at 2 m intervals for five transects across each palaeochannel. The depth to the gravel base of the channel was also recorded along intermediate transects between the five main sections, allowing construction of a detailed three-dimensional image of the former river bed. To reconstruct the bed morphology of each palaeochannel, the original channel characteristics have been calculated from the surface morphology measurements to allow incorporation of a theoretical riffle spacing ratio of 5–7 times the palaeochannel width. On the basis of this ratio, the length of channel sampled within each palaeochannel incorporated one riffle–pool–riffle sequence.

Palaeoecological methods

The waterlogged and anaerobic conditions that exist in palaeochannels are suitable for the preservation of organic remains, though the preservation of such peaty sediments is rare in northern Britain because of extensive reworking of the valley floor throughout Holocene time (Passmore & Macklin 1997; Moores *et al.* 1999). The reconstruction of fossil communities can be undertaken using a number of fossil types, including insects, pollen remains and plant macrofossils (Lowe & Walker 1997). Pollen data indicate biotic change across the whole floodplain and have been shown to be of importance in the evaluation of human activity on the valley floor (Moores *et al.* 1999). Detailed changes in vegetation communities, however, cannot be seen in the pollen record, because of the long-distance transport of pollen grains and the taxonomic accuracy of identification (Birks 1973, 1993); therefore, such analyses are more effective using plant and insect macrofossil remains (Wasylikowa 1986; Baker & Drake 1994). Insect remains are very useful in providing detailed information on the environmental conditions present at that time (Coope 1986; Greenwood *et al.* 1991).

Plant macrofossils consist of seeds, plant fragments, wood and moss (Wasylikowa 1986). Although such organic remains allow accurate reconstruction of wetland and aquatic environmental change (Field 1992), and have frequently been used as a palaeoenvironmental tool with lake sediments and ombrotrophic bogs, they have rarely been utilized in studies of alluvial systems (Amoros & Van Urk 1989). This is despite the fact that many species characteristic of these environments produce large numbers of resilient seeds, which are well preserved when not germinated (Dickson 1970; Mannion 1986). The main advantage of their use is that they travel only short distances before deposition (Birks 1993) and therefore they are representative of *in situ* vegetation communities. In addition, such macrofossils can often be identified to species level, allowing derivation of accurate images of past communities, and facilitating comparison with modern analogues.

For the macrofossil analyses a core was taken from the deepest point of each palaeochannel, as indicated by the reconstruction of the former channel bed, using both Russian and piston corers, and sampled at 2 cm intervals. The material was prepared as described by Dickson (1970) and Watts (1978). All of the material was used for analysis, with the full volume being used to calculate the concentration of each species encountered. The macrofossil counts were then rounded to the same volume concentration to allow comparison between species and over time. Identification was carried out using published material and the seed collections at Durham University and the Hancock Museum, Newcastle upon Tyne.

Sediment coring and stratigraphic logging methods

The palaeochannel infills were cored and logged at the resolution described in the palaeohydrological methods above. At each point the structure, texture, colour and organic content of the sediment were noted, as were the depth and nature of changes within the stratigraphic sequence. The results were then extrapolated from single coring points to the whole transect to create a 2D image of the channel fill, to allow comparison of the channel fills with respect to the proximity of cut-off, the former channel-bed morphology and the fossil vegetation communities.

Results

Terrace and floodplain evolution

A minimum of seven discrete Holocene terraces have been identified within the study site. The Holocene terraces are inset at least 2 m below the high-elevation terraces provisionally assumed to be of late Pleistocene age. Detailed investigations were undertaken on palaeochannels developed in terraces T4 and T6 (see Fig. 2). Radiocarbon dates were obtained from the peaty sediments at the base of both channels. A date of 3750 ± 80 BP was obtained from the base of the palaeochannel corresponding to Terrace Unit 4 (Fig. 2), hereinafter referred to as palaeochannel DC2. This date calibrates to 2440–1920 BP (Stuiver *et al.* 1993). The radiocarbon date obtained from the base of the associated former river channel is much later, and dated at 460 ± 50 BP, which calibrates to AD 1410–1620 (Stuiver *et al.* 1993). This channel is referred to as palaeochannel DC4. Continued incision between the two dates is evident from the significant terrace height difference down to Terrace Unit 6, whereas a smaller vertical drop accompanied by an increase in the lateral expansion of terrace units occurs below Terrace Unit 6 (Fig. 3). Recent historical channel migration is also evident from the first edition Ordnance Survey map of the area, which shows the palaeochannel running along Terrace Unit 8 as being connected to the main channel during the early part of the 19th century.

Channel-bed morphology and stratigraphy

Palaeochannel DC2 (c. 2440–1920 cal BC). Figure 4a shows the reconstructed morphology of DC2. The left bank of the channel is poorly defined, though the steeper right bank has been well preserved. Evidence exists of the presence of bar forms, as well as a possible secondary channel in the upstream section.

Alluviation within the former channel appears to have been relatively uniform. The shallow infills (Fig. 4b) consist of very fine grained sediment, dominated by silt and clay. Organic sediments predominate in the deeper sections, seen in Transects 3 and 4. All of the transects display a fining upwards of grain size, though thin layers of coarse-grained material exist within the finer material in Transects 2, 3 and 5.

Palaeochannel DC4 (c.1410–1620 cal AD). In contrast to DC2, palaeochannel DC4 exhibits a well-preserved morphology (Fig. 5a). Both sides of the channel are well defined, and a depression at the apex of the meander succeeded by a higher section further downstream is interpreted as a meandering pool–riffle sequence.

From the sedimentary data, shown in Fig. 5b, a sediment plug is evident in Transect 1, immediately upstream of the channel-bed depression. In the lee of this plug, sediments consist of organic-rich silts and fine sands. Downstream of the meander apex in the higher part of the former channel bed, sandy horizons have been deposited within the organic sediments.

Analysis of the palaeosurface reconstructions and sediment infills allowed a variety of channel variables to be compared between the palaeochannels investigated (see Table 1).

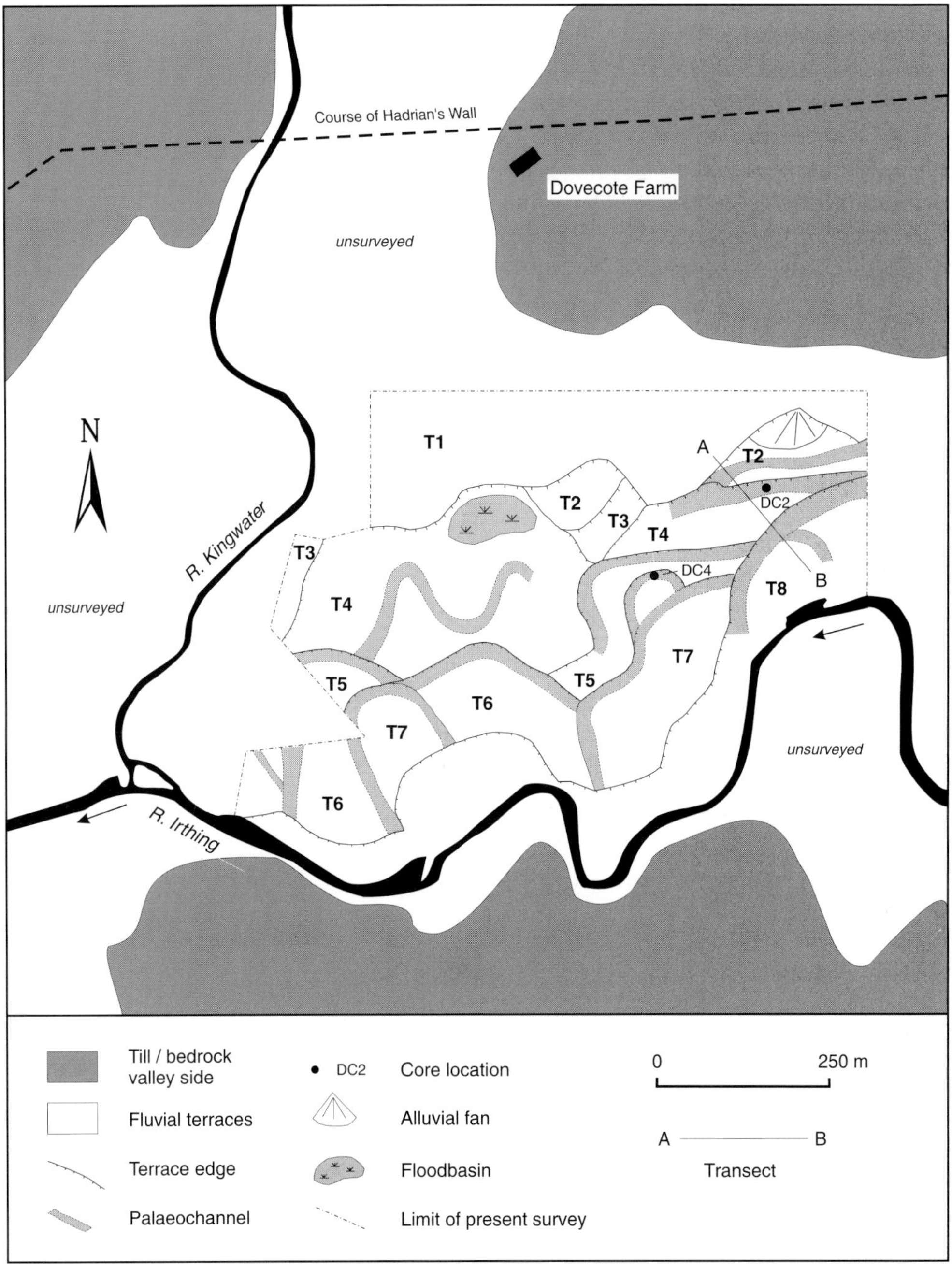

Fig. 2. Geomorphological map of the study site.

Plant macrofossil results

Plant macrofossils were well preserved within the organic silts and clays of DC2. As shown in Fig. 6, changes within the vegetation communities from the former river channel could be seen from the macrofossil record. Aquatic species dominate the fossil record throughout the sequence. Both open

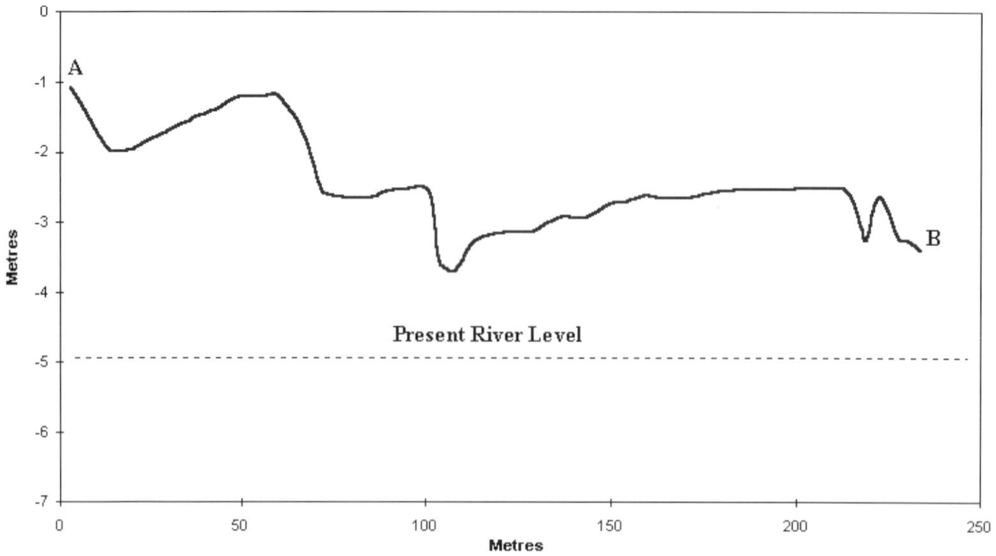

Fig. 3. Floodplain cross-section (location A–B in Fig. 2.)

water and wetland communities are evident. The presence of aquatic open water species such as *Daphnia ephibium* and *Nitella* spp. at the bottom of the core indicate the close proximity of the main channel following abandonment (this corresponds to the parapotamon stage *sensu* Amoros *et al.* (1987)). Species indicative of more stagnant open water conditions including *Potemogeton* spp. and *Alisma plantago-aquatica*, follow on from the initial pioneer communities. Up to 1.25 m depth a degree of succession is evident with the silting-up of the palaeochannel and an associated increase in wetland species. A recurrence of aquatics can be seen above the sandy horizon at 1.20 m depth, coinciding with an increase in laminated sediments indicative of more regular inundation.

In comparison with palaeochannel DC4, there is a marked absence of arboreal remains. Although *Alnus glutinosa* appears, it is present only in low numbers. The species have been broadly categorized, although it should be noted that many have wider environmental niches than the terms may suggest. The fossil record is cut short as a result of the effects of oxidation, which hinder identification of the remains. Overall there is little stratigraphic change throughout the core.

Figure 7 contains the macrofossil results for palaeochannel DC4, the younger of the two channels surveyed. The aquatic open water species at the base of the core indicate oxbow lake conditions following channel abandonment. Terrestrialization is evident above 1.80 m depth, with a decrease in aquatics accompanied by an increase of wetland species and overall organic matter content within the stratigraphy. This phase underlies a further stratigraphic change indicated by an accumulation of sandier material. The re-emergence of aquatic and moisture-loving species including *Daphnia ephibium* and *Sphagnum* spp. in the fossil record parallels the stratigraphic change. Colonization by some ruderal species continues until 1.30 m depth, at which point the accumulation of peaty sediments commences. Ecological succession continues through to primary woodland, shown by the significant presence of *Betula*, *Salix* and *Alnus*. This woodland, however, has not developed into what could be classified as the final seral stage of established woodland. Near the top of the core, a greater dominance of wetland species, with fewer arboreal remains, can be seen to correspond to a stratigraphic change to coarser material.

Discussion

Holocene alluvial terrace sequence

In common with many Holocene fluvial terrace sequences documented elsewhere in river basins in northern England (e.g. Harvey 1985; Macklin & Lewin 1993; Passmore & Macklin 1997), the past 4000 years has seen net incision of the Irthing valley with younger terraces inset below older fluvial units (Fig. 3). Correlations with regional

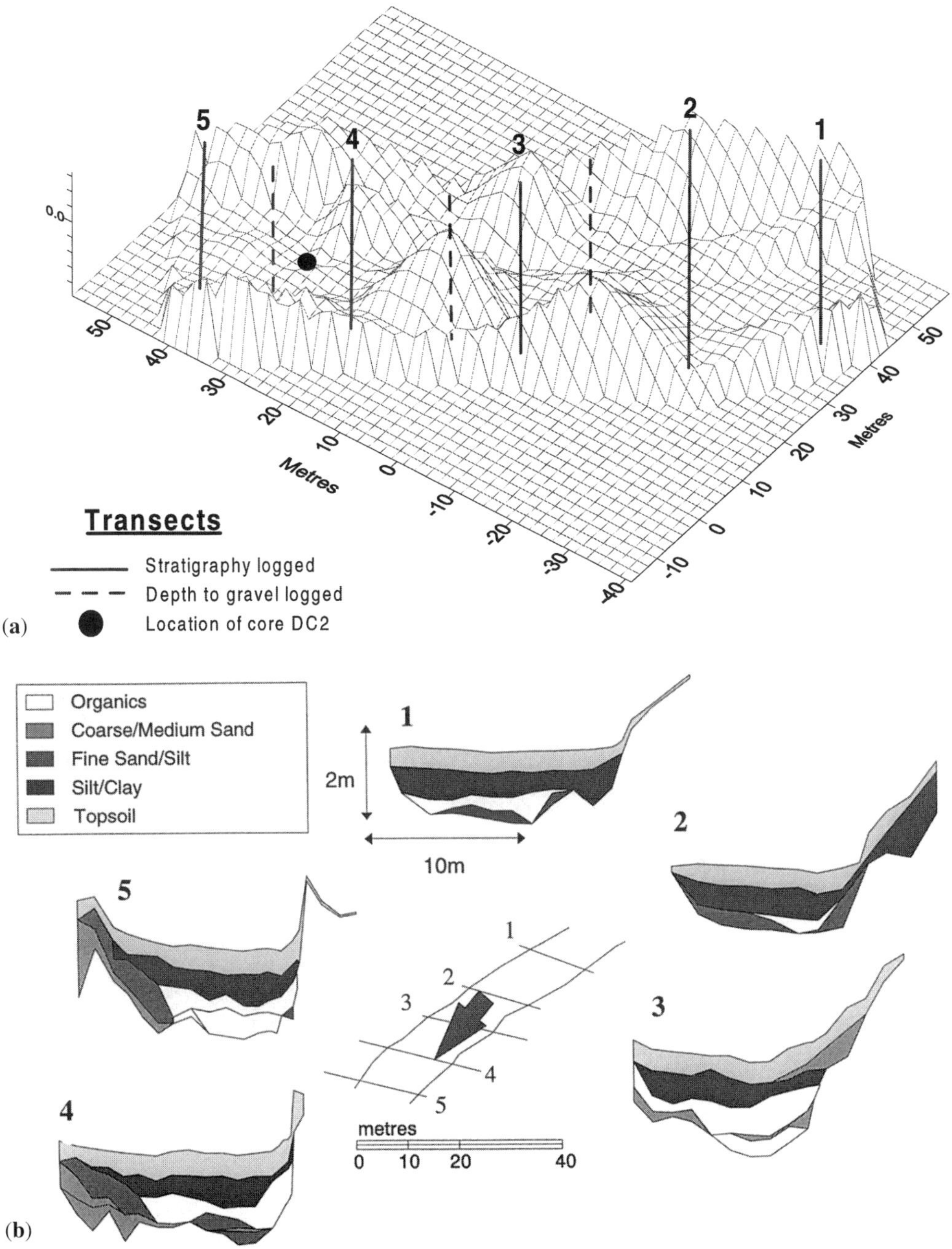

Fig. 4. (**a**) Three dimensional reconstruction of the former channel bed of DC2, also depicting location of cored transects and core taken for palaeoecological analysis. (**b**) DC2 palaeochannel infill stratigraphy along cored transects.

alluvial histories must be tentative at this stage of the investigation, but it is interesting to note that phases of channel abandonment (and most probably channel incision) dating to *c.* 2440–1920 cal BC and *c.* cal AD 1410–1620 (implied by dated basal channel fills in DC2 and DC4, respectively) at the study reach broadly coincide with increased rates of alluviation and/or incision recorded in the adjacent

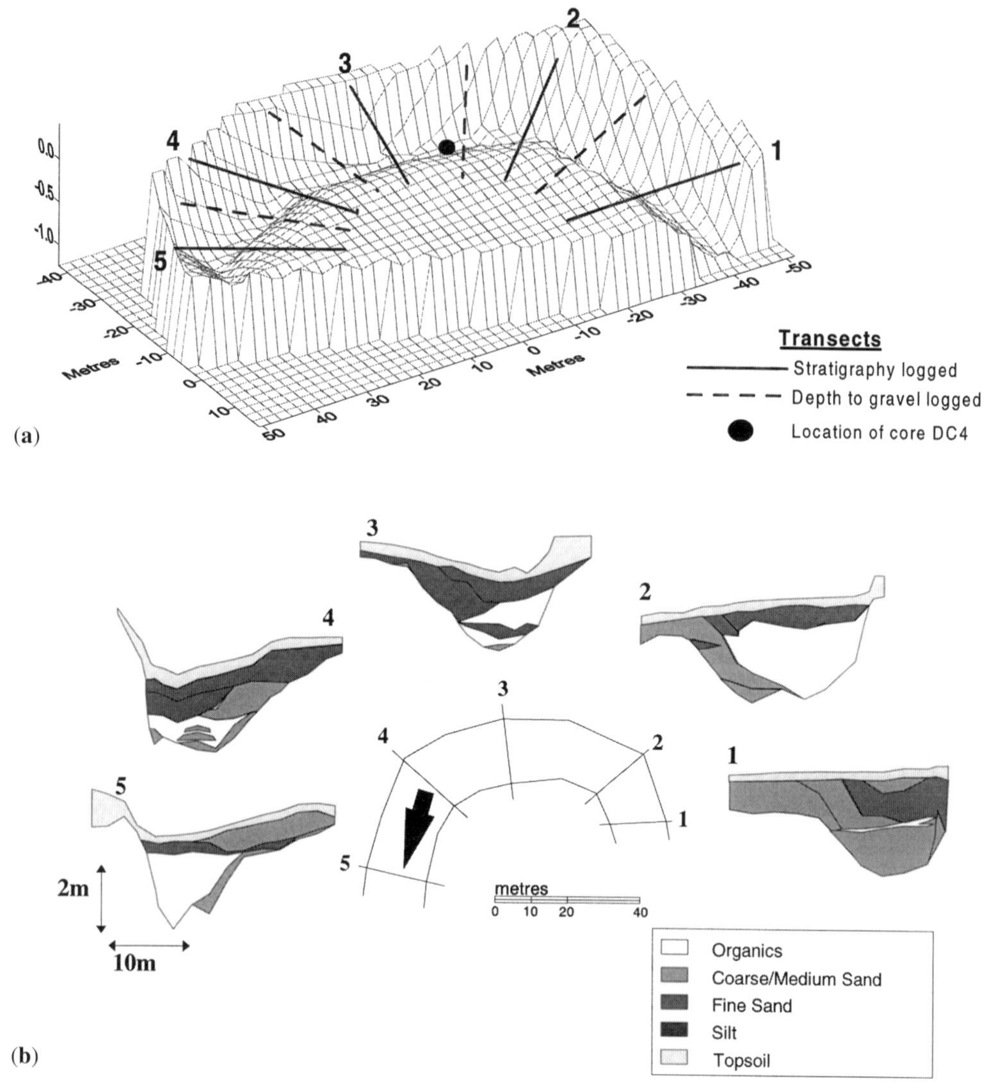

Fig. 5. (**a**) Three dimensional reconstruction of the former channel bed of DC4, also depicting location of cored transects and core taken for palaeoecological analysis. (**b**) DC4 palaeochannel infill stratigraphy along cored transects.

Tyne basin (Passmore *et al.* 1993; Passmore & Macklin 1997). Both periods correspond to Europe-wide climatic shifts to colder and wetter conditions (Roberts 1998), including the neoglacial 'Little Ice Age', which are recorded in the peat stratigraphy of Bolton Fell Moss (Barber *et al.* 1993, 1994), 2 km northwest of Dovecote. However, an important precursor to accelerated fluvial activity, as is proposed in the Tyne and other regional catchments (e.g. Macklin *et al.* 1992; Tipping 1992, 1998; Passmore & Macklin 1997), is likely to have been anthropogenic disturbance of catchment vegetation and soil cover.

Palaeochannel palaeohydrology and stratigraphy

In this study, sedimentary and geomorphological evidence has been used in combination to determine the palaeoenvironment of the Dovecote study site. Many previous studies have suggested

Table 1. *Variables used to define palaeochannels DC2 and DC4 for the River Irthing floodplain at Dovecote, Cumbria*

Factor	Palaeochannel DC2	Palaeochannel DC4
Bed morphology		
Bank preservation	Good only on right bank	Good on both sides
Characteristic units	Indeterminate	Point bar, pool–riffle
Cross-section morphology	Variable	Asymmetric
Bed material	Gravel	Gravel
Stratigraphy		
Abandoned channel deposits	Fining upwards	Varied
Organic content	Non-peaty sediment; occasional organic fragments	Peaty sediments; frequent organic fragments

particular sedimentary deposits or facies that can be used to infer a particular fluvial environment; however, such an over simplified approach has been subject to considerable criticism (Miall 1977; Jackson 1978; Reineck & Singh 1975). The use of some of the distinctions between river type in contemporary systems is often restricted by the poor preservation of evidence in the sedimentary record. However, when channel-bed morphology is well preserved, several ratios can be used to distinguish channel type, including channel sinuosity (see Friend & Sinha 1993) and channel gradient, which allows a calculation of the energy level (Jackson 1978). These ratios were calculated for each channel using the reconstructed bed morphology to estimate bankfull depths and water surface slope (Table 2). Overall, the two channels possessed different geomorphological characteristics, and particularly with regard to stream power. In the reaches analysed, energy levels provide a significant distinction between the channels.

Finally, channel preservation is also seen as a distinguishing attribute and, indeed, provides possibly the best distinction between the two channels studied in the Irthing system. In such floodplain systems, abandoned channel preservation is often better in meandering systems than in braided (Jackson 1978), as the latter have a propensity to rework older deposits over a short time scale (Passmore *et al.* 1993). Many researchers including Miall (1994) and Reineck & Singh (1975) have noted the lack of preservation of cohesive bank margins in braided systems, in contrast to the well-preserved palaeochannel infills of many meandering system deposits. Such a pattern was clearly evident in the two palaeochannels examined as part of this investigation.

Floodplain ecological analysis

From the fossil record contained within the older channel DC2, evidence is seen of open water communities developing into wetland habitat. Post-abandonment flood incursions into the cut-off channel were limited, as very fine grained material accumulated and the initial stages of succession can be seen to correspond to this infilling. However, a number of sand lenses have accumulated within the channel, emphasizing the degree of periodic connectivity via flooding. As discussed above, connectivity can have the effect of enhancing floodplain diversity, therefore it can be concluded that the disturbance evident from the stratigraphy was helping the wetland community species to persist. Dominance of *Juncus* spp. within the fossil record may relate to the excessive seed production of the species, rather than the dominance of the species.

Within the older channel DC2, the presence of occasional organic inclusions, along with peaty and sandy lenses, implies the presence of smaller habitat niches within the study reach. From the palaeosurface reconstruction, the surface variability of the channel would appear to indicate differential rates of terrestrialization, with preservation of more moisture-loving species within the surface depressions. To some extent, this has the effect of complicating the fossil record, although successional trends can still be seen within the overall pattern. Amoros & Wade (1996) highlighted the fact that braided systems are dominated by allogenic processes, and thus are more prone to reversion of the successional processes. With the morphological data pointing towards a high-energy environment, the effects on the biota are evident within the fossil record. Stratigraphic evidence points to flood inundation disturbance affecting the vegetation community above a core depth of 1.20 m. The increase in aquatic species above this level would appear to indicate more rapid water-table fluctuations, and subsequent interruption of the general ecological succession towards more terrestrial conditions.

The limit of habitat reconstruction occurs above 0.75 m depth, where oxidation of organic matter

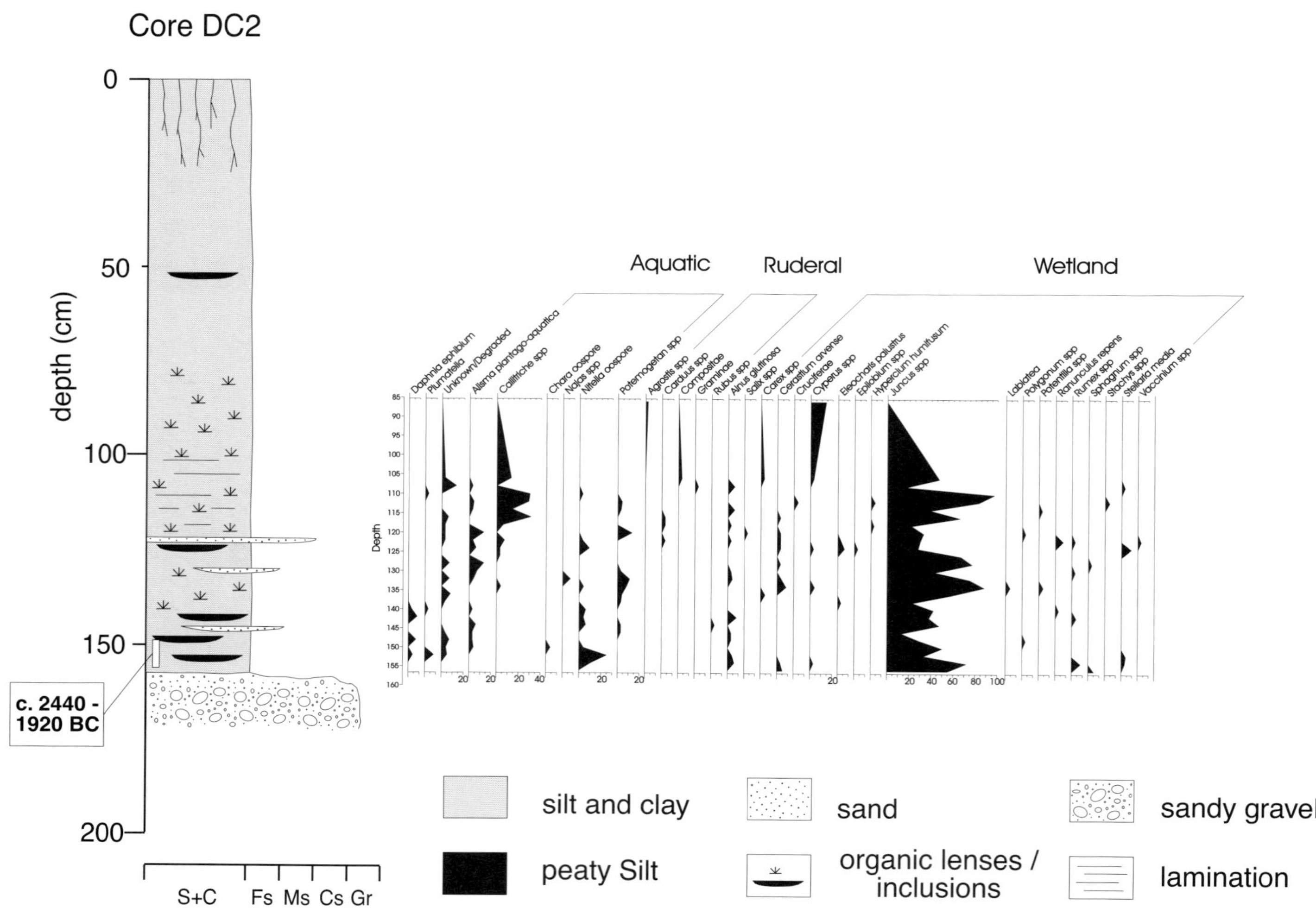

Fig. 6. DC2 sediment log and macrofossil data.

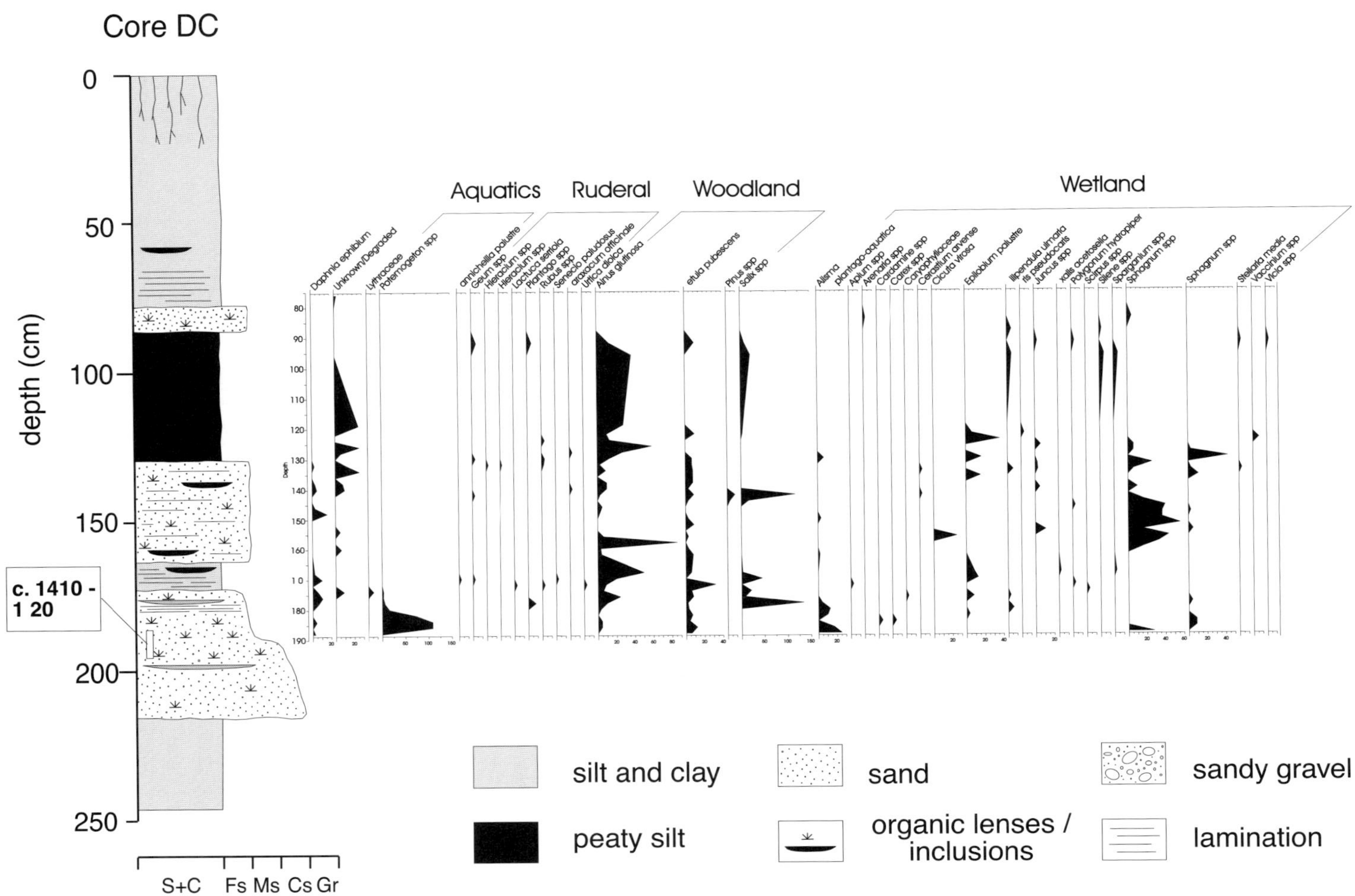

Fig. 7. DC4 sediment log and macrofossil data.

Table 2. *Calculated ratios of the former fluvial regimes of each channel*

Ratio	Measurement	DC2	DC4
Sinuosity	100 m reach	<1.3	>1.3
Past energy levels	Bankfull discharge/slope	High energy	Low energy

degrades the macrofossil remains and thus hinders identification. This results in the loss of the upper part of the sequence. Despite this limitation, the limited progression of ecological succession is notable throughout the studied profile. The absence of arboreal remains within the channel DC2 exemplifies the persistence of a wetland environment. A number of *Betula* and *Alnus* remains were found but not in quantities sufficient (compared with those found in DC4) to imply primary woodland development. Therefore, ecological succession within the channel does not appear to have progressed beyond the wetland stage.

The core from the younger channel DC4 was taken from the section of the reach interpreted as being the pool section (see Fig. 5a), which the sedimentary evidence shows to be immediately downstream of a sandy sediment plug deposited after abandonment. The open water aquatic species at the base of the core suggest that oxbow lake conditions may have persisted in the lee of this plug. A fining upwards of the sediments parallels ecological succession from aquatic to predominantly wetland communities. The rapid increase in organic matter indicates a progressive change, led by autogenic processes, with organic matter accumulation potentially altering the trophic status and increasing the nutrient content. Such developments are considered to be characteristic for meander cut-offs, as the influence of the contemporary channel is thought to be minimal (Amoros & Wade 1996).

Above 1.60 m, however, the change to a coarser sediment and the return of aquatic species indicates an increased influence of the contemporary main channel. In the downstream section of this reach, flood horizons are evident, as sediment inputs have been preferentially deposited in the former riffle (see Fig. 5b, transect 4). This contrasts with the theory put forward by Carrel & Juget (1987) whereby floods cause less disturbance in former meanders. Colonization by ruderal species following disturbance is evident at 1.30 m depth (see Fig. 9, below). Although anthropogenic disturbance during historical times is well documented in floodplain environments, further radiocarbon dates are required to elucidate a more precise cause of the change in species dominance. Despite the discontinuity in the macrofossil record, succession continues to the primary woodland stage (although the presence of secondary woodland is not seen in the fossil record). Grazing, deforestation and drainage may all account for both such limited progression and for the suppressed number of herbaceous species.

Modelling succession

Conceptual models linking temporal biotic, geomorphological and hydrological change have been constructed on the basis of ecological successions evident within each palaeochannel (Figs 8 and 9). The model from the prehistoric channel, DC2, is dominated by aquatic and wetland species, and reflects the succession pathways following channel abandonment, and subsequent development of an open water environment. Termed 'stable' biotic development, the change from open water to a more terrestrial wetland environment results from the predominance of autogenic processes. For this limited period, periodic inundation of the channel appears to be minimal. Contrasting biotic change is often related to abandoned channels in higher-energy environments, as allogenic inputs frequently destabilize the system by retarding terrestrialization processes and reducing species dominance (Bornette *et al.* 1998). Variation in abandoned channel-bed morphology and spatial variability of substrate causes differential biotic development, thus allowing the coexistence of vegetation communities. Therefore the autogenic processes and allogenic influences combine to create a diverse wetland environment. Unfortunately, the limit to the fossil record resulting from reduced sedimentation and a lowering of the water table since abandonment prevents further modelling of biotic change beyond the wetland phase.

In contrast to the older channel, the conceptual model for palaeochannel DC4 exhibits a greater degree of terrestrialization. Autogenic processes are evidently dominant, as the aquatic communities rapidly disappear with colonizing wetland and subsequent change to primary woodland communities. Affirmation of the limited influence of the contemporary channel up to the woodland stage is possible when external processes impress upon succession, thus causing the re-emergence of

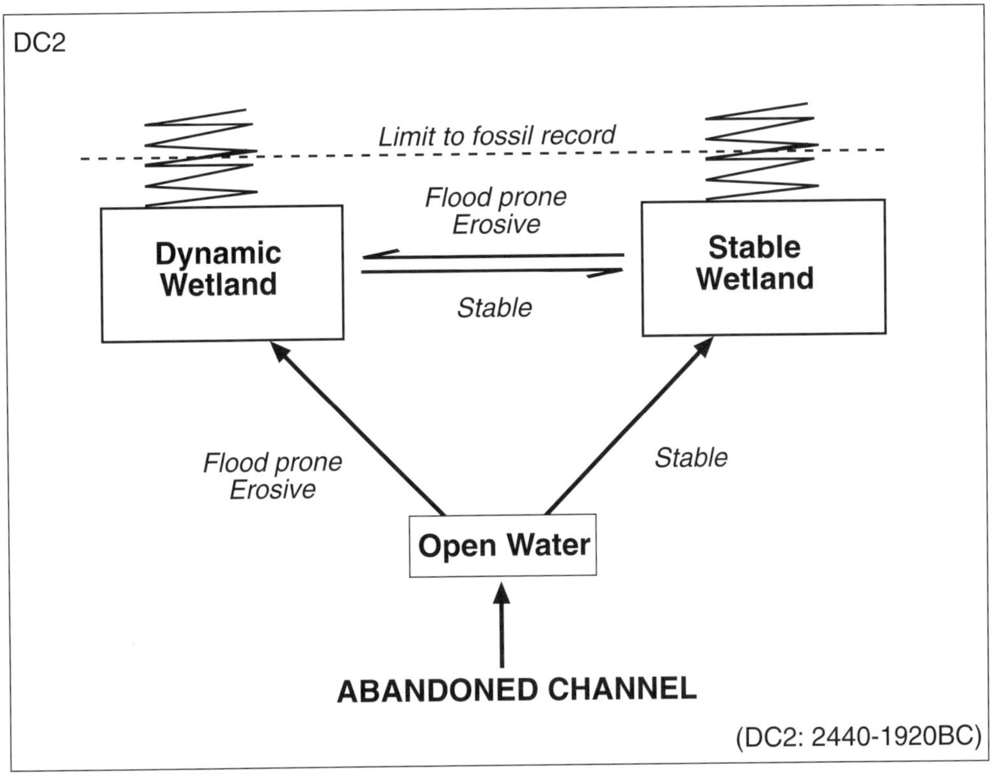

Fig. 8. Model of biotic change within palaeochannel DC2.

wetland communities. Disturbance also retards succession, leading to increased colonization by ruderal species. Though a greater degree of succession is evident within DC4, possibly because of the lower-energy environment and reduced connectivity to the channel, a secondary woodland stage is not reached. It is likely that this decline of woodland to herbaceous communities is a result of human intervention.

Comparison of the two conceptual models emphasizes the effect of differential external and internal influences upon community development. The contrasting geomorphological settings, evident from the morphological reconstruction and stratigraphic data, allow an assessment of the nature of the hydrological regime and subsequent effects upon the biota. The models also show that connectivity between floodplain and contemporary river for both channels was highly influential, and emphasize the conclusions put forward by Bravard *et al.* (1986) regarding the importance of regeneration within fluvial systems. An important implication of this, also discussed by Large & Prach (1998), is the fact that the role of allogenic influences in systems development must be considered when formulating strategies for the restoration of floodplain environments.

Conclusions

This research demonstrates the importance of combining data from as many components of the floodplain system as possible to allow a holistic appraisal. Although the 3D morphological reconstruction proved most useful in defining channel palaeoenvironments, it was still not conclusive in terms of categorizing the palaeochannels studied as either meandering or braided. Further research is being undertaken to evaluate the study reaches in relation to floodplain-scale palaeochannel and terrace morphology. The relative differences in bed morphology and calculated channel ratios for this part of the Irthing system proved invaluable in providing information on channel and floodplain variability, which conditions subsequent biotic characteristics. This technique therefore represents a potentially useful tool in palaeoenvironmental reconstruction of these dynamic systems. The different dynamics and stability of the palaeochannels investigated has appeared to significantly

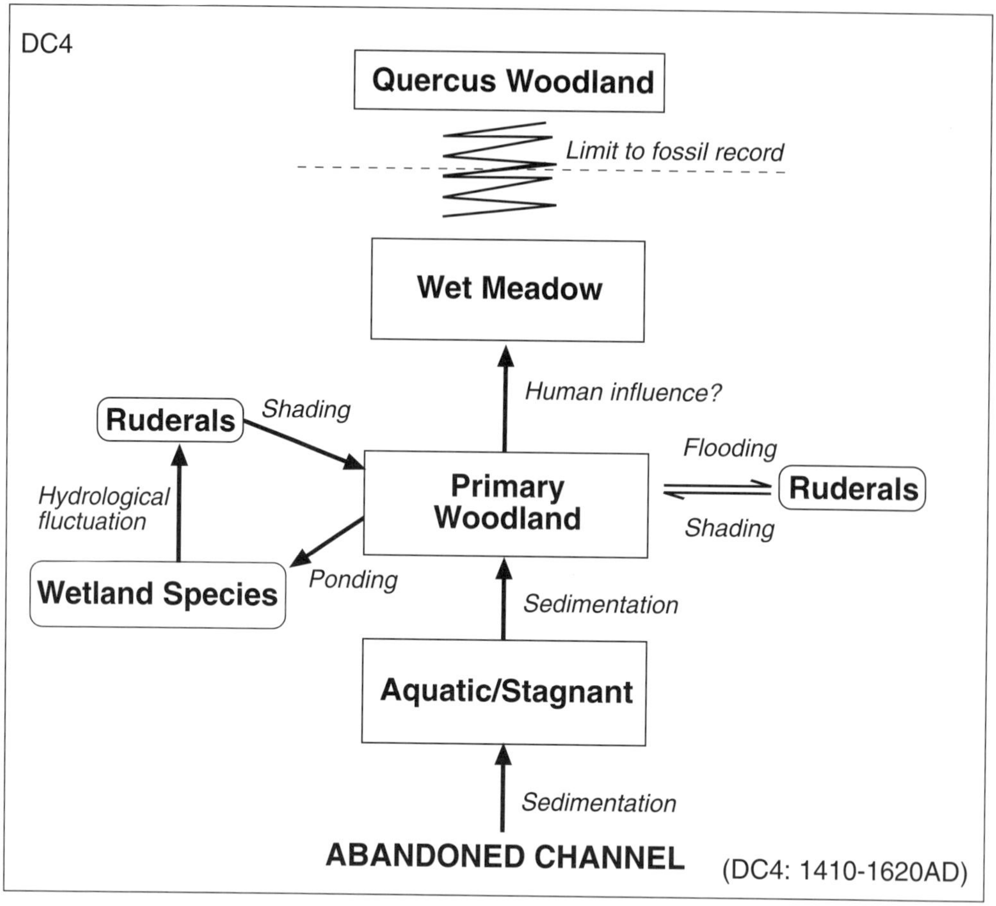

Fig. 9. Model of biotic change within palaeochannel DC4.

influence the extent and direction of vegetation development, with community succession principally affected by the changes in sedimentological inputs, hydrological fluctuations, flood histories and anthropogenic activity.

The advantages of macrofossil data in reconstructing local biotic change are shown. However, further research into the taphonomic processes affecting the fossil record will allow more detailed interpretation of the data. The potential of macrofossils as a tool for alluvial palaeoenvironmental reconstruction is emphasized, and it is proposed that a programme of research designed to test their usefulness in a variety of fluvial settings is required to further determine their value in palaeoenvironmental research.

This work was carried out as part of a wider study funded by a university research grant from the University of Newcastle upon Tyne. Acknowledgement is due to A. Rooke and L. Burns for assistance with the figures. We would like to thank J. Dobson for granting access to his land. We would also like to thank everybody who braved the Cumbrian weather and helped with field work.

References

AMOROS, C. & VAN URK, G. 1989. Palaeoecological analyses of large rivers; some principles and methods. *In*: PETTS, G. E., MÖLLER, H. & ROUX, A. L. (eds) *Historical Changes of Large Alluvial Rivers in Western Europe*. Wiley, Chichester, 143–165.

—— & WADE, P. M. 1996. Ecological successions. *In*: PETTS, G. E. & AMOROS, C. (eds) *Fluvial Hydrosystems*. Chapman & Hall, London, 211–241.

——, GIBERT, J. & GREENWOOD, M. T. 1996. Interactions between units of the fluvial hydrosystem. *In*: PETTS, G. E. & AMOROS, C. (eds) *Fluvial Hydrosystems*. Chapman & Hall, London, 184–210.

——, ROUX, A. L., REYGROBELLET, J. L., BRAVARD, J. P. & PAUTOU, G. 1987. A method for applied ecological studies of fluvial hydrosystems. *Regulated Rivers: Research and Management*, **1**, 17–36

ARMITAGE, P. D. 1996. Prediction of biological responses. *In*: PETTS, G. E. & CALOW, P. (eds) *River Biota: Diversity and Dynamics*. Blackwell Science, Oxford, 231–252.

BAKER, R. G. & DRAKE, P. 1994. Holocene history of prairie in mid western United States: pollen versus plant macrofossils. *Ecoscience*, **1**, 333–339.

BARBER, K. E., CHAMBERS, F. M., DUMAYNE, L., HASLAM, C. J., MADDY, D. & STONEMAN, R. E. 1994. Climatic change and human impact in North Cumbria: peat stratigraphic and pollen evidence from Bolton Fell Moss and Walton Moss. *In*: BOARDMAN, J. & WALDEN, J. (eds) *The Quaternary of Cumbria: Field Guide*. Quaternary Research Association, Oxford, 20–49.

——, DUMAYNE, L. & STONEMAN, R. 1993. Climate change and human impact during the late Holocene in northern Britain. *In*: CHAMBERS, F. M. (ed.) *Climate Change and Human Impact on the Landscape*. Chapman & Hall, London, 225–236.

BIRKS, H. H. 1973. Modern macrofossil assemblages in lake sediments in Minnesota. *In*: BIRKS, H. J. B. & WEST, R. G. (eds) *Quaternary Plant Ecology*. Blackwell, London, 173–189.

—— 1993. The importance of plant macrofossils in late-glacial climatic reconstructions: an example from western Norway. *Quaternary Science Reviews,* **12**, 719–726.

BORNETTE, G. & HEIHER, G. 1994. Environmental and biological responses of former channels to river incision: a diachronic study on the Rhone River. *Regulated Rivers: Research and Management,* **9**, 79–92.

—— AMOROS, C. & LAMOUROUX, N. 1998. Aquatic plant diversity in riverine wetlands: the role of connectivity. *Freshwater Biology,* **39**, 267–283.

BRAVARD, J.-P. & GILVEAR, D. J. 1996. Hydrological and geomorphological structure of hydrosystems. *In*: PETTS, G. E. & AMOROS, C. (eds) *Fluvial Hydrosystems*. Chapman and Hall, London, 98–117.

——, AMOROS, C. & PAUTOU, G. 1986. Impact of civil engineering works on the successions of communities in a fluvial system. A methodological approach applied to a section of the Upper Rhône River, France. *Oikos,* **47**, 92–111.

BROWN, A. G. 1997. *Alluvial Geoarchaeology. Floodplain Archaeology and Environmental Change*. Cambridge Manuals in Archaeology. Cambridge University Press, Cambridge.

BURT, T. P. 1993. Nutrient dynamics in riparian ecotones. *Global Ecology and Biogeography Letters*, **3**(3), 91–92.

CARREL, G. & JUGET, J. 1987. La Morte du Sauget, un ancien meandre du Rhône: bilan hydrologique et biochimique. *Schweizerische Zeitschrift für Hydrologie*, **49**, 102–125.

COOPE, G. R. 1986. Coleoptera analysis. *In*: BERGLUND, B. E. (ed.) *Handbook of Holocene Palaeoecology and Palaeohydrology*. Wiley, Chichester, 703–713.

DE WAAL, L. C., LARGE, A. R. G. & WADE, P. M. (eds) 1998. *Rehabilitation of Rivers: Principles and Practice*. Wiley, Chichester.

DICKSON, C. A. 1970. The study of plant macrofossils in British Quaternary deposits. *In*: WALKER, D. & WEST, R. G. (eds) *Studies in the Vegetational History of the British Isles*. Cambridge University Press, Cambridge, 233–255.

FIELD, M. H. 1992. *Study of plant macrofossil taphonomy in lakes and rivers and its application for interpreting some Middle Pleistocene assemblages*. PhD thesis, Cheltenham and Gloucester College of Higher Education, Cheltenham.

FRIEND, P. F. & SINHA, R. 1993. Braiding and meandering parameters. *In*: BEST, J. L. & BRISTOW, C. S. (eds) *Braided Rivers*. Geological Society, London, Special Publications, **75**, 105–111.

GREENWOOD, M. T., BICKERTON, M. A., CASTELLA, E., LARGE, A. R. G. & PETTS, G. E. 1991. The use of Coleoptera (Arthropoda: Insecta) for floodplain characterisation on the River Trent, UK. *Regulated Rivers: Research and Management,* **6**, 321–332.

GURNELL, A. M. 1995. Vegetation along river corridors: hydrogeomorphological interactions. *In*: PETTS, G. E. & GURNELL, A. M. (eds) *Changing River Channels*. Wiley, Chichester, 237–260.

HARVEY, A. M. 1985. The river systems of North-west England. *In*: JOHNSON, R. H. (ed.) *The Geomorphology of North-West England*. Manchester University Press, Manchester, 122–142.

HIGHAM, N. 1986. *The Northern Counties to AD 1000*. Longman, Harlow.

HUPP, C. 1996. Plant ecological aspects of flood geomorphology and paleoflood history. *In*: ANDERSON, M. G., WALLING, D. E. & BATES, P. D. (eds) *Floodplain Processes*. Wiley, New York, 335–356.

JACKSON, R. G. 1978. Preliminary evaluation of lithofacies models for meandering alluvial streams. *In*: MIALL, A. D. (ed.) *Fluvial Sedimentology*. Canadian Society of Petroleum Geologists Memoirs, **5**. 543–573.

LARGE, A. R. G. & PETTS, G. E. 1994. Restoration of river margins. *In*: CALOW, P. & PETTS, G. E. (eds.) *The Rivers Handbook*, Vol. 2. Blackwell Scientific, London, 401–418.

—— & —— 1996. Historical channel–floodplain dynamics along the River Trent: implications for river rehabilitation. *Applied Geography,* **16**, 191–201.

—— & PRACH, K. 1998. Floodplain ecology of the regulated River Trent: implications for rehabilitation. *In*: BAILEY, R. G., JOSÉ, P. V. & SHERWOOD, B. (eds) *United Kingdom Floodplains*. Westbury, Otley, 409–421.

LOWE, J. J. & WALKER, M. J. C. 1997. *Reconstructing Quaternary Environments*. Longmans, Harlow.

MACKLIN, M. G. & LEWIN, J. 1993. Holocene river alluviation in Britain. *Zeitschrift für Geomorphologie,* **88**, 109–122.

——, PASSMORE, D. G. & RUMSBY, B. T. 1992. Climatic and cultural signals in Holocene alluvial sequences: the Tyne Basin, Northern England. *In*: NEEDHAM, S. & MACKLIN, M. G. (eds) *Alluvial Archaeology in Britain*. Oxford, Oxbow Monograph, **27**, 123–140.

MANNION, A. M. 1986. Plant macrofossils and their significance in Quaternary palaeoecology. Part I: Introduction. *Progress in Physical Geography,* **10**, 194–214.

MARRIOTT, S. B. 1998. Channel–floodplain interactions and sediment deposition on floodplains. *In*: BAILEY, R. G., JOSÉ, P. V. & SHERWOOD, B. (eds) *United Kingdom Floodplains*. Westbury, Otley, 43–62.

MCCARTHY, M. R. 1995. Archaeological evidence for the Roman impact on vegetation near Carlisle. *The Holocene*, **5**, 491–495.

MIALL, A. D. 1977. A review of the braided-river depositional environment. *Earth-Science Reviews*, **13**, 1–62.

—— 1994. Reconstructing fluvial macroform architecture from two-dimensional outcrops: examples from the Castlegate Sandstone, Book Cliffs, Utah. *Journal of Sedimentary Research*, **B64**, 146–158.

MOORES, A. J., PASSMORE, D. G. & STEVENSON, A. C. 1999. High resolution palaeochannel records of Holocene valley floor environments in the North Tyne basin, northern England. *In*: BROWN, A. G. & QUINE, T. A. (eds) *Fluvial Processes and Environmental Change*. Wiley, Chichester, in press.

O'KEEFFE, J. 1997. Methods of assessing conservation status for natural freshwaters in the Southern hemisphere. *In*: BOON, P. J. & HOWELL, D. L. (eds) *Freshwater Quality: Defining the Indefinable?* Stationery Office, Edinburgh, 369–386.

PASSMORE, D. G. & MACKLIN, M. G. 1997. Geoarchaeology of the Tyne Basin: Holocene river valley environments and the archaeological record. *In*: TOLAN-SMITH, C. (ed.) *Landscape Archaeology in Tynedale*. Tyne–Solway Ancient and Historic Landscapes Research Programme Monograph, **1**, 37–52.

——, ——, BREWER, P. A., LEWIN, J., RUMSBY, B. T. & NEWSON, M. D. 1993. Variability of Late Holocene braiding in Britain. *In*: BEST, J. L. & BRISTOW, C. S. (eds) *Braided Rivers*. Geological Society, London, Special Publications, **75**, 205–232.

PAUTOU, G. & DECAMPS, H. 1985. Ecological interactions between the alluvial forests and hydrology of the Upper Rhône. *Archiv für Hydrobiologie*, **105**, 13–37.

PETTS, G. E., LARGE, A. R. G., GREENWOOD, M. T. & BICKERTON, M. A. 1992. Floodplain assessment for restoration and conservation: linking hydrogeomorphology and ecology. *In*: CARLING, P. A. & PETTS, G. E. (eds) *Lowland Floodplain Rivers: Geomorphological Perspectives*. Wiley, Chichester, 217–234.

——, MÖLLER, H. & ROUX, A. L. (eds) 1989. *Historical Changes of Large Alluvial Rivers in Western Europe*. Wiley, Chichester.

PRACH, K., JENÍK, J. & LARGE, A. R. G. (eds) 1997. *Floodplain Ecology and Management: the Luznice River in the Trebon Biosphere Reserve*. SPB Academic Publishing, Amsterdam.

REINECK, H. E. & SINGH, I. B. 1975. *Depositional Sedimentary Environments*. Springer-Verlag, Berlin.

ROBERTS, N. 1998. *The Holocene: an Environmental History*. Blackwell Scientific, London.

SCHUMM, S. A. 1979. Geomorphic thresholds: the concept and its applications. *Transactions of the Institute of British Geographers*, **4**, 485–515.

STUIVER, M., LONG, A. & KRA, R. S. (eds) 1993. 1993 Calibration issue. *Radiocarbon*, **35**, 1.

TIPPING, R. 1992. The determination of cause in the generation of major prehistoric valley fills in the Cheviot Hills, Anglo-Scottish border. *In*: NEEDHAM, S. & MACKLIN, M. G. (eds) *Alluvial Archaeology in Britain*. Oxford, Oxbow Monograph, **27**, 111–121.

—— 1998. The chronology of Late Quaternary fluvial activity in part of the Milfield Basin, northeast England. *Earth Surface Processes and Landforms*, **23**(9), 845–852.

WASYLIKOWA, K. 1986. Analysis of fossil fruits and seeds. *In*: BERGLUND, B. E. (ed.) *Handbook of Holocene Palaeoecology and Palaeohydrology*. Wiley, Chichester, 571–590.

WATTS, W. A. 1978. Plant macrofossils and Quaternary palaeoecology. *In*: WALKER, D. & GUPPY, J. C. (eds) *Biology and Quaternary Environments*. Australian Academy of Science, Canberra, A.C.T., 53–67.

The contribution of a multiproxy approach in reconstructing floodplain development

MARK DINNIN[1] & BARBARA BRAYSHAY[2]

[1]*School of Geography and Archaeology, University of Exeter, Exeter EX4 4QH, UK (e-mail:m.h.dinnin@sussex.ac.uk)*

[2]*Department of Archaeology and Prehistory, University of Sheffield, Sheffield S1 4ET, UK (e-mail:b.brayshay@sheffield.ac.uk)*

Abstract: This paper discusses the results of palynological, fossil insect and sedimentological investigations of floodplain deposits in the lower reaches of the River Trent. The results demonstrate the value of using multiproxy records by enhancing the resolution of a model for the development of the Trent floodplain during Holocene time. In particular, these data provide evidence for ecological and hydrological process–response relationships within the catchment. The combined results highlight the role of anthropogenic disturbance in major changes in floodplain ecology and hydrology during the later prehistoric and historical periods. The results and their implications for nature conservation management of floodplain natural resources are discussed in their national contexts and with reference to the implications for habitat restoration.

With a catchment of *c.* 10 450 km^2, and a length of 247 km the River Trent ranks as one of Britain's largest rivers (Fig. 1). In common with most other rivers in Northwest Europe, the natural floodplain habitats have been significantly fragmented and degraded by human activities that include canalization, drainage and agricultural reclamation (Brown *et al.* 1997). Historical and documentary sources suggest that intense habitat change occurred during the last few centuries (Baldock 1984; Cory 1985; Purseglove 1988; Dinnin 1997*a*). The palaeoenvironmental archive preserved within the floodplain sediments of the middle and upper Trent and its tributaries provides evidence for major changes in floodplain habitats and sedimentation regimes during the historical and prehistoric periods (e.g. Smith 1973; Greig *et al.* 1979; Straw & Clayton 1979; Buckland & Sadler 1985; Greenwood *et al.* 1991; Brown & Keough 1992; Gaunt *et al.* 1992; Brown *et al.* 1994; Gaunt 1994; Knight & Howard 1994; Lillie & Grattan 1994). In contrast to the upper and middle Trent, the lower Trent has until recently received relatively little palaeoenvironmental attention. This may be explained in part by the practical constraints imposed by thick minerogenic Holocene aggradation deposits, which are in places in excess of 30 m deep (Gaunt *et al.* 1992), and relatively few opportunities for clear sections afforded by commercial mineral extraction.

The relative roles of human activity and natural processes in triggering major changes in floodplain development in the Trent and Humberhead Levels continue to be debated. For example, Lillie & Grattan (1994) attributed a suite of so-called 'flood gravels', containing large numbers of possibly reworked trees, in the middle Trent at Langford and Girton to a phase of landscape instability triggered by intensified forest clearance during the Bronze Age. Knight & Howard (1994) and Lillie & Grattan (1994) *inter alia* questioned both the timing and anthropogenic role in the proposed phase of enhanced flooding. In many of the peri-estuarine floodplains of the region the uppermost sedimentary unit comprises a metre or more of reddish oxidized fine-grained alluvium (Buckland & Sadler 1985; Riley *et al.* 1995; Dinnin 1997*b*, *c*). Buckland & Sadler (1985) attributed this change in overbank sedimentation to accelerated soil erosion resulting from agricultural developments during the late Roman period. However, the exact timing and origin of this sediment input remains uncertain (Long *et al.* 1998).

Sampling and analysis

The sample site of Bole Ings is a low-lying area of former carr land 2 km south of Gainsborough (Fig. 1). At this point the Trent floodplain is bounded to the east by the Jurassic escarpment of the Lincoln Edge and to the west by the low hills of Mercia Mudstone (Keuper Marl). The floodplain deposits

From: MARRIOTT, S. B. & ALEXANDER, J. (eds) 1999. *Floodplains: Interdisciplinary Approaches*. Geological Society, London, Special Publications, **163**, 179–195. 1-86239-050-9/99/$15.00

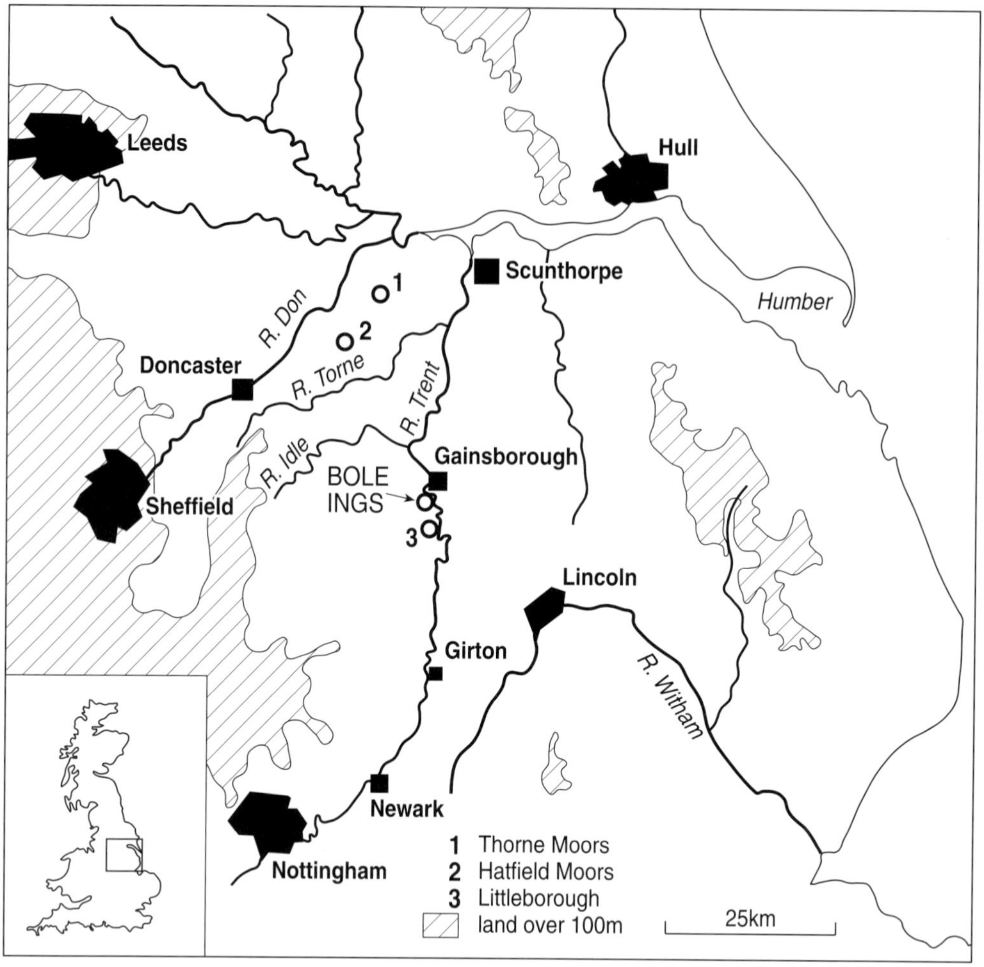

Fig. 1. The location of Bole Ings (Nottinghamshire, UK).

comprise Holocene alluvium overlying Devensian (last glaciation) First Terrace sands and gravels (Fig. 2). The post-glacial rise in sea level led to channel and floodplain aggradation during early to mid-Holocene time, resulting in the deposition of alluvium in the deeply incised sand and gravel flanked Devensian drainage network (Gaunt *et al.* 1992; Gaunt 1994). Reconnaissance boreholes indicate a maximum of *c.* 9 m of Holocene alluvium in the vicinity of Bole Ings. The intercalated organic and minerogenic sediments provided the potential for detailed palaeoenvironmental reconstruction and a chronological framework of radiocarbon dates (Fig. 3). The stratigraphic sequences recorded from these boreholes are referenced to Ordnance Datum (OD), Newlyn. A virtually continuous sediment sequence was recovered using a Pilkon percussion drilling rig from the site with the most complete organic-rich sediment sequence (borehole C). The sequence of fine-grained minerogenic sediments (–8.69 to –5.58 m OD) and peaty–silty clay with abundant *Alnus* macrofossils (–5.59 to +0.59 m OD) implies a low-energy, poorly drained depositional environment, such as a backswamp or floodbasin. Five radiocarbon date determinations indicate that the palaeoenvironmental record spans the early to later Holocene period, *c.* 8300–2700 BP (Fig. 3).

Samples for pollen analysis were prepared using standard KOH digestion and acetolysis procedures (see Berglund & Ralska-Jasiewiczowa 1989); identifications were made with reference to standard keys and texts (Punt 1976; Punt & Clarke 1980; Moore *et al.* 1991) and the University of Manchester departmental reference collection. Pollen taxa and plant nomenclature follow Stace

(1991) and Bennett (1994). A minimum 300 terrestrial pollen grains (excluding *Alnus*) per sample were counted. A pollen diagram has been constructed by calculating pollen percentages based on total land pollen (tlp) and tlp + pteridophytes (Fig. 4). A second pollen diagram presents tlp excluding wetland taxa, to provide a clearer picture of changes in vegetation beyond the floodplain (Fig. 5). Both pollen diagrams were constructed using TILIA and TILIA GRAPH (Grimm 1993). The oxidized clay–silt that composed the uppermost 2.5 m of core C contained no identifiable pollen or insect remains and is therefore not included in Figs 4 and 5. The sedimentary record was sampled in 100 mm slices for insect analysis (Table 1). Fossil beetle fragments representing a minimum of 1871 individuals belonging to at least 374 taxa were extracted from 29 samples using the 'paraffin flotation' technique of Coope & Osborne (1968). A full list of insects identified from the samples, together with an indication of their conservation status (*sensu* Shirt 1987) is provided in Table 2. This table can be obtained from the Society Library or the British Library Document

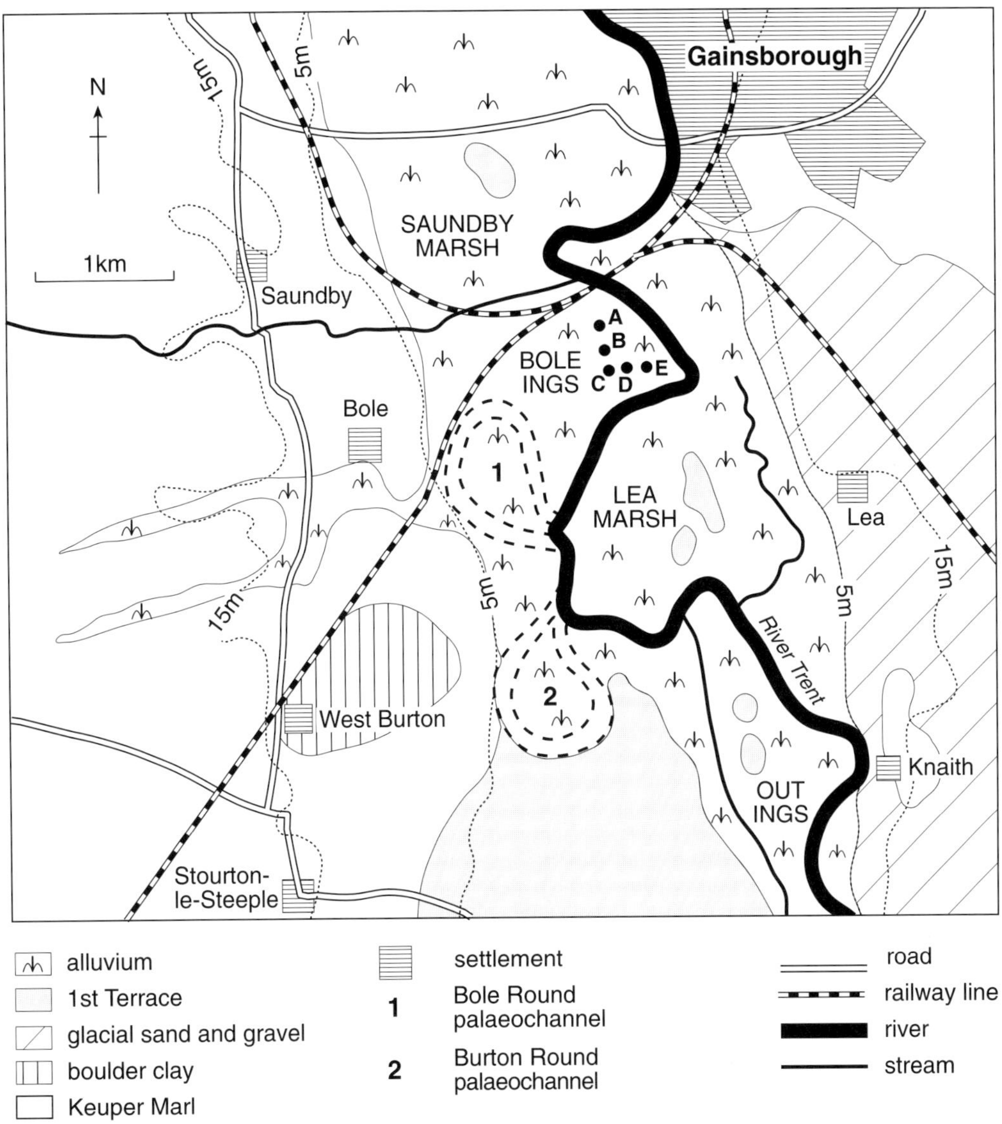

Fig. 2. Summary of surface geology, geomorphology and borehole locations around Bole Ings.

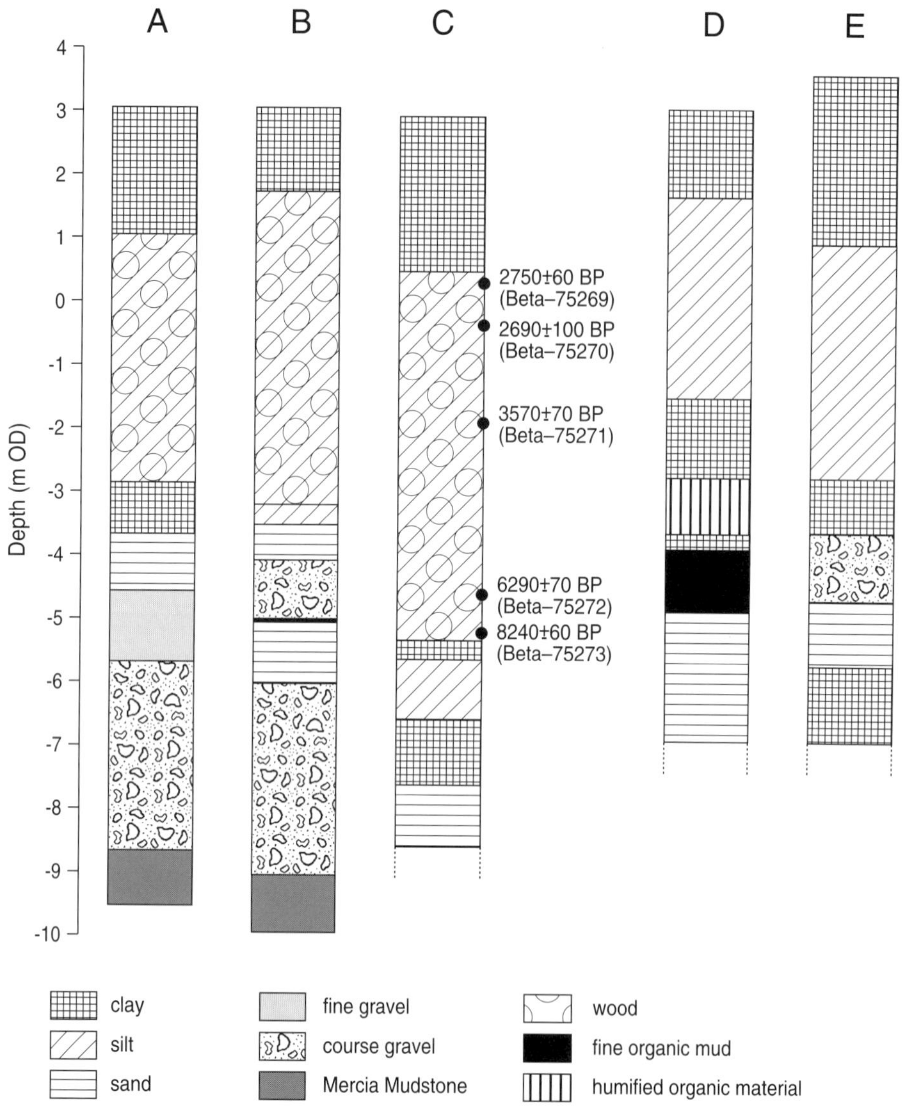

Fig. 3. Sedimentary sequences and radiocarbon dated horizons recorded from boreholes across Trent floodplain at Bole Ings.

Supply Centre, Boston Spa, Wetherby, West Yorkshire LS23 7BQ, UK, as Supplementary Publication No. SUP 18135 (21 pages). Taxonomy follows Kloet & Hinks (1977).

Floodplain development

Fossil beetle and pollen concentration was generally higher in the more organic or fine-grained sediments than in the coarser sands and minerogenic clay–silt. Low numbers of both insect and botanical fossil remains were recorded between −2.87 and −1.27 m OD. This is probably a consequence of poor preservation resulting from the drying out of the humified silty peat, discussed below. Clearly discernible changes occur in the faunal and floral assemblages through time (Fig. 6).

Early–mid-Holocene period: c. *8240–*c. *6300* BP

The basal pollen zone (LPAZ BI/1) spans a period of almost 2000 years between 8240 ± 60 BP (Beta-75272) and 6290 ± 70 BP (Beta-75271) (Figs 4 and

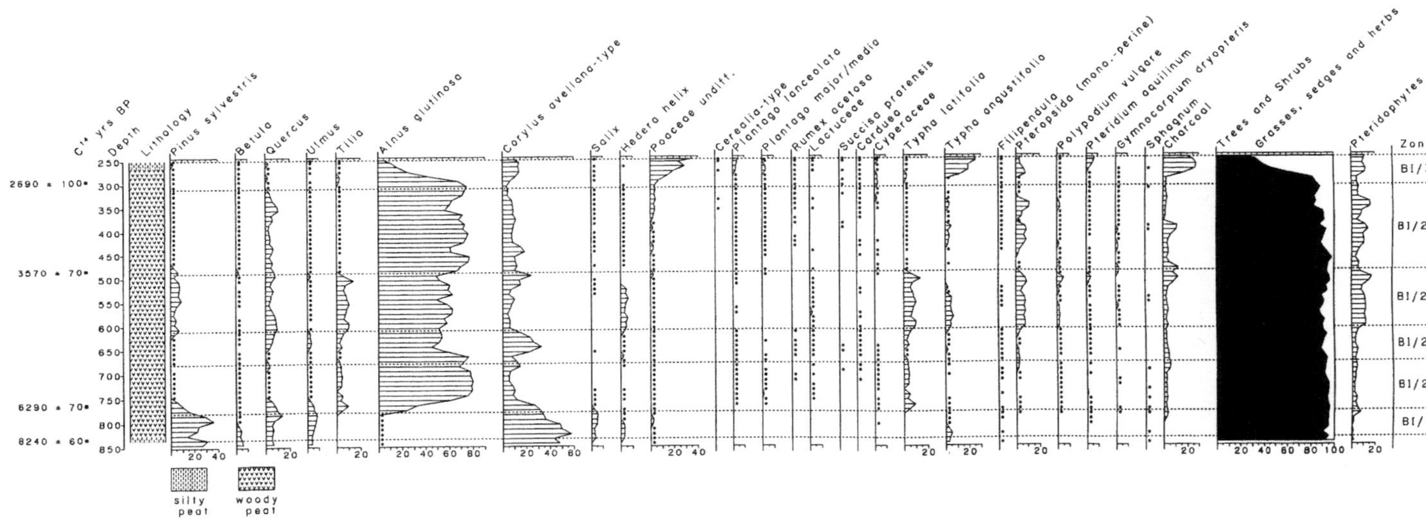

Fig. 4. Pollen percentage diagram for selected taxa, Bole Ings borehole C. Depth in centimetres.

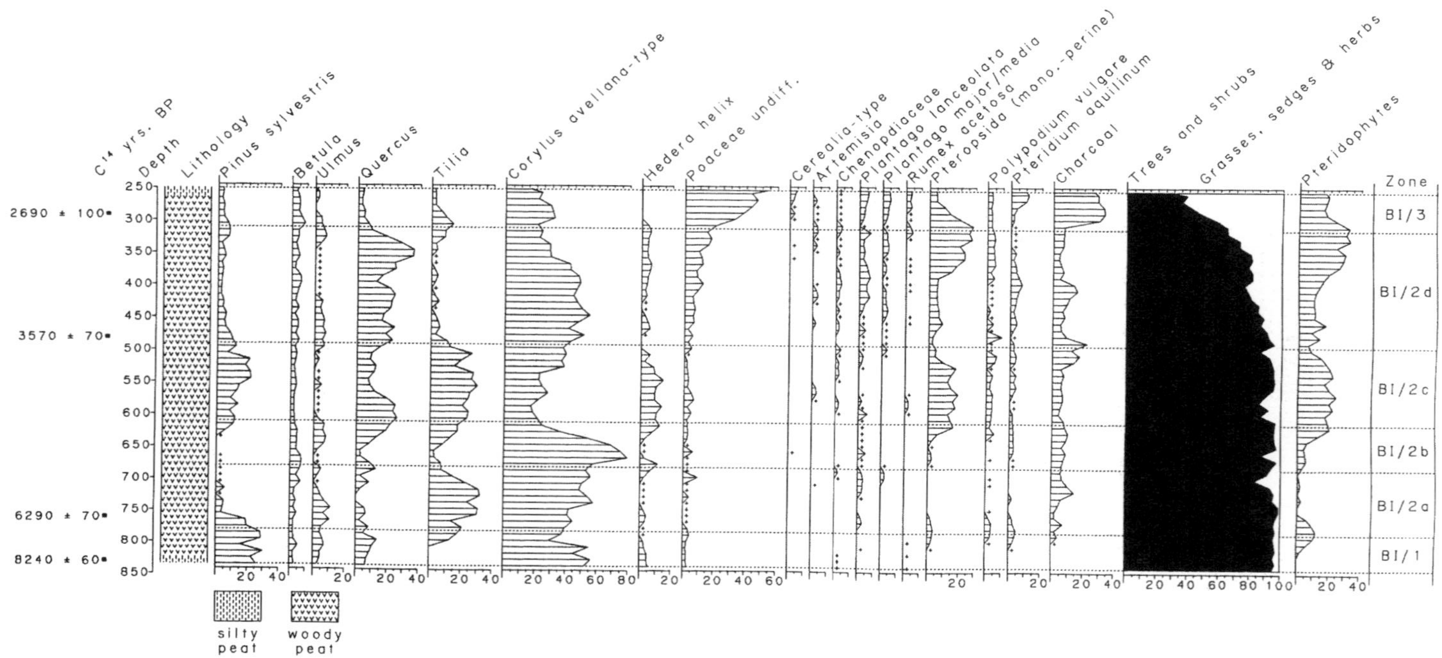

Fig. 5. Pollen percentage diagram for selected dryland taxa, Bole Ings borehole C. Depth in centimetres.

Table 1. *Fossil insect sample numbers and depths in borehole C*

Sample number	Depth(m OD)	Sample number	Depth (m OD	Sample number	Depth(m OD)
1	0.80–0.90	11	−1.73 to −1.63	21	–4.05 to –3.98
2	0.44–0.54	12	−1.87 to −1.77	22	–4.23 to –4.13
3	0.30–0.40	13	–2.17 to –2.07	23	–4.58 to –4.48
4	–0.05–0.05	14	–2.37 to –2.27	24	–4.73 to –4.63
5	–0.20 to –0.1	15	–2.72 to –2.62	25	–5.08 to –4.98
6	–0.56 to –0.46	16	–2.87 to –2.77	26	–5.23 to –5.13
7	–0.70 to –0.60	17	–3.27 to –3.17	27	–5.60 to –5.50
8	−1.06 to –0.96	18	–3.37 to –3.27	28	–6.63 to –5.63
9	−1.27 to −1.17	19	–3.78 to –3.67	29	–7.63 to –6.63
10	−1.37 to −1.27	20	–3.82 to –3.72		

5). The pollen data indicate a predominantly wooded floodplain with the developing woodland containing three main canopy-forming elements. Pollen and records of the pine shoot feeder *Hylobius abietis* (L.) indicate that *Pinus sylvestris* expanded during this period, perhaps colonizing suitable habitats on the gravel islands and ridges of the floodplain, apparently uninhibited by competition from deciduous species such as *Quercus* and *Ulmus*. A second softwood component comprised *Salix* and *Populus*. These fast-growing pioneer taxa tolerate periodic flooding and probably occupied wetland margins along the river bank. The abundance of this vegetation type is further suggested by numerous records of *Phyllodecta vulgatissima* (L.), *Chalcoides fulvicornis* (F.) and *Curculio salicivorus* (Payk.), phytophagous beetles that are dependent on these tree taxa. A third minor component consisted of deciduous hardwoods such as *Quercus, Tilia* and *Ulmus*, which were most probably migrating into the area and establishing small local populations, or were growing some distance from the site, possibly as part of woodland establishing in drier areas of the floodplain. Low pollen percentages of *Alnus* also suggest establishment in the wider region. Such an interpretation would explain the dearth of beetles associated with these tree species.

Low pollen percentage frequencies of *Betula* and the phytophagous beetle *Ramphus pulicarius* (Herbst) probably reflect the remnants of a once more substantial early post-glacial pioneer *Betula* population, out-competed by later arriving tree species. The leaf roller beetle *Rhynchaenus avellanae* (Donov.) and high levels of *Corylus avellana*-type pollen indicate the local abundance of this tree taxon, either as a dominant understorey shrub or in woodland edges on drier areas of the wetland margin and the surrounding landscape. The increasingly diverse saproxylic (tree-dependent) component of beetle assemblages during this period indicates an abundance of dead wood within the floodplain forest (e.g. *Hylecoetus dermestoides* (L.), *Cis bidentatus* (Ol.), *Tritoma bipustulata* (F.), *Opilo mollis* (L.), *Anobium* spp., *Cerylon histeroides* (F.), *Melasis buprestoides* (L.), *Agrilus* spp., *Scolytus* spp., *Xyloborus* spp., *Dromeolus barnabita* (Villa & Villa). Woodland similar in plant species composition appears to be typical of the early to mid-Holocene period in the Humber lowlands (e.g. Bartley 1962, pers. comm.; Beckett 1981; Gilbertson 1984). The relatively limited range of tree species present at this time explains the somewhat restricted arboreal phytophagous beetle fauna; similar patterns have been noted at other early Holocene sites (e.g. Osborne 1974, 1980; Dinnin 1992).

Plants associated with grassland and disturbed habitats make a persistent but minor contribution to the pollen assemblage during early Holocene time, most notably *Plantago lanceolata, Rumex acetosella, Artemisia, Cirsium*, Euphorbiaceae, Gentianaceae, Geraniaceae and Asteraceae. Likewise, the beetle assemblages contain sporadic records of xerophilous and phytophagous taxa dependent on disturbed or open ground plants (e.g. *Chaetocnema concinna* (Marsh.), *Gastrophysa viridula* (Deg.), *Ceuthorrynchus troglodytes* (F.), *C. assimilis* (Payk.), *Cidnorhinus quadrimaculatus* (L.), *Serica brunnea* (L.)). Background levels of disturbed land or grassland, ground and dung elements in assemblages with large saproxylic–woodland components are not unusual in early Holocene floodplain sequences (e.g. Robinson 1991; Allen & Robinson 1993). For example, analysis of an early to mid-Holocene palaeochannel sequence at Girton (25 km upstream of Bole Ings) indicates a relatively open and unstable floodplain environment during the first two millennia of the Holocene period (Dinnin 1992).

During this time the floodplain was characterized by a dynamic braided channel system that allowed

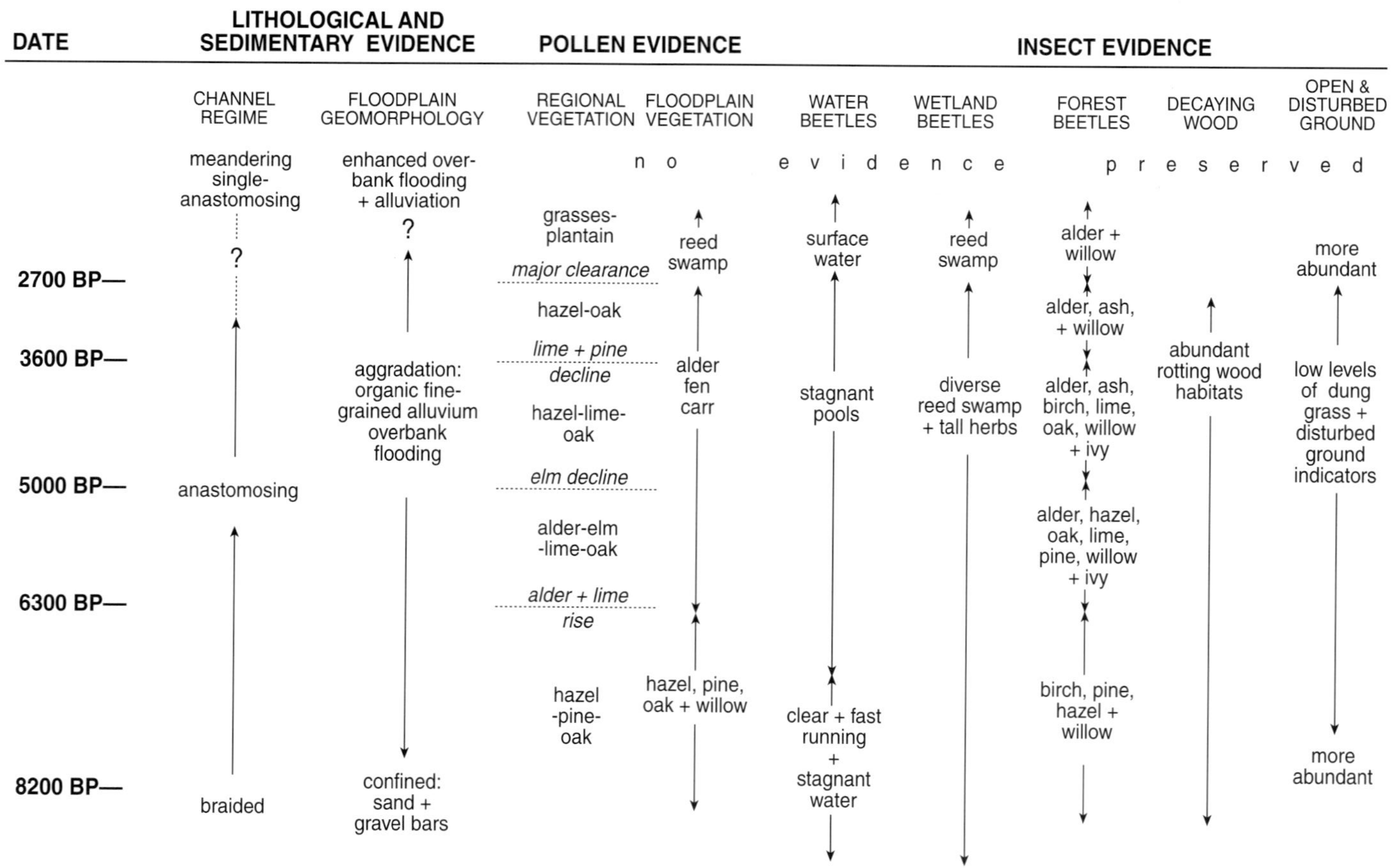

Fig. 6. Summary of environmental changes reflected by plant, insect and sedimentary evidence from Bole Ings.

only a sparse, early successional vegetation (e.g. *Rumex*, Chenopodiaceae) to become established, with *Pinus* perhaps restricted to adjacent gravel terraces (Green 1991). Contemporary floodplain invertebrate communities at Girton comprised mainly open ground, xerophilous taxa. Apart from occasional specimens of *Salix*- and *Corylus*-dependent beetles (e.g. *Chalcoides* spp.), there is a notable dearth of invertebrate taxa associated with floodplain woodland (Dinnin 1992).

Sedimentary evidence indicative of a relatively high-energy, unstable channel system is provided by substantial units of sands and gravels in boreholes across Bole Ings (between *c.* –6.00 to –9.00 m OD in boreholes A and B and below –8.70 m OD in borehole C). A sandy unit within core B (–4.00 to –5.10 m OD) is interpreted as an avulsion event. During early Holocene time, the confinement of the river within the channel created by the phase of Devensian incision would probably have resulted in erosion and reworking of the First Terrace and Devensian sands and gravels. Extensive exposures created by quarrying on the floodplain of the middle Trent at sites such as Girton, Langford Lowfield and Colwick (Brown *et al.* 1994; Knight & Howard 1994; Lillie & Grattan 1994) include units of cross-stratified channel trough sands and gravels, indicative of braided channel regimes during early Holocene time. At both Girton and Bole Ings this dynamic and unstable floodplain environment, with bank erosion cutting into gravel terraces, and bar and spit formation, would have created a shifting mosaic of riparian grassland. This, together with natural openings in the developing forest, for example, created by windthrow, would have ensured a continuous supply of suitable early successional habitats.

The aquatic beetle assemblages from Bole Ings provide specific information about river water quality during early Holocene time. Relatively fast-flowing, clear and well-oxygenated river conditions are suggested by the riffle beetles *Macronychus quadrituberculatus* (Mull.), *Normandia nitens* (Mull.) and *Stenelmis canaliculata* (Gyll.). Similar suites of elminthids occur in early Holocene palaeochannel deposits at Girton and in the River Aire (Dinnin 1992), as well as a series of other early to mid-Holocene sites in the English Midlands (Osborne 1997). The absence of elminthids in overlying deposits at Bole Ings can be attributed to loss of suitable habitat resulting from a shift in water flow away from the sample site. Evidence for this is provided by a change in lithostratigraphy towards increasingly fine-grained peaty sediment and palynological evidence for a development of riparian woodland. There is, however, evidence from later Holocene sites in the UK for an anthropogenic role in the decline of many of these taxa. Osborne (1988, 1997) attributed their disappearance to habitat destruction caused by anthropogenically triggered soil erosion during the Late Bronze Age.

Both pollen and insect records provide evidence for stagnant weed-choked water with rich fringing reedswamp (e.g. *Typha angustifolia, Phragmites australis, Glyceria aquatica, Scirpus lacustris*), tall herb (e.g. *Lythrum salicaria, Rorripa* spp., *Iris pseudacorus, Hydrocharis morsus-ranae, Lysimachia vulgaris, Veronica* spp., *Ceratophyllum demersum, Hippuris vulgaris, Myosotis* spp., *Apium inundatum, Heracleum sphondylium, Filipendula ulmaria*) and aquatic vegetation. For example, Cruciferae (specifically *Rorippa amphibia* or *R. sylvestris*), aquatic Ranunculaceae, *Myriophyllum spicatum, Littorella uniflora* and *Lemna* (*cf. minor*) are represented in the pollen record or indicated by host-specific beetle taxa such as *Hydrothassa marginella* (L.) and *Tanysphyrus lemnae* (Payk.). Stagnant or shallow water conditions are also implied by records of beetles dependent on the water lilies *Nymphaea alba* and *Nuphar lutea*.

By about 7000 BP the middle and lower reaches of the Trent had evolved an anastomosing or single-channel regime in response to hydrological changes associated with rising sea level, catchment soil stabilization and forest. Geomorphological evidence for this includes changes in sedimentation from crudely bedded gravels and sands to finer-grained silts and clays, and there is sedimentological evidence for abandonment and silting-up of secondary channels (Brown *et al.* 1994; Gaunt 1994; Knight & Howard 1994). Floodplain stabilization and extensive forest cover is reflected by the subsequent decline in disturbed ground and open floodplain plant and insect communities.

Mid–late Holocene period: c. *6300–*c. *2700* BP

A major change is recorded in the floodplain vegetation at 6290 ± 70 BP (Beta-75271, –4.76 m OD): a massive increase in *Alnus* pollen, defining the lower limit of LPAZ BI/2. Local domination is corroborated by abundance in the sediments of *Alnus* seeds and dependent beetle taxa (e.g. *Chrysomela aenea* (L.), *Dryocoetes alni* (Georg)). This probably reflects regional expansion, establishment and domination of the local floodplain vegetation by *Alnus* carr (Figs 4 and 5). This expansion appears to have been at the expense of other arboreal species, particularly *Corylus*. The marked change may be related to rising water table and waterlogging of floodplain soils, which would favour *Alnus*, triggered by rising sea level and resulting floodplain aggradation. Although it has been argued that *Alnus* expansion may have been aided by climatic amelioration (Chambers & Elliot 1989), there is nothing to suggest this was the case at Bole Ings. Once established in the fen carr, *Alnus* formed a dense canopy woodland that persisted for *c*. 3600 years.

A second significant vegetational event occurs with the arrival and establishment of *Tilia*, shortly after the expansion of *Alnus* (*c*. 6290 ± 70 BP (Beta-75271)). The relatively low amounts of *Tilia* pollen recorded at Bole Ings may reflect its establishment in the landscape beyond the fen carr. However, consistent records of the lime-borer *Ernoporus caucasicus* (Lind.) imply local growth of *Tilia* and the dearth of *Tilia* pollen could reflect a taphonomic bias (see Greig 1982). Furthermore, as Waller (1994) has stressed, aggradation in response to rising sea level would have meant an increasingly broad floodplain, and a corresponding increase in the distance between the sample site and drier soils favoured by *Tilia*.

Woodland also becomes more diverse in this zone, with light-demanding species such as *Fraxinus excelsior, Prunus spinosa*, *Ilex* and *Sambucus* present for the first time. The changes recorded in the pollen and beetle data during this period suggest that there were significant successional changes in the woodland stands of the floodplain. As the pioneering softwood species *Pinus, Salix* and *Populus* declined there was a corresponding rise in *Alnus*, which, once established, dominated the local fen carr. The increase in pollen frequencies of *Quercus, Ulmus* and *Tilia* also suggest a more stable catchment with less overbank flooding. In the fen carr understorey, the ground flora included ferns such as *Polypodium vulgare*, *Pteridium aquilinum*, *Gymnocarpium dryopteris* and woody climbing plants such as *Lonicera periclymenum*, *Hedera helix* and *Solanum dulcamara*, which were probably limited to the woodland margins, shaded out by a dense *Alnus* canopy. *Juniperus communis* disappears from the pollen record probably as a result of habitat loss; it may have been part of an understorey in the gravel ridge *Pinus* stands and was unable to find a new niche once the river stabilized.

There is a notable deterioration in pollen and insect fossil concentration and quality between *c*. 5350 and 3600 BP in borehole C at Bole Ings (Fig. 4). This could result from a phase of drier floodplain conditions, which would adversely affect fossil preservation. This degradation could be a result of aeration associated with root penetration and water-table draw-down by *Alnus* growing in the fen carr (see Bennett 1990). There is, however, both pollen and insect evidence for the continued presence of aquatic and riparian vegetation similar to that recorded for the early Holocene period.

Following an initial decline at the expense of *Alnus*, trees and shrubs such as *Quercus, Ulmus* and *Pinus* re-established throughout LPAZ BI/2, perhaps in new populations in drier parts of the landscape (Fig. 5). Declines in *Ulmus, Pinus, Tilia* and *Corylus* recorded during this period suggest changes to the vegetation in the wider landscape, beyond the floodplain. A distinct *Ulmus* decline occurs at –3.16 m OD (LPAZ BI/2b-2c). Elsewhere this widely recorded palynological event has been dated to *c*. 5100–5300 BP (Huntley & Birks 1983; Peglar 1993). There is little evidence to support a purely anthropogenic clearance hypothesis for the *Ulmus* decline at Bole Ings. However, neither is there entomofaunal evidence for the forest pathogen vector (e.g. *Scolytus scolytus* (F.)) responsible for spreading Dutch elm disease. A marked *Tilia* decline at *c*. 3579 ± 70 BP (Beta-75271, –1.95 m OD) is associated with increased representation of species typical of grassland and/or disturbed habitats, notably *Plantago lanceolata*, *Rumex conglomeratus* type, and species of Caryophyllaceae, Polygonaceae, Chenopodiaceae and Lactuceae. There is a concomitant peak in particulate charcoal, signalling an increase in the magnitude and/or frequency of fire in the landscape at this time. These features imply significant anthropogenic forest disturbance beyond the floodplain. Similar Bronze Age agricultural clearances have been recorded in pollen signals from a wider region (Turner 1962, 1965; Greig 1982; Smith 1985), and there is a wealth of archaeological evidence for human exploitation of dryland areas during this period (Knight & Howard 1994).

Despite evidence for regional landscape disturbance, the combined palynological and entomological data indicate stability in the floodplain environment. *Alnus* continued to dominate the fen carr on this part of the Trent floodplain, typified by further records of the *Alnus*-feeder *Chrysomela aenea*. Furthermore, the plethora of mature forest or dead wood habitat indicator beetles in these deposits (e.g. *Aderus brevicornis* (Per.), *Hylecoetus dermestoides*, *Ptilinus* spp., *Anobium* spp., *Xyloborus* spp.) suggests that the floodplain woodland community remained relatively undisturbed by anthropogenic activity.

The combined palaeoecological data show that during this period the floodplain and regional vegetation underwent a complex series of successional changes, some of which may have been induced by anthropogenic activity. The palaeoecological evidence for lack of appreciable disturbance of the broad floodplain is corroborated by the sedimentary records in boreholes A–E. The continuous sequence of low-energy organic-rich deposits provides no significant change in sedimentary regime. The intercalated clay silts and woody peat suggest continued overbank flooding within a wooded floodplain environment. Brown *et al.* (1994) pointed out that the confluence of the Trent, Dove and Soar results in the middle Trent receiving a large sediment input. Prehistoric catchment disturbance (either natural or anthropogenic) may have triggered relatively dramatic changes in river sedimentation. There is, however, no evidence at Bole Ings for the high-energy Bronze Age flood event(s) sediments described by Lillie & Grattan (1994) some 30 km upstream. It would appear that whatever the magnitude and frequency of these events, the floodplain ecosystem in this part of the Trent was able to buffer most of the effects until *c.* 2690 ± 100 BP (Beta-75270, –0.46 m OD). The dissipation of sediment in the middle reaches of the Trent may have been facilitated by the anastomosing channel in this reach of the river system.

Late Holocene period: c. *2700–*c. *2200* BP

The uppermost LPAZ BI/3 (Figs 4 and 5) is characterized by reductions in pollen representation of woodland and scrub taxa and a corresponding increase in grassland and reedswamp indicators (e.g. *Typha angustifolia*). The decline in floodplain woodland corresponds to a sedimentary change from woody, silty peat to organic silty clay. Fossil insect assemblages are dominated by aquatic and wetland taxa including *Hydrochara caraboides* (L.), *Hydrophilus piceus* L. and *Hydraena palustris* Erich. Continued records of taxa such as *Chrysomela aenea*, *Leperesinus varius* (F.), *Hylesinus oleiperda* (F.), *Grynobius* sp. and *Anobium* sp., *Leptura* sp. imply the local presence of some *Alnus*, *Fraxinus* woodland with dead wood. There is no increase in beetles associated with pasture, disturbed or open ground in these levels. These sedimentary and palaeoecological changes suggest increased wetness at the site rather than local expansion of grassland or cultivation. Disturbance indicator plants, including *Plantago lanceolata*, *P. major* or *P. media* and *Rumex* spp., could have found a niche in a changing river floodplain environment in which increased water-logging or flooding created open or disturbed habitats.

These changes could have been forced by a combination of natural and anthropogenically driven allochthonous processes. There is evidence for a contemporary episode of rapid coastal change between about 3000 and 2000 BP with a widespread expansion and contraction of marine conditions in the Humber (Long *et al.* 1998). Such an episode of positive sea-level tendency would have had a significant effect on the floodplain water table. It would have increased the tidal influence in the Trent and triggered the backing-up of runoff, leading to enhanced floodplain inundation and wetness. There is also pollen evidence for intensified arable and pastoral agriculture on land beyond the floodplain after 2690 ± 100 BP (Beta-75270, –0.46 m OD). This includes significant declines in arboreal taxa, especially *Quercus*, and corresponding increases in Poaceae, *Plantago*, *Rumex* spp. and Cerealia pollen indicating the cultivation of crops. Charcoal representation also increases dramatically during this period. This supports archaeological evidence for the exploitation of freely draining soils of the Lincoln Edge, Lincolnshire Wolds, Sherwood Sandstone and gravel terraces of the Trent (Holland 1975; May 1976; Loughlin & Miller 1979; Knight & Howard 1994). Such significant vegetational changes would have had an effect on catchment hydrology and soil stability, and the impacts of this have been reported from lowland rivers elsewhere in the UK (e.g. Burrin & Scaife 1984; Burrin 1985).

At Bole Ings there is a marked change in sedimentation above +0.9 m OD in borehole C, from organic mud to a reddish brown partially oxidized weakly laminated silty clay (2.0 m thick). A similar sedimentary unit occurs in all boreholes across the floodplain at this site (Fig. 3). The sediment has the characteristics of one deposited by overbank flooding. Unfortunately, the inorganic unit did not contain any preserved pollen or insect remains, and consequently the timing of this sedimentary change and duration of the phase are uncertain. The radiocarbon date of 2750 ± 60 BP (Beta-7526) obtained from a bulk sample of organic mud at +0.24 m OD is discordant with that of 2600 ± 100 BP (Beta-75270) obtained from an *Alnus* twig at –0.46 m OD. This could result from bulk sample contamination by reworked carbon, eroded from catchment sediments. The change in sedimentation occurs some 0.5 m above the palynological signal for regional intensification of agricultural deforestation during the Late Bronze Age. The stratigraphic position of the oxidized alluvium at Bole Ings would imply that this change in sedimentation occurred during the Romano-British period or later. This could reflect a significant process–response lag between forest clearance, soil erosion and overbank deposition in the catchment. Alternatively, a more recent, post-forest clearance, environmental change may be responsible (Buckland & Sadler 1985). A lithologically similar unit seals a series of Roman sites on the lower Trent floodplain, including that 3 km upstream at Littleborough (Riley *et al.* 1995), and elsewhere in the Humberhead Levels (Buckland & Sadler 1985; Dinnin 1997*b*). It seems reasonable to suggest that the deposit at Bole Ings is coeval with that at Littleborough. Buckland & Sadler (1985) attributed the change in overbank sedimentation to accelerated soil erosion resulting from agricultural developments, more specifically, a switch to winter wheat production, during the late Roman or early post-Roman period. Although Buckland & Sadler (1985) pointed out that this may have been enhanced by an episode of positive sea-level tendency, there is no clear evidence for such a trend in the Humber at this time (Long *et al.* 1998).

Thus, the demise of the floodplain forests in this part of the Trent appears to have occurred in the context of regional deforestation within the catchment (e.g. Smith 1985). During this period, when, regionally, vegetation changed from the original forest mosaic to an agricultural landscape characterized by open grassland plant communities, soil erosion in the catchment brought to an end almost 3600 years of floodplain stability.

Implications for conservation and restoration of floodplain habitats

The results of this and similar palaeoenvironmental studies provide a temporal perspective on the development and demise of natural floodplains. The results suggest that the restoration of natural floodplain communities is challenged by the problems of national extinction, or virtual extirpation, of many species that were once significant components of the ecosystem. The present-day restriction of these taxa to isolated remnant habitats, coupled with apparently poor dispersal ability among many of the key invertebrates (Foster 1991; Warren & Key 1991), suggests that many elements (e.g. *Hydrophilus piceus*, *Aderus brevicornis*, *Ernoporus caucasicus*) may not benefit from floodplain habitat restoration unless careful consideration is given to enhancing habitat continuity. The palaeoenvironmental record can help to resolve this problem by providing indicator species for evaluating restoration, identifying biogeographical linkages and identifying processes that are important in maintaining high floodplain diversity.

This palaeoenvironmental investigation contributes baseline information about the ecological characteristics of unmodified floodplain ecosystems and biogeomorphology. This information is of potential value in habitat restoration and assessment of changes in species diversity (e.g. Amoros *et al.* 1987; Greenwood *et al.* 1991; Amoros & Van Urk 1992; Petts *et al.* 1992; Rosenzweig 1995; Hughes 1997). A series of methodological and interpretational factors create difficulties when using the pollen record to reconstruct species composition and spatial distribution of past vegetation communities (Webb *et al.* 1978; Jacobsen & Bradshaw 1981; Bradshaw & Webb 1985; Prentice 1985, 1988). This study demonstrates that combining the pollen and insect fossil data allows more detailed assessment of local and regional habitat communities. For example, the coincidence of arboreal pollen and fossils of their host-specific beetle taxa in an assemblage suggests a local source, demonstrated here by records of *Tilia* pollen and an associated lime-borer (*Ernoporus caucasicus*). Additional advantages arise from records of phytophagous (plant-feeding) insects indicative of plants that are poorly represented in the pollen record (e.g. *Juncus* spp.) and, in particular, monophagous (plant-specific) insects, which may increase the taxonomic resolution of the vegetation record. For example, in this study phytophagous beetles have indicated specific aquatic plants (e.g. *Apium inundatum*, *Rorripa amphibium*, *Caltha palustris*) that are palynologically difficult to resolve beyond family level. Stenotopic insect taxa may also provide information about the structural and functional diversity of vegetation and habitats (*sensu* Franklin 1988) that cannot be inferred from the pollen record. For example, the diverse dead wood beetle assemblages recorded from Bole Ings indicate the survival of relatively undisturbed mature floodplain forest during later prehistory.

Floodplain restoration and management aims to restore the vegetation associated with natural floodplain features, such as wooded islands (Brown *et al.* 1997). This study demonstrates how palaeoenvironmental data can provide an indication of the community composition in these natural habitats. This includes identification of biotic elements that might not generally be considered in relation to floodplain restoration. During much of the Holocene period the floodplain forests, riparian vegetation and aquatic habitats at Bole Ings supported insect communities that contained many taxa that are now considered relic species, or indicators of relict habitats in the UK. The fossil records from Bole Ings add to a growing picture that by the Bronze Age species-rich primary forest beetle faunas (e.g. the nationally extinct *Dromaeolus barnabita* (Villa & Villa)) had become increasingly restricted to relatively inaccessible and little disturbed woodland in agriculturally marginal areas and extensive wetlands (Buckland 1979; Girling 1985; Dinnin 1992; Whitehouse 1993). Forested floodplains may have served as the last refuges for many of the denizens of primary forest during the later prehistoric and historical periods. Floodplain forest management and clearance has destroyed all but a few of these refugia (Brown *et al.* 1997; Harper *et al.* 1997) so that today many of

Britain's rarest forest invertebrates (e.g. *Ernoporus caucasicus*) are restricted to ancient pasture woodland and medieval deer parks (Warren & Key 1991). At these sites over-mature centuries old pollarded trees have provided the habitat continuity necessary for the continued survival of these specialist taxa (Warren & Key 1991; Harding & Rose 1987). The recognition that floodplain forests in the past supported what are today considered 'primary forest' invertebrate communities highlights the potential of floodplain woodland restoration in expanding the habitats available for this now nationally rare component of the fauna. This, however, necessitates considering restoration on a catchment scale, to ensure that suitable networks or migration corridors exist within the landscape to enable these late-successional forest taxa to colonize restored floodplain forest from their current refuges (see Warren & Key 1991).

Records from Bole Ings and later Holocene sites in the UK identify a series of relict floodplain fen indicators, such as *Hydraena palustris*, *Limnebius aluta* (Bedel), *Hydrochara caraboides* and *Hydrophilus piceus*, *Dromius longiceps* (Dej.) and *D. sigma* (Rossi). The evidence suggests that they remained relatively widespread and common until the medieval and early post-medieval period (e.g. Girling & Greig 1989; Dinnin 1991). Their decline must be mainly attributed to drainage and reclamation of floodplain wetlands, since the mid-16th century, and mainly during the last 150 years or so (Ratcliffe 1984; Williams 1990). Similarly, the palaeontological record demonstrates that the Red Data Book elminthids *Macronychus quadrituberculatus*, *Normandia nitens* and *Stenelmis canaliculata* were present in the lower and middle Trent until at least the 11th century AD (Smith & Greenwood pers. comm.). Pollution and habitat destruction has greatly affected these taxa, with the result that only *M. quadrituberculatus* has been recorded in modern times from the Trent, last recorded at Burton-on-Trent during the 1950s (Holland 1980). Recently, attention has been drawn to the importance of woody debris in floodplain ecological and geomorphological systems (e.g. Gurnell *et al.* 1995; Gurnell 1997). The practice of scouring and removing obstructing wood has probably contributed to the virtual extirpation from Britain of *M. quadrituberculatus*, a species that clings to and probably feeds on submerged rotten wood (Holland 1980). These taxa not only indicate the need to retain woody debris in channels but also the need to retain or restore coarse-bedded reaches within lower parts of fluvial systems. The suite of elminthids could be used as indicators of the success of this aspect of habitat restoration.

The identification of potential biogeographical modern analogues is useful in providing a basis for habitat restoration. This study indicates that during the early to mid-Holocene period the floodplain forest vegetation of the middle to lower Trent had much in common with the modern-day alluvial hardwood forests found in large European floodplains (Karpati & Toth 1961; Schnitzler 1994). This is further emphasized by records of several invertebrate taxa with a markedly continental aspect to their distribution (e.g. *Dromaeolus barnabita*, *Rhynchites auratus* or *R. bacchus*). Similar affinities have been noted from later prehistoric forest insect faunas preserved in the peats of Thorne and Hatfield Moors, to the west of the Trent floodplain (Buckland 1979; Whitehouse *et al.* 1998). The modern floodplains of the Rhine, Rhône, Loire, Po, Danube and Oder contain areas of *Quercus–Ulmus–Fraxinus* forest (*Querco-Ulmetum minoris*), which together with stands of *Salix* and *Populus,* form a mosaic of floodplain forest habitats that are associated with periodically flooded but well-drained and very fertile soils (Karpati & Toth 1961; Schnitzler 1994). Possibly before the inundation of the North Sea basin, *c.* 8500 BP (Preece 1995), the Trent and the Rhine were part of a more extensive palaeohydrological and palaeo ecosystem, hence the similarities in the species composition of the potential alluvial forest vegetation. The palaeoenvironmental record thus reaffirms the value of the surviving fragments of semi-natural floodplain woodland in these major European rivers as analogues for habitat restoration in lowland reaches of rivers in this part of Britain.

Perhaps the greatest contribution of the palaeoenvironmental record is the significant information provided about turnover rates and processes within floodplains at a variety of temporal and spatial scales. It is possible in the record to distinguish between changes in floodplain resulting from natural processes such as post-glacial relaxation phenomena and those superimposed on this pattern by anthropogenic disturbance and management (see Hughes 1997, fig. 1). Long-term turnover patterns (i.e. $>10^3$ years) on a large scale (i.e. $>10^5$ m) are most readily identifiable in the early Holocene period. These changes are linked to succession on floodplain and terraces associated with a complex interplay between climate change, rising base level, the influence of these on hydrological and sedimentological inputs to the floodplain, biogeographical range-associated response to these factors, and the resulting inter-species competition (see Hughes 1997). Examples of this include the transformation in early Holocene time from a braided, unstable floodplain regime, dominated by open ground and early successional taxa, to an anastomosing channel regime with extensive floodplain woodland. This is intimately linked with the stabilization of catchment soils, and the replacement of *Pinus*-dominated floodplain woodland by mixed deciduous forest (*c*. 8300 BP) and ultimately *Alnus*-dominated woodland (*c*. 6300 BP). The abruptness with which these major biogeomorphological changes may occur are illustrated by the rapid expansion of *Alnus*, *c*. 6300 BP, and change in floodplain sedimentation to a regime dominated by fine-grained overbank deposits during early Holocene time (Figs 4 and 6). The apparent low turnover rate at this scale within the floodplain during mid- to later Holocene time (*c*. 6300–2700 BP) demonstrates the stability of established *Alnus* floodplain woodland. The marked change in floodplain biogeomorphology after *c*. 2700 BP resulted from a phase of enhanced overbank flooding, which had a marked effect on floodplain communities and soils (Fig. 6). This demonstrates the potential response of the river channel and floodplain regime to longitudinal variations in catchment land use and perhaps minor adjustments in base level, resulting from projected climate change.

In contrast, at lower temporal and spatial scales (i.e. $<10^3$ years and $<10^5$ m) there is evidence for high turnover. This is mainly reflected in high biodiversity of the insect and plant assemblages from the Trent floodplain. This diversity is linked to floodplain habitat heterogeneity and the close association in space and time of early and late successional habitats, ranging from gravel bars to undisturbed mature forest (see Hughes 1997; Harper *et al*. 1997). This heterogeneity can be attributed to natural disturbance and succession regimes associated with overbank flooding, erosion and windthrow. On natural floodplains this allowed the existence close in space and time of communities containing taxa that today are geographically disparate. Within this system overbank flooding plays an important role in creating opportunities for regeneration, turnover, creating a heterogeneous age structure and variety of early to late successional habitats, addition of soil nutrients and dispersal of plants and animals (Hughes 1997). The role of beavers in altering floodplain hydrology and successional trends is recognized. The creation of floodplain wetlands and early successional habitats that arise from beaver activity is likely to play an important role in creating opportunities for plant and invertebrate communities associated with forest clearings and edges. The extinction of beavers from Britain and their increasing rarity in mainland Europe may also be a significant factor in the decline of dead wood insect communities, and especially those associated with submerged dead wood (e.g. *Macronychus quadrituberculatus*).

The importance of enhancing the coupling between river channel and floodplain hydrology and biogeomorphology is increasingly recognized in restoration. In a floodplain with a long history of management, such a change is likely to be feasible only in selected reaches, using controlled flooding and disturbance (Hughes 1997). Harmon *et al*. (1986) estimated that 17% of trees in undisturbed woodland are dead or moribund, of which perhaps only one in a hundred provide a niche suitable for a late successional forest species such as *Dromaeolus barnabita* or *Ernoporus caucasicus*. Speight (1989) suggested that a minimum forest reserve size of 400–500 ha is needed to allow a return to natural forest conditions with a full range of successional habitats. This figure provides some indication of targets necessary for restoration of the late successional forest insect taxa found in undisturbed floodplain forests in Britain.

Conclusion

The contribution of multiproxy approaches to reconstructing floodplain development can be seen to lie in three main areas.

(1) These records provide baseline data for species and community composition of undisturbed floodplain habitats. These data demonstrate that forested floodplains have in the past provided important habitats for a range of forest communities no longer associated with these habitats. This provides potentially important indicator groups for assessing the success of future habitat restoration.

(2) In broader terms, comparisons can be made between unmodified floodplain environments and surviving semi-natural habitats in different regions. Biogeographical linkages may potentially be used to evaluate these habitats as analogues for restoration planning.

(3) Integrated analyses allow identification of ecological and hydrological process–response relationships within the catchment, and the recognition of processes that are important in maintaining high floodplain diversity on a variety of spatio-temporal scales. This may be used, for example, to model channel and floodplain sedimentation responses to changes in catchment vegetation and the resulting influence on floodplain communities in the past. This information provides a basis for identifying key requirements of future habitat restoration schemes, in particular the creation of a heterogeneous mosaic of successional floodplain habitats.

This research was assisted by funding from National Power. We are grateful to P. C. Buckland, J. Symonds and D. Gilbertson for project support and co-ordination. We thank Doncaster Museum and Art Gallery for providing access to insect reference collections, and H. Mendell and P. Whitehead for assistance with certain identifications.

References

ALLEN, T .G. & ROBINSON, M. A. 1993. *The Prehistoric Landscape and Iron Age Enclosure Settlement at Mingies Ditch Hardwick-with-Yelford, Oxon.* Thames Valley Landscapes: the Windrush Valley, **2**.

AMOROS, C. & VAN URK, G. 1992. Palaeoecological analyses of large rivers: some principles and methods. *In*: PETTS, G. E., MOLLER, H. & ROUX, H. (eds) *Historical Change of Large Alluvial Rivers.* Wiley, Chichester, 143–166.

——, ROUX, A. L., REYGROBELLET, J. L., BRAVARD, J. P. & PATOU, G. 1987. A method for applied ecological studies of fluvial hydrosystems. *Regulated Rivers*, **1**, 17–36.

BALDOCK, D. 1984. *Wetland Drainage in Europe. The Effect of Agricultural Policy in Four EEC Countries.* Russel Press, Nottingham.

BARTLEY, D. D. 1962. The stratigraphy and pollen analysis of lake deposits near Tadcaster, Yorkshire. *New Phytologist*, **61**, 277–287.

BECKETT, S. C. 1981. Pollen diagrams from Holderness, North Humberside. *Journal of Biogeography*, **8**, 177–198.

BENNETT, K. D. 1990. Postglacial history of alder (*Alnus glutinosa* (L) Gartn.) in the British Isles. *Journal of Quaternary Science*, **5**, 123–133.

—— 1994. *Annotated Catalogue of Pollen and Pteridophyte Spore Types of the British Isles.* Department of Plant Sciences, University of Cambridge.

BERGLUND, B. E. & RALSKA-JASIEWICZOWA, M. 1989. Pollen analysis and pollen diagrams. *In*: BERGLUND, B. E. (ed.) *Handbook of Holocene Palaeoecology and Palaeohydrology.* Wiley, Chichester.

BRADSHAW, R. H. W & WEBB, T. 1985. The relationship between contemporary pollen and vegetation data from Wisconsin and Michigan, USA. *Ecology*, **66**, 721–737.

BROWN, A. G. & KEOUGH, M. K. 1992. Palaeochannels, palaeoland-surfaces and the three-dimensional reconstruction of floodplain environmental change. *In*: CARLING, P. A. & PETTS, G. E. (eds) *Lowland Floodplain Rivers: Geomorphological Perspectives.* Wiley, Chichester, 185–202.

——, HARPER, D. & PETERKEN, G. F. 1997. European floodplain forests: structure, functioning and management. *Global Ecology and Biogeography Letters*, **6**, 169–178.

——, KEOUGH, M. K. & RICE, R. J. 1994. United Kingdom: the Lateglacial and Flandrian alluvial record from the Soar and Nene valleys. *Philosophical Transactions of the Royal Society of London, Series A*, **348**, 261–293.

BUCKLAND, P. C. 1979. *Thorne Moors: a Palaeo-ecological Study of a Bronze Age Site (a Contribution to the History of the British Fauna).* Department of Geography, University of Birmingham, Occasional Publications, **8**.

—— & SADLER, J. P. 1985. The nature of late Flandrian alluviation in the Humberhead Levels. *East Midlands Geographer*, **11**, 239–251.

BURRIN, P. J. 1985. Holocene alluviation in South-East England and some implications for palaeo-hydrological studies. *Earth Surface Processes and Landforms*, **10**, 257–271.

—— & SCAIFE, R. G. 1984. Aspects of Holocene valley sedimentation and floodplain development in southern England. *Proceedings of the Geologists Association*, **95**, 81–96.

CHAMBERS, F. M. & ELLIOTT, L. 1989. Spread and expansion of *Alnus* Mill in the British Isles: timing, agencies and possible vectors. *Journal of Biogeography*, **16**, 542–550.

COOPE, G. R. & OSBORNE, P. J. 1968. Report on the coleopterous fauna of the Roman Well at Barnsley Park, Gloucestershire. *Transactions of the Bristol and Gloucester Archaeological Society*, **86**, 84–87.

CORY, V. 1985. *Hatfield and Axholme: an Historical Review.* Providence Press, Ely.

DINNIN, M. H. 1991. The sub-fossil occurrence of the greater silver water-beetle *Hydrophilus piceus* (L.) (Col: Hydrophilidae) at Shirley Pool, South Yorkshire. *Naturalist (Hull)*, **116**, 57–59.

—— 1992. *Islands within islands: Holocene development of the British entomofauna and the implications for conservation.* PhD thesis, University of Sheffield.

—— 1997*a*. The drainage history of the Humberhead Levels. *In*: VAN DE NOORT, R. & ELLIS, S. (eds) *Wetland Heritage of the Humberhead Levels.* Humber Wetlands Project, Hull, 19–30.

—— 1997*b*. The palaeoenvironmental survey of the Rivers Idle, Torne and Old River Don. *In*: VAN DE NOORT, R. & ELLIS, S. (eds) *Wetland Heritage of the Humberhead Levels.* Humber Wetlands Project, Hull, 81–155.

—— 1997*c*. Holocene beetle assemblages from the Lower Trent floodplain at Bole Ings, Nottinghamshire, UK. *Quaternary Proceedings*, **5**, 83–104.

FOSTER, G. N. 1991. Conserving insects in aquatic and wetland habitats with special reference to beetles. *In*: COLLINS, M. N. & THOMAS, J. A. (eds) *The Conservation of Insects and their Habitats.* Blackwell, London, 235–262.

FRANKLIN, J. F. 1988. Structural and functional diversity in temperate forests. *In*: WILSON, E.O. (ed.) *Biodiversity.* National Academy Press, Washington, DC, 166–175.

GAUNT, G. D. 1994. *Geology of the Country around Goole, Doncaster and the Isle of Axholme.* Memoir for 1:50 000 geological sheets 79 and 88 (England and Wales). HMSO, London.

——, FLETCHER, T. P. & WOOD, C. J. 1992. *Geology of the Country around Kingston upon Hull and Brigg.* Memoir for 1:50 000 geological sheets 80 and 89 (England and Wales). HMSO, London.

GILBERTSON, D. D. 1984. Early Neolithic utilisation and management of alder carr at Skipsea Withow Mere, Holderness. *Yorkshire Archaeological Journal*, **56**, 17–22.

GIRLING, M. A. 1985. An old forest beetle fauna from a Neolithic and Bronze Age peat deposit at Stileway. *Somerset Levels Papers*, **11**, 80–83.

—— & GREIG, J. R. A. 1989. Ecological interpretations from the beetle fauna. *In*: HAYFIELD, C. & GREIG, J. R. A. (eds) *Environmental Samples, Excavation and Salvage work on a Moated Site at Cowick, South*

Humberside, 1976. Yorkshire Archaeological Journal, **61**, 47–70.

GREEN, F. 1991. *Girton Quarry Archaeological Watching Brief. Palaeobotany Preliminary Report.* Trent and Peak Archaeological Trust, Nottingham.

GREENWOOD, M. T., BICKERTON, M. E., CASTELLA, E., LARGE, A. R. G. & PETTS, G. E. 1991. The use of Coleoptera (Arthropoda: Insecta) for patch characterisation on the floodplain of the River Trent, UK. *Regulated Rivers*, **6**, 321–332.

GREIG, J. R. A. 1982. Past and present lime woods of Europe. *In*: BELL, M. & LIMBREY, S. (eds) *Archaeological Aspects of Woodland Ecology*. British Archaeological Reports, Oxford, **S146**, 23–56.

——, OSBORNE, P. J., SMITH, C. & WILLIAMS, P. 1979. Part IV—the landscape of the Fisherwick area during the Iron Age. *In*: SMITH, C. (ed.) *Fisherwick: the Reconstruction of an Iron Age Landscape*. British Archaeological Reports, Oxford, **B61**, 93–103.

GRIMM, E. 1993. *TILIA 2.0 b4. Pollen Analysis Software*. Illinois State Museum, Springfield.

GURNELL, A. 1997. The hydrological and geomorphological significance of forested floodplains. *Global Ecology and Biogeography Letters*, **6**, 219–229.

——, GREGORY, K. J. & PETTS, G. E. 1995. The role of coarse woody debris in forest aquatic habitats: implications for management. *Aquatic Conservation*, **5**, 143–166.

HARDING, P. T. & ROSE, F. 1987. *Pasture–Woodland in Lowland Britain*. Institute of Terrestrial Ecology, Huntingdon.

HARMON, M. E., FRANKLIN, J. F., SWANSON, F. J., CROMACK, K. & CUMMINS, K. W. 1986. The ecology of coarse woody debris in temperate ecosystems. *Advances in Ecological Research*, **15**, 133–302.

HARPER, D., MEKOTOVA, J., HULME, S., WHITE, J. & HALL, J. 1997. Habitat heterogeneity and aquatic invertebrate diversity in floodplain forests. *Global Ecology and Biogeography Letters*, **6**, 275–285

HOLLAND, D. G. 1980. Distribution of Elmidae/Elminthidae. *Balfour–Browne Club Newsletter*, **16**, 1–12.

HOLLAND, S. M. 1975. Pollen analytical investigations at Crosby Warren, Lincolnshire, in the vicinity of the Iron Age and Romano-British settlement at Dragonby. *Journal of Archaeological Science*, **2**, 353–364.

HUGHES, F. M. R. 1997. Floodplain biogeomorphology. *Progress in Physical Geography*, **21**, 501–529.

HUNTLEY, B. & BIRKS, H. H. 1983. *An Atlas of Past and Present Pollen Maps for Europe: 0–13 000 years ago*. Cambridge University Press, Cambridge.

JACOBSEN, G. L. & BRADSHAW, R. H. W. 1981. The selection of sites for palaeovegetational studies. *Quaternary Research*, **16**, 80–96.

KARPATI, I. & TOTH, I. 1961. Die Auenwaldtypen Ungarns. *Acta Agronomica Academiae Scientiarum Hungaricae*, **11**, 421–452.

KLOET, G. S. & HINKS, W. D. 1977. *A Checklist of British Insects: Coleoptera and Strepsiptera*, 2nd edn. Handbooks for the Identification of British Insects, V(ii). Royal Entomological Society, London.

KNIGHT, D. & HOWARD, A. J. 1994. *Archaeology and Alluvium in the Trent Valley, an Archaeological Assessment of the Floodplain and Gravel Terraces*. Trent and Peak Archaeological Trust, University of Nottingham.

LILLIE, M. C. & GRATTAN, J. P. 1994. Geomorphological and palaeoenvironmental investigations in the Lower Trent Valley. *East Midlands Geographer*, **18**, 12–24.

LONG, A. L., INNES, J. B., KIRBY, J. R., LLOYD, J. M., RUTHERFORD, M. M., SHENNAN, I. & TOOLEY, M. J. 1998. Holocene sea-level change and coastal evolution in the Humber estuary, eastern England: an assessment of rapid coastal change. *The Holocene*, **8**, 229–247.

LOUGHLIN, N. & MILLER, K. R. 1979. *A Survey of Archaeological Sites in Humberside*. White & Farrell, Hull.

MAY, J. 1976. *Prehistoric Lincolnshire*. History of Lincolnshire Committee, Lincoln.

MOORE, P. D, WEBB, J. A. & COLLINSON, M. 1991. *Pollen Analysis*, 2nd edn. Blackwell Scientific, London.

OSBORNE, P. J. 1974. An insect assemblage of early Flandrian age from Lea Marston, Warwickshire, and its bearing on the contemporary climate and ecology. *Quaternary Research*, **4**, 471–486.

—— 1980. The Late Devensian–Flandrian transition depicted by serial insect faunas from West Bromwich, Staffordshire, England. *Boreas*, **9**, 139–147.

—— 1988. A Late Bronze Age fauna from the River Avon, Warwickshire, England: its implications for the terrestrial and fluvial environment and for climate. *Journal of Archaeological Science*, **15**, 715–727.

—— 1997. Insects, man and climate in the British Holocene. *Quaternary Proceedings*, **5**, 193–198.

PEGLAR, S. M. 1993. The mid-Holocene *Ulmus* decline at Diss Mere, Norfolk, UK: a year-by-year pollen stratigraphy from annual laminations. *The Holocene*, **3**, 1–13.

PETTS, G. E., LARGE, A. R. G., GREENWOOD, M. T. & BICKERTON, M. A. 1992. Floodplain assessment for restoration and conservation: linking hydrogeomorphology and ecology. *In*: CARLING, P. A. & PETTS, G. E. (eds) *Lowland Floodplain Rivers: Geomorphological Perspectives*. Wiley, Chichester, 217–234.

PREECE, R. C. (ed.) 1995. *Island Britain: a Quaternary Perspective*. Geological Society, London, Special Publications, **96**.

PRENTICE, I. C. 1985. Pollen representation, source area and basin size, towards a unified theory of pollen analysis. *Quaternary Research*, **23**, 70–86.

—— 1988. Records of vegetation in time and space: the principles of pollen analysis. *In*: HUNTLEY, B. & WEBB, T. (eds) *Vegetation History III*. Kluwer, Dordrecht, 17–32.

PUNT, W. 1976. *The Northwest European Pollen Flora, I (Parts 1–7)*. Elsevier, Amsterdam.

—— & CLARKE, G. C. S. (eds) 1980. *The Northwest European Pollen Flora, II (Parts 8–20)*. Elsevier, Amsterdam.

PURSGLOVE, J. 1988. *Taming the Flood*. Oxford University Press, Oxford.

RATCLIFFE, D. A. 1984. Post-Medieval and recent changes in British vegetation: the culmination of human influence. *New Phytologist*, **98**, 73–100.

RILEY, D. N., BUCKLAND, P. C. & WADE, J. S. 1995. Aerial reconnaissance and excavation at Littleborough-on-Trent, Notts. *Britannia*, **26**, 253–284.

ROBINSON, M. A. 1991. The Neolithic and Bronze Age insect assemblages. *In*: NEEDHAM, S. (ed.) *Excavations and Salvage Work at Runnymede Bridge, 1978: the Late Bronze Age Waterfront Site*. British Museum, London, 277–326.

ROSENZWEIG, M. L. 1995. *Species Diversity in Space and Time*. Cambridge University Press, Cambridge.

SCHNITZLER, A. 1994. European alluvial hardwood forests of large floodplains. *Journal of Biogeography*, **21**, 605–625.

SHIRT, D. B. 1987. *British Red Data Books: 2. Insects*. Nature Conservancy Council, Peterborough.

SMITH, B. 1985. *A palaeoecological study of raised mires in the Humberhead Levels*. PhD thesis, University of Wales, Cardiff.

SMITH, E. G. 1973. *Geology of the County around East Retford, Workshop and Gainsborough*. Memoir for sheet 101 (England and Wales). HMSO, London.

SPEIGHT, M. C. D. 1989. *Saproxylic Invertebrates and their Conservation*. Council of Europe, Strasbourg, Nature and Environment Series, **42**.

STACE, C. 1991. *New Flora of the British Isles*. Cambridge University Press, Cambridge.

STRAW, A. & CLAYTON, C. 1979. *The Geomorphology of the British Isles, Eastern and Central England*. Methuen, London.

TURNER, J. 1962. The *Tilia* decline: an anthropogenic interpretation. *New Phytologist*, **61**, 328–341.

—— 1965. A contribution to the history of forest clearance. *Proceedings of the Royal Society of London*, **161**, 91–104.

WALLER, M. 1994. Paludification and pollen representation: the influence of wetland size on *Tilia* representation in pollen diagrams. *The Holocene*, **4**, 430–434.

WARREN, M. S. & KEY, R. S. 1991. Woodland: past, present and potential for insects. *In*: COLLINS, M. N. & THOMAS, J. A. (eds) *The Conservation of Insects and their Habitats*. Blackwell, London, 155–211.

WEBB, T., III, LASESKI, R. A. & BERNABO, J. C. 1978. Sensing vegetation patterns with pollen data: choosing the data. *Ecology*, **59**, 1151–1163.

WHITEHOUSE, N. J. 1993. *A mid-Holocene forested site from Thorne Moors: the fossil insect evidence*. MSc thesis, University of Sheffield.

——, DINNIN, M. H. & LINDSEY, R. A. 1998. Conflicts between palaeoecology, archaeology and nature conservation: the Humberhead Peatlands SSSI. *Landscape Archaeology and Ecology* **3**, 70–78.

WILLIAMS, M. 1990. *Wetlands: a Threatened Landscape*. Blackwell, Oxford.

A mechanism for the formation of overconsolidated horizons within estuarine floodplain alluvium: implications for the interpretation of Holocene sea-level curves

STEPHEN CROOKS

Postgraduate Research Institute for Sedimentology, University of Reading, Whiteknights, Reading RG6 6AB, UK

Present address: Centre for Social and Economic Research on the Global Environment, School of Environmental Sciences, University of East Anglia, Norwich NR4 7TJ, UK (e-mail:s.crooks@uea.ac.uk)

Abstract: The stratigraphic sequence held in accumulating estuarine floodplain alluvium provides a detailed record of past sea-level movement. However, contrasting Holocene sea-level curves have been derived for two basins in southern Britain. The sea-level curve of the Thames Estuary Basin describes up to five possible regressive cycles superimposed onto a long-term transgressive trend. In the Severn Estuary no regressive cycles have been, as yet, identified. These differences in interpretation are, in part, due to the identification of 'overconsolidated' horizons, found within Essex alluvium, which have been argued to represent prolonged periods of desiccation. No such horizons have been found in the Severn Estuary. By examining the sedimentological and geotechnical characteristics of a number of active, land-claimed and regenerated saltmarshes from these regions, this paper identifies the mechanism for the formation of overconsolidated horizons to be geochemically controlled and regionally variable. The presence of suitable quantities of detrital calcium carbonate is important in preventing the formation of these horizons. The carbonate-deficient alluvium in the Thames Estuary region appears to be particularly sensitive to the effects of a lowered saline water table, inducing the deflocculation of clays and formation of dense horizons. By contrast, the sediments in the carbonate-bearing alluvium of the Severn Estuary are insensitive to pore-water salinity changes so an absence of overconsolidated horizons in the floodplain alluvium does not preclude past sea-level falls.

Estuaries lie at the boundary between marine and fluvial systems and are extremely sensitive to the effects of sea-level movements. This sensitivity is reflected within the vertical stratigraphy of the accumulating floodplain alluvium, which records past sea-level movements as a sequence of interdigitated marine and fresh-water facies. The UK has extensive, though now mostly embanked, estuarine floodplains. Two such areas are the coastal plain of southeast England, covering the estuaries of Essex and north Kent, and in southwest Britain, the floodplain of the Bristol Channel–Severn Estuary. To researchers investigating Holocene relative sea-level movements, these sequences provide a detailed sedimentological, palaeontological and geotechnical record of coastal evolution stretching back at least 6000 years (Hawkins 1971; Kidson & Heyworth 1973; Greensmith & Tucker 1976; Devoy 1977, 1979, 1982; Heyworth & Kidson 1982; Allen 1991; Hewlett & Birnie 1996; Haslett *et al.* 1998). However, even though these regions are separated geographically by only a very short distance, roughly 150 miles (240.59 km), considerable differences exist in terms of the form of derived sea-level curves (Fig. 1). Sea-level curves derived from southern Essex and the inner Thames Estuary suggest up to five sea-level regressions or still-stands superimposed on a long-term sea-level rise. However, the transgressive sea-level curve for the Bristol Channel–Severn Estuary does not pick up the signature of any major regressive or stillstand periods.

The differences in sea-level curve interpretation are, in part, influenced by the identification of what have been termed 'overconsolidated horizons', beds of sediments with significantly higher shear strength and density than the ambient sediment within the Essex alluvium. (The term overcon-solidated is part of the geological engineering

From: MARRIOTT, S. B. & ALEXANDER, J. (eds) 1999. *Floodplains: Interdisciplinary Approaches*. Geological Society, London, Special Publications, **163**, 197–215. 1-86239-050-9/99/$15.00

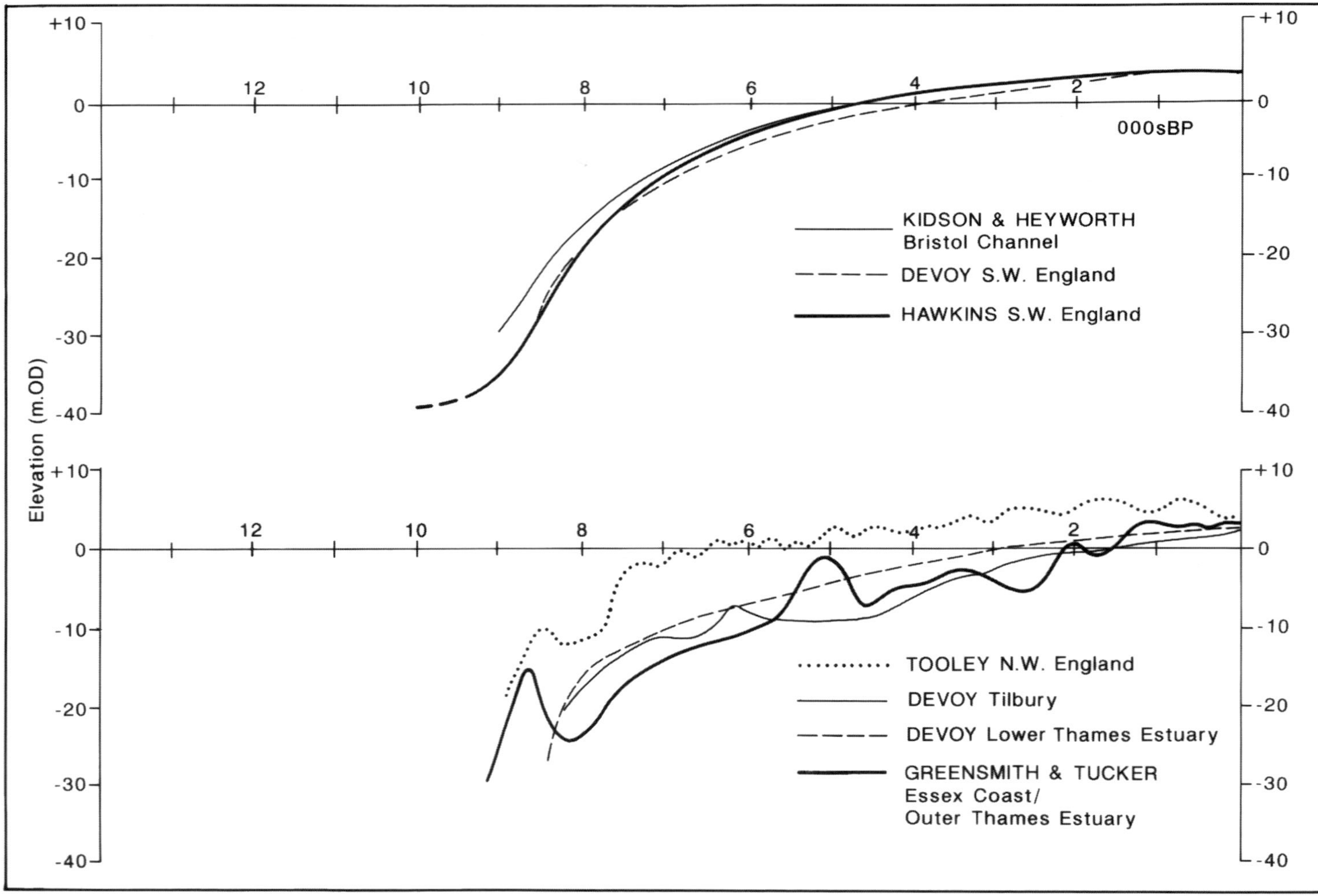

Fig. 1. Holocene sea-level curves for the Severn Estuary, Thames Estuary and northwest England (after Hawkins, 1984).

vocabulary, and refers to a sediment that has been subject to an 'apparent' load greater than the weight of the existing overburden (Skempton 1964). Loading can be due to previous thick cover of younger sediment or a to an overriding icesheet; effects similar to loading are also induced by prolonged periods of desiccation (Terzaghi & Peck 1969). The primary characterisitics of such sediment are high bulk densities associated with high shear strengths. (Evidence that overconsolidation is due to effects of desiccation rather than loading pressure is provided by the physiographical nature of the bed.) These are believed to indicate prolonged desiccation associated with a fall in relative sea level (Greensmith & Tucker 1971*a*, *b* 1973, 1976). Failure to identify such overconsolidated horizons in the southwest of England has led some to question their validity and further strengthen the argument that sea-level has risen without fluctuation within the Severn Estuary (Hawkins 1984). Clearly, therefore, it is important to determine why the overconsolidated horizons described by Greensmith and Tucker (1973) were not evident in the floodplain sediments of the Severn Estuary and what the implications may be for the determination of Holocene sea level curves.

To determine the validity and possible mechanisms by which overconsolidated horizons might form and the reasons why their distribution is regionally inconsistent then we must, relying on the Principle of Uniformitarianism, turn to a system which, in a modern-day environment, models a fall in sea level. Agricultural land-claim and subsequent drainage of estuarine saltmarshes provides a possible analogy for the cessation of tidal flooding and lowering of the saline watertable which would be associated with a fall of relative sea level. If overconsolidated horizons are a surface phenomenon related to a sea level fall then comparable geotechnical and pedological characteristics should be identifiable in near surface sediments of reclaimed clay-rich marshes. Moreover, there are a number of examples along the Essex coastline of reclaimed saltmarshes having been reintroduced into the intertidal environment following a storm breach in the protecting sea-embankment, thus providing an opportunity to examine the preservation potential of any structurally altered former reclaimed marsh sediments.

This paper draws upon clay-dispersion theory and sedimentological and geotechnical analysis of short cores from both regions, to suggest a mechanism for the formation of overconsolidated horizons. This is used to clarify the reasons for regional variability in structural behaviour of the fine-grained alluvium and to discuss the implications for sea-level curve interpretation.

Exchangeable cations and clay particles dispersion

The effect of exchangeable cations on the geotechnical properties and stability of sodic clay-rich sediments is now well documented (Norrish 1954; Quirk & Scholfield 1955; Rowell 1963; McNeal & Coleman 1966; Rimmer & Greenland 1976; Shanmuganathan & Oades 1982; Renasamy 1983; Dexter *et al.* 1984; Kjellander *et al.* 1988; Dexter & Chan 1991; Hodgkinson & Thorburn 1995; Skene & Oades 1995; Regea *et al.* 1997;Anson & Hawkins 1998). The interaction of clay particles in aqueous pore waters is an interplay between attractive and repulsive forces. Of the attractive forces between the clay particles, interatomic Van der Waals forces are the most significant, though the strength of these is strongly inversely related to particle separation. Repulsive forces include interactions of diffuse double-layers surrounding the clay particles, hydration of exchangeable cations and hydration of clay surfaces (Churchman *et al.* 1993).

The role of exchangeable cations lies in their ability to influence the dimensions of the diffuse double-layer and hence determine the magnitude of the repulsive forces. The diffuse double-layer consists of hydrated cations and water molecules, which are drawn to the negative charge on the clay mineral surface. The dimensions of the diffuse double-layer are most extensive when monovalent cations, particularly sodium, occupy exchange sites, as more hydrated molecules are required to balance the negative charge on the clay than if di- or trivalent molecules are present at any given pore fluid concentration. The dimensions of the diffuse double-layer are also affected by osmotic forces and, as the concentration of the pore waters falls, water molecules are drawn to the clay, leading to an expansion of the water film. In pore waters of low salinity, complete dispersion of sodium-saturated illites and smectites takes place as the diffuse double-layer expands and repulsive forces overcome attractive forces.

In sea water divalent cations are present in only low concentrations whereas sodium is abundant. However, the high ionic concentrations of sea water is sufficient to diminish the dimensions of the diffuse double-layer, so allowing flocculation. Crooks (1996) has shown that on coastal floodplains, Ca^{2+} may be made available through the dissolution of detrital bioclastic and lithoclastic $CaCO_3$. However, this distribution is locally and regionally variable, dependent on local geology and hydrodynamic conditions at the site of deposition. A strong correlation was found between the presence of $CaCO_3$ and consolidation of intertidal saltmarsh deposits, which led to the assertion that

Ca^{2+} may be an influencing factor in the erosion resistance of such deposits.

Agricultural land management and clay dispersion of sodic floodplain alluvium

Given that clay dispersion is a process dependent upon pore-water chemistry, and supply of stabilizing calcium ions is related to the supply of detrital carbonate particles, then it is possible that clay dispersion effects are also regional or local phenomena and are not ubiquitous to all coastal floodplains. Agriculturalists are particularly sensitive to the behaviour of soils, so agricultural land management practices provide some indication of the spatial distribution of any possible sites prone to clay dispersion.

The traditional agricultural practice on land-claimed marshland in southeast England and the Severn has been dominantly towards grazing pasture, but where arable cultivation has taken place drainage has been required to lower the saline water table. During the mid-20th century an intensification of arable production led to an increase in the extent of land drainage through the placement of ditches and internal drains. In north Kent and Essex, however, the drainage programme was not found to be wholly successful because within less than 10 years severe waterlogging occurred on a number of drained marshes and crop yield suffered. An investigation by the Soil Survey of the problem in Kent identified the drainage problems to be specific to those soil formations low in $CaCO_3$, most notably the Wallasea series (Hazelden *et al.* 1986). Waterlogging of sediment was not found to be a problem on nearby grasslands, in which a high saline water table was permanently maintained, nor on drained soil formations that contained $CaCO_3$. Further, laboratory analysis determined the cause to be dispersion of the sodium–clay particles induced by the reduction in salinity of surface pore waters as mobilized particles accumulated to block pores and cracks lower in the sediment profile where pore waters were more saline, as well as filling the drainage pipes and channels. As a result, permeability was reduced and surface waterlogging occurred. It was the presence of suitable quantities of pore-water Ca^{2+} in unaffected $CaCO_3$-bearing soil formations that protected the soil structure from dispersion. Thus Hazelden *et al.* identified the occurrence of $CaCO_3$ within sodic coastal marshes as being important in determining whether dispersion of clays occurs following land drainage. Similar problems and conclusions have been reported on the estuarine alluvium of Essex (Hodgkinson & Thorburn 1995). The application of gypsum (a source of calcium ions) to coastal floodplains affected by salt-water intrusion is now the standard treatment to prevent the detrimental effects of clay dispersion. In the Severn Estuary artificial drainage of the floodplain has been common practice over past centuries but to date no such problems of clay dispersion have been reported to the author's knowledge.

Site descriptions

Severn Estuary

Tidally derived post-glacial, fine-grained alluvium blankets an area of about 840 km^2 across the Inner Bristol Channel and Severn Estuary and attains a depth of 10–15 m in many places (Allen 1990). The main bulk of alluvium has been assigned, by Allen (1987) and Allen & Rae (1987), to a generally upward-fining lithostratigraphic unit called the Wentlooge Formation. Between depths of about –5 m OD (Ordnance Datum) to about 1.5 m OD (Hawkins 1984) the clayey silts are interbedded with continuous and discontinuous organic horizons. These organic horizons, which vary in thickness and extent, have been dated to between 6500 and 2500 ^{14}C years BP (Heyworth & Kidson 1982), and represent a range of non-saline wetland habitats including woodland, carr and raised bog (Smith & Morgan 1989). Seaward of defences, a stepped sequence of three younger morphostratigraphical units, in addition to localized Wentlooge outcrops, have been identified (Allen & Rae 1987). From the oldest to the youngest they are the Rumney, Awre and Northwick, which are believed to represent marsh initiations during the late 17th to early 18th century, late 19th century and mid-20th century, respectively.

A description of the geotechnical properties of floodplain alluvium in the Severn Estuary has been published by Skempton (1970), Cook & Roy (1984) and Hawkins *et al.* (1989). These researchers described the stratigraphy to consist of soft to firm, lightly consolidated blue–green clay with silty seams becoming common at the base and intermittent peats at mid-depth. Skempton (1970), for example, described a core collected from near Avonmouth as having plasticity that is remarkably constant throughout the sediment column, with plastic and liquid limits averaging between 27% and 71% and liquidity indices generally exceeding 0.5%.

Two sites were visited and at each, three marsh terraces sampled (Fig. 2A). At Littleton Warth, just north of Aust Cliff on the east bank of the estuary, three of the four marsh terraces are found. These are referred to as the Lower, Intermediate and Upper Marsh terraces, which, on the basis of

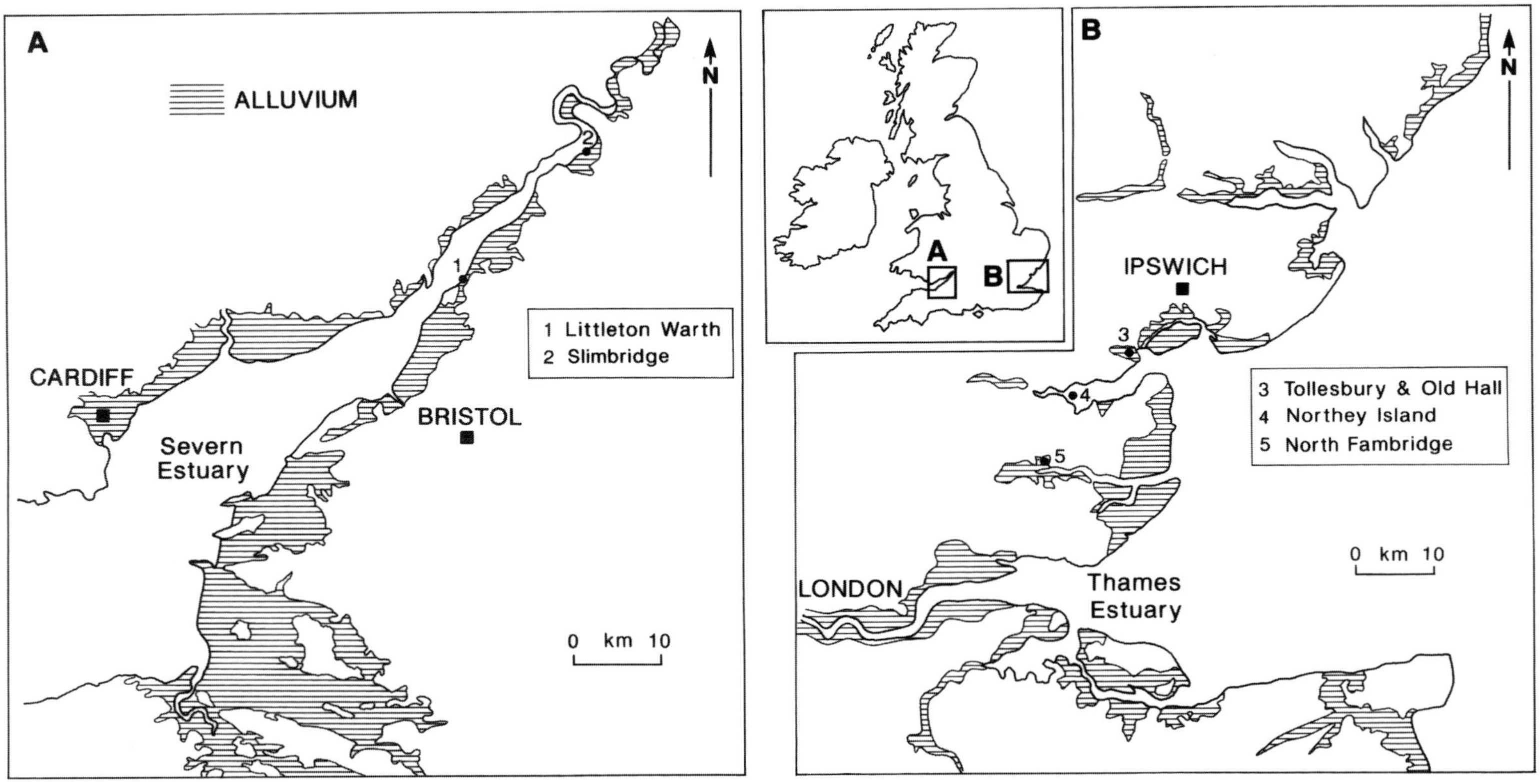

Fig. 2. Location of field sites in the Severn Estuary and across the Essex coast.

stratigraphic, sedimentological and geochemical evidence, are believed to represent the Northwick, Rumney and Wentlooge Formations, respectively. The Awre Formation was not present at this location. Each narrow terrace, typically less than 30 m in width, terminates seaward at, in the case of the two younger formations, a short clifflet 600–800 mm in height or, in the case of the oldest formation, a gentle ramp.

Further up the estuary near Slimbridge Warth three sequentially land-claimed marshes were sampled. The history of the post-Romano-British agricultural reclamation between Slimbridge Warth and Frampton-on-Severn, as described by Allen (1986), has taken place piecemeal over the past 700 years as local saltmarsh has built up and ground suitable for enclosure presented itself. The oldest land claim to be sampled is that of Katherine Cook's Leyes, dated to AD 1335–1336 on documentary evidence. Two subsequent land-claims were sampled, including that undertaken in the mid-18th century, to the northwest of the 14th-century land-claim, and the 19th-century embankment to the west. The fields are drained by peripheral ditches as well as by a series of mediaeval wooden culverts which underlie the area. The outfalls of these can be found along the shores of the estuary. At the time of sampling, nearby excavation by the National Rivers Authority (the predecessor of the Environment Agency) uncovered one of these wooden culverts, which was found to be still functioning to drain the hinterland (Ordnance Survey Grid Reference: SO 742066).

Southern Essex

The estuarine and embayment topography of the southern Essex region reflects the existence of a network of proto-Thames and Medway channels cut into the London Clay by periglacial rivers (Sheldon 1968; Greensmith & Tucker 1971*a*, *b*, 1976; Gibbard 1977; Green *et al.* 1982; Conway *et al.* 1984; Bridgland 1988). Surficial alluvium consisting of marine sands giving way to intertidal clays now blankets this basement, and reaches a maximum thickness of 39 m in buried channels (Greensmith & Tucker 1976). At a number of sites, the London Clay crops out to form low islands fringed by saltmarsh, such as Mersea Island, Northey Island and Osea Island. There is a large amount of archaeological evidence that Neolithic and Romano-British settlement took place on the floodplain throughout north Kent and southern Essex, which now lie several metres below mean sea level (Hazzeldene-Warren *et al.* 1936; Evans 1953; Ackeroyd 1972; Kirby 1990).

It is from some 130 cores collected across the floodplain between the Blackwater and Thames estuaries that Greensmith & Tucker (1976) based their interpretation of general Flandrian sea-level rise with episodic minor sea-level falls. These regressions were identified by the seaward extension of peat layers and the more widespread development of supratidal overconsolidated layers and partially gleyed soils (Greensmith & Tucker 1971*a*, *b*, 1973). The overconsolidated horizons possessed high densities and shear strengths, akin to the underlying Eocene basement and in stark contrast to the soft alluvium (Table 1). The natural water contents, liquid limits and plastic limits were also significantly lower than in the remainder of the alluvium. Physiographically, these horizons were marked by: a sharp vertical change in lithology at the surface of the unit (e.g. often from silt clay to sand); a change in faunal content (e.g. brackish-water fauna to open marine); increased carbon content usually as peat traces (reflecting slowing in minerogenic sedimentation); mottled structure in silty clay caused by rootlets; surface fissuring of silty clays (reminiscent of desiccation cracks); colour changes interpreted as a change in oxidation state.

This stretch of coastline, along with much of eastern England, is susceptible to storm surges, the results of which have been periodic breaches in embankments. Within Essex alone there are at least 30 sites of sea-embankment failure and natural marsh regeneration on former land-claimed marshland initiated since the 19th century, which make up a significant proportion of the existing saltmarsh habitat (IECS 1992). In all, four sites were investigated (Fig. 2B): the active marsh at Old Hall (age unknown but continually accreting for at least 400 years); the adjacent land-claimed section called Tollesbury Marsh, and two regenerated marshes, at Northey Island (Blackwater estuary) and North Fambridge (Crouch estuary), both taken by the sea in the storm of November 1997.

Field and laboratory methods

Sampling involved the characterization of sedimentological and geotechnical properties of the sediments

Table 1. *The undrained shear strength of sediments found at Bradwell, Essex, as recorded by Grennsmith & Tucker (1971*a*)*

Deposit	Shear strength (kPa)
Soft alluvium	13.7–29.4
Firm horizons	43.2–121.7
London Clay	49.1–147.2

throughout the upper 1 m of selected marshes in each of the study areas. Where possible, marshes of differing ages were selected to identify changes in sedimentological and geotechnical characteristics that take place as both active and reclaimed marshes mature. Sampling sites were chosen purposefully, on the basis of being representative of the marsh as a whole, following air photograph interpretation and ground walkover survey.

Undisturbed cores were collected for laboratory analysis, and *in situ* undrained shear strength measurements were made in the field. Cores were collected by driving short lengths of plastic pipe (250 mm length; 100 mm diameter), with a bevelled edge, vertically into the marsh surface. After insertion of the first core tube, a pit was dug to allow gentle removal of the core, which was sealed with water-tight end-caps and stored upright. A second tube was then placed in the exact location from which the first core had been removed and driven into the marsh in a similar manner. This procedure was repeated to give a total core length of 1 m. This method, although laborious, was found to be most effective in providing undisturbed samples with minimal compaction. At each stage of core extraction five undisturbed, undrained shear strength measurements were taken using an using an ELE Field Inspection Vane (range 0–200 kPa).

In the laboratory the sediments were extruded in short measured lengths (generally 30–40 mm; accurate to 0.5 mm) and the colour was recorded. A section from the centre of each sediment section was removed using a square cutter (area 360 mm^2) and used to determine bulk density and moisture content according to the methods described in BS 1377 (British Standards Institution 1990).

Sub-samples from the sediment slice were prepared for microfabric analysis by scanning electron microscopy (SEM), determination of clay mineralogy by X-ray powder diffraction (XRD), grain-size analysis by laser granulometry (Coulter LS130; range 0.1–900 μm) and determination of Atterberg limits (according to BS 1377 1990). For SEM examination of the sediment microfabric, small sub-samples were removed from the parent sub-sample, critical-point dried and then coated with a thin layer of gold in preparation in a JEOL 5300 scanning electron microscope equipped with a LINKED SYSTEM AN10085 X-ray analyser.

At depth intervals of *c.* 150 mm, 100–150 g of sediment were collected from the excess cutting, dried at 45°C for 3 weeks then ground in an agate mill for 10 min to a particle size of 5 μm. These powdered sub-samples were used to determine the whole-rock mineralogy by XRD, carbonate content by calcimetry, organic matter by loss on ignition (LOI), specific gravity as well as sodium absorption ratio (SAR) and electrical conductivity (EC) from a saturated paste by methods outlined by Rowell (1994).

Results: core descriptions

Littleton Warth: active floodplain, Severn Estuary (Table 2)

Marsh deposits consisted of firm mottled brown clayey silts finely laminated with sandy material. The degree of lamination was greatest in the youngest (lowest) marsh unit and decreased with elevation to being less common in the older marshes. Bulk densities were high (up to 2.1 g cm^{-3}) in all units, and increased with age, although they were comparable between the older two terraces; this suggests that full consolidation was approached within 300 years (the age of the mid-terrace). Liquid limits were high but moisture content generally approached the plastic limit (40%), resulting in very high sediment undrained

Table 2. *Geotechnical characteristics of Littleton Warth active saltmarsh units*

Property	Littleton lower marsh	*n* *n*	Littleton intermediate marsh	*n*	Littleton upper marsh	*n*
Shear strength (kPa)	41.5 (23.0; 76.1)	5	95.4 (75.1; 109.4)	4	83.9 (57.9; 106.2)	4
Moisture content (%)	47.6 (35.0; 64.4)	13	34.6 (31.3; 37.7)	9	40.7 (27.1; 78.9)	14
Bulk density (g cm^{-3})	1.65 (1.53; 1.76)	12	1.72 (1.57, 1.93)	10	1.71 (1.25; 2.09)	14
Dry density (g cm^{-3})	1.12 (1.00; 1.28)	12	1.25 (1.15; 1.37)	10	1.26 (0.70; 1.64)	14
Porosity (%)	56.8 (50.6; 61.4)	12	51.9 (48.1; 55.6)	10	51.7 (36.7; 73.1)	14
Liquid limit (%)	67.5 (57.3; 75.4)	7	56.4 (52.4; 59.3)	6	61.8 (50.6; 93.3)	7
Plastic limit (%)	39.4 (36.1; 47.0)	4	32.3 (27.1; 36.8)	4	34.8 (27.1; 53.2)	4
Plasticity index	27.1 (22.6; 34.0)	4	23.0 (19.3; 27.5)	4	28.0 (22.2; 40.1)	4
Liquidity index	0.24(-0.04;0.39)	4	0.09 (-0.12; 0.37)	4	0.24 (0.07; 0.52)	4
Carbonate content (%)	12.1 (10.0; 14.3)	8	12.49 (6.19; 15.9)	8	2.98 (0.23; 7.13)	8
Loss on ignition (%)	7.84 (7.12; 8.53)	8	5.99 (2.00; 10.6)	8	3.09 (1.19; 5.16)	8
Percentage sand (%)	4.56 (2.15; 8.03)	8	5.56 (2.78;9.25)	8	2.26 (1.99; 3.60)	6
Percentage silt (%)	73.1 (66.6; 76.4)	8	73.6 (69.9;77.5)	8	74.4 (70.4; 79.6)	6
Percentage clay (%)	22.4 (19.0; 28.5)	8	20.8 (14.1; 24.3)	8	23.4 (16.8; 27.9)	6
EC (dS m^{-1})	15.7 (15.5; 15.9)	3	12.9 (9.8; 17.8)	4	13.4 (10.6; 15.5)	4
SAR	37.7 (36.0; 39.4)	2	35.5 (35.4; 36.0)	4	35.6 (32.7; 38.2)	4

Average (minimum; maximum).

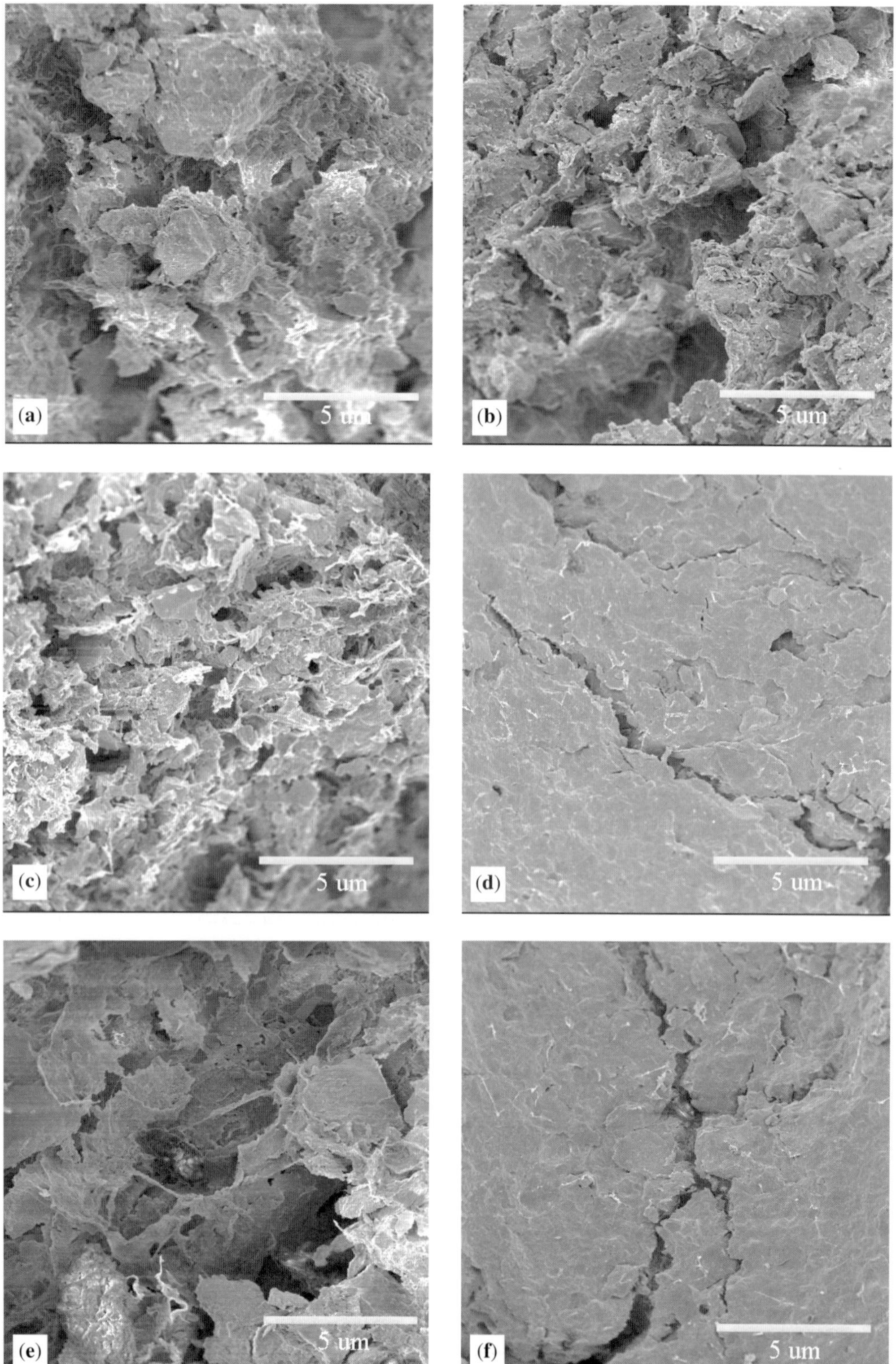
(a)
5 um
(b)
5 um
(c)
5 um
(d)
5 um
(e)
5 um
(f)
5 um

Table 3. *Geotechnical characteristics of active land-claimed marshes at Slimbridge Warth*

	Age of reclamation					
Property	14th century	*n*	18th century	*n*	19th century	*n*
Shear strength (kPa)	65.9 (58.4; 70.0)	4	89.2(71.2; 109.7)	4	99.8(66.4; 122.4)	5
Moisture content (%)	37.2 (31.0; 58.6)	17	38.2 (25.3; 65.2)	17	32.8 (24.9; 59.3)	21
Bulk density (g cm^{-3})	1.73 (1.58; 1.83)	17	1.73 (1.36; 1.89)	17	1.70 (1.17; 1.93)	21
Dry density (g cm^{-3})	1.36 (1.11; 1.43)	16	1.27 (0.62; 1.44)	17	1.30 (0.74; 1.53)	21
Porosity (%)	46.6 (43.9; 56.3)	16	51.7 (45.4; 76.6)	17	49.8 (40.8; 71.5)	21
Liquid limit (%)	56.1 (50.1; 68.5)	4	54.2 (50.1; 65.3)	3	46.9 (42.6; 53.7)	4
Plastic limit (%)	25.4 (18.5; 32.0)	4	29.2 (25.2; 33.5)	3	24.5 (22.5; 28.5)	4
Plasticity index	26.2 (17.6; 36.5)	4	25.0 (21.1; 31.8)	3	22.4 (20.0; 25.2)	4
Liquidity index	0.40 (0.21; 0.64)	4	0.28 (0.05; 0.51)	3	0.27 (0.15; 0.45)	4
Carbonate content (%)	4.57 (0.06; 10.7)	7	7.77 (0.00; 12.6)	7	8.93 (2.98; 12.7)	9
Loss on ignition (%)	3.89 (0.90; 12.4)	7	3.86 (1.12; 11.1)	7	3.28 (1.64; 9.97)	9
Percentage sand (%)	9.12 (7.00; 16.5)	6	6.31 (3.86; 11.4)	9	7.58 (3.40; 17.2)	6
Percentage silt (%)	75.9 (74.4; 79.4)	6	73.8 (72.7; 74.7)	9	74.5 (73.0; 77.2)	6
Percentage clay (%)	14.1 (9.15; 17.0)	6	19.9 (14.4; 22.9)	9	17.9 (8.95; 23.6)	6
EC (dS m^{-1})	0.5 (0.4; 0.6)	3	0.8 (0.7; 1.0)	4	1.2 (0.7; 1.5)	4
SAR	4.8 (0.9; 11.4)	3	7.2 (3.6; 11.4)	3	5.5 (2.0; 7.4)	4

Average (minimum; maximum).

shear strengths (up to 109 kPa). Carbonate was present in all units (up to 15%); however, the near-surface sediments of the older terraces exhibited considerable dissolution of carbonate particles. Fine rootlet remains were present through the sequence, with living root material abundant in near surface. Microfabric analysis found an aggregated floc structure with moderate inter-aggregate porosity (Fig. 3a). Electrical conductivity and sodium adsorption ratios were high, reflecting the saline nature of pore waters.

Slimbridge Warth: land-claimed floodplain, Severn Estuary (Table 3)

Land-claimed marsh units were well-drained, mottled brown clayey silts finely laminated with sandy material. Bulk densities were high (up to 1.93 g cm^{-3}), liquid limits moderate (decreasing with age of land-claim from average values of 56 to 47%) and moisture contents low (roughly 35%), inducing high undrained sediment shear strengths (up to 122 kPa). Carbonate was present throughout, although in reduced quantities in organic near-surface deposits of older units. Scanning electron microscopy identified the microfabric to consist of aggregated clay particles with a moderately open pore structure (Fig. 3b). Electrical conductivity and sodium adsorption ratios were low, reflecting replacement of saline pore waters by meteoric waters.

Old Hall Marsh: active floodplain Blackwater Estuary (Table 4)

The deposit consisted of soft mottled grey and brown clayey silts with occasional fine laminae of coarser material. Moisture contents and liquid limits were extremely high, reaching a maximum of 135% and 137%, respectively. Bulk densities were low, ranging between 124 and 1.51 g cm^{-3}. The site was subsequently resampled by Crooks & Garcia San Leon (January 1998; unpublished data) to check that the high moisture contents were not due to untypical conditions. Moisture contents were found to be consistently high, exceeding 100% across the cycle and falling only by 20% during the neap tide period. Measured undrained shear strengths were low (typically 27 kPa) as a result of the high moisture content. Carbonate was not detected. High electrical conductivity values and sodium adsorption ratios reflect the saline marsh

Fig. 3. SEM images of soil microfabric: (**a**) Littleton (intermediate marsh); (**b**) Slimbridge Warth (19th-century land-claim); (**c**) Old Hall Marsh; (**d**) Tollesbury land-claim; (**e**) Northey Island active marsh; (**f**) Northey Island former land-claim. All active marshes and the land-claim at Slimbridge exhibit an open porous fabric. By contrast, both Tollesbury land-claim and the former land-claim unit at Northey Island show evidence of a breakdown in soil structure with reorientation of clay platelets and loss of porosity.

Table 4. *Geotechnical characteristics of active Old Hall Marsh and Tollesbury land-claim*

Property	Old Hall Marsh	*n*	Tollesbury land-claim	*n*
Shear strength (kPa)	19.5 (14.7; 27.8)	3	116.8 (57.5; 186.3)	5
Moisture content (%)	84.2(64.0; 136.9)	17	34.3 (24.0; 48.3)	23
Bulk density (g cm^{-3})	1.45 (1.24; 1.51)	17	1.65 (1.59; 1.73)	3
Dry density (g cm^{-3})	0.88 (0.55; 0.92)	17	1.15 (1.09; 1.24)	3
Porosity (%)	66.9 (65.2; 79.1)	17	56.7 (53.5; 59.3)	3
Liquid limit (%)	117.0(99.5;134.5)	2	74.1 (69.5; 77.3)	4
Plastic limit (%)	65.0 (61.0; 66.0)	2	35.8 (35.0; 37.0)	4
Plasticity index	52.0 (35.5; 68.5)	2	38.4 (33.5; 40.5)	4
Liquidity index	0.64 (0.57; 0.71)	2	0.08 (0.01; 0.21)	4
Carbonate content (%)	Not detected	6	Not detected	9
Loss on ignition (%)	6.81 (1.69; 12.7)	6	3.7 (8.5; 1.9)	9
Percentage sand (%)	7.55 (3.79; 11.40	5	10.8 (6.64; 18.6)	7
Percentage silt (%)	77.3 (75.3; 79.2)	5	72.4 (70.9; 74.7)	7
Percentage clay (%)	15.2 (11.6; 20.20	5	16.8 (16.8; 19.5)	7
EC (dS m^{-1})	34.0 (22.3; 45.6)	2	3.2 (1.0; 6.5)	3
SAR	60.2 (53.4; 66.9)	2	12.4 (4.4; 20.6)	3

Average (minimum; maximum).

pore waters. Microfabric analysis found an porous aggregate structure (porosity of 57%) with a high degree of edge-to-edge interparticle contacts (Fig. 3c).

Tollesbury Marsh: land-claim floodplain, Blackwater Estuary (Table 4)

The land-claim sediments consisted of very stiff mottled grey and brown clayey silts with strong brown coloration around fine rootlet remains. Moisture contents were low, but increased with depth from 24% at the surface to 48% at 0.98 m depth, and tended towards the plastic limit. Liquid limits were high (70–79%). Because of the extremely stiff nature of the sediment and problems in sampling, only three bulk density values were recorded at the base of the core (159–173 g cm^{-3}). Carbonate was not detected within the sequence. Shear strengths were extremely high at the surface (186 kPa), but fell at the base of the core, where electrical conductivity and sodium adsorption ratios identified brackish pore waters (Table 3). In the upper 0.50 m of the profile pore waters were of low salinity. SEM investigation found a dispersed fabric with aggregate breakdown, reoriented clay particles and greatly contracted pore structure (Fig. 3d).

Northey Island: regenerated floodplain: Blackwater Estuary (Table 5)

The profile at Northey Island marsh was found to consist of 0.66 m of very soft, grey mud (regenerated marsh) overlying a firm mottled grey and black sediment of the former land-claimed unit. Both units consisted predominantly of clayey silt, though the regenerated unit was slightly coarser. The former land-claim surface was marked by an organic-rich horizon containing plant stems and root material. Moisture contents of the regenerated marsh were extremely high, increasing to a maximum of 156% just above the land-claim surface then decreasing to 45% in the older unit. A similar trend in liquid limits was detected. Bulk densities were low (1.27 g cm^{-3}) in the upper unit, contrasting with a higher value in the lower unit (1.62 g cm^{-3}). Carbonate was not detected throughout the profile. Undrained shear strengths were very low in the soft, upper mud (11–20 kPa) but much higher in the former land-claim deposit (65–77 kPa). Electrical conductivity and sodium adsorption ratios were high throughout, reflecting high salinity of pore waters. Scanning electron microscopy found a highly open and porous structure within the regenerated marsh unit (Fig. 3e), similar to that of Old Hall, but a consolidated and deflocculated structure in the lower deposit (Fig. 3f).

North Fambridge: regenerated floodplain, Crouch Estuary (Table 5)

Two cores were collected from this marsh, the findings from which will be discussed together. As at Northey Island a soft, grey mud was found to overlie a firm, grey and black former land-claim deposit. The depth of the soft mud was variable depending on the underlying topography, but within the cores was some 330 and 570 mm thick,

Table 5. *Geotechnical properties of active and former land-claim sediments at Northey Island and North Fambridge*

	Northey Island				North Fambridge			
Property	Active	*n*	Former land-claim	*n*	Active	*n*	Former land-claim	*n*
Shear strength (kPa)	15.6 (11.6; 20.0)	3	78.5 (77.5; 79.6)	2	13.8 (10.5; 17.5)	5	71.3 (65.0; 77.3)	5
Moisture content (%)	130.4 (116.1; 156.0)	14	47.6 (44.8; 52.3)	7	93.9 (84.1; 157.5)	5	50.3 (41.1; 56.8)	5
Bulk density (g cm^{-3})	1.27 (1.22; 1.36)	14	1.62 (1.45; 1.67)	7	1.35 (1.32; 1.40)	5	1.63 (1.51-1.70)	5
Dry density (g cm^{-3})	0.55 (0.48; 0.60)	14	1.10 (1.00; 1.17)	7	0.65	2	1.09 (1.05; 1.13)	2
Porosity (%)	79.2 (77.4; 80.7)	14	59.0 (56.4; 62.6)	7	75.7	2	60.6 (59.2; 60.0)	2
Liquid limit (%)	138.8 (128.5; 149.0)	2	87.5 (86.5; 88.5)	2	128.8 (125.3;135.1)	5	88.6 (78.9; 99.1)	4
Plastic limit (%)	64.5 (61.0; 68.0)	2	39.0 (37.5; 40.5)	2	42.1 (37.4; 46.9)	5	39.7 (26.1; 53.3)	4
Plasticity index	73.9 (60.5; 87.2)	2	50.5 (46.0; 55.0)	2	84.4 (80.6; 90.2)	5	48.9 (37.9; 60.6)	4
Liquidity index	0.97 (0.88; 1.05)	2	0.19 (0.12; 0.26)	2	0.86 (0.52; 1.37)	5	0.24 (0.13; 0.29)	4
Carbonate content (%)	Not detected	5	Not detected	4	0.94 (0; 3.7)	6	0.03 (0; 0.1)	3
Loss on ignition (%)	9.2 (7.5; 11.1)	5	4.5 (4.1; 5.2)	3	8.4 (7.8; 9.3)	6	5.2 (3.3; 6.5)	3
Percentage sand (%)	17.94 (16.9; 19.9)	4	7.62 (6.6; 8.7)	3	10.1(8.1; 11.8)	3	4.60 (2.3; 5.7)	6
Percentage silt (%)	71.90 (70.7; 72.9)	4	69.90 (68.3; 71.2)	3	73.60 (68.3; 76.4)	3	69.36 (67.1; 73.1)	6
Percentage clay (%)	10.16 (9.39; 11.7)	4	22.48 (22.1; 23.0)	3	16.25 (13.1; 23.6)	3	25.04 (19.5; 30.7)	6
EC (dS m^{-1})	39.7 (23.9; 50.5)	3	48	1	na		na	
SAR	60.5 (53.9; 64.4)	3	63	1	na		na	

Average (minimum; maximum). na, not analysed.

respectively. The regenerated marsh consisted of poorly consolidated clayey silty sediment, with very high moisture content (84–158%) and low bulk density (1.32–1.40 g cm^{-3}). The underlying unit was again marked by an organic horizon with desiccation cracking apparent on the surface. Moisture contents in the lower unit were low (41–57%) and bulk densities correspondingly high (1.51–1.70 g cm^{-3}). No electrical conductivity or sodium adsorption data were available. Microfabric analysis found the upper unit to possess a highly open porous structure but a consolidated and deflocculated fabric in the former land-claim unit.

Discussion

Although possessing similar granulometric characteristics, the 'behaviour' of the sediments on land-claim and drainage at sampled sites in Essex does appear to be in stark contrast to those in the Severn Estuary (Fig. 4). Destructuring of the sediment fabric at the Tollesbury land-claim, and the buried former surfaces at Northey Island and North Fambridge, is in line with the findings of Hazelden *et al.* (1986) and Hodgkinson & Thorburn (1995) and their description of unstable saline soils in the region. As discussed above, the survey of Hazelden *et al.* (1986) identified sodium-saturated clays on coastal and estuarine floodplains to be potentially 'unstable' and prone to deflocculation and mobilization of clay particles on drainage of the saline water table. Clay particle flocculation led to the formation of a dense, low-permeability horizon with a very high shear strength. Hazelden *et al.* found that the distribution of soils subject to clay dispersion and waterlogging was specific to non-calcareous soils and that the process of deflocculation could be prevented by maintaining a high saline water table that counteracted the soil instability. The North Kent survey (Hazelden *et al.* 1986) further identified that only a small quantity of carbonate <2% by volume) was sufficient to prevent deflocculation. Following from these findings, the ability of the clay fabric in the Severn Estuary marshes to resist deflocculation on drainage, and to maintain an aggregated structure, appears to reflect the presence of significant quantities of carbonate.

Turning to the intertidal zone, although deflocculation is not a process that occurs in saltmarshes, because of the high pore-water salt content, the presence of carbonate may have an effect on the consolidation rate of upper intertidal sediments. It has been found in this study that marshes in the Severn consolidate at a greater rate than those in Essex. This is not simply an age phenomenon, as the Old Hall marsh and others identified by Crooks (1996) are older than the lower and mid-terraces at Littleton Warth, yet possess lower bulk densities. The increased consolidation rate in the Severn may, in part, be aided by exchangeable cation effects. It has been established that at any given pore-water salt concentration Ca^{2+}-saturated clays will have more compact diffuse double-layer than Na^{+}-saturated clays (Rowell 1994). This would allow closer and stronger interaction between clay particles. It is postulated that Ca^{2+} is being supplied in sufficient quantities in the Severn marsh sediments to displace Na^{+} from exchange sites, so causing this effect. Other factors may also be coming into play to cause the increased consolidation. The Severn has an exceptionally large tidal range (14.6 m spring tide range at Avonmouth), which may cause down-draw effects, and the sediments include laminae of coarse silts that aid drainage.

Regional variations in the geotechnical properties of fine-grained estuarine alluvium

Several published investigations exist reporting geotechnical analysis of fine-grained coastal alluvium from a number of sites in the Severn region (Avonmouth, Newport, Brean, the length of the M5 corridor across the Somerset Level) and across north Kent and Essex (Tilbury Dock, Long Reach, Shell Haven, the Swale and a suite of boreholes across the Essex coastal plain). Collectively, these investigations describe alluvium in the Severn region to have typically lower moisture contents, lower liquid limits, higher bulk densities and higher undrained shear strengths than sediments in southeast England (Greensmith & Tucker 1967, 1969, 1971*a*, *b*, 1973, 1976; Skempton 1970; Cook & Roy 1984; Marsland 1986; Hawkins *et al.* 1989). Unfortunately, the carbonate content of the sediments was not published in these studies.

Beyond these regions of southern Britain, further investigation, by Hawkins *et al.* (1989), shows that coastal alluvium at Bothkennar in the Firth of Forth possesses geotechnical properties more in line with that of the southeast of England than with that of the Severn. The work of Hawkins *et al.* (1989) also indicates that the organic content at sampled sites within each of the three regions is comparable and thus is not the dominant control on sediment plasticity. Again, the carbonate content of the sediment was not reported. However, Paul *et al.* (1992) provided geochemical details of pore waters collected from another core at the same site. It was found that alluvial pore-fluid chemistry was dominated by marine-derived ions Na^{+}, Cl^{-} and

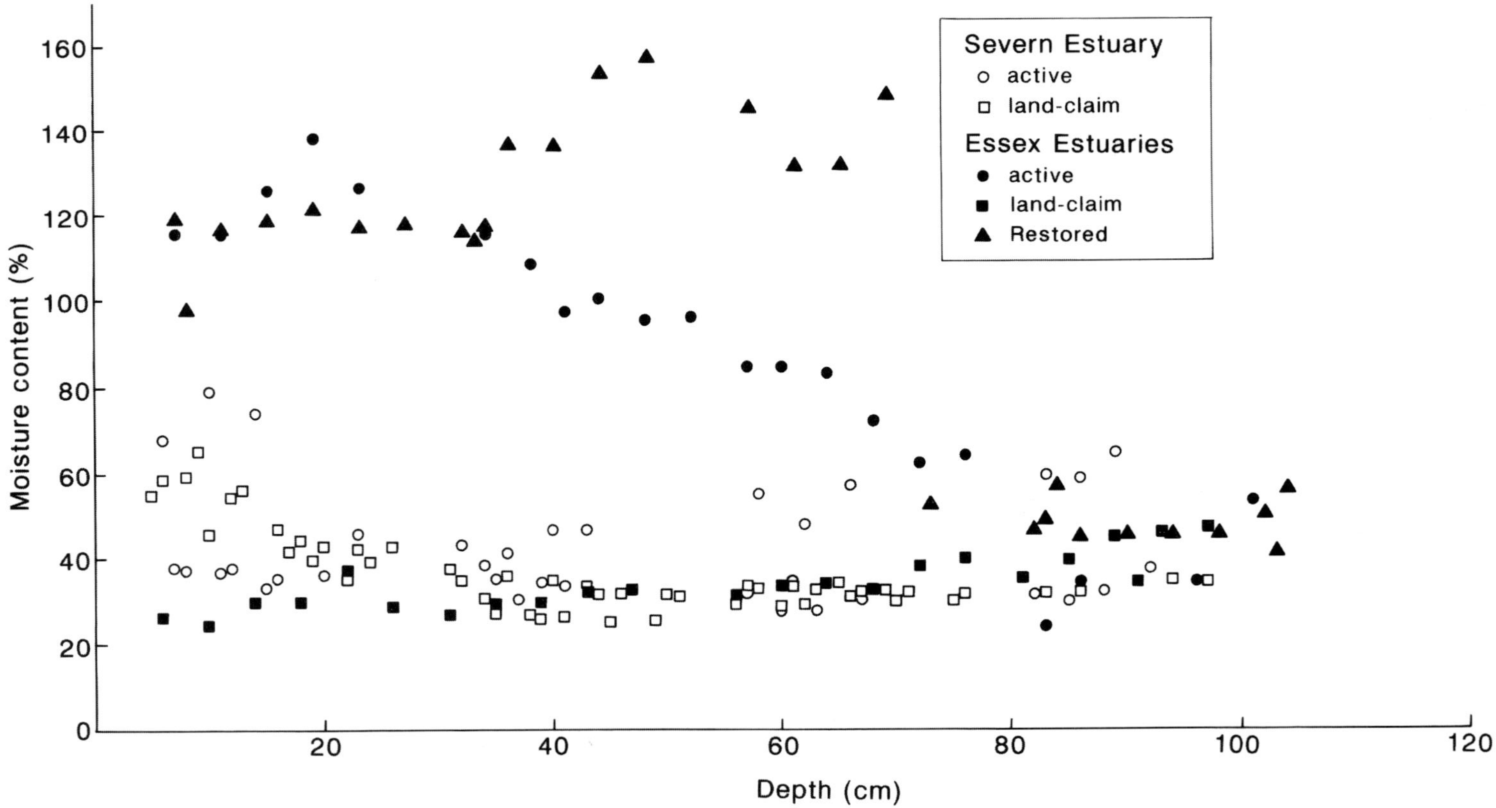

Fig. 4. Moisture content profiles through the upper 1 m of sampled marshes. The similarity in moisture content levels between the active and land-claim marshes in the Severn estuary should be noted; in contrast, those of the active marshes in Essex are considerably higher than their land-claim counterpart. Interestingly, the increase in moisture content with depth suggests that the former land-claim surface of Northey Island (at 0.68 m depth) may be acting as an aquiclude, maintaining high water levels in the developing marsh above.

SO_4^{2-}, in similar proportions to the adjacent estuary. Exchangeable cations were dominated by Na^+ (*c.* 5000 mequiv kg^{-1}) with much lesser quantities of K^+ (*c.* 1000 mequiv kg^{-1}), Mg^{2+} (*c.* 1000 mequiv kg^{-1}), and Ca^{2+} (*c.* 500 mequiv kg^{-1}). The Bothkennar clays are therefore sodium dominated, and the results presented by Paul *et al.* (1992) appear to show a relationship between the concentration of pore-water sodium content and the liquid limit, although no direct correlation was undertaken during their study.

However, a more conclusive relationship between high plasticities and cation content was identified by Crooks & Graham (1976) within the alluvium underlying Belfast. They described a stratigraphic sequence consisting of 6 m of soft, lightly fissured, slightly overconsolidated, silty clays of medium to high sensitivity, overlain by 2 m of sands and silty sands. The clay mineralogy was dominated by illite, with kaolinite, chlorite and swelling chlorite also present. Shear strength increased from a minimum of 10 kPa at the top of the clay stratum to 22 kPa at a depth of 8 m. A significant variation in Atterberg limits was recorded. Plasticity indices decreased from 70% at 4.0 m depth to about 45% at 7.5 m depth, and liquid limit generally exceeded 80%, often approaching 100%. Organic content, measured by loss on ignition, was found to be roughly 4% and constant throughout the tested horizon.

Analysis of exchangeable cation content (Na^+, K^+, Ca^{2+} and Mg^{2+}) found sodium content to be high (15–20 mequiv per 100 g), decreasing with depth, and calcium to be constant but low (2–3 mequiv per 100 g). The Na^+:Ca^{2+} ratio decreased with depth from 12–15 at 2–4 m to about six at 7.5 m depth. In plotting sodium cation concentration against plasticity index, an approximate linear relationship was found, high plasticities being related to high sodium concentrations and consequently high Na^+:Ca^{2+} ratios. Crooks & Graham (1976) therefore showed that the variation in liquid limit and plasticity index, and hence sediment interactions, was due to changes in exchangeable cation content. Their findings are supported within this study, as a strong positive relationship between SAR and sediment plasticity has been identified (Fig. 5).

Overconsolidated horizons as indicators of a fall in saline water table

The deflocculation of calcium-deficient clays does appear to provide a mechanism for the formation of overconsolidated horizons as identified by Greensmith & Tucker (1971*a*). Moreover, these horizons form only under specific conditions, namely $CaCO_3$ deficiency within clay-rich alluvial sediments and the diluting or downward displacement of a saline water table. If such horizons are identified in $CaCO_3$-depleted alluvium, as described by Greensmith & Tucker (1971*a*) for the Essex coastal floodplain, then their presence does strongly suggest that a fall in the saline water table has occurred at some time during the evolution of the coastline. In addition, examination of the former land-claim surfaces buried at Northey Island and North Fambridge shows a range of characteristics indicative of a period of reduced sedimentation and extended subaerial exposure (developed organic surface horizon, desiccation features, etc.).

It is also apparent from the nature of the former land-claim units that reintroduction to the intertidal zone, even after 100 years, has little or no impact on sediment structure or a range of physiographic features. Apart from a slight increase in moisture content, and accompanying slight lowering in undrained shear strength, the geotechnical properties of the former land-claim unit appear to have altered little with return to tidal inundation. The change in sediment coloration from grey–brown to black suggests anaerobic conditions after saline flooding. A walkover survey of the experimental managed retreat site at Tollesbury, only a few months after breaching of the sea-wall, found the former surface to be blackened and discoloured, and buried beneath several centimetres of grey–brown soft sediment. Anoxia therefore takes hold rapidly within the buried surface, aided by the covering of saturated sediment, the decay of glycophyte vegetation and supply of sulphate by sea water. Overconsolidated horizons therefore appear to a have a very high preservation potential and make good markers of a fall in the saline water table. In Essex coastal alluvium identification of overconsolidated horizons is aided by the marked geotechnical contrast with the bulk soft sediment from accreting tidal deposits. Of course, whether this fall in water table is due to a relative sea-level regression or some local factor, such as isolation from tidal waters behind a shingle barrier, it cannot be discerned from the geotechnical measurements alone.

In contrast to the Essex marshes, examination of the stratigraphy of $CaCO_3$-bearing alluvium of the Severn Estuary, the sediment fabric analysis, bulk density, plasticity indices and undrained shear strength measurements provide little information to assist the distinction between mature active marsh and supratidal marsh deposits. A palaeontological investigation of any organic horizons will be required to delineate mature saline and non-saline marsh environments.

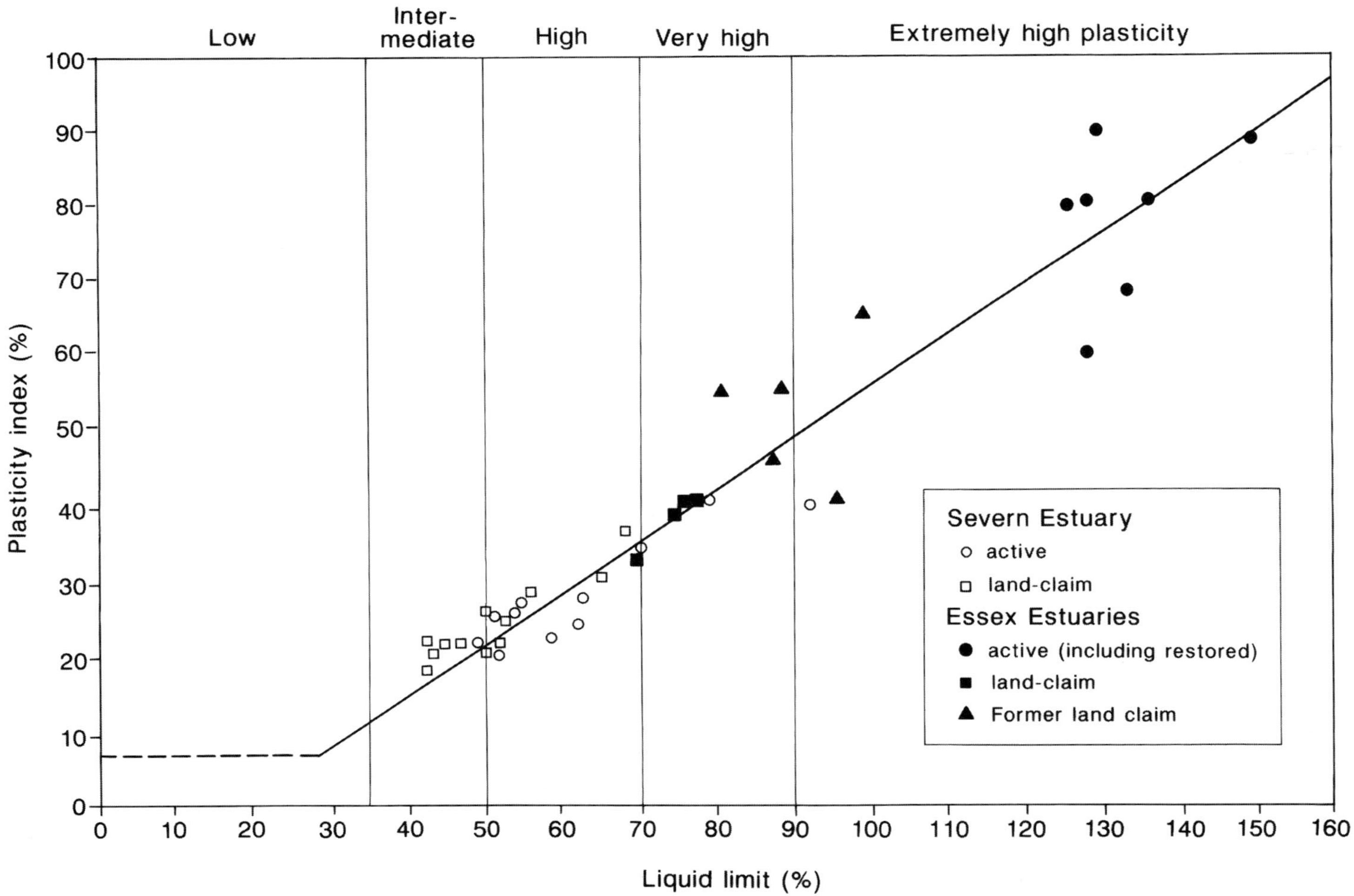

Fig. 5. Plasticity chart for all sampled marshes.

Implications for the interpretation of Holocene sea-level curves

Given that overconsolidated horizons strongly reflect a fall in saline water table, then if identified they provide a reasonable marker for past sea-level regressions. However, knowledge about their geographical extent is limited. As far as the author is aware, apart from the work of Greensmith & Tucker on the Essex coastal floodplain, overconsolidated horizons have not been described within alluvial stratigraphy elsewhere. This may reflect either that the identification of overconsolidated horizons in Essex was not correct or that the Essex coastal area is particularly sensitive to clay dispersion. This study has derived a mechanism for the formation of overconsolidated horizons, which would support the observations of Greensmith & Tucker. Failure to encounter these markers elsewhere may perhaps suggest, however, that detrital carbonate is commonly present within muddy coastal areas in sufficient quantities (>2% by volume) to 'protect' clay aggregates from dispersion under low-salinity conditions. Further research is required to determine whether overconsolidated horizons occur in other coastal areas around the globe. As detrital carbonate distribution is largely hydrodynamically controlled, prime areas to investigate may be large floodplain regions subject to a microtidal regime or back-barrier system. Of course, alluvium sequences consisting of a coarse silt- or sand-supported matrix (such as NW England, areas of Northern France, Netherlands, etc.) would be insensitive to clay dispersion and would not exhibit overconsolidation.

It has not been possible here to quantify the implications of clay dispersion and sediment consolidation on the accuracy of sea-level index points. Typically, these points possess four characteristics: a known location, age, an altitude relative to a past sea level and a tendency relating to whether there is an increase or decrease in marine influence (Shennan 1982). It is particularly problematic to assign with certainty the past elevation of the index point given syndepositional consolidation within alluvial deposits (Haslett *et al.* 1998). Uncertainty in dealing with this issue is reflected in the use of rough estimates of elevation change as a result of autocompaction (Heyworth & Kidson 1982; Long 1992). It is clear from the findings in the present study that the amount of consolidation can, depending on sedimentology, be variable from region to region. Consolidation rates are notably high in the Essex region, where drainage may reduce the moisture content of saltmarsh sediments from over 100% to the order of 35–40%. In the Severn Estuary, given the greater state of consolidation of the upper intertidal deposits, lower amounts of further consolidation appear to take place on drainage and so elevation errors in this area are likely to be correspondingly less. (Although the effects of reclamation on the consolidation of alluvium made up of granular material have not been discussed within this study it might be suggested that these sediments would be the least succeptible to autocompaction and so sea-level index points will be displaced from their depositional elevation to a lesser degree.) Moreover, the observation that bulk density of estuarine alluvium in the studied regions increases with depth and overburden pressure suggests that sea-level index points within the main body of alluvium now exist at an elevation lower than that at which they were originally deposited.

An indication of the amount of consolidation that occurs with substrate dewatering can be provided by reference to the 0.75 m thick saltmarsh deposit overlying the London Clay at Old Hall. The (reasonable) assumption can be made that change in surface elevation is due to the loss in volume of the pore water as consolidation takes place. Thus by calculating pore volume change associated with reducing the moisture content throughout the active marsh profile down to, say, 35% to reflect that of the land-claimed marsh equates to a surface elevation fall of some 320 mm. This calculation is possible only by knowing the mass of sediment and the mass change in water per given unit volume calculated for a complete sequence of depth interval throughout the core (derived from bulk density and moisture content measurements). It is believed that given these variables for cores through alluvial sequences it may be possible, in future studies, to estimate the consolidation experienced beneath peat horizons as a response to drainage and autocompaction induced by sea-level fall.

Conclusions

Within the present study, the effect of embankment construction and reclamation on the geotechnical properties of saltmarsh sediments has been considered to be analogous in effect to a fall in relative sea level, in that surface minerogenic sedimentation ceases and the saline water table is lowered, leading to leaching of salts and soluble minerals by meteoric waters. It has been found that the formation of dense horizons, such as the overconsolidated horizons described by Greensmith & Tucker (1971*a*), is dependent on the mineral and geochemical composition of the sediment and appears to be regionally specific, determined by the supply of detrital carbonate particles. This may give conflicting indications of relative sea-level movement even across relatively small distances.

Calcium-deficient active marshes possess a very open porous sediment fabric and maintain this structure, even at high intertidal elevation. Only reclamation of these sediments and lowering of the saline water table results in dispersion of the clay particles and the formation of a dense, low-permeability surface with a very high shear strength (Hazelden *et al.* 1986). Such horizons do not form in $CaCO_3$-bearing marshes following a fall in the saline water table. The identification of such overconsolidated horizons in $CaCO_3$-depleted sediments therefore strongly indicates a fall in saline water table, which in turn most likely reflects a fall in relative sea level if local factors can be discounted. Overconsolidated horizons are readily identified in the Essex marches described here because of the stark geotechnical contrasts with the soft adjacent alluvium.

The absence of overconsolidated horizons within the Severn Estuary alluvium does not preclude sea-level fluctuations. These $CaCO_3$-bearing sediments consolidate heavily whilst in the intertidal zone as a function of marsh age, and after land-claim, the sediments are not prone to dispersion and show little further consolidation. This leads to the formation of a relatively uniform alluvium with high density and shear strength. The identification of horizons of particularly high bulk density does not necessarily suggest a former supratidal deposit, as these horizons can also reflect a saltmarsh unit that can form under conditions of sea-level rise as well as fall. As such, careful palaeontological investigation is required to determine whether organic horizons associated with such horizons are derived from marine or fresh-water assemblages.

This work was supported partly by an NERC research studentship held by the author while at the Postgraduate Research Institute for Sedimentology, University of Reading. I am grateful to K. Pye, J. R. L. Allen and D. L. Rowell for helpful discussion, and to D. Thornley, G. Patterson, M. Andrews, S. Bennett and P. Judge for technical and cartographic assistance. I am grateful to the anonymous reviewers for comments on this manuscript. I would also like to thank Slimbridge Wildlife Trust, J. Cullimore of Cullimore farm, English Nature and North Fambridge Marina for permission to access sampling areas.

References

ACKEROYD, A. 1972. Archaeological and historical evidence for subsidence in southern Britain. *Philosophical Transactions of the Royal Society of London, Series A*, **272**, 151–169.

ALLEN, J. R. L. 1986. A short history of saltmarsh reclamation at Slimbridge Warth and neighbouring areas, Gloucestershire. *Transactions of the Bristol and Gloucestershire Archaeological Society*, **104**, 139–155

—— 1987. Late Flandrian shoreline oscillations in the Severn Estuary: the Rumney Formation at its typesite (Cardiff area). *Philosophical Transactions of the Royal Society of London, Series B*, **135**, 157–184.

—— 1990. The Severn Estuary in southwest Britain: its retreat under marine transgression, and fine-sediment regime. *Sedimentary Geology*, **96**, 13–28.

—— 1991. Salt-marsh accretion and sea-level movement in the inner Severn Estuary, Southwest Britain: the archaeological and historical contribution. *Philosophical Transactions of the Royal Society of London, Series B*, **148**, 485–494.

—— & RAE, J. E. 1987. Late Flandrian shoreline oscillations in the Severn Estuary: a geomorphological and stratigraphical reconnaissance. *Philosophical Transactions of the Royal Society of London, Series B*, **315**, 185–230.

ANSON, R. W. W. & HAWKINS, A. B. 1998. The effect of calcium ions in pore water on the residual shear strength of kaolinite and sodium montmorillonite. *Géotechnique*, **48**, 787–800.

BRIDGLAND, D. R. 1988. The Pleistocene fluvial stratigraphy and palaeogeography of Essex. *Proceedings of the Geologists' Association*, **99**, 291–314.

BRITISH STANDARDS INSTITUTION 1990. *Methods of Testing for Soils for Civil Engineering Purposes*, BS 1377. British Standards Institution, London.

CHURCHMAN, G. J., SKJEMSTAD, J. O. & OADES, J. M. 1993. Influence of clay-minerals and organic matter on effects of sodicity on soils. *Australian Journal of Soil Research*, **31**, 779–800.

CONWAY, B. W., MCCANN, D. M., SARGINSON, M. & FLOYD, R. A. 1984. A geophysical survey of the Crouch/Roach river system in south Essex with special reference to buried channels. *Quarterly Journal of Engineering Geology, London*, **17**, 269–282.

COOK, D. A. & ROY, M. R. 1984. A review of the geotechnical properties of Somerset Alluvium using data from the M5 motorway and other sources. *Quarterly Journal of Engineering Geology*, **17**, 235–242.

CROOKS, J. K. A. & GRAHAM, J. 1976. Geotechnical properties of the Belfast estuarine deposits. *Géotechnique*, **26**, 293–315.

CROOKS, S. 1996. *Sedimentological controls on the geotechnical properties of intertidal saltmarsh and mudflat deposits.* PhD thesis, University of Reading.

DEVOY, R. J. N. 1977. Flandrian sea-level changes in the Thames estuary and the implications for land subsidence in England and Wales. *Nature*, **270**, 712–715.

—— 1979. Flandrian sea-level changes and vegetation history of the Lower Thames Estuary. *Philosophical Transactions of the Royal Society of London, Series B*, **285**, 355–407.

—— 1982. Analysis of the geological evidence for Holocene sea-level movements in the southeast of England. *Proceedings of the Geologists' Association*, **93**, 65–90.

DEXTER, A. R. & CHAN, K. Y. 1991. Soil mechanical properties as influenced by exchangeable cations. *Journal of Soil Science*, **42**, 219–226.

——, KROESBERGEN, B. & KUIPERS, H. 1984. Some mechanical properties of aggregates of top soils from the Ijsselmeer polder. 2. Remoulded soil aggregates and the effects of wetting and drying cycles. *Netherlands Journal of Agricultural Science,* **32**, 215–227.

EVANS, J. H. 1953. Archaeological horizons on the north Kent Marshes. *Archaeologia Cantiana*, **66**, 103–146.

GIBBARD, P. L. 1977. Pleistocene history of the Vale of St Albans. *Philosophical Transactions of the Royal Society, Series B*, **280**, 445–483.

GREEN, C. P., MCGREGOR, D. F. M. & EVANS, A. H. 1982. Development of the Thames drainage system in early and mid Pleistocene times. *Geological Magazine*, **119**, 281–290.

GREENSMITH, J. T. & TUCKER, M. V. 1967. Morphology and evolution of inshore shell ridges and mud-mounds on modern intertidal flats, Bradwell, Essex. *Proceedings of the Geologists' Association*, **77**, 329–346.

—— & —— 1969. The origin of the Holocene shell deposits in the chenier plain facies of Essex, Great Britain. *Marine Geology*, **7**, 403–425.

—— & —— 1971*a*. Overconsolidation in some fine-grained sediments, its nature, genesis and value in interpreting the history of certain English Quaternary deposits. *Geologie en Mijnbouw*, **50**, 743–748.

—— & —— 1971*b*. The effects of Late Pleistocene and Holocene sea-level changes in the vicinity of the River Crouch, East Essex. *Proceedings of the Geologists' Association,* **82**, 301–321.

—— & —— 1973. Holocene transgression and regressions on the Essex coast outer Thames Estuary. *Geologie en Mijnbouw*, **52**, 193–202.

—— & —— 1976. Major Flandrian transgressive cycles, sedimentation, and palaeogeography in the coastal zone of Essex, England, *Geologie en Mijnbouw*, **55**, 131–146.

HASLETT, S. K., DAVIES, P., CURR, R. H. F., CAVIES, C. F. C., KENNINGTON, K., KING, C. P. & MARGETTS, A. J. 1998. Elevating late-Holocene relative sea-level change in the Somerset Levels, southwest Britain. *The Holocene*, **8**, 197–207.

HAWKINS, A. B. 1971. Sea-level changes around South West England. *Colston Symposium on Marine Archaeology*. Colston Papers, **23**, 67–88.

—— 1984. Depositional characteristics of the estuarine alluvium: some engineering implications. *Quarterly Journal of Engineering Geology,* **17**, 219–234.

——, LARNACH, W. J., LLOYD, I. M. & NASH, D. F. T. 1989. Selecting the location, and the initial investigation of the SERC soft clay test bed site. *Quarterly Journal of Engineering Geology*, **22**, 281–316.

HAZELDEN, J., LOVELAND, P. J. & STURDY, R. G. 1986. *Saline Soils in North Kent*. Soil Survey of England and Wales, Special Soil Survey, **14**.

HAZZELDINE-WARREN, S., PIGGOT, S., CLARK, J. G. D., BURKITT, M. C., GODWIN, H. & GODWIN, M. E. 1936. Archaeology of the submerged land-surface of the Essex coast. *Proceeding of the Prehistorical Society*, **2**, 178–210.

HEWLETT, R. & BIRNIE, J. 1996. Holocene environmental change in the inner Severn estuary, UK: an example of the response of estuarine sedimentation to relative sea-level change. *The Holocene*, **6**, 49–61.

HEYWORTH, A. & KIDSON, C. 1982. Sea level changes in SW England and Wales. *Proceedings of the Geological Association,* **93**, 91–111.

HODGKINSON, R. A. & THORBURN, A. A. 1995. Factors influencing the stability of salt affected soils in the UK—criteria for identifying appropriate management options. *Agricultural Water Management*, **29**, 327–338.

IECS 1992. *Historical Study of Sites of Natural Sea Wall Failure in Essex*. Institute of Estuarine and Coastal Studies, unpublished report to English Nature.

KIDSON, C. & HEYWORTH, A. 1973. The Flandrian sea-level rise in the Bristol Channel. *Proceedings of the Ussher Society*, **2**, 565–584.

KIRBY, R. 1990. The sediment budget of the erosional intertidal zone of the Medway estuary. *Proceedings of the Geologists' Association*, **101**, 63–77.

KJELLANDER, R., MARCELJA, S., PASHLEY, R. M. & QUIRK, J. P. 1988. Double-layer ion correlation forces restricted calcium–clay swelling. *Journal of Physical Chemistry,* **92**, 6489–6492.

LONG, A. J. 1992. Coastal response to changes in sea-level in the East Kent Fens and southeast England, UK, over the last 7500 years. *Proceedings of the Geologists' Association*, **103**, 187–199.

MARSLAND, A. 1986. The flood plain deposits of the Lower Thames. *Quarterly Journal of Engineering Geology*, **19**, 223–247.

MCNEAL, B. L. K. & COLEMAN, N. T. 1966. Effect of solution composition on soil hydraulic conductivity. *Soil Science Society of America Proceedings*, **40**, 308–312.

NORRISH, K. 1954. The swelling of montmorillonite. *Discussions of the Faraday Society*, **18**, 120–134.

PAUL, M. A., PEACOCK, J. D. & WOOD, B. F. 1992. The engineering geology of the Carse Clay at the national soft clay research site, Bothkennar. *Géotechnique*, **42**, 183–198.

QUIRK, J. P. & SCHOFIELD, R. K. 1955. The effect of electrolyte concentration on soil permeability. *Journal of Soil Science*, **6**, 163–178.

REGEA, M., YANO, T. & SHAINBERG, I. 1997. The response of low and high swelling smectites to sodic conditions. *Soil Science*, **162**, 299–307.

RENASAMY, P. 1983. Clay dispersion in relation to changes in electrolyte composition of dialysed red–brown earths. *Journal of Soil Science*, **34**, 723–732.

RIMMER, D. L. & GREENLAND, D. J. 1976. Effects of calcium carbonate on the swelling behaviour of a soil clay. *Journal of Soil Science*, **27**, 129–139.

ROWELL, D. L. 1963 Effect of electrolyte concentration on the swelling of orientated aggregates of montmorillonite. *Soil Science*, **96**, 368–374.

—— 1994. *Soil Science: Methods and Applications*. Longman, Harlow.

SHANMUGANATHAN, R. T. & OADES, J. M. 1982. Effect of

dispersible clay on the physical properties of the B horizon of red–brown earth. *Australian Journal of Soil Research,* **20**, 315–324.

SHELDON, R. W. 1968. Sedimentation in the estuary of the River Crouch, Essex. *Limnology and Oceanography*, **13**, 72–83.

SHENNAN, I. 1982. Interpretation of Flandrian sea-level data from the Fenland, England. *Proceedings of the Geologists' Association*, **93**, 53–63.

SKEMPTON, A. W. 1964. Long-term stability of clay slopes. *Géotechnique*, **14**, 77–102.

—— 1970. The consolidation of clays by gravitational compaction. *Quarterly Journal of the Geological Society of London*, **125**, 373–411.

SKENE, T. M. & OADES, J. M. 1995. The effect of sodium adsorption ratio and electrolyte concentrations on water quality: laboratory studies. *Soil Science,* **159**, 65–73.

SMITH, A. G. & MORGAN, L. A. 1989. A succession to ombrotrophic bog in the Gwent Levels and its demise: a Welsh parallel to the peats of the Somerset Levels. *New Phytologist*, **112**, 145–167.

TERZAGHI, K. & PECK, R. B. 1969. *Soil Mechanics and Engineering Practice.* Wiley, New York.

Historical background to floodplain morphology: examples from the East European Plain

A. V. PANIN, A. YU. SIDORCHUK & A. V. CHERNOV

Geographical Faculty, Moscow State University, Vorobyovy Gory, Moscow 119899, Russia (e-mail: panin@morpho.geogr.msu.su)

Abstract: Floodplain morphology is described in relation to channel pattern changes in the past. The oldest segments of the present-day floodplains were formed in the Late Valdai (Weichselian). Their morphology is inherited from large palaeomeanders (macromeanders) formed under extremely high discharge conditions. A map of the spatial distribution of Late Valdai macromeander relics on the East European Plain is given. Various floodplain segments have a spectrum of ages that may reflect only selected portions of the river history. The most widespread morphological units correspond to the periods when the river was undergoing its widest lateral migration. Periods of stable channel position may result in a gap in the floodplain age spectrum. The channel pattern and rates of river migration are controlled both by hydrological conditions and by valley floor morphology inherited from the preceding river regime. Heterogeneity of floodplains is illustrated from three key sites: the Khoper River (Southern Russia), the Protva River (Central Russia) and the Vychegda River (Northern Russia).

This paper contributes to the understanding of floodplain morphology from the point of view of hydrological changes in the past. In particular, two main issues are discussed: (a) the inheritance of recent floodplains from former river regimes, with a focus on the extreme hydrological conditions in the Late Valdai (Weichselian); (b) the age of floodplain segments and contribution of river regime at various time periods to contemporary floodplain composition and form.

The data reported have been collected from river valleys of the East European Plain. There are at least two reasons for this region to be of interest in the context of floodplain research. First, it is the largest area in Europe characterized by rather uniform morphology and geological composition. It also has a well-pronounced recent landscape zone pattern. Therefore it provides a wide range of sites for investigation of river and floodplain history under a changeable climate with minimal influence of 'local noise'. On the other hand, Late Valdai and Holocene fluvial history within the borders of the former USSR is still poorly covered in publications (see Starkel 1995*a*, fig. 2.1).

Methods of investigation

The spatial scale covered by this research is both of regional and local (key sites) character. To reveal the Late Valdai fluvial topography of the region the analysis of large-scale (1:25 000) maps for all the rivers with a length over 200 km was completed. Detailed field investigations of river and corresponding floodplain development were conducted at three key sites, which were chosen because of their distinctly different valley floor morphology. At these key sites aerial photographs were used to determine the relative age of floodplain segments and to compile corresponding geomorphological maps. The geological composition of the main morphological units was studied by examination of borehole cores and of natural exposures. Absolute age was determined by radiocarbon analysis of organic matter found in the alluvium. All dates referred to below are non-calibrated. The time of abandonment of palaeochannels was in most cases derived from dating the basal layer of the channel filling. Such dating techniques probably diminish the real age of the palaeochannels, especially if they were incorporated into the active channel pattern after abandonment (Starkel 1995*a*). Nevertheless, the lack of organic matter in within-channel deposits gives no alternative. The reliability of dating was also controlled by geomorphological features, and several dates from different locations within the same palaeochannel system were obtained where it was possible.

In wider terms, the age of the floodplain at a given site is considered here as the period since the site was a part of the active channel. Channel bars become components of a floodplain when they are fixed by vegetation. Former channels become parts of the floodplain at the time of their abandonment, which may be either abrupt or gradual. In both cases, changes in flow regime at the site are reflected in the vertical sediment stratification, and the floodplain as a geomorphological surface is considered

From: MARRIOTT, S. B. & ALEXANDER, J. (eds) 1999. *Floodplains: Interdisciplinary Approaches*. Geological Society, London, Special Publications, **163**, 217–229. 1-86239-050-9/99/$15.00

younger than the upper layer of channel sediment and older than the base of overbank deposits or palaeochannel infill.

Geomorphological evidence of high fluvial activity in the Late Glacial period

The past few decades have been marked by a growing interest in floodplains as a source of palaeohydrological information. Dury (1964) probably stimulated this interest. His concept of river underfitness based on the existence of large palaeomeanders ('valley meanders') with wavelengths several times greater than that of the present river, assumed a dramatic increase in river discharge at some time during the Quaternary period, including at the end of the last glaciation. The overall geographical distribution of giant palaeomeanders in the temperate zone is an argument for global climatic and hydrological causes for their development rather than for local or infrequent high-magnitude events. Since the 1970s, the postglacial development of river valleys has been studied in numerous locations in Europe. Transformation from a braiding pattern into large meanders during the Late Valdai seems to be a common feature for many European valleys. It has been recognized in England (Rose *et al.* 1980), the Netherlands (Vandenberghe 1987; Bohncke & Vandenberghe 1991), Poland (Kozarski & Rotnicki 1977; Mycielska-Dowigiallo 1977; Schumanski 1983; Kozarski 1991; Starkel 1995*b*), Belorussia (Kalicki 1995) and in other regions, though few examples of other river behaviour have been described (Starkel 1991).

The analysis of valley morphology confirms that the East European Plain was not an exception from the rest of Europe. Large relic meanders, referred to here as macromeanders, which are up to 15 times larger than the recent ones of the same rivers, are common on the floodplains and low terraces of small and medium-sized rivers through the whole region. Table 1 shows typical examples of macromeander parameters in relation to recent channels. The study sites were selected because of the well-conserved former fluvial forms and valley topography that allows derivation of palaeochannel parameters from topographic maps. Another selection criterion was the complete isolation of the selected valleys from glacial melt-water flows during deglaciation, to ensure that macromeander formation had resulted from climatically caused hydrological conditions. The dimensions of the palaeochannels (Table 1) show the dramatic increase of maximum discharges during the periods of macromeander formation.

Three main regions may be determined in relation to the distribution and the prevalent type of macromeanders and their role in floodplain structure (Fig.1):

(I) The region where macromeanders are virtually absent from the river valleys corresponds to the territory that was under the Late Valdai icesheet. The reasons for the absence of macro-

Table 1. *Macromeander parameters in relation to the contemporary channel dimensions*

River	F (km^2)	Q_a (m^3 s^{-1})	λ_{pr} (m)	W_{pr} (m)	λ_p (m)	W_p (m)	λ_p/λ_{pr}	W_p/W_{pr}
Dnieper River Basin								
Seym	10700	37.1	780	60	5600	400	7.2	6.7
Ubort'	5260	10.7	250	40	1700	120	6.8	3.0
Sula	14200	28.5	340	30	5000	400	14.7	13.3
Uday	6120	11.4			3000	300		
Psyel	11300	30.5	660	40	2500	330	3.8	8.3
Khorol	1740	4.5	350	30	1800	250	5.1	8.3
Orel'	9400	21.6	500	50	3600	350	7.2	7.0
Don River Basin								
Khoper	19100	67.0	840	100	5500	1000	6.5	10.0
Buzuluk	9200	13.0	920	75	2400	250	2.6	3.3
Ilovlya	8730	8.0	250	40	1700	150	6.8	3.8
Volga River Basin								
Dyema	12500	42.3	480	50	1700	200	3.5	4.0
Ikk	7660	31.9	350	40	2900	290	8.3	7.3
B. Kinel'	5970	19.4	400	25	1800	200	4.5	8.0
M. Irgiz	2110	3.3	320	30	1600	100	5.0	3.3
B. Uzen'	9490	6. 8	250	40	2000	200	8.0	5.0

F, Basin area; Q_a, present mean annual discharge; λ_{pr} and W_{pr}, present meander wavelength and bankfull channel width; λ_p and W_p, meander wavelength and bankfull channel width of the past (time of macromeander formation).

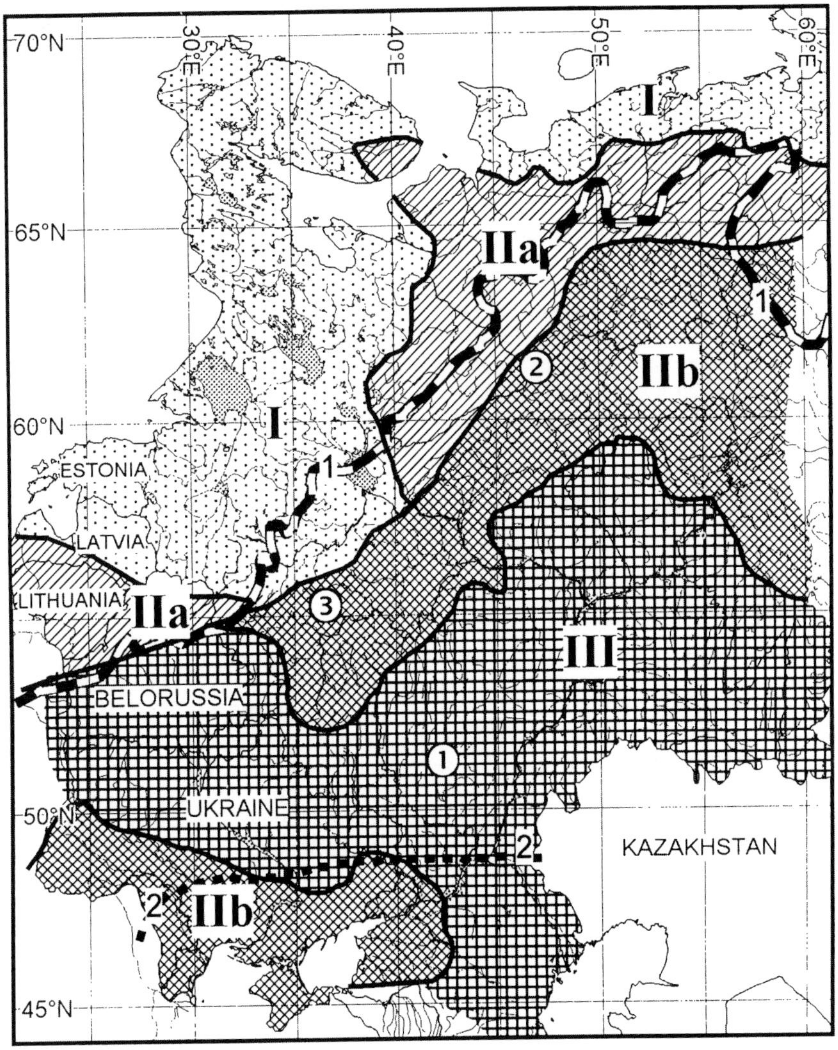

Fig. 1. Spatial distribution of relic Late Valdai macromeanders in river valleys over the Russian Plain. I, No macromeanders; II, macromeanders on the low tarraces prevalent (IIa, northern subregion, IIb, southern region); III, widespread floodplain macromeanders. 1, Last ice-sheet boundary (after Velichko 1982); 2, maximum permafrost extent (after Velichko 1982). Location of the case studies described in the text is shown by numbers inside circles: 1, the Khoper River; 2, the Vychegda River; 3, the Protva River.

meanders are different across the region. The northwest of the region (the Baltic Sheet) is composed of nearly bare magmatic and metamorphic rocks, and most of the river valleys were stable during the Late Glacial and Holocene periods. River basins of the northeastern part of the Russian Plain are still composed of sediments with continuous permafrost. The hydrological conditions of the channels have not changed dramatically here during the last 12 000 years. The prolonged existence of ice-dammed lakes in river valleys must also be taken into account. In the northwest some of these lakes were drained only in the late Holocene, and river floodplains inherit their flat bottoms. The reasons for the absence of macromeanders in the northern-west of the region are less clear.

(II) The region with macromeanders situated on low terraces of modern river valleys. In this region modern channels are mainly incised into the bottoms of palaeochannels due to various reasons, such as glacioisostatic uplift, changes of the local erosion base (e.g. draining of ice-dammed lakes), etc. The region is divided into two subregions. The northern subregion (IIa) occupies the north–eastern part of the forest zone. Macromeanders are found in some, but not all river valleys. The southern subregion (IIb) is characterized by overall distribution of macromeanders. It occupies two separated areas in forests and the southern part of steppes near the Black Sea coast.

(III) The region with macromeanders at the level of the modern floodplain forms a wide belt on the steppe–forest and steppe landscapes of the East European Plain. To the west it continues into Poland (Schumanski 1983; Starkel 1995*a*). River valleys of this region are characterized by very high floodplain width–channel width ratio. More than a hundred years ago, Dokuchaev (1878) pointed out that the floodplains of the rivers of Southern Russia are 30–300 times wider than the river channels. Dokuchaev suggested that this phenomenon is related to the influence of ancient lakes. Recent investigations, however, show that this feature is inherited from the Late Valdai valley widening as a result of the migration of extremely large palaeochannels.

The following example represents a rather typical situation.

Late Glacial relics in floodplain morphology: the Khoper River, Southern Russia

A 25 km valley stretch was investigated in the middle course of the Khoper River, a left tributary of the middle Don (Fig.1, Table 2). At this site the valley bottom, 10–11 km wide, includes a floodplain 2–3.5 m high (here and elsewhere, floodplain height is measured above the low-flow level), and fragments of low, sandy terrace (7–15 m). Most river activity is concentrated during the spring flood, which is 3–4 m high. Stream deposits are represented by fine sands. The riverbanks are composed mainly of highly erodible fine sands and silt. The channel, 0.07–0.12 km wide, has free meanders with a mean wavelength of 0.840 km. Abandoned meanders preserved in valley floor topography differ noticeably in size. Three generations of palaeomeanders and corresponding alluvial surfaces may be distinguished (Fig. 2).

Oxbows, which have parameters similar to those of the present channel (width 20–150 m, wavelength 200–800 m) mark a 3–6 km strip along the river course and indicate hydrological conditions close to contemporary ones. The ^{14}C dating 4670 ± 120 (Ki-6169) derived from the base of a palaeomeander infill shows that this part of the floodplain was formed during the second half of the Holocene period, including probably the Atlantic period. As seen from the abandoned palaeochannel pattern, the river formed either a meandering or an anastomosed planform, the latter probably being developed during high-flood periods or as a result of individual events of high magnitude.

The second generation of palaeomeanders is characterized by wavelengths of 0.75–1.5 km and bankfull channel width of 0.15–0.25 km. Belousova (1997) obtained regime equations connecting meander wavelengths and discharges for a number of rivers in the middle reaches of the Don Basin. Her estimations show that the Khoper peak discharges necessary for formation of such meanders are two times greater than at present. Radiocarbon dates from other alluvial segments allow us to place these palaeomeanders and the corresponding floodplain area into the early Holocene period. The bottom of oxbows have the same level as the young floodplain (2.5–3.0 m). Their relief was smoothed by subsequent sedimentation, although, in some places, curved alluvial levees are seen marking the successive meander growth. Old channels are filled by loamy sediment 2.5–3.0 m thick, and now lie 1–2 m above the water table. The top of stream-bed sediment corresponds to the present river-bed level, indicating the absence of any significant incision or aggradation during Holocene time.

The oldest and the largest meanders (macromeanders) with a wavelength of 5.0–6.0 km, were formed by a channel 0.8–1.2 km wide (Fig. 2). Macromeanders cut a sandy terrace, 7–15 m high,

Table 2. *Brief characteristics of the key sites*

Characteristic	Khoper	Vychegda	Protva
Basin area (km^2)	19100	121000	1450
Mean July temperature (°C)	+21	+17	+18
Mean January temperature (°C)	−10	−14	−10
Mean annual precipitation (mm)	460	700	600
Landscape zone	Steppe	Middle taiga	Mixed forest
Mean annual discharge (m^3 s^{-1})	68	1160	8.0
Mean maximum discharge (m^3 s^{-1})	991	7520	234

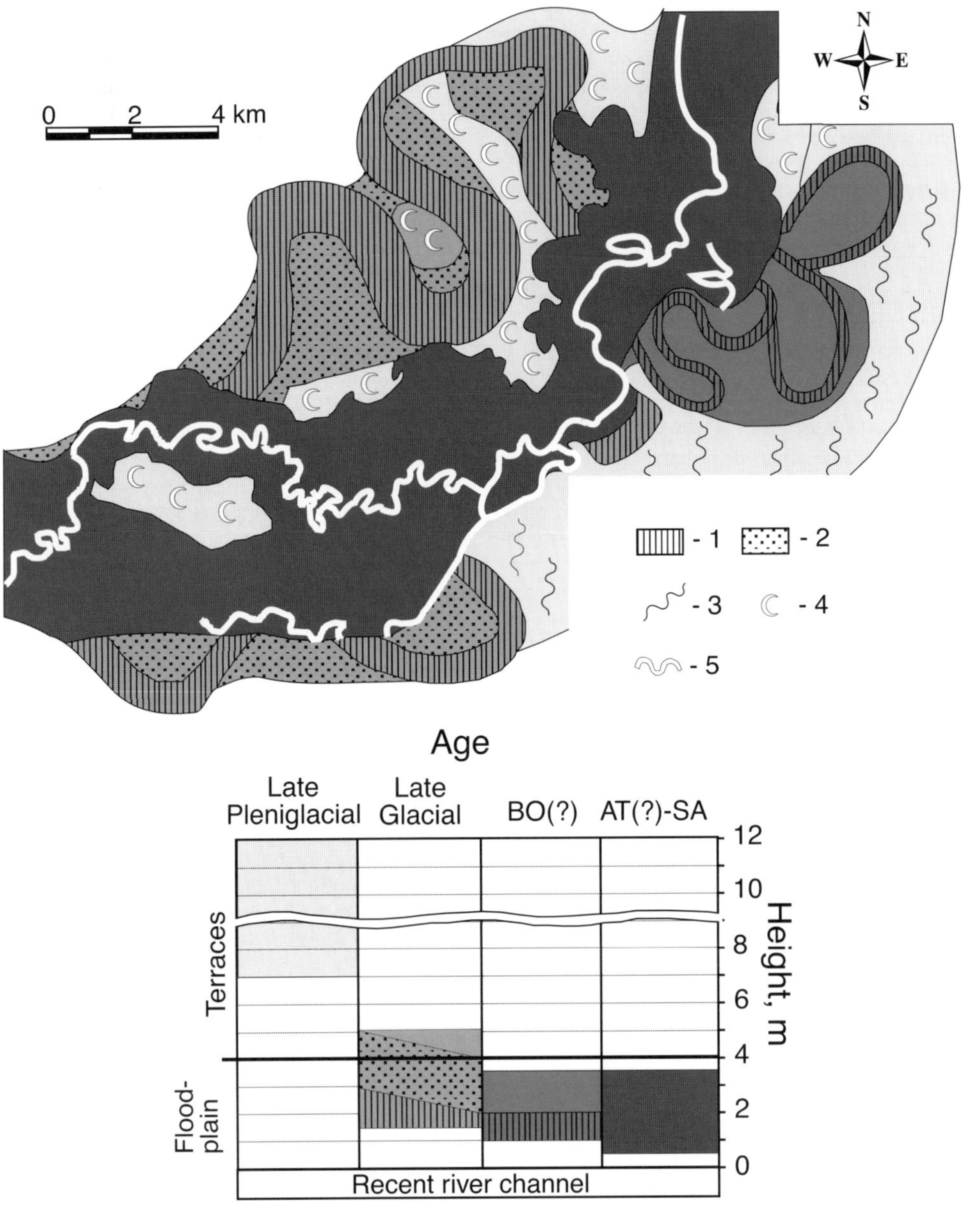

Fig. 2. Main morphological units of the Khoper River valley near Povorino. 1, Old channels; 2, old bars and islands; 3, solifluction cover; 4, aeolian forms; 5, recent river and floodplain lakes.

which has been considerably reworked by aeolian processes. Estimations based on palaeomeander wavelengths reveal that the required mean maximum discharges for their formation are nearly six times as much as the present values (Belousova 1997). Radiocarbon dating of the base of the palaeochannel infill (11 900 ± 120, Ki-5305) points to a Late Glacial age for this generation of floodplain. This corresponds to the data on large palaeomeanders in the Vistula Basin, which existed between 13 and 10 ka BP (Schumanski 1983; Starkel 1995*b*). The top of the stream alluvium inside the giant oxbows corresponds to the modern stream level in palaeoriffles and lies 4–6 m below it

(2–4 m below the modern river bed) in palaeopools. It may point to some aggradation after abandonment of the macromeander, or result from high vertical relief of the palaeochannel bed. During Holocene time, palaeochannels have been partially filled and now form rather low (1.5–2.5 m) bend segments inherited by the recent floodplain. Old bars that form alluvial surfaces 2–5 m high with their top parts lying sometimes above the modern flood level represent the other kind of inherited floodplain.

Holocene floodplain development in European valleys was characterized by relative morphological stability but with much spatial diversity, which results from variability of climatic, geological and geomorphological conditions on local and regional scales (Starkel 1995*a*; Vandenberghe 1995). A remarkable difference in floodplain structure is demonstrated by rivers that have different stream power. According to the Nanson & Croke (1992) classification, the Holocene floodplain of the Khoper River may be ranked as a low-energy one. The development of medium- and high-energy floodplains is illustrated by the following two cases.

Holocene floodplain development on a large meandering sand-bed river: the Vychegda River, Northern Russia

The River Vychegda is the right tributary of the Severnaya Dvina River (Fig. 1). The field investigations were carried out for a 40 km valley stretch (Table 2). Channel sediments are fine or medium sands, which are very easily transported by the stream. Consequently, the river is very mobile: sand bars move at rates of 100–150 m a^{-1}, and the annual retreat of highly erodible banks reaches 5–10 m. More than half of the annual discharge passes through during the spring flood (April–June), and maximum flood levels reach 7–8 m above low-flow level.

The valley is 30–40 km wide, and 60% of its area is occupied by a 20–30 m terrace. The latter was formed by the valley filling with sediments as a result of the existence of a glacier dam in the middle reaches of the Severnaya Dvina River. The ice-dammed lake was drained 12.5–13 ka BP (Val'chik *et al.* 1994). In the course of the subsequent incision, the River Vychegda formed a number of alluvial surfaces that are grouped into four generations (Fig. 3). The ages of different floodplain segments were obtained by radiocarbon dating of the base of palaeochannel infills or overbank alluvium.

The first generation of the valley floor is represented by a 3–4 km wide segment lying 7–14 m above the river. An abandoned meander with 6-7 km half-wavelength is formed by a palaeochannel 1.0–1.5 km wide with well-developed bars and islands. The palaeochannel surface (partially inundated recently during high floods) is 7–9 m high. The basal layers of the palaeochannel infill are dated at 8400 ± 70 (Ki-6407) and 8630 ± 60 (Ki-6405) thus indicating the approximate time of the meander abandonment. The 10–12 m high palaeo-floodplain created during the meander shift is now represented by a low terrace dated at 8650 ± 60 (Ki-6413) and 8230 ± 50 (Ki-6400). The branch channel with a series of small meanders located upstream is dated at 9260 ± 70 (Ki-6406). The whole fluvial unit may be referred to between 8.5 and 9.5 ka BP (the early Boreal). The dimensions of the palaeochannel and its morphology (braiding features) show that discharge at that time could have been somewhat higher than at present (Sidorchuk *et al.* 1999). The upper surface of stream sands at old riffles within the palaeochannel lies 3–4 m above the present stream, so during early Boreal time incision was still in progress.

The next floodplain generation, 7–10 m high, contains a distinct series of curved levees that mark the development of meanders noticeably smaller than at previous and all following stages. Nowadays it is submerged partially in depressions only, but the levees remain higher than the maximum flood level. One of the oldest palaeomeanders is dated as 8120 ± 50 (Ki-6404), so it is clear that further development of this generation of floodplain continued through the Atlantic period. A small palaeomeander generation dated at 7700 ± 80 (Ki-6411) is found at the Vyled' River Valley (left tributary of Vychegda). So the low-discharge period covers the Boreal–Atlantic transition and the early Atlantic period. It probably lasted throughout the Atlantic period as surfaces of this age are very scarcely represented in the floodplain. This implies negligible lateral channel migration, and consequently a 'gap' of dates corresponding to the Atlantic period exists.

The Sub-Boreal stage is represented by a meander series whose development started not later than 4.2–4.7 ka BP (4670 ± 60 Ki-6409, 4470 ± 60 Ki-6402, 4200 ± 50 Ki-6401). The increase in meander dimensions, when compared with the previous period, indicates an increase in discharge. Levee pattern shows that channel curvature increased down the valley eventually resulting in loop cut-off. The date of 1900 ± 50 (Ki-6390) derived from the palaeomeander infill seems to underestimate the age of the abandonment, as it overlaps the dates of the subsequent floodplain generation. The base of the abandoned channel (1–2 m below the river water surface), the heights

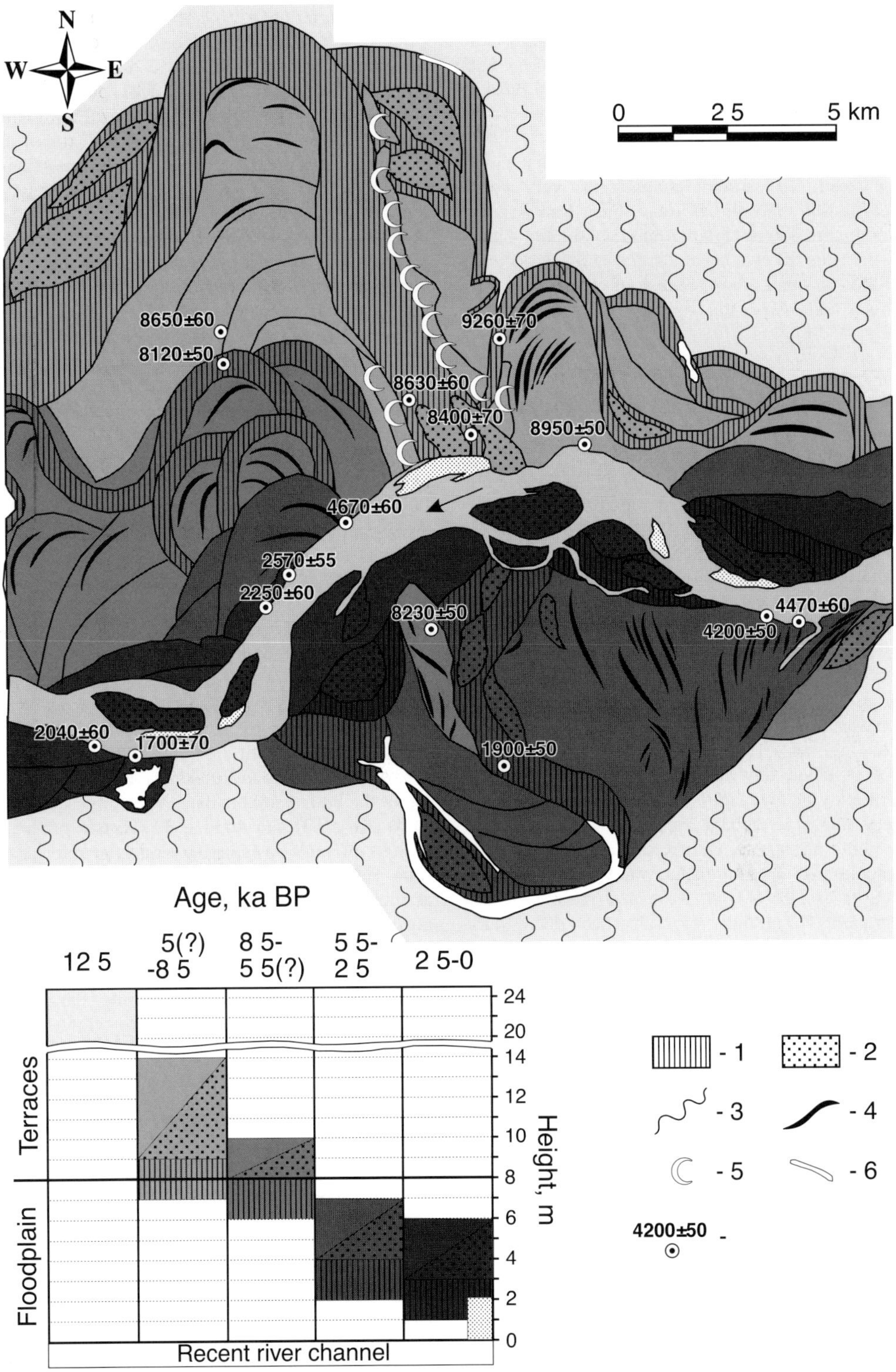

Fig. 3. Geomorphological map of a key site in the Vychegda River valley. 1, Old channels; 2, old bars and islands; 3, solifluction cover; 4, levees; 5, aeolian dunes; 6, floodplain lakes; 7, radiocarbon dates.

of the levees (5–7 m) and inter-levee depressions (2–5 m) confirm that by the beginning of Sub-Boreal time river incision had ceased. The Sub-Boreal surface as a whole is inundated by present-day floods and consequently represents the typical floodplain of a meandering river.

During the Sub-Atlantic period discharges increased. The current channel has very gentle bends and simple braids. The Sub-Atlantic floodplain, 3–6 m high, is found as isolated islands both inside the channel and at the banks. It occupies relatively small areas because of channel straightening and narrowing of the channel migration belt.

Floodplain formation on a lowland gravel-bed river: the Protva River, Central Russia

The Protva River is a left tributary of the River Oka (Upper Volga Basin, see Fig. 1). The key site is located in the middle reach (Table 2). Some 70% of water yield comes from snowmelt, and during the spring flood (March–April) the water rises by 4.0–5.5 m. The stream width is 25–40 m (40–55 m bankfull). The channel is rather straight; at some reaches it bends very gently or is split by isolated islands. Channel sediment is gravel (mainly 2–5 cm) mixed with coarse sand. It is transported during flooding, but a 10 year record of observations has not revealed any movement of bars or riffles (Antonov & Rychagov 1996). As the channel is very stable, river banks are also stable: zones of washing are sparse, with retreat rates only locally reaching 0.3 m a^{-1}. At the study site the valley bottom is 500–600 m wide and there is a low terrace (8–12 m high), which has been dated to the end of the Last Glacial period (Antonov & Rychagov 1996). It is only locally present, like the Late Glacial floodplain surfaces (6–8 m high); both are covered by colluvial mantles and fans. As the top of the basal gravels indicates, the post-glacial incision lasted till early Holocene time, but since at least 8.5 ka BP no incision and no significant aggradation has occurred. Three main units of the floodplain are detected differing in morphology and relative height (Fig. 4).

The 4–5 m high floodplain is the widest. It is characterized by a rather smooth surface, which nevertheless preserves a system of interrelated linear depressions left by a channel with multiple braids. Some of the individual palaeobranches are twice as wide as the present channel. Though it is almost impossible to determine which of the braids were active simultaneously, it is evident that the palaeostream was characterized by significantly higher discharges than the present stream. The channel migration belt occupied the whole valley floor. Three radiocarbon dates were derived from the basal gravels: 8500 ± 75 (Ki-5307), 7910 ± 90 (Ki-6155) and 6840 ± 230 (MGU-1483). The bottom layers of the palaeochannel infill were dated at 6200 ± 85 (Ki-5217), 6150 ± 70 (Ki-6156) and 4970 ± 100 (Ki-6175). By the end of the Atlantic period all the channel branches were separated from the main channel. It follows from the chronology of the branch abandonment and the formation of the next generation of floodplain that concentration of the stream in a single channel was completed between 5.5 and 6 ka BP. During the second half of the Holocene period the early Atlantic floodplain was subject only to vertical sediment accretion and has therefore been covered by overbank alluvium 2–3 m thick (sandy loam and silt sediments).

The second generation of floodplain forms a strip along the channel and commonly reaches 50–70 m in width (locally up to 100–120 m). It has a more pronounced relief than earlier generations, with distinct old islands and bars with tops 2–3.5 m above the river. The large size of palaeobars indicates that they formed under much higher energy conditions than at present. Three radiocarbon dates were derived from the top of the basal gravels: 5370 ± 80 (Ki-6471), 3560 ± 65 (Ki-6463) and 2980 ± 80 (MGU-584). The overbank facies of the early Atlantic floodplain contain a buried humus horizon at the depth interval 1.2–1.6 m giving evidence of low flood frequency. The base and the top of the buried soil are dated respectively at 4570 ± 70 (Ki-6467) and 4020 ± 80 (Ki-6466), so the periods of high river activity may be referred to time intervals of 4.6–5.5 and 3.0–4.0 ka BP assuming that floodplain stripping could take place during individual flood events of high magnitude.

During the Sub-Atlantic period the stream power was still insufficient to provide active transport of coarse-grained material; consequently, the associated floodplain is revealed as low (0.5–1.0 m) separated fragments on rare, small bars and islands.

Discussion

As Brown (1996) has pointed out, contemporary processes of floodplain formation may be influenced by forms inherited from a preceding river regime. Extraordinary hydrological conditions in the Late Valdai have resulted in three main aspects that could influence the Holocene river dynamics and floodplain formation: (a) creation of large alluvial forms, which are conserved in floodplain morphology; (b) considerable river valley widening; (c) creation of valley slopes corresponding to discharges much higher than those during Holocene time. Therefore, at least the

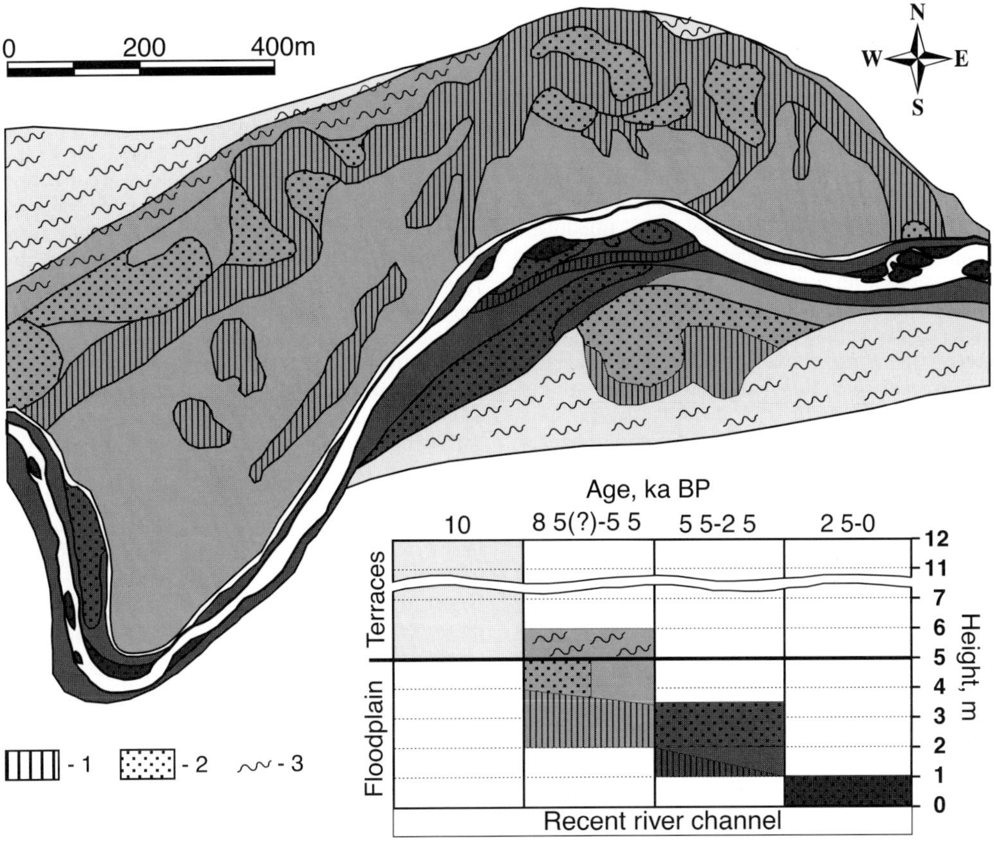

Fig. 4. Geomorphological map of the Protva River floodplain near Satino. 1, Old channels; 2, old bars and islands; 3, colluvial cover.

early Holocene floodplains seem to show some disequilibrium with river discharge, and this has been reported from other locations in northwest Europe (Brown 1995, 1996). However, in many cases river floodplains are relatively recent and thus reflect only the latest periods of fluvial history.

In the temperate zone most of the lowland rivers, which are controlled mainly by climatic factors, reveal common behaviour features during the past 15 millennia (i.e. intensive incision during the deglaciation period and relative stability in Holocene time). River incision, which started at the beginning of the Late Glacial period, is reported for most European regions (Vandenberghe, 1995). The timing of long profile stabilization indicates the beginning of the construction of recent floodplains. In the Baltic Sea region incision has lasted throughout Holocene time as a result of glacioisostatic crustal movement, and recent floodplain development is attributed to the second half of the Holocene period (Dvareckas 1990; Miidel & Raukas 1991). A similar situation is demonstrated by river valleys in the northern part of the central regions of the East European Plain (regions IIa and IIb in Fig. 1). The Vychegda River was undergoing incision until no later than the end of the Atlantic period. Its early Holocene fluvial surfaces experience present-day flood inundation only in palaeochannel depressions whereas palaeobars and floodplain levees remain dry. Such partially flooded units illustrate the statement by Dawson & Gardiner (1987) that the frequency of inundation is not a strict criterion for distinguishing between the floodplain and higher terraces. The Protva River was incising until the middle of the Boreal period, and its early Holocene fluvial belt makes up a significant part of the present floodplain. In such cases, the Late Valdai fluvial surfaces now form terraces above the current inundation level and therefore influence floodplain construction only as a component of the valley sides that confine the river channel.

In the southern half of the East European Plain, river incision stopped simultaneously with the decrease in discharge at the end of the Late Valdai, and the topographic position of Late Valdai surfaces allows their incorporation into the modern floodplains (as in the case of the Khoper River). Floodplain segments inherited from the Late Valdai hydrological regime create certain morphological units that differ from the Holocene floodplain. Large palaeochannels (macromeanders) form broad depressions with a width comparable with that of the Holocene meander belt. Old point-bars now make up vast upland areas, which are completely or partly (the tops) above the contemporary inundation level. Such floodplains are characterized by a width that significantly exceeds the potential amplitude of migration of the Holocene channel. For example, about a third of the present-day Khoper River floodplain is occupied by fluvial surfaces originating from the Late Valdai.

The low valley slope created by powerful Late Valdai flows predetermines the low-energy conditions of channel development during Holocene time. In addition, the high cohesiveness of bank material might also have prevented the formation of a meandering pattern as illustrated by Richards (1972) for the River Severn. Alternatively, low gradients, coupled with extreme floods, are favourable for the formation of anastomosing channels (Knighton & Nanson 1993). The width of the Holocene channel migration belt of the Khoper River is five times as much as that of the Vychegda and Protva rivers (Table 3). Such a large amplitude of channel migration could hardly be achieved by lateral erosion alone. It is assumed therefore that widening of the Holocene floodplain occurred to a large extent as a result of frequent channel avulsion superimposed on the previously created floodplain.

Present-day floodplain morphology depends also on the destruction of previously created units by river migration. In the model for long-term meandering channel evolution presented by Howard (1996) the simulated floodplain consists only of fragments formed as a result of development of individual channel loops. Combination of the fragments results in a continuum of ages, but the older surfaces occupy smaller portions of the whole floodplain than the younger ones. The following simple model may illustrate the simulation of floodplain age spectra. A further simplification of procedure was used: (a) lateral channel shift is of a stochastic character; (b) rate of lateral migration W_0 is constant in time; (c) channel migration occurs within a belt of limited width. Under these assumptions the area of floodplain destroyed during a given time span equals that of newly created floodplain. At a given time moment t the probability of destruction of segments created at a moment τ in the past is proportional to their area $W_\tau(t)$ expressed as a proportion of the whole floodplain area against unit valley stretch W_b. Then the change in area W_τ over time is described by the equation:

$$\frac{dW_\tau(t-\tau)}{dt} = -\frac{W_0}{W_b} W_\tau(t-\tau) \qquad (1)$$

Table 3. *Areas occupied by floodplain units of different ages*

	Period (ka BP)	Area F_T (km²)	Mean width W_T (km)	Occupied portion of the valley floor: for the period	per 1000 years	W_T/W_c
Khoper	14–10	55.1	2.6	0.32	0.080	26.3
	10–8	20.4	0.97	0.12	0.059	9.71
	8–0	92.7	4.4	0.54	0.067	44.1
	Holocene	113.1	5.4	0.65	0.065	53.9
	Present channel	4.5	0.10	0.03		
Vychegda	9.5–8.5	107.3	5.4	0.39	0.390	4.51
	8.5–5.5	37.0	1.9	0.13	0.045	1.56
	5.5–2.5	69.1	3.5	0.25	0.084	2.91
	2.5–0	34.1	1.7	0.12	0.050	1.43
	Holocene	247.5	12.5	0.90	0.09	10.4
	Present channel	27.4	1.2	0.10		
Protva	8.5–5.5	0.52	0.34	0.71	0.237	8.6
	5.5–2.5	0.11	0.076	0.16	0.053	1.9
	2.5–0	0.010	0.006	0.01	0.005	0.16
	Holocene	0.64	0.42	0.88	0.09	10.7
	Present channel	0.080	0.040	0.12		

W_c, present-day mean channel width.

As $t-\tau$ is the age T of given floodplain portions, the solution of (1) describes the floodplain age spectra, i.e. the occupied portion of floodplain area as a function of age:

$$W_\tau(T) = W_0 \exp\left(-\frac{W_0}{W_b}T\right) \quad (2)$$

The assumption $W_b(t)$ = const. makes this model valuable for rivers confined by firm valley sides or having a very wide floodplain. Howard (1996) obtained a logarithmic growth function for meander belt width. Inclusion of a time depending function for W_b in (1) leads to an expression that does not permit finite integration. In this case, approximation from numeric simulations is more preferable, although it is obvious that floodplain age distribution would also be a time-decreasing function.

In reality, the assumption of non-changing river activity may be acceptable only for relatively short time periods. An alternating rate of river lateral migration is a causative factor in the more or less wide presence of fluvial units formed during different epochs in the past. To illustrate this, the areas of floodplain units of different age were calculated for the three case studies (Table 3). The calculations include partially flooded surfaces as well. Areas were estimated based on geomorphological maps (Figs 2–4), and mean width (Table 3, column 4) was calculated as a ratio between the area and the length of a valley stretch. To make data comparable, they are normalized in relation to the total floodplain area (Table 3, column 5) and to the duration of the corresponding formation period (Table 3, column 6).

None of the three cases gives the expected decrease in area with increase in age of floodplain sites (Table 3, Fig. 5). The Khoper River demonstrates the relatively even distribution of floodplain area–age relationship and the significant role of pre-Holocene surfaces in the current floodplain structure. The entire floodplains of the Vychegda and Protva rivers were formed in Holocene time, and the youngest surfaces (Late Holocene) occupy noticeably less area than the Early Holocene ones (Table 3, Fig. 5). It is natural to suppose, therefore, that the floodplain age spectrum is defined mainly by preceding changes in the width of the strip wherein lateral deformations occur. The units that are distributed most widely should be those that were formed during the epochs when the highest amplitude of channel migration took place.

It is worth noting that the floodplain age spectrum cannot be interpreted distinctly from a palaeohydrological point of view. Most frequently, an increase in discharge causes an increase in the rate of bank erosion, and likewise an increase in the amplitude of horizontal channel migration. In some cases, however, the meander belt decreases in width. The Vychegda and Protva manifest both variants.

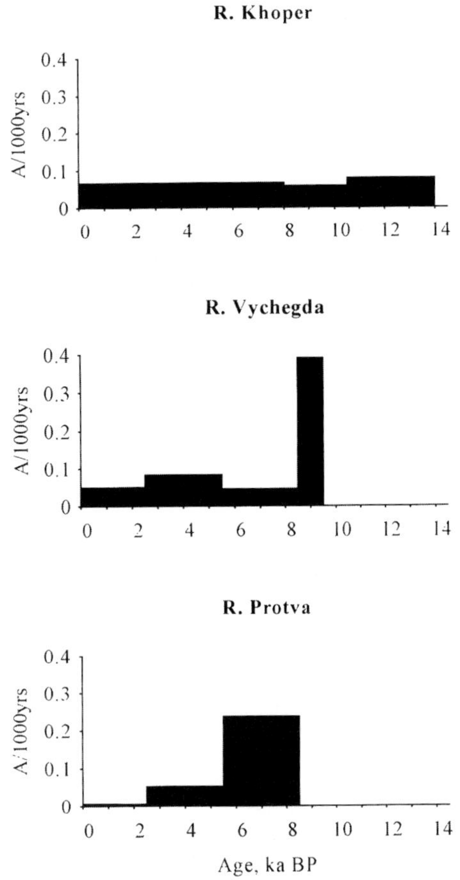

Fig. 5. Floodplain age spectra of presented case studies. *A*, Portion of floodplain area occupied by surfaces of given age.

The Vychegda River is characterized by a considerable increase in the channel width, a decrease in sinuosity and the appearance of braiding features during the Sub-Atlantic period as compared with the Sub-Boreal time. This shows that discharge reached the threshold values separating meandering and braiding channels for given valley slopes (Alabyan & Chalov 1998). This channel transformation led to a narrowing of the meanderbelt. In such cases, the period associated with high discharge is represented only on minor areas of a floodplain and has little potential to survive during further channel migration.

In the case of the Protva River, the present period is characterized by too low a rate of sediment transport to create point bars. This results in stable channel position, low rates of bank erosion and a low rate of formation of new floodplain areas. The existing broad floodplain was formed during the two high-water periods, the early Atlantic and the late Sub-Boreal. Evidence of the low river activity in earlier Holocene periods (similar to that at present), is not found in the floodplain morphology. Such contrasting changes in river activity alongside the alternation of discharge are caused by the coarse-grained composition of the channel sediment. This floodplain type is similar to the high-energy floodplains subject to widespread stripping during large floods and a predominance of vertical accretion during low flood periods (Nanson 1986; Nanson & Croke 1992). In this case, floodplain morphology stores information only from high flood periods, and periods of low river activity may be studied only from the sedimentary record stored in overbank alluvium. A similar example was reported by Brown & Keough (1992) from the Nene River in the UK.

Conclusions

The following conclusions can be reached.

(1) Floodplains, both as individual segments and as a whole, may be dated within a wide age range. There are two main reasons for this: destruction of old surfaces by channel migration and chronology of late- and post-glacial incision–aggradation cycles.

(2) In most cases the floodplain surface may not be regarded as a continuous chronicle of river history, the latter being represented only selectively. Surfaces created during periods with wider channel migration belts are more likely to be preserved in the present-day floodplain, whereas periods with low amplitude of lateral channel shift may make a gap in the floodplain age spectrum.

(3) Previously created valley floor characteristics, namely, width, slope and morphology, may, to a greater or lesser extent, influence further channel development and associated floodplain construction. Such kinds of intrinsic control should be taken into consideration within the framework of a process–response approach to studying river history.

Investigations are supported by the Russian Foundation for Basic Research (Project 97-05-64708).

References

ALABYAN, A. M. & CHALOV, R. S. 1998. Types of river channel patterns and their natural controls. *Earth Surface Processes and Landforms*, **23**, 467–474.

ANTONOV, S. I. & RYCHAGOV, G. I. (eds) 1996. *Geology and Fluvial History of the Protva River Valley.* Moscow University, Moscow [in Russian].

BELOUSOVA, E. E. 1997. The Khoper River floodplain in the middle reaches - morphology and some problems of palaeohydrology. *Geomorphologia*, **1**, 54–58 [in Russian].

BOHNCKE, S. J. P. & VANDENBERGHE, J. 1991. Palaeohydrological development in the Southern Netherlands during the last 15000 years. *In*: STARKEL, L., GREGORY, K. J. & THORNES, J. B. (eds) *Temperate Palaeohydrology*. Wiley, Chichester, 253–281.

BROWN, A. G. 1995. Holocene channel and floodplain change: a UK perspective. *In*: GURNELL, A. & PETTS, G. (eds) *Changing River Channels.* Wiley, Chichester, 43–64.

—— 1996. Floodplain palaeoenvironments. *In*: ANDERSON, M. G., WALLING, D. E. & BATES, P. D. (eds) *Floodplain Processes*. Wiley, Chichester, 95–137.

—— & KEOUGH, M. 1992. Palaeochannels, palaeolandsurfaces and the three-dimensional reconstruction of floodplain environmental change. *In*: CARLING, P. A. & PETTS, G. E. (eds) *Lowland Floodplain Rivers: Geomorphological Perspectives*. Wiley, Chichester, 185–202.

DAWSON, M. R. & GARDINER, V. 1987. River terraces: the general model and a palaeohydrological and sedimentological interpretation of the terraces of the Lower Severn. *In*: GREGORY, K. J., LEWIN, J. & THORNES, J. B. (eds) *Palaeohydrology in Practice.* Wiley, Chichester, 269–305.

DOKUCHAEV, V. V. 1878. *Types of River Valley Formation in European Russia.* S Petersburg Vasily Dermakor Publishing House [in Russian]

DURY, G. H. 1964. *Principles of Underfit Streams.* US Geological Survey Professional Papers, **452-A**.

DVARECKAS, V. 1990. The development of the Lithuanian river valleys in Late- and Post-Glacial times. *Quaternary Studies in Poland*, **10**, 41–45.

HOWARD, A. D. 1996. Modelling channel evolution and floodplain morphology. *In*: ANDERSON, M. G., WALLING, D. E. & BATES, P. D. (eds) *Floodplain Processes*. Wiley, Chichester, 15–62.

KALICKI, T. 1995. Lateglacial and Holocene evolution of some river valleys in Byelorussia. *In*: FRENZEL, B., VANDENBERGHE, J., KASSE, K., BOHNCKE, S. & GLASER, B. (eds) *European River Activity and Climatic Change During the Lateglacial and Early Holocene. Palaoklimaforschung/Palaeoclimate Research*, **14,** 89–100.

KNIGHTON, A .D. & NANSON, G. C. 1993. Anastomosis and the continuum of channel pattern. *Earth Surface Processes and Landforms*, **18**, 613–625.

KOZARSKI, S. 1991. Warta—a case study of a lowland river. *In*: STARKEL, L., GREGORY, K. J. & THORNES, J. B. (eds) *Temperate Palaeohydrology*. Wiley, Chichester, 189–215.

——, & ROTNICKI, K. 1977. Valley floors and changes of river channel patterns in the north Polish Plain during the Late-Würm and Holocene. *Questiones Geographicae*, **4**, 51–93.

MIIDEL, A. & RAUKAS, A. 1991. The evolution of the river systems in the east Baltic. *In*: STARKEL, L., GREGORY, K. J. & THORNES, J. B. (eds) *Temperate Palaeohydrology*, Wiley, Chichester, 365–380.

MYCIELSKA-DOWIGIALLO, E. 1977. Channel pattern changes during the last glaciation and Holocene in the northern part of the Sandomierz basin and the middle part of the Vistula valley, Poland. *In*: GREGORY, K. J. (ed.) *River Channel Changes*. Wiley, Chichester, 75–87.

NANSON, G. C. 1986. Episodes of vertical accretion and catastrophic stripping: a model of disequilibrium floodplain development. *Geological Society of America Bulletin*, **97**, 1467–1475.

—— & CROKE, J. C. 1992. A genetic classification of floodplains. *Geomorphology*, **4**, 459–486.

RICHARDS, K. S. 1972. Meanders and valley slope. *Area*, **4**, 288–290.

ROSE, J., TURNER, C., COOPE, G. R. & BRYAN, M. D. 1980. Channel changes in a lowland river catchment over the last 13000 years. *In*: CULLINGFORD, R. A., DAVIDSON, D. A. & LEWIN, J. (eds) *Timescales in Geomorphology*. Wiley, New York, 159–176.

SCHUMANSKI, A. 1983. Palaeochannels of large meanders in the river valleys of the Polish Lowland. *Quaternary Studies in Poland*, **4**, 207–216.

SIDORCHUK, A. YU., BORISOVA, O. K., KOVALUKH, N. N., PANIN, A. V. & CHERNOV, A. V. 1999. Palaeohydrology of the Lower Vychegda River in Late Glacial and Holocene. *Vestnik Moskovskogo Universiteta, Seria 5, Geografia*, **5**, in press [in Russian].

STARKEL, L. 1991. Long-distance correlation of fluvial events in the temperate zone. *In*: STARKEL, L., GREGORY, K. J. & THORNES, J. B. (eds) *Temperate Palaeohydrology*. Wiley, Chichester, 473–496.

—— 1995*a*. Changes of river channels in Europe during the Holocene. *In*: GURNELL, A. & PETTS, G. (eds) *Changing River Channels*. Wiley, Chichester, 27–42.

—— 1995*b*. The place of the Vistula river valley in the late Vistulian—early Holocene evolution of the European valleys. *In*: FRENZEL, B., VANDENBERGHE, J., KASSE, K., BOHNCKE, S. & GLASER, B. (eds) *European River Activity and Climatic Change During the Lateglacial and Early Holocene. Palaoklimaforschung/Palaeoclimate Research*, **14**, 75–88.

VAL'CHIK, M. A., MAKKAVEEV, A. N., FAUSTOVA, M. A. & SHUPRICHINSKY, YA. 1994. Hydrographic net developing in Poland and European Russia during deglaciation. *In*: VELICHKO, A. A. & STARKEL, L. (eds) *Palaeogeographic Background of Recent Landscapes*. Nauka, Moscow, 40–53 [in Russian].

VANDENBERGHE, J. 1987. Changing fluvial processes in a small lowland valley at the end of the Weichselian Pleniglacial and during the Late Glacial. *In*: GARDINER, V. (ed.) *International Geomorphology 1986, Part I*. Wiley, Chichester, 731–744.

—— 1995. Postglacial river activity and climate: state of the art and future prospects. *In*: FRENZEL, B., VANDENBERGHE, J., KASSE, K., BOHNCKE, S. & GLASER, B. (eds) *European River Activity and Climatic Change During the Lateglacial and Early Holocene. Palaoklimaforschung/Palaeoclimate Research*, **14,** 1–9.

VELICHKO, A. A. (ed.) 1982. *Palaeogeography of Europe During the Last One Hundred Thousand Years (Atlas–Monograph)*. Nauka, Moscow [in Russian].

Palaeochannels and ground-water storage on the North China Plain

YINGKUI ZHAO[1], CHEN WU[2] & XIUQING ZHANG[2]

[1]*School of Geography and Environmental Management, University of the West of England, Bristol BS16 1QY, UK (e-mail: yingkui.zhao@uwe.ac.uk)*

[2]*Geography Institute of Hebei Academy of Sciences, Shijiazhuang, People's Republic of China*

Abstract: The North China Plain, on which there are many palaeochannels as a result of frequent river channel changes, consists of a complex of floodplains formed by the Yellow, the Haihe and the Luanhe rivers. The palaeochannels act as a significant store of fresh ground water. Most of the storage is within sand and gravel layers. The buried depth of these layers varies among palaeochannels formed in different periods. For example, those formed between the last principal stage of glaciation and early Holocene time are buried 20–40 m deep. Those formed in mid-Holocene time are found at a depth of 10–20 m and those formed in late Holocene time at 0–10 m depth. Ground-water storage and palaeochannel discharge was calculated for the North China Plain. Ground-water in palaeochannels is recharged mainly by rainfall. With good management practice, the water withdrawn can be balanced against recharge to achieve a sustainable level of use. In addition, the shallow-buried palaeochannels provide a suitable environment for water regulation and storage and can be used to develop underground reservoirs.

The North China Plain is located in the eastern part of China, between latitudes 35°00′N and 40°30′N, and longitudes 113°00′E and 119°30′E. It is bordered by the Taihang Mountains to the west, Yanshan Mountains to the north and the Bohai Sea to the east. The southern boundary is the Yellow River (Fig.1).

The North China Plain, including all the plain area of Hebei Province, Beijing and Tianjin, and the northern part of the plain areas of Henan Province and Shandong Province, has an area of 136 000 km^2 with a population of 112 million people. This extensive feature was constructed from sediment deposited by the Yellow River, the Haihe River and its main branches and the Luanhe River. The plain provides an important economic base for northern China, including the capital, Beijing, China's political, economic and cultural centre.

Investigation of the palaeochannels on the North China Plain has been undertaken since the 1950s. The initial aim of the studies was to provide a database for pumping ground water stored in the palaeochannels (Wu & Zhao 1985), and then for the development of ground-water storage using the porosity of the sand in the palaeochannels. By mapping the area of sand and studying the sand properties, information would then be available for agricultural planning and to guide sand extraction for building purposes. To undertake palaeochannel research, historical documents covering a period of more than 2000 years and the large-scale contour maps made in 1921 provided the foundation.

The North China Plain is located in the warm, temperate zone and experiences a semi-humid monsoon climate. Uneven rainfall distribution, both spatially and temporally, and storage of runoff by the reservoirs in the upper reaches, limits the water supply for agriculture and industry. In addition, it is uneconomic and unsustainable to tap ground water at a depth greater than 100 m. Therefore, the fresh water stored in the sand and gravel of the shallow-buried palaeochannels (depth less than 50 m) is an important water supply for the North China Plain. Palaeochannels within the North China Plain also provide a suitable environment for water regulation for overcoming the uneven distribution of rainfall or for regulation and storage of water transferred from South China. Understanding the distribution of palaeochannels, palaeochannel deposit characteristics and their relationships with ground-water storage is thus essential for the North China Plain. This paper describes surface and shallow-buried palaeochannels and palaeochannel zones and their relationship with ground-water storage.

Methods

The Daming–Qinghe–Jingxian–Qingxian palaeochannel zone (Fig. 2) is typical of the palaeochannel zones on the

From: MARRIOTT, S. B. & ALEXANDER, J. (eds) 1999. *Floodplains: Interdisciplinary Approaches*. Geological Society, London, Special Publications, **163**, 231–239. 1-86239-050-9/99/$15.00

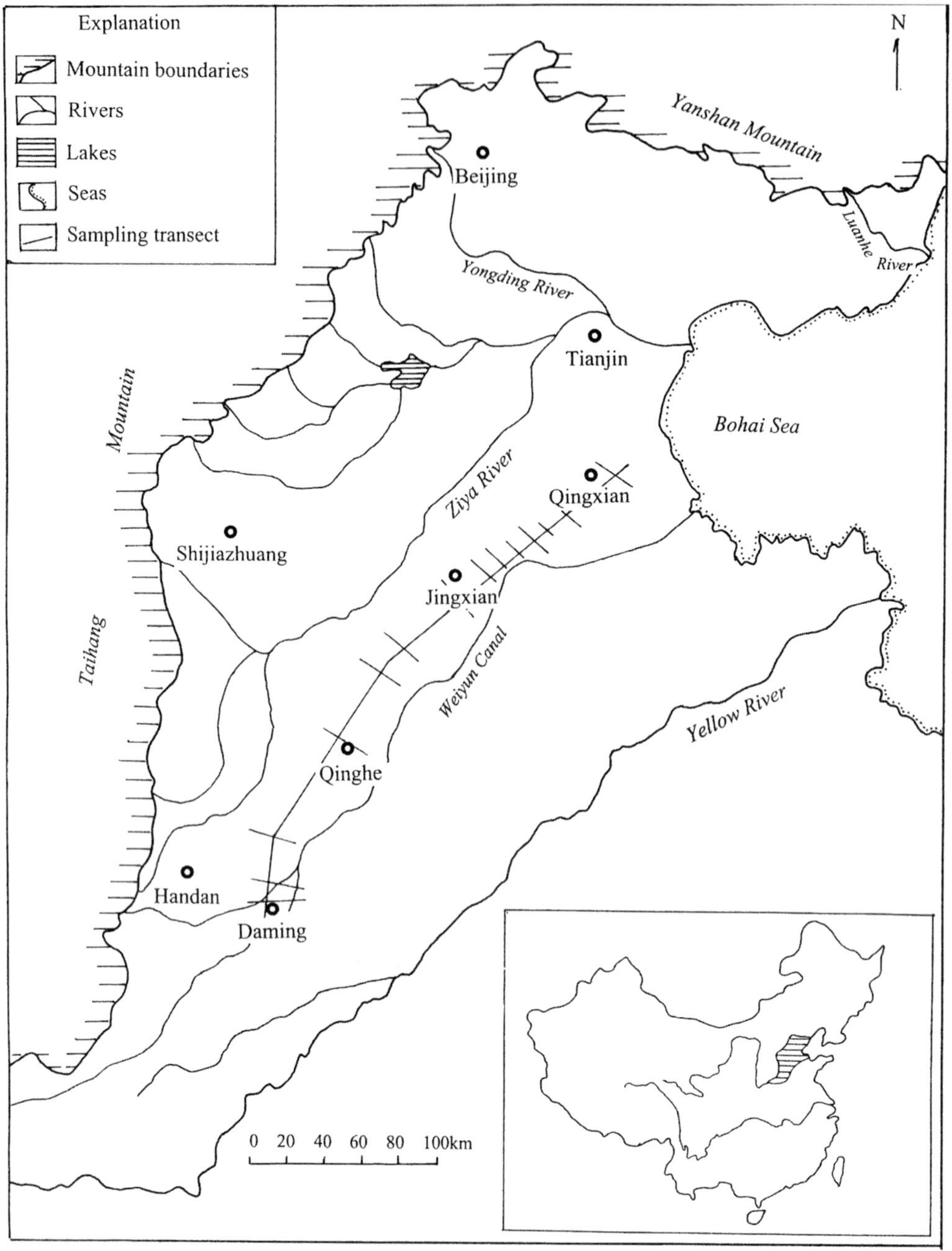

Fig. 1. Location of the North China Plain, showing the sampling transects.

North China Plain and was chosen for detailed research particularly to consider the variations in ground-water storage. This palaeochannel zone is about 340 km in length. Fourteen cross-sections along this length were chosen for detailed research. Along each section 5–10 boreholes were drilled to a depth of up to 60 m. Samples were collected at 1 m intervals and taken back to the laboratory for particle-size analysis, ^{14}C dating, pollen analysis and mineralogical analysis.

Results of particle-size analysis were plotted on a cumulative frequency diagram so that coarse termination and fine termination of all the samples could be

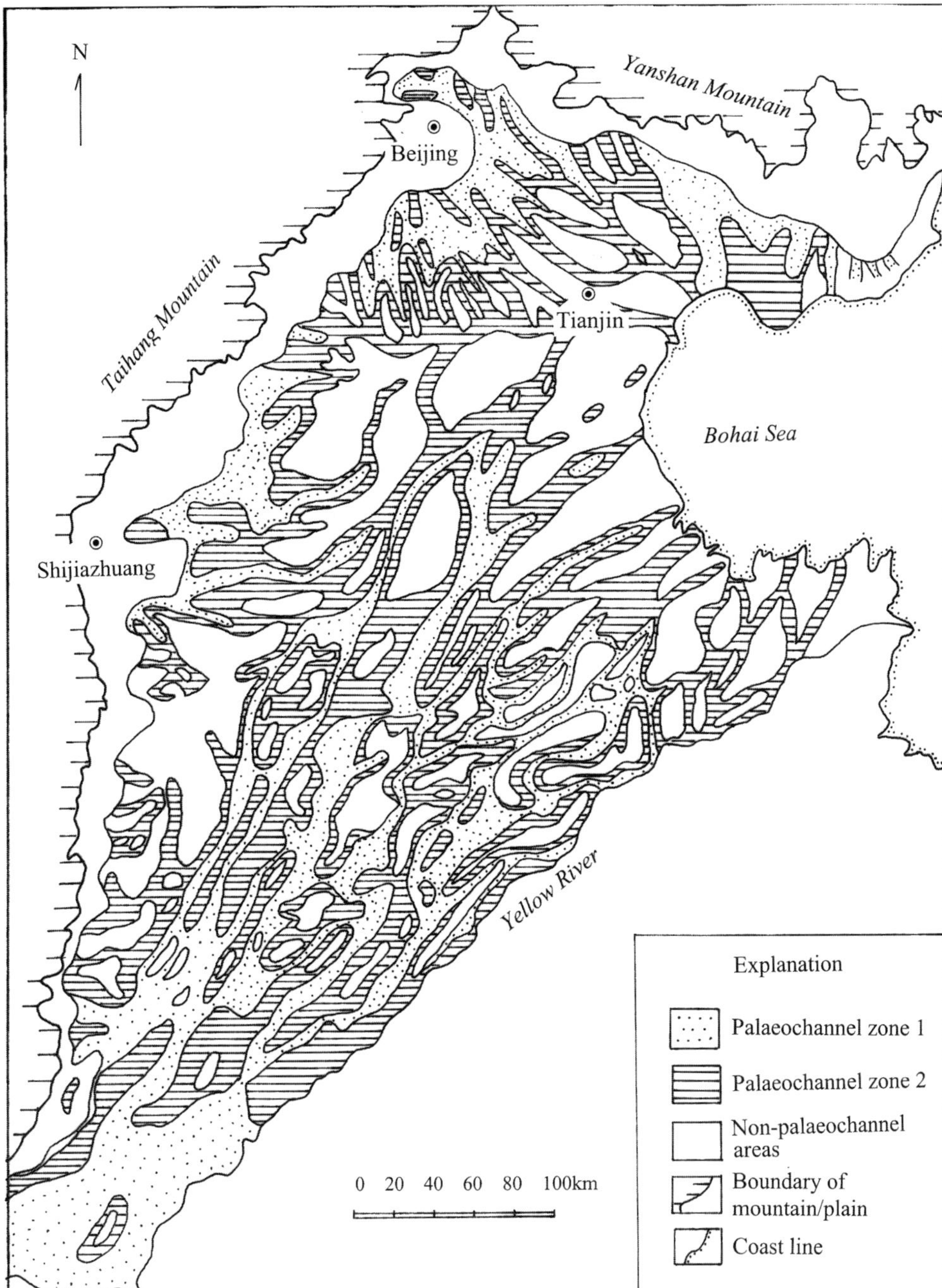

Fig. 2. Distribution of shallow-buried palaeochannel zones on the North China Plain.

calculated. Particles distributed between the coarse termination (CT, 1–1.25ϕ) and fine termination (FT, 2–3.5ϕ) are deemed to be related to saltation. Particles coarser than CT are related to traction. Those finer than FT are related to suspension.

Palaeochannel discharge is calculated using Schumm's (1960) and Chezy equations (Starkel et al. 1991) on the basis of results of the field survey and laboratory analysis. The field survey at each section included the recording of sand belt width (*W*, represented by coarser materials), number of channels within the belt (N), channel width (Wd, represented by main stream deposits), water depth (D, represented by the thickness of point bars) and river channel gradient (S). Drilling samples were taken back to the laboratory for particle-size analysis. The data were used for calculation of traction loads (Qs), suspension loads (M) and mean grain size (d):

(1) width/depth ratio (F)

$$F = \frac{\sqrt{W}}{D}$$

$$F = 225M^{-1.08}$$

(2) mean velocity (V)

$$V = C\sqrt{(RS)}$$

$$C = \frac{87}{1 + \frac{r}{\sqrt{D}}}, \qquad r = 0.25$$

$$R = \frac{\bar{A}}{P}$$

where R is the hydraulic radius, $\bar{A}$ is mean cross-sectional area and P is the wetted perimeter;

(3) mean discharge (Qm)

$$Qm = WDV$$

(4) annual discharge (Qw)

$$Qw = Qm\,t$$

(5) Sinuosity (Ω)

$$\Omega = 3.5F^{-0.27}$$

(6) index of river-bed stability (R_b)

$$R_b = \frac{d}{S}$$

Distribution of palaeochannels

On the North China Plain, six stages of late Pleistocene and Holocene palaeochannels can be identified (Xu *et al.* 1996). According to their depth of burial, palaeochannels on the North China Plain are classified into surface palaeochannels (with a buried depth of 0–10 m) and shallow-buried palaeochannels (with a buried depth of 10–40 m) (Wu *et al.* 1996*a*). In general, stage 6 palaeochannels on the surface were formed in late Holocene time, whereas stages 1–5 shallow-buried palaeochannels formed during late Pleistocene to mid-Holocene time. A number of palaeochannels cross or run parallel to each other to form palaeochannel zones. A palaeochannel zone can stretch to a width of several to tens of kilometres, within which palaeochannels are densely distributed; on a vertical scale, palaeochannels of various ages come into mutual contact or are superimposed. There are more than 20 palaeochannel zones on the North China Plain (Wu *et al.* 1996*b*), and several of them exceed 100 km in length and 10 km in width and lead to the Bohai Sea.

In considering ground-water storage, the shallow-buried palaeochannel zones provide the greatest ground-water resource. Figure 2 shows the distribution of shallow-buried palaeochannel zones on the North China Plain. The palaeochannels run in a northeasterly direction in the south and in an easterly or southeasterly direction in the north, with a starting point at the mountain front and ending at the coastline. This is a similar pattern to that of the modern river system (Fig. 1) and corresponds to the ground-water distribution on the North China Plain. Investigation of the palaeochannels shows that the shallow-buried palaeochannel zones on the North China Plain cover an area of about 55 000 km^2. The aquifer unit is between 10 and 20 m in thickness. The estimated water storage, on the basis of an average thickness of 15 m and average effective porosity of 20% (Wu & Zhao 1985), is about 16.5×10^9 m^3. This amount is about 90% of the total ground-water storage on the North China Plain.

Characteristics of palaeochannel sediments

The sedimentary characteristics of deposited materials are the main indicators of palaeochannels. In general, deposited materials in palaeochannels are coarser than those in areas where palaeochannels are absent (Wu *et al.* 1996*a*) and become finer downstream. Across the channels the sediments tend to be finer on natural levees; however, mean grain size varies between individual palaeochannels as a result of source differences. For example, the sediments in the palaeochannels left by the Yellow River are smaller (mean particle size 2ϕ–5ϕ) because in the upper reaches this river runs through the Loess Plateau and then runs for a long distance through the plain area. In contrast, the sediments of the Luanhe River are larger (mean grain size 1ϕ–3ϕ) because this river runs through mountains in the upper reaches and has a shorter course through the plain area.

The sand in the palaeochannels is usually well sorted (sorting coefficient 0.5–1.0). The result of calculations of the traction, saltation and suspension fractions of the sediments shows that 80–89% of the distribution relates to saltation, 10–20% relates to suspension and less than 1% to traction. This is similar to Visher's (1969) classification of fluvial sediments. Within the palaeochannel zones, ground-water storage varies from one site to another because of the differences in sedimentary characteristics.

The Daming–Qinghe–Jingxian–Qingxian palaeochannel zone was chosen for consideration of the variations in ground-water storage. The results

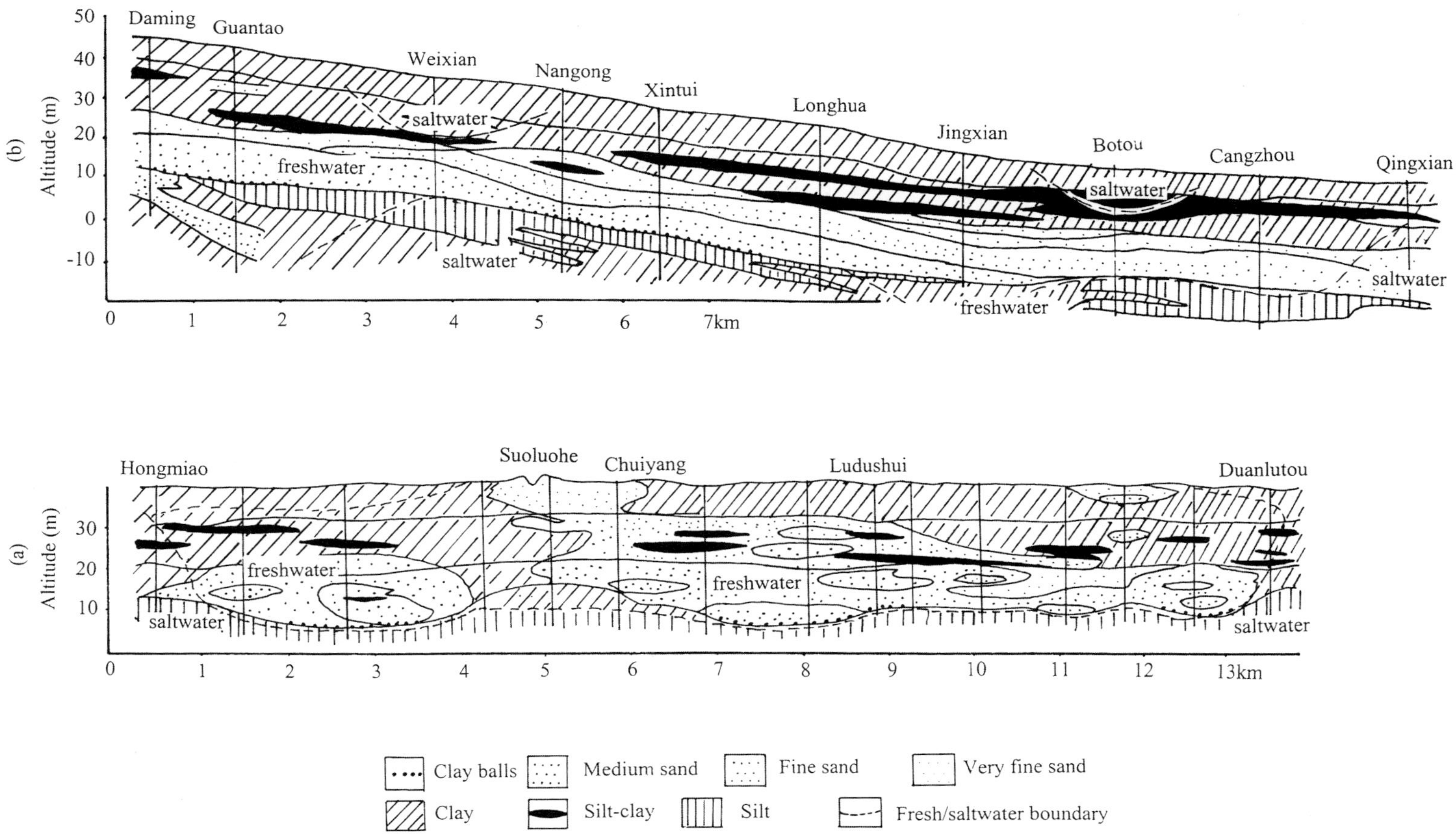

Fig. 3. Vertical profiles of Daming–Qinghe–Jingxian–Qingxian palaeochannel zones: (**a**) downstream; (**b**) cross-section.

Table 1. *Changes in fluvial deposits in a downstream direction of Daming–Qinghe–Jingxian–Qingxian palaeochannel zone*

Location (reaches)	Deposits	Thickness (m)	Traction load (%)	Suspension load (%)	Ratio of suspension load to traction load
Upper	Medium sand	20	18.29	6.54	0.36
	Fine sand	15–20	11.45	19.28	1.68
	Very fine sand	10–15	5.13	13.59	2.65
Lower	Silt	3–10	0.48	14.57	30.35

show that the sand materials have typical fluvial-facies sedimentary characteristics, as follows.

Erosion surface and river-bed facies sediments

Within the drilling depth generally four erosion surfaces are observed, at 30 m, 25 m, 20 m and 8 m depth. On the first erosion surfaces, elliptical calcium carbonate nodules and/or brownish clay balls are found, which suggests that the sediments are river-bed facies. In some areas, the river-bed deposits on the erosion surfaces at 30 m depth have been slightly calcified. The silt–clay deposits between the top two erosion surfaces form an aquifer, above which salt water is perched as a result of sea-water intrusion. For example, the shallow ground-water in the Botou and Weixian areas contains salt water above a thick silt–clay layer (Fig. 3).

Fining upwards

Within the drilling depth, the sediments in which fresh ground-water is stored change from medium sand in the bottom layer, to fine sand in the middle layer, and very fine sand in the top layer. This indicates ordinary grading of the deposition processes in general. Within this sequence are three cycles distinguished by erosion surfaces at 20 m and 8 m depth. Each cycle can be further divided into two sub-units, which can be easily recognized by the two layers of the deposits with sand on the bottom and very fine sand or silt on the top (Fig. 3).

Fining downstream

Downstream the particle size becomes finer and the thickness of the sediments decreases; also, the ratio of suspension load to traction load becomes larger (Table 1). The vertical profile in the downstream direction (Fig. 3) also shows that the thickness of the layer of coarser materials decreases downstream and that of finer materials increases.

Changing width of sand belt

In general, the sand belt is wider in the upper reaches (7–12 km) and narrower in the lower reaches (1–4.5 km), with a SW–NE orientation, although the width varies along the channel (Table 2). In cross-section, the sand bodies are lenticular with thicker centres and thinner sides (Fig. 3).

Fossils

Within the sand bodies there are many fossils. For example, there are fossils of a large mammal community represented by *Coelodonta antiquitatis*, *Palaeoloxodon namadicus*, *Equus przewalskii*, *Equus hemianus*, *Bos primiqenius*, etc. Many fossils of *Lamprotula antiqua* and *Ostracoda*, as well as tree trunks, stone implements and micro-fossils exist in the palaeochannels.

At the base of the sand layer there is usually an erosion surface, over which are deposits consisting of rounded calcium carbonate nodules, elliptical clay balls and thick-shelled fossils of *Lamprotula antiqua*. Underneath is a continuous layer con-

Table 2. *Changes of sand belt width along the channel length (Fig. 3a)*

Location	Site	Width (km)
Upper reaches	Daming	6.8
	Guantao	10.5
	Weixian	11.8
	Nangong	11.5
	Xintui	4.4
	Longhua	4.5
	Jingxian	6.5
	Botou	1.0
	Cangzhou	1.5
Lower reaches	Qingxian	4.5

sisting of clay, which forms the base of palaeochannels, that have developed since late Pleistocene time.

Palaeochannel type and palaeoriver hydrology

Four cross-sections known as Daming, Qinghe, Jingxian and Qingxian, across the palaeochannel zone shown in Fig. 3, were chosen for detailed research on palaeochannel types and palaeohydrology. Within the drilling depth, and according to their sediment texture and structure, the palaeochannels can be divided into three stages. The Stage 1 palaeochannels, with a buried depth of 0–10 m, were formed during late Holocene time. The Stage 2 palaeochannels, with a buried depth of 10–30 m, were formed during mid-Holocene time. The Stage 3 palaeochannels, with a buried depth of 30–50 m, were formed during the late Pleistocene time.

Table 3 shows the calculated palaeoriver parameters for the three stages of the palaeochannels. The ground-water storage within these palaeochannels varies because of the different palaeoriver types. For example, the Stage 1 palaeochannels have the greatest annual discharge (about 8×10^8 m^3). The channels in the upper reaches above Qinghe have a steeper longitudinal gradient (0.29‰) and are of braided type; the deposit contains more traction load (32.44%) and less suspension load. The channels in the lower reaches below Qinghe are of a braided–straight type showing decreased longitudinal gradient and decreased traction loads. Therefore, overall Stage 1 palaeochannels have thick and wide sand bodies, with large particle size, and thus the ground-water storage potential is large. Compared with Stage 1 palaeochannels, the Stage 2 palaeochannels have a gentler longitudinal gradient and less annual ground-water discharge (about 2.5×10^8 m^3). The deposits contain a smaller traction load (5.92%) and a greater suspension load (11.73%); the channels are of braided type in the upper reaches above Jingxian and have a meandering platform in the lower reaches. Both the sand body geometry and particle size are smaller than that in the upper reaches. Therefore, there is less ground-water storage with poor water quality, especially where lake-facies silt deposits exist within the sand bodies. The Stage 3 palaeochannels have the lowest annual ground-water discharge (about 5×10^7 m^3). The deposits consist of less than 1.00% traction loads and more than 40.71% suspension loads. The channel is of straight–meandering type. The sand bodies are thinner and widespread with finer particles. Therefore, the ground-water is lower in quantity but is rechargeable by rainfall and surface runoff.

Palaeochannel geomorphology and ground-water occurrence

On the North China Plain, palaeochannel deposits provide a suitable environment for abundant ground-water storage, with the ground-water restricted within the palaeochannel zones. Investigation of the 47 counties and cities on the North China Plain found that 80% of fresh-water supply is from the shallow-buried palaeochannels (Wu *et al.* 1996*b*).

Although palaeochannel zones are closely associated with ground-water distribution, within the zones ground-water storage varies in terms of quantity and quality because of the variations in deposit characteristics, combinations of different stages of palaeochannels, micro-geomorphology and channel formation. Generally, within the Daming–Qinghe–Jingxian–Qingxian palaeochannel zone, Stage 1 palaeochannels (formed in late Holocene time) contain salt water in the coastal area and brackish water in the floodplain area (Wu *et al.* 1996*a*). Stage 2 palaeochannels (formed in mid-Holocene time) contain good-quality fresh water up to 30 m deep, except in the coastal area where salt water replaces fresh water as a result of rising sea level during mid-Holocene time. The Stage 3 palaeochannels (formed during late Pleistocene to early Holocene time) contain fresh water up to 50 m deep in the southwest area (away from the sea) and salt water in the northeast area (near the sea). This was possibly influenced by seawater intrusion at a time of high sea level during the last interglacial of the late Pleistocene period (Wu *et al.* 1996*b*).

Combinations of palaeochannels have a significant influence on ground-water quality and quantity. A single palaeochannel has only a small capacity for ground-water storage, because of the limited volume of sand. Where palaeochannels have overlapped, the interconnected sand layers become much thicker and provide a greater capacity for ground-water storage. The upper sand layer provides a pathway for leading surface runoff recharge to lower layers. Thus there is a greater capacity for water infiltration storage where palaeochannels overlap.

As well as the influence of palaeochannel combinations on ground-water occurrence, the micro-geomorphology of palaeochannels also has a significant influence, because of differing sediment characteristics. Generally, the upper reaches of palaeochannels can store more water than the lower reaches. For example, the water storage in Daming

Table 3. *Palaeoriver types and palaeoriver hydrology*

Stage	Site	*Wd* (m)	*N*	*W* (m)	*D* (m)	*S* (‰)	Q_s (%)	M (m)	*d* (mm)	*F*	*V* ($m\ s^{-1}$)	Q_m ($m^3\ S^{-1}$)	*Qw* ($\times 10^9\ m^3$)	Ω	R_b	Channel type
1	DM	6800	4	4725	2.2	0.32	42.38	4.82	0.19	31.24	1.97	20470	6450	1.38	.63	*
	QH	11500	4–5	6900	3.0	0.27	22.51	3.50	0.15	27.69	2.16	43350	13670	1.43	0.55	*
	JX	6500	3–4	4700	3.1	0.17	8.81	7.02	0.13	22.12	1.75	25490	8040	1.52	0.74	†
	QX	4500	3–4	3150	3.2	0.07	0.90	7.39	0.10	17.54	1.14	11490	3620	1.62	1.40	†
	Mean	7325	4	4869	2.9	0.21	16.65	5.68	0.14	24.65	1.75	25200	7940	1.49	0.83	†
2	DM	4750	2	2450	2.3	0.14	11.20	8.27	0.13	21.52	1.34	7550	2380	1.53	0.89	†
	QH	6300	3	4100	3.3	0.11	7.41	9.50	0.10	19.40	1.46	19750	6220	1.57	0.91	†
	JX	1500	2	950	3.5	0.07	6.52	7.19	0.09	8.01	1.20	5990	1250	1.99	1.20	‡
	QX	1100	2	700	3.6	0.02	0.06	21.96	0.07	7.35	0.65	1630	510	2.04	3.70	‡
	Mean	3412	2	2050	3.2	0.08	5.92	11.73	0.09	14.07	1.16	8230	2590	1.78	1.40	‡
3	DM	1600	1	480	2.8	0.13	1.28		0.07	7.82	1.44	1930	610	2.01	0.58	‡
	QH	2000	1–2	430	3.0	0.13	0.40	46.80	0.05	6.91	1.50	1930	610	2.07	0.40	‡
	JX	1300	1	340	3.0	0.10	0.06	26.56	0.03	6.14	1.32	1340	420	2.14	0.30	‡
	QX					0.05										
	Mean	1633	1	416	2.95	0.10	0.58	40.71	0.05	6.22	1.42	1730	540	2.07	0.43	‡

* Braided channels with sediments of mainly traction loads

† Channels of straight and braided mixture

‡ Straight–meander channels.

DM, Daming; QH, Qinghe; JX, Jingxian; QX, Qingxian.

(upper reaches) is 8–10 $t\,h^{-1}\,m^{-1}$, that between Guantao and Zaoqiang (middle reaches) is 5–8 $t\,h^{-1}\,m^{-1}$ and that in the area north to Xintui (lower reaches) is 2.5–5 $t\,h^{-1}\,m^{-1}$. River-bed deposits contain more water than other riverine deposits such as point bars. For example, the water storage in river-bed deposits in Daming is 6–10 $t\,h^{-1}\,m^{-1}$ and that in the point-bar deposits is less than 5 $t\,h^{-1}\,m^{-1}$. Also, in Cangzhou (further downstream) the water storage in river-bed deposits is 5 $t\,h^{-1}\,m^{-1}$ and that in the point-bar deposits is less than 2.5 $t\,h^{-1}\,m^{-1}$. Crevasse splay deposits, because of the coarser grain sizes, contain more water than nearby palaeochannels. For example, the water storage in the crevasse splay area in Nangong is 6–10 $t\,h^{-1}\,m^{-1}$ whereas that in the palaeochannel is less than 5 $t\,h^{-1}\,m^{-1}$.

Conclusion

(1) On the North China Plain, palaeochannel zones confine ground-water distribution. More than 90% of ground-water is stored in palaeochannels.

(2) Each stage of palaeochannel has different sediment characteristics and this influences the ground-water storage both in quantity and quality. Palaeochannels formed during the Holocene period, especially during mid-Holocene time, have a large amount of fresh-water storage within the sand and gravel layers. Palaeochannels formed during late Pleistocene time contain fresh water in the upper reaches and salt water in the lower reaches.

(3) A single palaeochannel has limited water storage capacity but combinations of palaeochannels produce large interconnected sand bodies, which are effective for ground-water storage. Where palaeochannels are superimposed, the upper sand layer provides a pathway for rainfall and surface runoff to recharge the lower aquifer beds. Interconnected palaeochannels comprise the principal ground-water storage environment on the North China Plain.

(4) Palaeoriver type and palaeohydrology have significant influences on ground-water storage by palaeochannels. Braided channels have a wider sand belt and contain more ground-water than straight channels but meander channels have the least water storage capacity. Therefore, palaeochannels in upper reaches have a greater capacity for ground-water storage than lower reaches.

We acknowledge the financial support of the China Natural Science Foundation, and R. Mourne and S. B. Marriott for their great help with English language.

References

SCHUMM, S. A. 1960. *The Shape of Alluvial Channels in Relation to Sediment Type*. US Geological Survey Professional Papers, **352-B**, 17–30.

STARKEL, L., GREGORY, K. J. & THORNES, J. B. 1991. *Temperate Palaeohydrology*. Wiley, Chichester.

VISHER, G. S. 1969. Grain size distribution and depositional processes. *Journal of Sedimentary Petrology*, **39**.

WU, C. & ZHAO, Y. 1985. Shallow-buried palaeochannels on Hebei Plain. *Quarterly Research*, **6**(2) [In Chinese].

——, XU, Q., MA, Y. & ZHANG, X.1996*a*. Palaeochannels on the North China Plain: palaeo-river geomorphology. *Geomorphology*, **18**, 37–45.

——, ZHU, X., HE, N. & MA, Y. 1996*b*. Compiling the map of shallow-buried palaeochannels on the North China Plain. *Geomorphology*, **18**, 47–52.

XU, Q., WU, C., ZHU, X. & YANG, X. 1996. Palaeochannels on the North China Plain: stage division and palaeo-environments. *Geomorphology*, **18**, 15–25.

Geochemical characteristics of overbank deposits and their potential for determining suspended sediment provenance; an example from the River Severn, UK

L. J. BOTTRILL[1], D. E. WALLING[1] & G. J. LEEKS[2]

[1]*Department of Geography, University of Exeter, Exeter EX4 4RJ, UK*

[2]*Institute of Hydrology, Wallingford OX10 8BB, UK*

Abstract: The sources of suspended sediment are an important factor controlling sediment yield and sediment budgets. Sediment provenance is an essential prerequisite for elucidating the overall sediment delivery system. Only a proportion of the suspended sediment transported by a river during floods may reach the river mouth, and lowland floodplains frequently represent important sediment sinks. The geochemical properties of floodplain sediments have been used in stratigraphic studies of long-term environmental changes in river basins, but their potential for investigating recent and contemporary sediment sources has not been fully exploited. This paper reports the results of a study that has used the geochemical properties of overbank deposits, including heavy metal, trace metal and cation exchange elements, to establish the main suspended sediment sources within the 10 000 km^2 basin of the River Severn, UK. The results confirm the importance of the upland catchments of the rivers Teme, Vyrnwy and Upper Severn as sources, providing 70% of sediment in the Basin, and of the River Avon as a source of 27% of the <63 μm material. The catchments of the rivers Perry, Stour and Tern, which drain the central areas of the Severn Basin, are shown to be insignificant sources of sediment, contributing only 0–4% to sediment deposited at Haw Bridge. An attempt was also made to establish historical sediment source contributions, although changes in sediment geochemistry during storage in floodplain sinks cause this to be largely unsuccessful.

Information on the source of the suspended sediment transported by a river is an essential prerequisite in any attempt to reduce or minimize the adverse effects of erosion and sedimentation (Walling 1990). The use of sediment tracers and signatures to fingerprint sediment sources, and thereby establish their relative importance, represents an important advance. Fluvial sediment fingerprinting isolates property signatures of sediment sources in a catchment area and relates these to the signatures of suspended sediment (Oldfield *et al.* 1979; Walling *et al.* 1993). The basis of the fingerprinting approach involves two steps: (1) the selection of physical or chemical properties that are diagnostic of particular sources and clearly differentiate such sources on a statistical basis; (2) the comparison of measurements of the same fingerprinting properties obtained for source materials with the corresponding properties of sediment samples. The quest for a single diagnostic property capable of distinguishing between sources is thought to be unrealistic (Walling & Woodward 1995) and several diagnostic sediment properties are now commonly used (Collins *et al.* 1996, 1997, 1998). The use of composite fingerprints has the benefit of permitting the discrimination of a greater range of sediment sources and is likely to prove more effective in establishing source–sediment linkages, by reducing the possibility of spurious matches, which may occur with the use of individual tracers (Collins *et al.* 1996).

Sediment fingerprinting for source identification has been widely used in small lake catchments, generally of less than 50 km^2 (e.g. Dearing 1992; Foster & Walling 1994), and in river basins of less than 300 km^2 (e.g. He & Owens 1995; Walling & Woodward 1995; Foster *et al.* 1996; Walden *et al.* 1997). These studies of small drainage basins have generally addressed sediment provenance in terms of source type, and thus, whether the sediment has originated primarily from erosion of topsoil by rilling or sheetwash, or of subsoil by gullying or channel scour; or from erosion of cultivated land, pasture or forest. In larger catchments there may be a need to identify the relative importance of different sub-catchments. There have been considerably fewer attempts to apply fingerprinting

From: MARRIOTT, S. B. & ALEXANDER, J. (eds) 1999. *Floodplains: Interdisciplinary Approaches*. Geological Society, London, Special Publications, **163**, 241–257. 1-86239-050-9/99/$15.00

techniques to establish spatial sediment sources in larger drainage basins. This paper presents the results of a study that utilizes fingerprinting techniques to establish the spatial provenance of fluvial suspended sediment in the 10 000 km^2 catchment of the River Severn, UK (Fig. 1).

Studies of spatial sediment sources in larger drainage basins (e.g. Collins *et al.* 1996, 1997, 1998) have examined the origins of contemporary suspended sediment, focusing on bulk suspended sediment samples collected during flood events, over a short duration and, therefore, representing essentially instantaneous samples of the suspended sediment load. To provide time-integrated and more generally representative samples of the suspended sediment transported by rivers during high-discharge events and therefore to provide more reliable information on the relative contributions of individual sediment sources to the overall suspended sediment load, this study used sediment samples collected from the surface of floodplains. This use of overbank floodplain sediment as a proxy for suspended sediment also overcomes the many safety and logistical considerations inherent in suspended sediment sampling during high flows. The floodplains of lowland rivers represent important sinks for fluvial suspended sediment and they are frequently characterized by extensive deposits of fine sediment resulting from the deposition of suspended sediment during overbank flood events (Walling *et al.* 1992). The geochemical properties of floodplain sediments have been widely used in stratigraphical studies of longer-term environmental and land-use changes in river basins (e.g. Macklin *et al.* 1992), and such information also offers considerable potential for elucidating changes in sediment sources over the past century. The key assumption in using floodplain sediment records is that down-core geochemical variability is

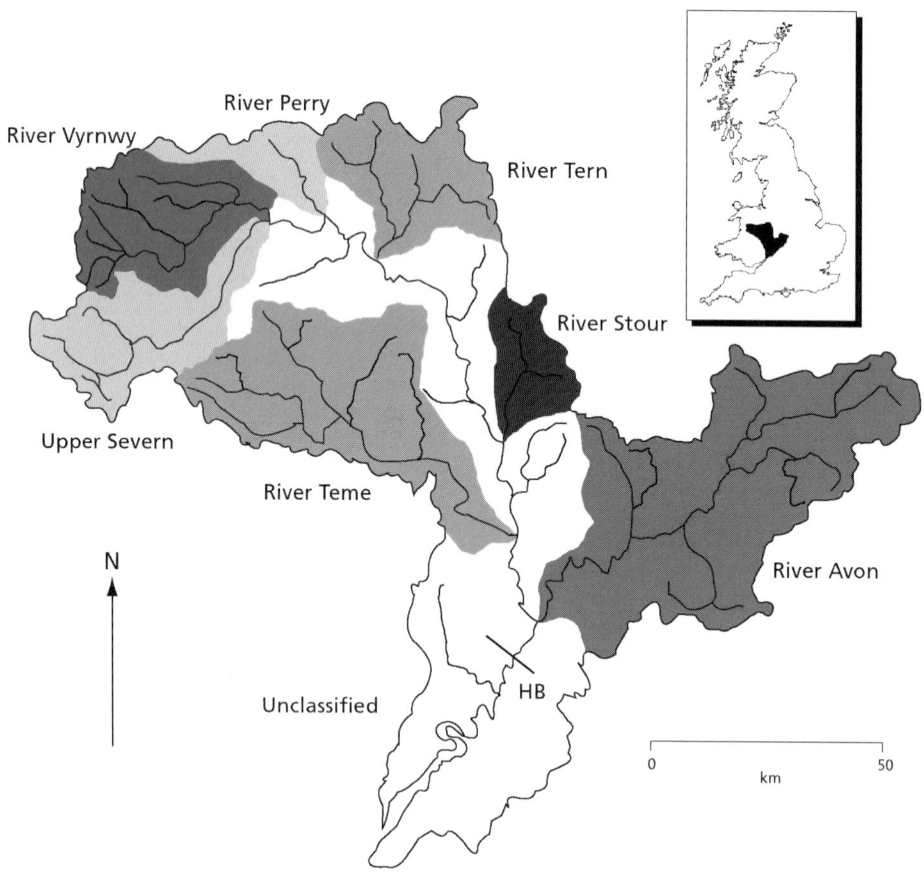

Fig. 1. The Severn Basin showing sub-catchment source areas and the location of the Haw Bridge floodplain (HB).

diagnostic of time-variant contributions from component source areas.

Previous fingerprinting studies have established the relative importance of sediment sources by comparing the properties of suspended sediment with the properties of samples collected from potential source areas (e.g. Walling & Woodward 1992, 1995; He & Owens 1995; Collins *et al.* 1996, 1997, 1998). However, the use of such source sediment comparison in tracing studies introduces uncertainties as a result of the preferential transport of certain size fractions caused by selective erosion of fines and selective deposition of the coarser fractions. It is generally accepted that particle size exerts an important influence upon element concentrations and it is therefore undesirable to attempt to compare fingerprint properties of samples with a different particle size composition. This study compares the properties of sediment collected from downstream floodplains with sediment collected from upstream floodplains. The degree to which the grain-size composition of the sediment sampled at the catchment bottom differs from that of the material used to characterize the potential sources is closely constrained.

Three applications of fingerprinting procedures to identify sediment sources in the basin of the River Severn, UK, are presented. The first investigates the relative contributions of upstream source areas to recent overbank sediment collected from the floodplain at Haw Bridge, which represents the non-tidal limit of the Severn Basin (Fig. 1). The second involves use of a simplified fingerprinting procedure to examine the contribution of the tributaries to the sediment load downstream of major confluences. The results of this application also allow the consistency of the source area sediment contributions apportioned by the more commonly applied fingerprinting approach to be assessed. The third application attempts to investigate changes in the sources of floodplain sediment deposited on the Haw Bridge floodplain over the past century.

Materials and methods

Study basin

The Severn Basin is the largest drainage area in England and Wales and is underlain predominantly by Ordovician (mudstones), Silurian (mudstones and sandstones), Devonian (sandstones), Carboniferous (limestones, mudstones and sandstones), Permo-Triassic (sandstones and mudstones) and Jurassic (sandstones, mudstones and clays) strata. The basin embraces a great variety of terrain types, ranging from mountainous uplands in the western headwaters, underlain by Ordovician and Silurian rocks, through rolling hills in the central areas, developed on Devonian and Carboniferous strata, to lowlands in the east of the catchment underlain by Triassic and Jurassic rocks. Relief generally decreases from west to east and from north to south, with elevations ranging from 752 m in the northwestern headwaters to <10 m in the southeastern valleys. Mean annual precipitation ranges from *c.* 2300 mm in the northwest of the catchment to less than 600 mm in the southeast. The western region is dominated by pastoral farming, mixed farming assumes greater importance in the central regions and arable farming dominates in the lowlands.

Sample collection

Collection of upstream floodplain sediment, to characterize the potential sources of downstream floodplain sediment, was based on representative sampling of the surface (0–20 mm) of floodplains distributed throughout the study catchment. Floodplain sediment representative of the suspended sediment output from the Severn Basin was characterized by two surface samples collected from the floodplain at Haw Bridge (Fig. 1). The samples were dried and screened to 2 mm and a sub-sample of the <63 μm particle size fraction was taken. Results of laboratory analysis on both size fractions were used to characterize potential sediment sources. To investigate the potential for deriving a historical record of sediment sources, two 1 m floodplain cores were collected from the Haw Bridge site using a motorized percussion corer. The cores were sectioned into 20 mm increments, dried and analysed for ^{137}C, to establish a core chronology. The samples were screened to obtain the <63 μm fractions that were analysed to characterize the floodplain core samples and establish their provenance.

The fingerprinting technique is based on the comparison of the geochemical properties of sediment collected from the catchment bottom with those of potential source sediments. It was therefore necessary to identify a range of properties that could be used to characterize the samples and discriminate different potential sources. The selection of these geochemical properties was based on the requirement for the suite to reflect different environmental controls, so as to maximize the discrimination between the potential sediment sources. The following properties were included: iron, manganese and aluminium; the heavy metals: chromium, copper, lead, zinc, nickel, cobalt and cadmium; the trace metals: arsenic and strontium; the cation exchange elements: potassium, magnesium, sodium and calcium; and the pyrophosphate, dithionite and oxalate extractable fractions of iron and manganese.

Following nitric/hydrochloric acid extraction, total concentrations were determined using atomic absorption spectrophotometry (AAS) with either an air–acetylene or nitrous oxide–acetylene flame. Particle-size composition was determined after sample pre-treatment with hydrogen peroxide to remove the organic fraction, and dispersion with sodium hexametaphosphate, using a Malvern Mastersizer. Carbon and nitrogen levels were determined by pyrolysis–thermal conductivity using a CE Instruments NA2500 elemental analyser. Analysis for ^{137}C was undertaken by gamma spectrometry using an HPGe coaxial detector (EG&G) coupled to a multi-channel analyser.

The fingerprinting technique of sediment source identification

To permit examination of the provenance of sediment from floodplain surfaces, the potential source areas were classified by sub-catchment (Fig. 1). The seven main tributary sub-catchments reflect variations in geology, topography, soils, land use and meteorology (see Table 1). To establish the relative contribution of these areas to the floodplain deposits at Haw Bridge, sediment from the individual sub-catchments was characterized by replicate samples collected from 58 floodplain sites. Two samples were collected from the floodplain of the River Severn at Haw Bridge to ensure that the sediment to be sourced was representative of the floodplain as a whole. One sample was taken from higher ground, 40 m from the channel, and a second from nearer the channel. Significant variations in properties of overbank sediments of the same age can occur laterally over relatively small distances (Macklin *et al.* 1994; Taylor 1996). More samples would be needed to assess this variation, but the available resources limited numbers.

Following selection of a range of fingerprint properties, statistical procedures are needed to provide an objective estimate of the relative contribution of the potential sediment sources to the suspended load of a stream. The statistical procedures involve two main stages: (1) statistical testing of the selected combination of fingerprint properties to confirm that individual sediment sources are clearly distinguished by their composite signatures; (2) application of an objective algorithm that is capable of comparing the composite fingerprint of a sediment sample with those of potential sources and will provide an estimate of the relative contributions of those sources to the overbank sediment deposit.

Selection of a composite fingerprint signature

It is important to confirm that each source area is characterized by a fingerprint, which is significantly different from those of the other areas. The properties that are not able to distinguish between source categories need to be identified and removed from the subsequent analysis. Various statistical methods have been used for this purpose (e.g. Walling *et al.* 1993; Yu & Oldfield 1993). Collins *et al.* (1996, 1997, 1998) proposed a two-stage verification procedure, which is used here, consisting of a Kruskal–Wallis *H*-test and discriminant function analysis.

The results of the Kruskal–Wallis *H*-test to analyse the variance within the <2 mm and <63 μm sub-sample datasets are presented in Table 2. Any variables that failed to reach the critical value for the 95% confidence limit of 12.59 would be removed from the analysis at this stage. However, all of the variables exceeded this value. The results of multivariate discriminant function analysis to identify the property subset that would provide a composite fingerprint capable of maximizing the statistical differentiation of the defined source areas are shown in Tables 3 and 4, for the <63 μm and <2 mm size fractions, respectively. The tables show the steps taken in each analysis, and the resulting percentage of samples that are correctly

Table 1. *Summary information for the sub-catchment sediment sources in the Severn Basin*

Catchment	Area (km^2)	Geology	Precipitation (mm a^{-1})	Relief	Land use
River Avon	2900	Jurassic clays, Triassic mudstones	600–672	Lowland, predominantly flat	Arable, grassland, urban
River Teme	1648	Silurian mudstones, Devonian sandstones	840–1200	Hilly, very little lowland	Upland pasture, arable, grassland
River Stour	373	Triassic sandstone	620–710	Lowland predominantly flat	Arable, urban, grassland
River Tern	852	Triassic sandstone	650–750	Lowland, predominantly flat	Arable, grassland, urban
River Perry	181	Triassic sandstone	660–770	Lowland, predominantly flat	Arable, grassland, urban
River Vyrnwy	778	Ordovician and Silurian mudstones	660–2500	Mountainous uplands, wide lowland valleys	Grassland, upland pasture, moorland
Upper Severn	1033	Ordovician and Silurian mudstones	660–2400	Mountainous uplands, wide lowland valleys	Grassland, upland pasture, moorland

Table 2. *Results of using the Kruskal–Wallis* H-*test to assess the ability of individual chemical properties to discriminate between the potential sources of <63 µm and <2 mm floodplain sediment*

Variable	*H* value <63 µm	*H* value <2 mm
Carbon	45.01*	55.81*
Nitrogen	46.18*	57.4*
Iron	39.33*	37.42*
Aluminium	38.42*	37.42*
Calcium	48.44*	50.34*
Chromium	44.99*	15.71*
Copper	45.22*	43.87*
Potassium	45.75*	30.16*
Magnesium	45.58*	25.34*
Manganese	24.58*	26.79*
Sodium	32.94*	30.36*
Nickel	44.27*	32.2*
Strontium	54.79*	33.93*
Lead	32.38*	33.86*
Zinc	39.48*	37.29*
Pyrophosphate iron	33.53*	17.5*
Pyrophosphate manganese	16.59*	31.45*
Pyrophosphate magnesium	42.15*	23.62*
Dithionite iron	34.57*	25.51*
Dithionite aluminium	18.42*	17.96*
Dithionite manganese	19.68*	20.61*
Dithionite magnesium	44.74*	24.44*
Oxalate iron	31.83*	23.24*
Oxalate manganese	22.05*	31.83*
Cobalt	35.17*	36.96*
Cadmium	32.17*	27.18*
Arsenic	25.93*	40.26*
Tin	42.81*	33.06*
Critical vValue	12.59	12.59

*Significant at $P = 0.05$.

apportioned to their source group. A composite signature consisting of seven variables, namely, magnesium, zinc, iron, manganese, pyrophosphate extractable iron, pyrophosphate extractable manganese and carbon, was identified as that most capable of distinguishing <63 µm sediment from the individual sub-catchment sources. This seven-variable signature was capable of correctly apportioning 81.9% of the overall set of samples to their source groups. The discrimination afforded by the selected signature is shown graphically in Fig. 2. To discriminate between <2 mm sediment representative of individual sub-catchments the analysis identified a seven-variable signature, nitrogen, strontium, cobalt, dithionite manganese, iron, lead and pyrophosphate manganese, that was capable of assigning 91.9% of samples to their correct sub-catchment origin (Fig. 3).

Use of a multivariate mixing model

To use the fingerprint signatures, identified by the multivariate discriminant function analysis, to establish the relative contribution of the different sub-catchments to the sediment collected from Haw Bridge and to maximize the accuracy of such source ascription, a mixing model is required. The mixing model optimizing algorithm proposed by Walling *et al.* (1993) was selected for this purpose. This mixing model seeks to satisfy the following set of linear constraints:

(1) the source type contributions must all sum to unity, i.e.

$$\sum_{s=1}^{S} P_s = 1$$

(2) the source type contributions must all be non-negative, i.e.

$$0 \leq P_s \leq 1$$

(3) the error term must be reduced to its minimal value, i.e.

$$E = \sum_{t=1}^{T} \left\{ \left[\left| B_t - \left(\sum_{s=1}^{S} V_{st} P_s Z_s \right) \right| \right] / B_t \right.$$

Table 3. *Results of the stepwise discriminant function analysis of sub-catchment source areas for <63 µm floodplain sediment*

	Enter*	Proportion of samples classified correctly (%) Avon	Teme	Stour	Tern	Perry	Vyrnwy	Upper Severn	Total
Step 1	Mg	68.4	16.7	0	30	50	62.5	20	38.9
Step 2	Zn	63.2	83.3	75	30	50	62.5	40	56.9
Step 3	Fe	63.2	83.3	75	40	25	87.5	40	59.7
Step 4	Mn	68.4	100	75	40	75	75	46.7	66.7
Step 5	Py. Fe	84.2	100	75	60	100	87.5	46.7	76.4
Step 6	Py. Mn	94.7	100	75	70	75	87.5	60	81.9
Step 7	C	94.7	100	75	70	75	87.5	60	81.9

Py, pyrophosphate. * At corresponding step, data relating to this parameter is added.

Table 4. *Results of the stepwise discriminant function analysis of sub-catchment source areas for <2 mm floodplain sediment*

	Enter*	Proportion of samples classified correctly (%)							
		Avon	Teme	Stour	Tern	Perry	Vyrnwy	Upper Severn	Total
Step 1	N	42.8	50	33.3	16.7	50	12.5	6.7	31.1
Step 2	Sr	68.4	50	50	33.3	66.7	25	6.7	43.2
Step 3	Co	68.4	71.4	50	50	66.7	37.5	46.7	58.1
Step 4	Di. Mn	73.7	78.6	83.3	33.3	83.3	75	60	70.3
Step 5	Fe	78.9	100	83.3	33.3	83.3	75	53.3	74.3
Step 6	Pb	73.7	92.9	83.3	33.3	100	100	80	81.1
Step 7	Py. Mn	94.7	100	100	66.7	100	100	80	91.9

Di., dithionite; Py, pyrophosphate. * At corresponding step, data relating to this parameter is added.

where B_t is the concentration of tracer t in sediment obtained from the catchment outlet, P_s is the percentage contribution from each source area, V_{st} is the mean concentration of tracer t for source area s and Z_s is a particle size correction factor for source area s.

The particle size of suspended sediment exerts a fundamental control over its mineralogy and geochemistry, and is hence the most important factor influencing trace element concentrations (Horowitz 1991). For this reason, particle size weighting factors based upon the ratio of the mean specific surface area of the Haw Bridge sediment samples to the mean specific surface area of the sediment samples from each of the sources were included in the mixing models (Table 5). The error term was minimized by using the optimization program available within the Quatro Pro database package. Initially, the source contributions (P_s) are set equal and the error term is calculated. New errors are then calculated for a series of changes in the values of P_s, and the status that minimizes the error term is

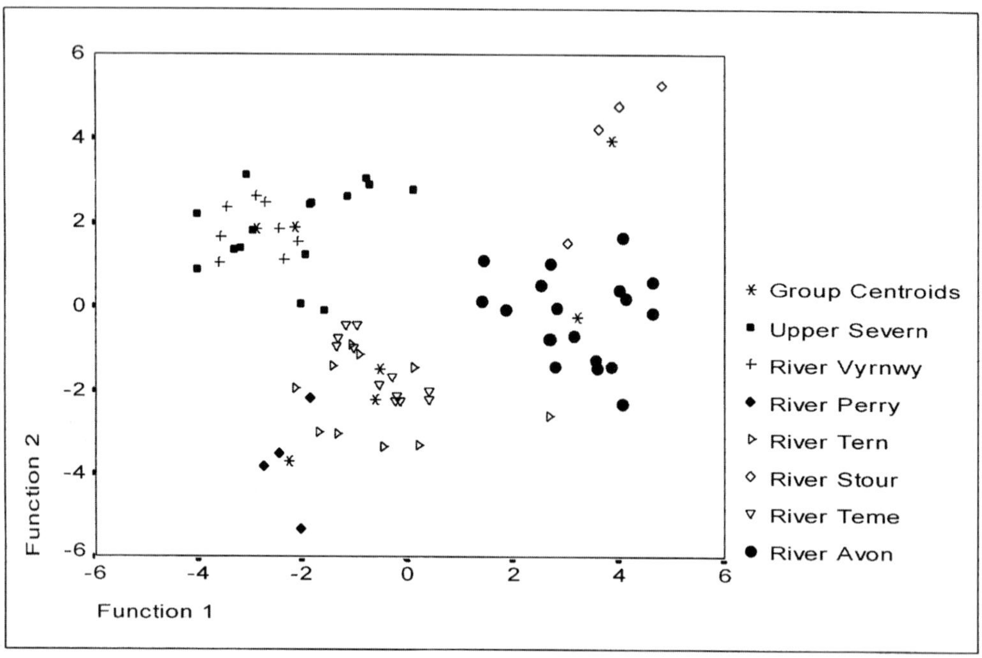

Fig. 2. Canonical discriminant functions constructed from the first and second discriminant functions calculated for sub-catchment source areas of <63 μm floodplain sediment.

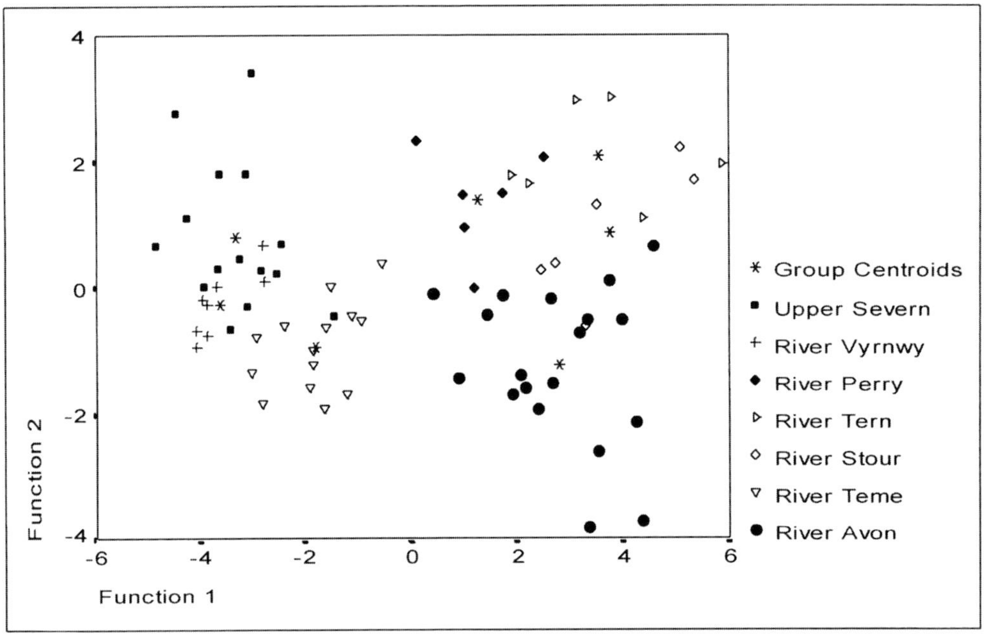

Fig. 3. Canonical discriminant functions constructed from the first and second discriminant functions calculated for sub-catchment source areas of <2 mm floodplain sediment.

selected as a solution and used to provide the percentage of source area contributions.

Sediment source identification using a confluence-based fingerprinting approach

Recent sediment fingerprint studies (e.g. Walling & Woodward 1995; Collins *et al.* 1996, 1997, 1998) have replaced qualitative interpretation of fingerprinting results with rigorous quantitative procedures, encompassing the statistical verification of multi-component signatures and the use of multivariate mixing models for ascription of sediment source contributions. However, such rigorous statistical selection of fingerprint signatures can be problematical in large river basins, where the properties of sediment may be similar over large areas. The possibility of developing a simplified source identification procedure for application in larger river basins, which does not require the development of composite signatures capable of discriminating between all potential sources simultaneously, is described below. It should be possible to consider a larger number of sources and to establish the relative contributions of smaller spatial sources by application of a mixing model at tributary confluences. This would establish the relative contributions of tributaries to sediment collected downstream. This approach was used to calculate the relative sediment contributions from two catchments at 17 major confluences in the Severn Basin. At each confluence, two samples collected from each of three floodplains were used to characterize overbank deposits below the confluence and in both contributing catchments. The characteristics of these samples were used with the mixing model proposed by Walling *et al.* (1993) to provide estimates of relative source contributions to the floodplain downstream of the confluence.

Table 5. *Mean particle size correction factors for sub-catchment source areas of <63 μm and <2 mm floodplain sediment*

Source area	Correction factor	
	<63 μm	<2 mm
River Avon	0.722	1.068
River Teme	1.308	1.455
River Stour	0.797	1.615
River Tern	0.922	1.597
River Perry	1.036	2.063
River Vyrnwy	1.057	1.299
Upper Severn	0.993	1.272

Although the simplified source identification procedure is founded upon two of the same fundamental requirements as the fingerprinting approach (see above), the procedures used for sediment characterization and statistical removal of properties that cannot singly explain sediment provenance are slightly different. In this simplified approach to source identification, the number of floodplains sampled in each of the two source areas is reduced to one, as it is assumed that material collected from the lowest floodplain is characteristic of a mixture of sediment from upstream areas (Darnley *et al.* 1995). Mixing of in-channel and overbank sediment during transport is therefore used as a surrogate for the mathematical averaging of sediment properties within source areas that are generally employed in fingerprinting studies.

A significant benefit of the confluence-based procedure is that, because it does not assign sediment contributions to all potential sources in one application, it is possible to include properties that are incapable of distinguishing between all of the potential sources at the same time but that may still afford a degree of discriminatory power. However, it is still important to exclude those variables where combinations of contributions from the two sources cannot explain the concentrations observed in the floodplain samples collected downstream of the confluences. As an example, if the level of copper in the outlet floodplain samples is ten and levels in the source areas are four and eight then there is no consistent relationship between the source sediment and the mixed sediment and, therefore, the mixing model cannot provide an accurate ascription of sediment source contributions. Hence, a key requirement is to remove such variables from consideration. The existence of these factors may indicate measurement error or alteration of the properties of the sediment during transport and in the period since deposition. Alternatively, it may be a result of a failure to represent accurately concentrations of elements in sediment from a particular source. This is perhaps more likely given that only two samples from one floodplain were used to characterize the sediment from each source area.

The property selection process was carried out for each confluence and the rejected properties for both the <63 μm and <2 mm size fractions included heavy and trace metals, cation exchange elements and various iron and manganese fractions. With refined suites of between 13 and 20 diagnostic properties, the simplified source identification procedure used the mixing model algorithm of Walling *et al.* (1993) to establish the relative sediment contributions of the tributary source areas at the confluences.

Use of fingerprinting to investigate changes in sediment sources during the last century

It is possible to apply the fingerprinting technique to deeper floodplain deposits to provide a longer-term record of the relative contributions from the various sediment sources than is offered by

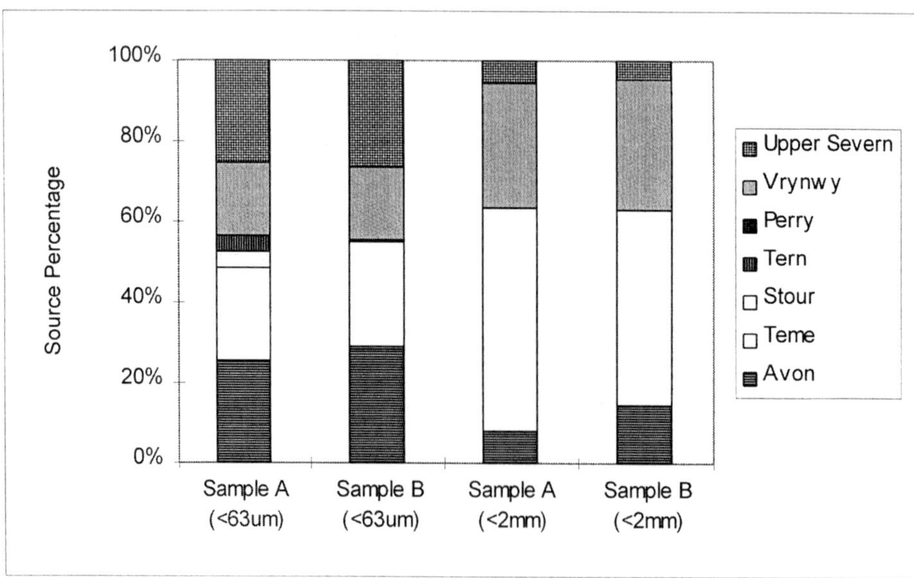

Fig. 4. Sub-catchment sources of Haw Bridge floodplain sediment established using the fingerprinting technique.

Table 6. *The percentage contribution of individual sub-catchment sources to <63 μm floodplain sediment samples collected at Haw Bridge*

Sample	Contribution (%)						
	Avon	Teme	Stour	Tern	Perry	Vyrnwy	Upper Severn
Haw Bridge A	25.7	22.8	3.9	4.3	0	18.1	25.3
Haw Bridge B	29.2	25.9	0.5	0	0	17.7	26.6

surficial deposits. Potentially, the geochemical record preserved in floodplain sinks offers considerable scope for identifying recent changes in sediment sources. This procedure has been employed in reconstructing the longer-term contributions of individual source types in small lake catchments (e.g. Dearing 1992; Foster & Walling 1994) but has been rarely employed in larger drainage basins. The source of <63 μm sediment obtained from the two Haw Bridge cores was interpreted using the traditional fingerprinting technique described above.

Results

Figure 4 presents the results of the application of the fingerprinting technique to the identification of the relative contributions of the seven source areas to the sediment collected from the surface of the floodplain at Haw Bridge, for both the <63 μm and <2 mm fractions. The <63 μm deposits indicate that the sub-catchments of the Rivers Avon, Teme, Vyrnwy and Upper Severn are the most important sources of sediment (Table 6). The catchments of the rivers Perry, Stour and Tern, which drain the Triassic sandstones in the central areas of the Severn catchment, are shown to be insignificant sources of <63 μm sediment. These relatively smaller central catchments are also insignificant sources of <2 mm sediment (Table 7). Most <2 mm sediment is ascribed to the catchments of the Vyrnwy and Teme. Figures 5 and 6 present the results of the simplified source identification procedure applied at major confluences in the Severn Basin. This approach also shows that the major sources of <63 μm material are the catchments of the rivers Avon, Vyrnwy and Teme. The most important sources of <2 mm sediment are shown to be the western upland catchments. Figures 7 and 8 present the results of the application of the fingerprinting technique to the down-core variation in sediment properties in the two sediment cores collected. The mixing model assigns significant sediment contributions from only the rivers Avon, Teme and Upper Severn to Core A (Fig. 7). The source contributions identified in the top 40 mm of the core are consistent with those reported for the Haw Bridge floodplain surface deposits. The application of the fingerprinting procedure to the sediment collected from the core below a depth of 40 mm shows the catchment of the River Avon to be the only significant sediment source. The record of sub-catchment provenance provided by Core B shows the same general trend in source area contributions as shown by Core A. The results show that the top increments of Core B are composed of sediment originating from the catchments of the Avon, Teme and Upper Severn. As in the case of Core A below 40 mm, the River Avon appears to become the only significant sediment source to Core B below 80 mm.

Discussion

As expected, the sub-catchments of the rivers Teme, Vyrnwy and Upper Severn on the steeply sloping western uplands of the Severn Basin, which receive higher levels of precipitation than the eastern areas, represent important sources of sediment. The Avon is a lowland river draining the lower-energy environment of the eastern Severn Basin. Its relative importance as a <63 μm sediment

Table 7. *The percentage contribution of individual sub-catchment sources to <2 mm floodplain sediment samples collected at Haw Bridge*

Sample	Contribution (%)						
	Avon	Teme	Stour	Tern	Perry	Vyrnwy	Upper Severn
Haw Bridge A	8.1	54.4	0	0	0	30.7	5.4
Haw Bridge B	13.9	47.5	0	0	0	31.5	4.5

Fig. 5. Sediment contributions (%) established by the tributary source identification technique from upstream catchments to the <63 μm fraction of surface samples collected from floodplains downstream of major confluences in the Severn Basin.

source is largely attributable to the underlying, easily erodible Jurassic mudstone, and the existence of unconsolidated Quaternary deposits which provide fine material. The insignificant contributions assigned to the Stour, Tern and Perry subcatchments, particularly of <2 mm sediment, are surprising given that all three drain sandstones and are therefore likely to generate sand-grade sediment. The River Vyrnwy is a more important source of <2 mm sediment than for <63 μm sediment. This is largely because the Upper Severn is ascribed a lower <2 mm contribution. The change in the relative importance of the northwestern upland source areas may be attributable to the coarser composition of the sediment that originates from the Vyrnwy catchment, or to conveyance losses within the Upper Severn as a result of deposition of coarse material in the extensive floodplain systems of the Vale of Powys. The higher contribution of the River Teme catchment to the <2 mm sediment fraction at Haw Bridge may also reflect the particle-size characteristics of the transported sediment. The catchment of the River Teme includes Devonian sandstones and the resulting sediment has a larger mean particle size than any of the other three major sediment contributing catchments. This suggests that it is likely to provide a more significant contribution to the coarser sediment fraction.

One of the aims of the development of the simplified, confluence-based source identification approach was to use the results as an independent guide to the reliability of the results obtained from the fingerprinting technique described above. However, the results of the simplified tributary mixing model and the results established by the fingerprinting technique are not directly comparable. This is because the former relate to relative contributions at the tributary junctions within the drainage basin, whereas the latter relate to the

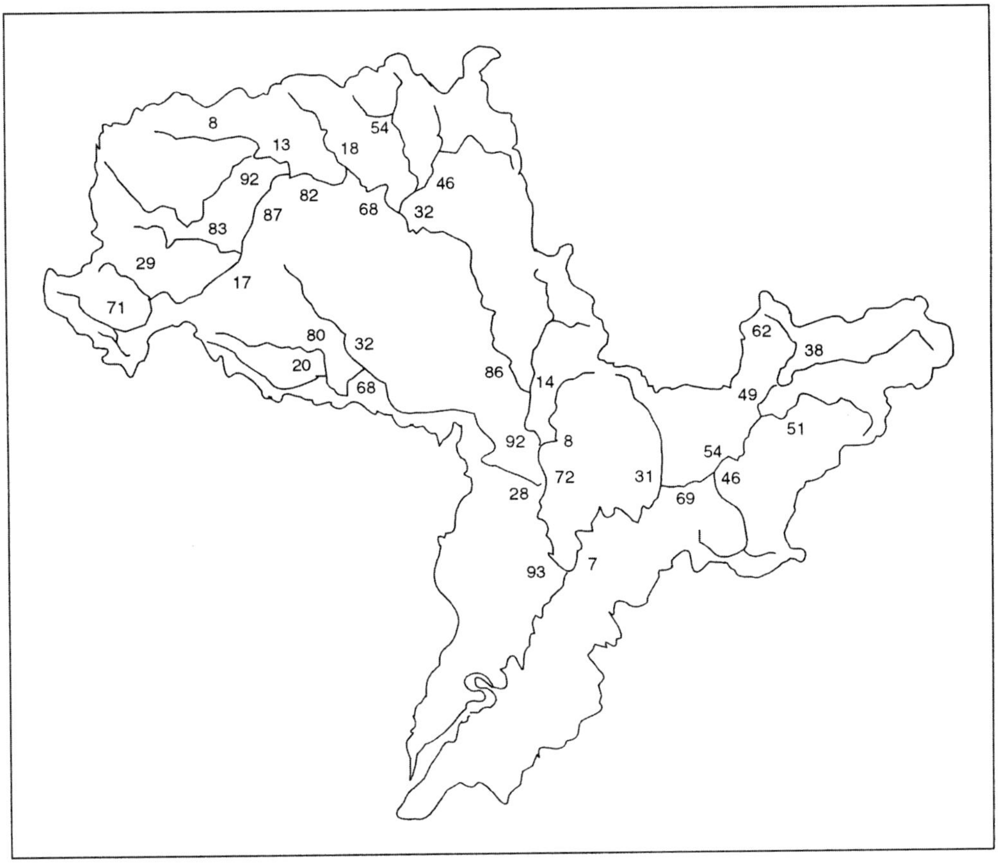

Fig. 6. Sediment contributions (%) established by the tributary source identification technique from upstream catchments to the <2 mm fraction of surface samples collected from floodplains downstream of major confluences in the Severn Basin.

catchment outlet and will therefore reflect conveyance losses in the sediment transportation system and additional downstream sediment inputs. Information on the relative sediment contributions at individual tributary confluences can nevertheless be used to assess some of the findings discussed above. A direct comparison of the results provided by the two source identification techniques is possible in relation to the sediment contribution from the Avon sub-catchment, for example, because the results provided by the tributary mixing model are also based on the sourcing of sediment deposited on the floodplain at Haw Bridge. Transmission losses are therefore of limited significance. The relative sediment contributions from the River Avon to both size fractions, indicated by both techniques, are very similar (Table 8).

A key finding of the application of the fingerprinting technique to Haw Bridge sediment was that the Middle Severn catchments of the rivers Stour, Tern and Perry were shown to be largely insignificant as sediment sources. In considering the sources of <63 μm sediment deposited on the floodplain at Haw Bridge, the fingerprinting technique assigned mean sediment contributions of 2% to the catchments of both the River Stour and Tern and 0% to the River Perry. The confluence mixing model provided estimates of sediment contributions of 8%, 21% and 10% from the Perry, Tern and Stour to the <63 μm fraction of sediment deposited on the floodplains of the River Severn below the respective confluences. These two sets of results are nevertheless consistent if both conveyance losses (estimated to be in the order of 23% (Walling & Quine 1993)), and the additional sediment inputs from smaller tributaries, located between these catchments and Haw Bridge, are

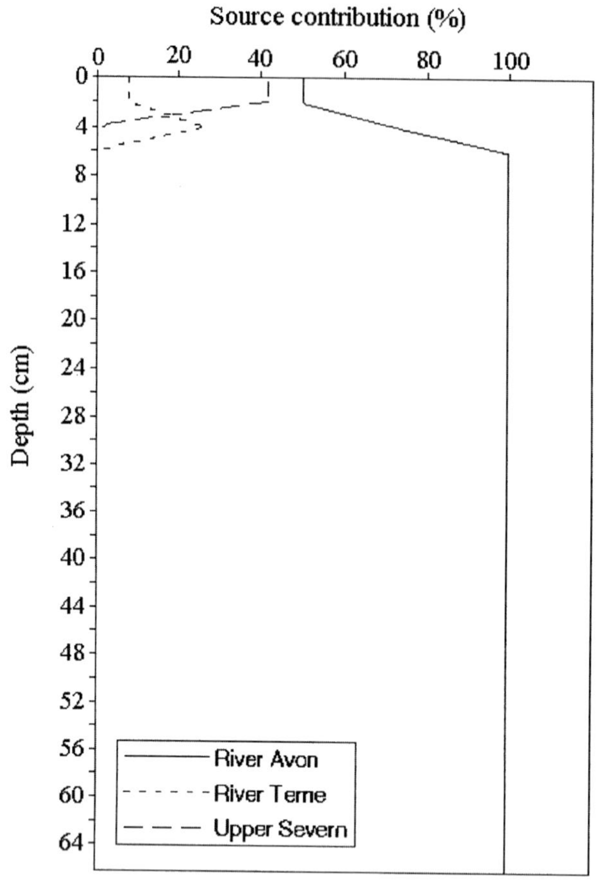

Fig. 7. Mixing model estimates of recent variations in the relative contributions of upstream sub-catchments to sediment from Core A.

taken into account. The simplified procedure indicated increased contributions from all three Middle Severn catchments to the <2 mm sediment fraction transported by the River Severn. The rivers Stour, Tern and Perry were shown to account for 14%, 32% and 18% of sediment deposited immediately downstream of their respective confluences. The more significant contributions of <2 mm sediment are a function of the sandstones that characterize the geology of the three catchments. Although the results from the two sediment source identification procedures are not directly comparable, the simplified procedures are successful in providing an independent guide to the reliability and consistency of the fingerprinting approach applied to the overall basin.

The apparent dominance of the River Avon sediment contribution below 40 mm in Core A and 80 mm in Core B is unlikely to reflect a dramatic change in source contributions within the Severn Basin during recent years. The difference in the depth at which this shift occurs in the cores may be attributable to differences in sedimentation rates at the two sites, as the mean annual sedimentation rate at site B (8.3 mm) is slightly higher than at site A (6.5 mm). Similar trends in both cores from Haw Bridge suggest that the errors in source identification for sediment below 40 mm and 80 mm are not due to the mixing model algorithm employed. This procedure has been shown to produce acceptable results for floodplain surface sediment and for the surface horizons of the cores; the errors produced by the application of the model are therefore likely to reflect the influence of post-depositional changes in the signature characteristics. A key principle underpinning the use of chemical properties in sediment fingerprinting is that they must behave conservatively (Walling & Woodward

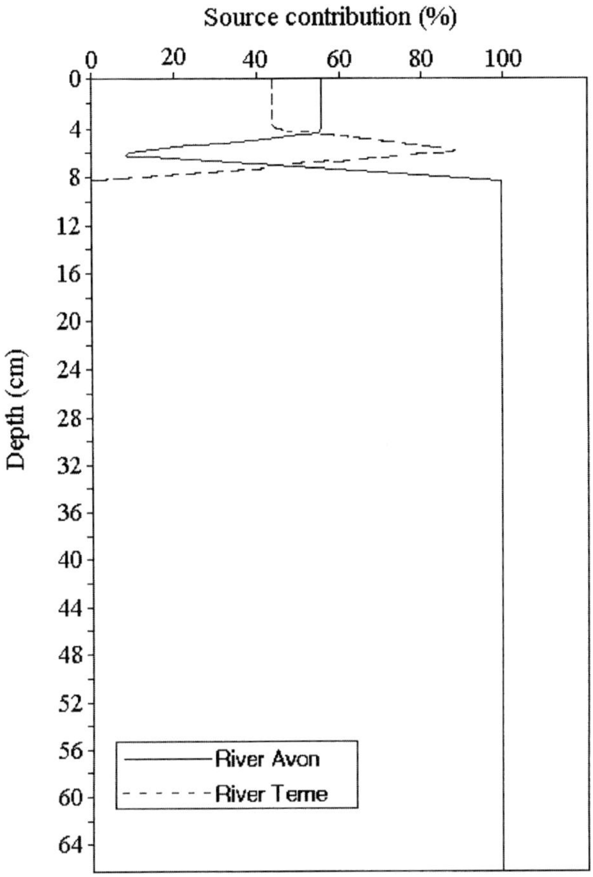

Fig. 8. Mixing model estimates of recent variations in the relative contributions of upstream sub-catchments to sediment from Core B.

1992, 1995). When examining contemporary suspended sediment samples and recent floodplain deposits, this effectively requires conservative behaviour during transportation from the source area to the point of sampling. However, in the examination of longer-term sediment source contributions a fundamental requirement is that fingerprint properties must also be physically and chemically stable in sediment sinks. Down-core variation in sediment signature property values must reflect variations in source rather than representing a function of processes (e.g. pedogenic, biogenic) occurring on the floodplain.

The existence of non-conservative behaviour of the fingerprint variables involved in the multi-component signatures of the two cores was investigated using Pearson's product-moment correlation tests to assess the importance of a number of controlling variables. Element concentrations in the <63 μm fraction are shown to be significantly related to depth (Table 9) and organic matter content (Table 10). The latter relationship suggests that concentrations of the various elements are not a function of depth *per se*, but are more likely to be a function of organic matter content, which is itself shown to be strongly depth dependent. This finding is consistent with that of Leenaers *et al.* (1988), who showed that metals in particular can be preferentially adsorbed to organic material, with a correlation of 0.810 between copper concentrations and organic matter content for flood deposits on the River Geul, Belgium. These two correlation tests demonstrate that the down-core variations in signatures are not a simple function of sediment source trends. It appears that where the organic carbon content in the floodplain profile declines with depth, reflecting the breakdown of organic matter, humic bound metals are released from the soil matrix and mobilized out of the profile. As a progressive down-core reduction

Table 8. *The contribution of the River Avon to floodplain sediment collected at Haw Bridge provided by the fingerprinting technique and the tributary source identification procedure*

Procedure	Sample	Sediment contribution from the River Avon (%)
<63 μm sediment		
Fingerprinting	Haw Bridge A	25.7
Fingerprinting	Haw Bridge B	29.2
Simplified	Haw Bridge	25.0
<2 mm seiment		
Fingerprinting	Haw Bridge A	8.1
Fingerprinting	Haw Bridge B	13.9
Simplified	Haw Bridge	7.0

in element concentrations occurs, the signatures displayed by the floodplain increments progressively reflect the signature of the source area with the lowest element concentrations. This has caused the optimization algorithm to focus often on a single contributing source area, the identity of which is dependent on the properties being considered in the characterizing signature. This situation is demonstrated in Fig. 9, which shows fingerprint signature property concentrations in Core A. Iron, magnesium, pyrophosphate extractable iron and pyrophosphate extractable manganese concentrations, which have been demonstrated to be related to depth and organic matter, decrease down the core profile as the organic carbon content decreases. As levels of iron, pyrophosphate extractable iron and pyrophosphate extractable manganese decline, they approach those characteristic of the River Avon samples and move further away from the concentrations in sediment associated with the other three major sub-catchments. Although the levels of magnesium that characterize the upland catchments of the rivers Vyrnwy and Upper Severn are lower than in the Avon catchment, concentrations in the floodplain only fall to the level found in the samples from the Avon source area.

For the elements manganese and zinc, which are positively correlated with depth and negatively correlated with organic matter content, concentrations increase down the profile to approach those characteristic of the Avon sub-catchment. However, the trends evident in both manganese and zinc concentrations may not be directly linked to those of organic carbon. Heavy metal concentrations in floodplains of the River Severn have a close association with the timing of sedimentation in relation to mining activity and erosion of associated spoil-heaps in the catchment (Lewin & Macklin 1987). Within the Haw Bridge floodplain profile, concentrations of lead and tin, as well as zinc, decline nearer the floodplain surface, and this may reflect the reduction in mining and erosion of spoil-heaps in the upper and middle reaches of the River Severn during the last century. Manganese concentrations are independent of organic carbon content in the upper 10 cm of the floodplain profile. The rise in manganese levels

Table 9. *Correlation of signature property concentrations with depth* (r)

Signature property	Core A	Core B
Carbon	–0.891	–0.934
Iron	–0.892*	–0.697*
Magnesium	–0.730*	–0.567*
Manganese	0.755*	0.555*
Zinc	0.928*	0.876*
Pyrophosphate iron	–0.868*	–0.907*
Pyrophosphate manganese	–0.689*	–0.641*
Nitrogen	–0.898*	–0.947*
Critical *r* value	0.365	0.381

* Significant at P =0.05.

Table 10. *Correlation of signature property concentrations with organic carbon content* (r)

Signature property	Core A	Core B
Iron	0.677*	0.506*
Magnesium	0.771*	0.633*
Manganese	–0.595*	–0.346
Zinc	–0.852*	–0.837*
Pyrophosphate iron	0.882*	0.936*
Pyrophosphate manganese	0.890*	0.801*
Critical *r* value	0.365	0.381

* Significant at P = 0.05

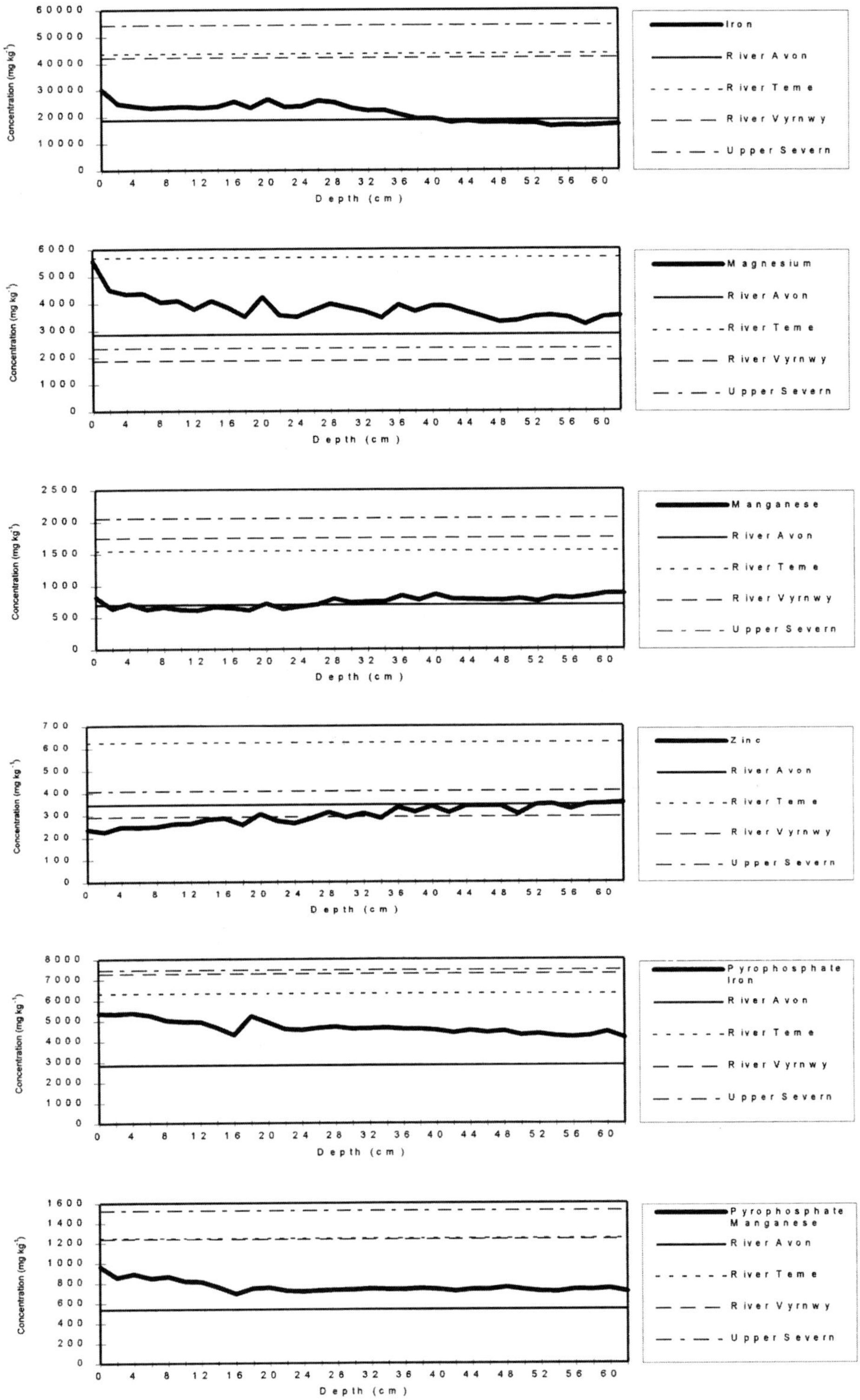

Fig. 9. Down-core variation in fingerprint signature property concentrations in Core A, and average fingerprint signature property concentrations for the four most significant source areas.

further down the profile could be a function of restricted drainage and poor aeration at depth, rather than organic matter decrease. In poorly drained and poorly aerated floodplain soils, with a high redox potential, manganese is reduced. Under moderate reducing conditions manganese is soluble and can therefore be leached down the soil profile (Patrick & Jugsujinda 1992). In consequence, the top of the profile would be depleted in manganese in relation to the bottom. Overall, with an increase in depth, the ascription of sediment sources by the mixing model is biased towards the Avon sub-catchment. This is as a result of the complex interaction of organic matter breakdown and a number of other processes that are occurring in the floodplain and not as a function of a shift in sediment supply in the Severn Basin to this sub-catchment.

Conclusions

The methods, results and discussion presented above clearly demonstrate the potential for using recent floodplain deposits in sediment source fingerprinting studies. The results of applying the fingerprinting technique to surface deposits from the Haw Bridge floodplain have revealed that there are only minor differences in the sources of <63 μm and <2 mm sediment in the Severn Basin. These have been attributed to both the particle-size composition of the sediment originating from the different sources and selectivity in the deposition of coarse particles transported from the most distant sub-catchments.

The simplified approach to sediment source identification at tributary confluences also provided consistent results. Overall, the sediment source contributions estimated by the simple tributary mixing model support the validity of the sediment source contributions established by the fingerprinting procedure and, in consequence, the applicability of the fingerprint technique for identifying the sources of sediment transported by, and deposited on the floodplains of the River Severn.

The examination of cores from the Haw Bridge floodplain has shown that identification of past sediment source contributions was not possible using the fingerprinting approach. This is attributable to the *in situ* alteration of fingerprint elements, and consequently the signature of floodplain sediment, by pedogenic processes including the breakdown of organic matter during storage in floodplain sinks. Further work is required to establish whether the problem of the chemical alteration of sediment in floodplain sinks is local to the Haw Bridge floodplain or is common to all River Severn floodplain sites. It may be worth while to continue to characterize source areas using floodplain surface deposits, but to fractionate individual chemical properties so as to isolate a fraction that is likely to behave in a conservative manner. This would develop further the applicability of the fingerprinting technique for establishing the historical source area contributions to sediment preserved in floodplains.

References

COLLINS, A. L., WALLING, D. E. & LEEKS, G. J. L. 1996. Composite fingerprinting of the spatial source of fluvial suspended sediment: a case study of the Exe and Severn river basins, UK. *Geomorphologie; Relief Processus Environnement*, **1996**(2), 41–54.

——, —— & —— 1997. Fingerprinting the origin of fluvial suspended sediment in larger river basins: combining assessment of spatial provenance and source type. *Geografiska Annaler*, **79(A)**, 239–254.

——, —— & —— 1998. Use of composite fingerprints to determine the provenance of the contemporary suspended sediment load transported by rivers. *Earth Surface Processes and Landforms*, **23**, 31–52.

DARNLEY, A. G., BJORKLUND, A., BOLVIKEN, B. *ET AL.* 1995. *A Global Geochemical Database for Environmental and Resource Management.* UNESCO Publications, 1–33.

DEARING, J. A. 1992. Sediment yields and sources in a Welsh upland lake-catchment during the past 800 years. *Earth Surface Processes and Landforms*, **17**, 1–22.

FOSTER, I. D. L. & WALLING, D. E. 1994. Using reservoir deposits to reconstruct sediment yields and sources in the catchment of the Old Mill Reservoir, South Devon, UK, over the past 50 years. *Hydrological Sciences*, **39**(4), 347–368.

——, OWENS, P. N. & WALLING, D. E. 1996. Sediment yields and sediment delivery in the catchments of Slapton Lower Ley, South Devon, UK. *Field Studies*, **8**, 629–661.

HE, Q. & OWENS, P. N. 1995. Determination of suspended sediment provenance using caesium-137, unsupported lead-210 and radium-226: a numerical mixing model approach. *In*: FOSTER, I. D. L, GURNELL, A. M. & WEBB, B. W. (eds) *Sediment and Water Quality in River Catchments.* Wiley, Chichester, 207–227.

HOROWITZ, A. J. 1991. *Sediment-Trace Element Chemistry,* 2nd edn. Lewis, Ann Arbor, MI.

LEENAERS, H., SCHOUTEN, C. J. & RANG, M. C. 1988. Variability of the metal content of flood deposits. *Environmental Geology and Water Science*, **11**, 95–106.

LEWIN, J. & MACKLIN, M. G. 1987. Metal mining and floodplain sedimentation in Britain. *In*: GARDINER, V. (ed.) *International Geomorphology 1986 Part 1.* Wiley, Chichester, 1009–1027.

MACKLIN, M. G., RIDGWAY, J., PASSMORE, D. G. & RUMSBY, B. T. 1994. The use of overbank sediment for geochemical mapping and contamination assessment: results from selected English and Welsh floodplains. *Applied Geochemistry*, **9**, 689–700.

——, RUMSBY, B. T. & NEWSON, M. D. 1992. Historical floods and vertical accretion of fine-grained alluvium in the Lower Tyne Valley, north-east England. *In*: BILLI, P., HEY, R. D., THORNE, C. D. & TACCONI, P. (eds) *Dynamics of Gravel-Bed Rivers*. Wiley, Chichester, 573–589.

OLDFIELD, F., RUMMERY, T. A., THOMPSON, R. & WALLING, D. E. 1979. Identification of suspended sediment sources by means of magnetic measurements: some preliminary results. *Water Resources Research*, **15**, 211–219.

PATRICK, W. H. & JUGSUJINDA, A. 1992. Sequential reduction and oxidation of inorganic nitrogen, manganese, and iron in flooded soil. *Soil Science Society of America, Proceedings*, **56**, 1071–1073.

TAYLOR, M. P. 1996. The variability of heavy metals in floodplain sediments: a case study from mid-Wales. *Catena*, **28**, 71–87.

WALDEN, J., SLATTERY, M. C. & BURT, T. P. 1997. Use of mineral magnetic measurements to fingerprint suspended sediment sources: approaches and techniques for data analysis. *Journal of Hydrology*, **202**, 353–372.

WALLING, D. E. 1990. Linking the field to the river: sediment delivery from agricultural land. *In*: BOARDMAN, J. & FOSTER, I. D. L. (eds) *Soil Erosion from Agricultural Land*. Wiley, Chichester, 129–152.

—— & QUINE, T. A. 1993. Using Chernobyl-derived fallout radionuclides to investigate the role of downstream conveyance losses in the suspended sediment budget of the River Severn, United Kingdom. *Physical Geography*, **14**, 239–253.

—— & WOODWARD, J. C. 1992. Use of radiometric fingerprints to derive information on suspended sediment sources. *In*: *Erosion and Sediment Monitoring Programmes in River Basins*. International Association of Hydrological Sciences Publications, **210**, 153–164.

—— & —— 1995. Tracing sources of suspended sediment in river basins: a case study of the River Culm, Devon, UK. *Marine and Freshwater Research*, **46**, 327–336.

——, QUINE, T. A. & HE, Q. 1992. Investigating contemporary rates of floodplain sedimentation. *In*: CARLING, P. A. & PETTS, G. E. (eds) *Lowland Floodplain Rivers: Geomorphological Perspectives*. Wiley, Chichester, 165–184.

——, WOODWARD, J. C. & NICHOLAS, A. P. 1993. A multi-parameter approach to fingerprinting suspended sediment-sources. *In*: *Tracers in Hydrology*. International Association of Hydrological Sciences Publications, 329–337.

YU, L. & OLDFIELD, F. 1993. Quantitative sediment source ascription using magnetic measurements in a reservoir–catchment system near Nijar, S.E. Spain. *Earth Surface Processes and Landforms*, **18**, 441–454.

The *CM* pattern as a tool for the classification of alluvial suites and floodplains along the river continuum

JEAN-PAUL BRAVARD[1] & JEAN-LUC PEIRY[2]

[1]*UFR de Géographie, Université Paris–Sorbonne, URA CNRS 141, 191 rue Saint-Jacques, 75005 Paris, France (e-mail:jean-paul.bravard@wanadoo.fr)*

[2]*Institut de Géographie Alpine, Université Joseph Fourier, UMR CNRS 5600, 17, rue Maurice Gignoux, 38031 Grenoble, France*

Abstract: The *CM* pattern is a technique used for describing textural patterns and depositional processes. Over the last 15 years in the Rhône watershed, the *CM* pattern has been intensively used to describe the actual functioning of aquatic and terrestrial environments of the river and its tributaries. Results from this research led us to propose a global *CM* pattern more adapted to floodplain environments, on which geomorphological units are mapped according to their position within the graph. In addition, data collected on a large range of fluvial environments from outwash plains to deltas confirm the mobility of the complete *CM* pattern within the graph. Each segment of the image can shift laterally or vertically in relation to the bottom turbulence of the river and/or the unit stream power. The complete *CM* pattern of river environments also provides a broad assessment of the cohesiveness of deposits in the channel and across the adjacent floodplain, and can be used to identify each river type in the continuum from torrents to deltas.

Research on floodplains is both more recent and less developed than river studies, but has met with growing interest among scientists, with floodplain classification being one of the recurrent topics. Much work has been done since Fisk's seminal research on the Mississippi River (1944). After Lewin (1978) and others, Nanson & Croke (1992) recently reviewed the literature and proposed the first genetic classification of floodplains which is based on three variables: (1) energy in the channel at bankfull discharge; three major classes are used, based on the unit stream power: >300 W m^{-2}, 300–10 W m^{-2} and <10 W m^{-2}; (2) the cohesion of floodplain sediments, described in two categories, 'cohesive' and 'noncohesive'; (3) nine geomorphological variables describing the behaviour of the river both laterally and vertically. This proposal is the most useful one because energy controls both channel dynamics and the conditions of sediment deposition on floodplains when bankfull discharge is exceeded. In the same way, the texture of floodplain sediments, which is related to cohesiveness, depends on flow energy.

This paper is based on Passega's *CM* image technique (1957, 1964), which has been widely used to relate the grain-size characteristics of sediments to the processes of their deposition. After reviewing applications of the *CM* diagram since Passega's papers, we will present the research aimed at reconstructing depositional environments in the Rhône watershed. Fluvial landforms are correlated with the textural characteristics of floodplain sediments. Finally, a model of downstream variation of texture is proposed, integrating the downstream distribution of energy as a variable.

A brief review of the use of *CM* pattern technique

Basic principles of the CM *pattern technique*

The first studies performed on sediment transport in rivers have provided evidence on three types of sediment movement: traction on the bed, saltation produced by high turbulence close to the bottom and homogeneous suspension in the water column (Doeglas 1946; Inman 1949). A vertical gradient in sediment grain size and sediment concentration is characteristic of sediment transport during floods, bottom sediments being coarser and more concentrated than sediments in the upper part of the water column. From data collected by the US Waterways Experiment Station along cross-sections of the Mississippi River, Passega (1957, 1964) and Passega & Byramjee (1969) made a

From: MARRIOTT, S. B. & ALEXANDER, J. (eds) 1999. *Floodplains: Interdisciplinary Approaches*. Geological Society, London, Special Publications, **163**, 259–268. 1-86239-050-9/99/$15.00

distinction between the following types of sediment movement:

(a) particles rolled on the bed, even when there is no turbulence.

(b) Graded suspension (the term is preferred to 'saltation'), composed of sand whose concentration (from 8.5 to 1 g l^{-1}) and coarseness decrease from the bottom of the river to an elevation of 2 m upwards. The size of the coarsest particles in the graded suspension (*Cs*) fluctuates according to the maximum turbulence when settling begins (Passega 1964).

(c) Overlying graded suspension, Passega (1964) described uniform suspension, defined by a constant concentration of particles at the top of the water column. The size of the coarsest suspended particles (*Cu*) is finer than 200 μm in the Mississippi River. Uniform suspension may be in direct contact with the bed when turbulence decreases below a threshold, allowing the graded suspension to settle on the bed. Slow-flowing rivers display uniform suspension, even during floods.

Deposits are described using only two variables extracted from the cumulative curve of the grain-size distribution in a sample: *C* is the coarsest one percentile (value at 99% on a cumulative curve), and *M* is the median of sediment size distribution. The values of *C* and *M* are plotted on a log–log graph (Fig. 1). The closer to the *C* = *M* line the sediments are, the better the sorting is. On the graph, RS is uniform suspension with a maximum value of *C* (*Cu*) at 200 μm, QR is graded suspension (parallel to the *C* = *M* line), PQ is a mixture of rolling grains and graded suspension, and NO and OP are composed of rolling particles.

Some further developments of the textural (CM) pattern technique

Since the papers by Passega (1957, 1964), the *CM* diagram technique has been presented in textbooks, but it has been used by few researchers. The *CM* patterns published belong to two categories, in-channel and river-floodplain patterns.

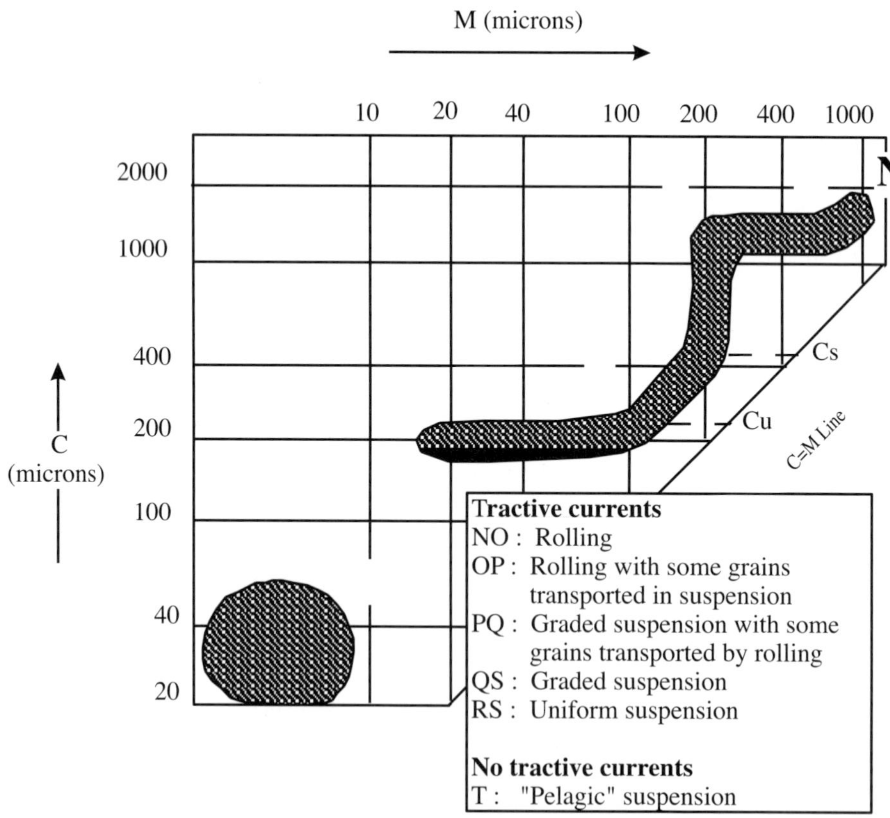

Fig. 1. Complete *CM* pattern (after Passega 1957).

River CM *patterns*. The most quoted in-channel *CM* pattern depicts the Mississippi at Mayersville (Fig. 2A). The river bed, the subaqueous bank and a protected section of the channel are plotted respectively on the three segments PQ–QR–RS separated by bends. This pattern was taken to be the standard image of a low-energy river with sandy bedload and loamy suspended load (Passega 1957). Passega also studied the *CM* patterns of the Adige River (Italy), the Enoree River (South Carolina) and the Niobrara (Wisconsin), plotted as the NO–OP segments (Fig. 2B). These patterns describe the conditions of deposition in high-energy rivers with gravel bedload (Passega 1957).

Bull (1962) studied the *CM* patterns of 11 alluvial fan deposits of Fresno County, California. Sediments from the main channels plot on the PQ segment (mixed rolled and suspended particles with poor sorting) whereas sediments from shallow braided channels and sand bars plot on segment QR (mostly graded suspension). Bull also described the *CM* pattern of poorly sorted mudflow deposits, which do not fit the fluvial *CM* pattern proposed by Passega (Fig. 2C). Williams & Rust (1969) presented the *CM* pattern of the braided Donjek River (Yukon), 40 km downstream of the glacial sediment sources. The different environments they studied plot to short segments in which they recognized segments OP (gravelly channel bars), QR and RS (mixed in-channel deposits); T is a thin layer of silt decanted in slough channels after flood recession (Fig. 2D). A wide range of sedimentation conditions explains the presence of low-energy environments: the *CM* pattern displays an alternation of low- and high-regime flow in flat channels.

CM patterns from rivers and adjacent flood-plains. The scientific literature provides at least three examples. Working on the *CM* pattern of Palaeocene deposits of Sentinel Butte and actual deposits of the Tongue River (North Dakota), Royse (1968) proposed the first graphical illustration of a continuum in deposits from the channel

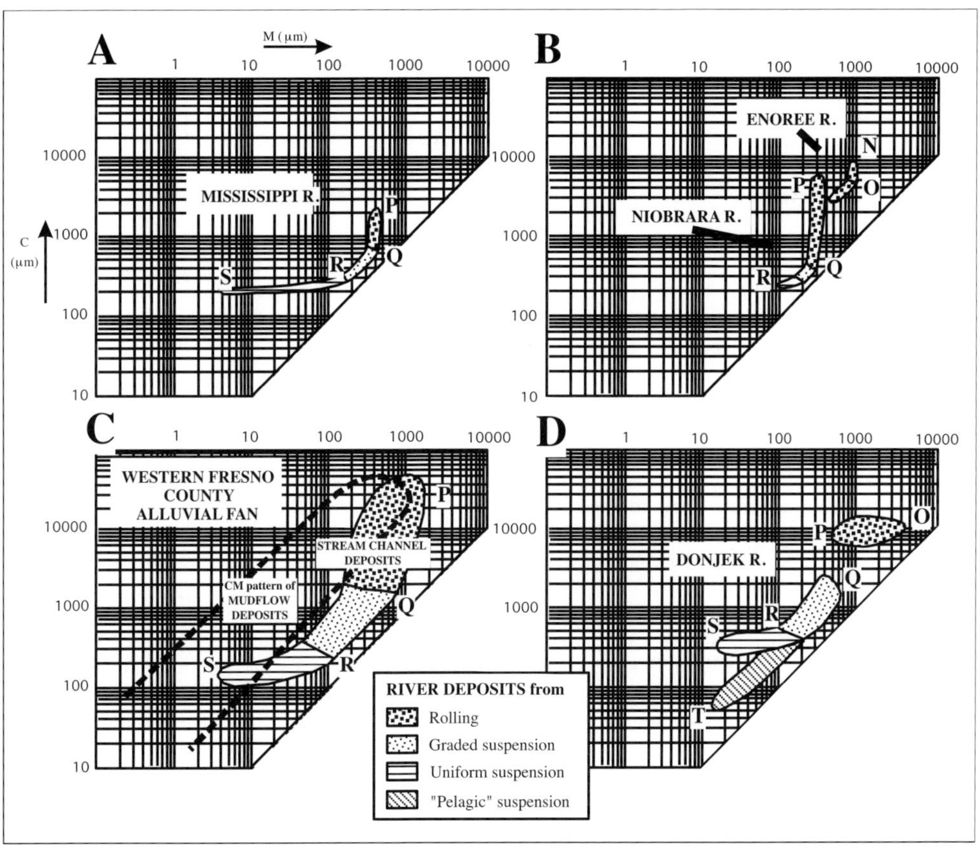

Fig. 2. In-channel *CM* diagrams (after Passega 1957; Bull 1962; Williams & Rust 1969).

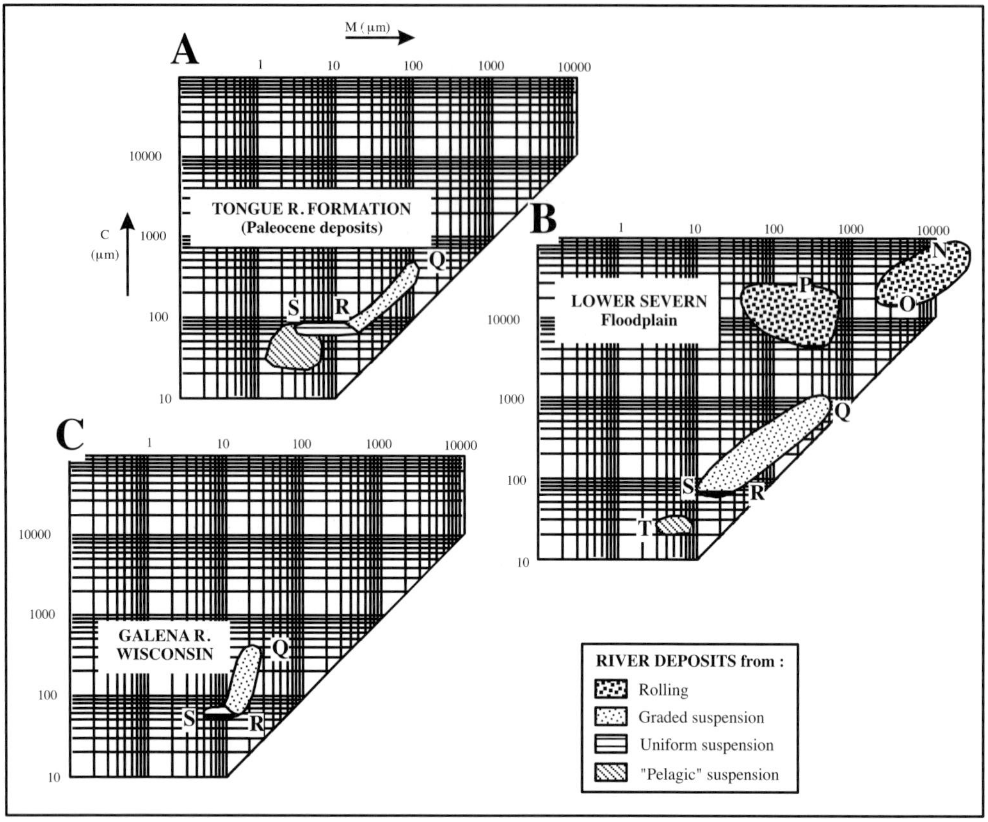

Fig. 3. River and floodplain *CM* diagrams (after Royse 1968; Brown 1985; Magilligan 1992).

to lateral backswamps (Fig. 3A). Channel deposits originate from tractive currents whereas backswamps slowly rise up from the deposition of 'pelagic' sediments. Brown (1985) confirmed the utility of *CM* patterns in differentiating landform units in the floodplain of the lower Severn (UK). Figure 3B shows deposits from tractive currents (P–O–N) and the upper 3 m of the deposits, which plot on RS and T segments. They allowed Brown (1985) to describe persisting overbank deposition during Holocene time with a deficiency in lateral channel migration. Lastly, the *CM* pattern of the Galena River floodplain, Wisconsin, was interpreted by Magilligan (1992). Comparing his results with the standard pattern (Passega 1957), Magilligan was surprised by the shift of the Galena samples towards the area of the fine sediments in the *CM* diagram, even for channel-fill facies, and by the fact that the magnitude of the coarsest one percentile represents the main sedimentological change across the valley bottom. This difference could result from the moderate scale of the Galena drainage basin and from the limited size range of the fine-grained source material within the loess-controlled basins (see Vandenberghe 1975).

The watershed of the Rhône River, an intensive research programme on the *CM* technique

CM *patterns as indicators of the actual functioning of aquatic and terrestrial environments*

An extensive programme of interdisciplinary ecological and geomorphological research has been conducted in the Upper Rhône valley from 1979 onward. Functional typologies of both terrestrial (levees, floodplains) and aquatic environments (different types of channels from the active channels to the remote palaeochannels) have been proposed for selected braided and meander reaches throughout the watershed of the Rhône River. In principle, each landform unit is also an ecological unit, which may be characterized by the texture of its sediments, by its physico-chemical composition, by the depth of the water table, by the frequency of flooding and by biological descriptors. Each unit has thus been classified using a so-called 'functional describer' (Amoros *et al.* 1987). Most interestingly, the *CM* pattern proved to be strongly related to other physico-chemical and biological descriptors of landform units, as the textural properties of sediments originate from the conditions of deposition and determine distinct pedological or substrate characteristics. An extensive programme of sediment sampling and analysis tested the pertinence of *CM* patterns as 'functional describers' in various fluvial environments. This research was integrated in a 3D approach for fluvial systems based on a hierarchical classification of 'functional units', 'functional sets' and 'functional sectors' (Amoros *et al.* 1987).

On this occasion, Passega's *CM* pattern of the Upper Rhône River was enriched by systematically integrating both river and floodplain environments (Bravard *et al.* 1986). Sediment samples were collected from recent deposits from single flood events and on cross-sections of distinct landforms such as levees, banks, channel bars, filled channels, etc. Peiry (1988) proposed a detailed *CM* image of the Arve valley bottom with a complex set of active and abandoned braided channels (Fig. 4). The global pattern of former channels (functional sets) was disaggregated into detailed patterns of landform units (functional units), such as alluvial plugs and in-channel deposits at different locations. In keeping with Passega's suggestions, landform units were mapped in reference to *CM* patterns obtained in a former braided area affected by embankments (Bravard 1983).

New perspectives on the CM *patterns of river reaches*

Shape of the CM *patterns and river energy.* As Royse (1968) stated, the *CM* pattern of a reach illustrates the continuum of deposition from the channel to the remote backswamps. The degree of dispersion of the samples in the graph represents the total amount of fluctuation in energy in the river responsible for the deposition of the sediments. In

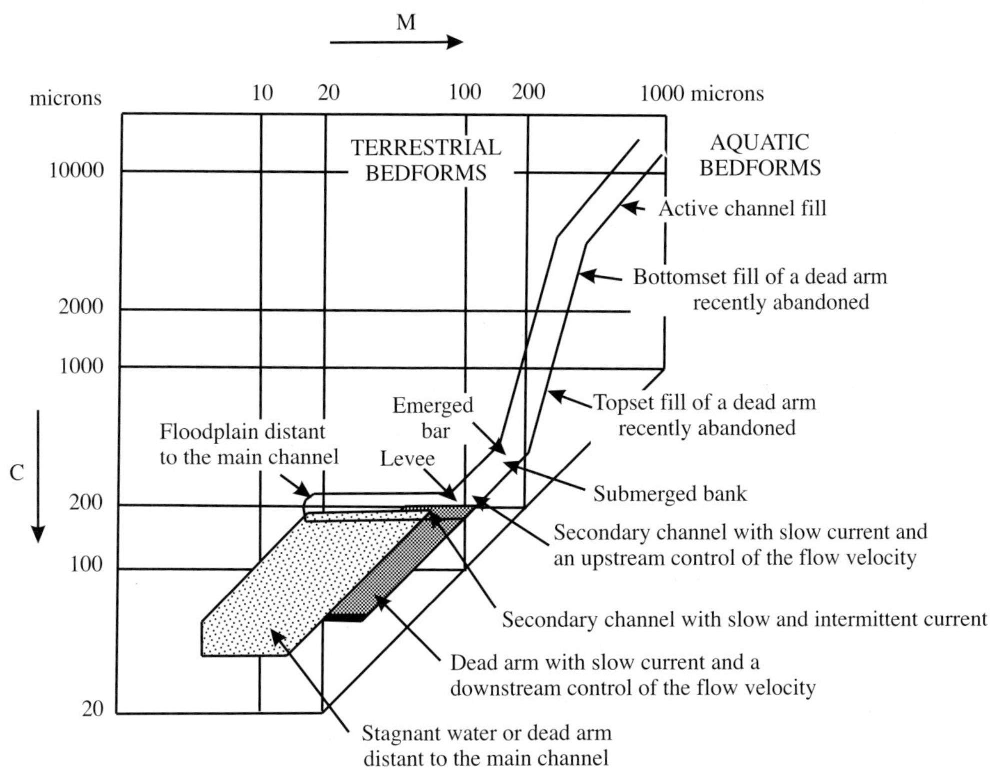

Fig. 4. *CM* pattern and associated terrestrial and aquatic bedforms (after Peiry 1988).

the example studied by Royse (1968), the 'maturity' of river environments, i.e. the downstream decrease of energy, is reflected by the development of backswamp type facies across wide floodplains. Thus, the complexity of *CM* patterns, deduced from their shape, depends on the nature of river and floodplain reaches. The more complex *CM* patterns (Fig. 4) describe gravel-bed piedmont rivers flowing amidst large floodplains prone to the preservation of flat surfaces and meanders or braided channels, because all the segments are present, from NO (cobbles transported on the river bed) to T (different types of decanted uniform suspensions).

Graded and uniform suspensions in the channel and over the floodplain. In fact, one should consider the vertical structure of the water column during floods: in the Mississippi River in flood at Mayersville (Passega 1964), the height of the graded suspension does not exceed 3 m underneath 10 m of uniform suspension. Any increase in discharge and bottom turbulence would simply raise the relative elevation of the graded suspension and decrease the thickness of the uniform suspension zone. In contrast, in a more turbulent river such as the Rhône, the flow during medium and large floods is composed of graded suspensions up to the water surface, implying that graded suspensions can be deposited on levees. The *Cu* value of the uniform suspension flowing over the floodplain is controlled by the turbulence of the graded suspension at the bankfull level discharge. In the upper Rhône valley, uniform suspension is not superimposed upon graded suspension but rather seems to be juxtaposed with it in space.

New data on deposition in cut-off channels. The study of several floodplains in the watershed of the Rhône River allowed us to elaborate the theoretical complete pattern of tractive current deposit (see Fig. 1). 'Pelagic suspension' T was not described in alluvial environments but in offshore marine environments, where particles float near the bottom in water that is not necessary turbulent. We were able to describe the sedimentation in several former meanders isolated from the main channel. All of them are positioned below the RS segment but cannot be defined as 'pelagic suspensions'. According to Passega (pers. comm.) this can be interpreted as a dispersion of particles in water: it could represent a transformation of distal uniform suspensions that would have lost their coarsest particles. Peiry (1988) made a distinction between the sediments deposited in stagnant water bodies remote from the main channel and the sediments deposited in former channels where the velocity of water fluxes is slowed by a downstream control. The former are finer because the coarser particles settle on the floodplain before the remaining sediment settles in stagnant water. In contrast, the cut-off channel still connected to the main channel can receive full graded suspensions that suddenly lose their turbulence and deposit the sand fraction. This is why their *CM* image extends the RS segment (graded suspension) in a way similar to the turbidity currents described by Passega (1964).

Position of RS segment (uniform suspension) in the CM *graph.* Of course, the most quoted *CM* pattern is the Mississippi pattern published in various papers by Passega (e.g. 1957, 1964). It is the only one by Passega to represent clearly uniform suspension, defining *Cu*, 'the maximum grain size' of the RS segment. On the Mississippi at Mayersville, *Cu* averages 200 μm and is considered as an indicator of the maximum turbulence in the current above the zone disturbed by bottom friction (Passega 1964). This statement implies that the value of *Cu* should increase in high-energy rivers, but a later study shows that the value of *Cu* should seldom be higher than 250 μm (Passega & Byramjee 1969) and is commonly 200 μm (Passega 1977). Moreover, Passega (1977) stated that the value of the median *M* increases in uniform suspension when the bottom turbulence increases, as the proportion of fines decreases. Therefore, the value of *M* in uniform suspension should be the correct indicator of bottom turbulence.

The Mississippi pattern has been considered as a standard pattern and all further published studies interpreting new patterns either refer to *Cu* as constant, i.e. having a value close to 200 μm, or do not mention the problem if the position of segment RS is higher (Williams & Rust 1969) or lower (Royse 1968; Magilligan 1992). In reality, there is no standard *CM* image for reference purposes and our findings demonstrate that parameters *Cs* and *Cu* are not constant but fluctuate according to the energy of river environments during floods. The value of *Cu* probably fluctuates between 100 μm for low-energy rivers and 400 μm for high-energy rivers.

An interesting point for discussion is the question of whether the *Cu* value of floodplain sediments deposited by different types of floods should be constant along the same river reach. One can assume that high, major floods can bring about higher turbulence in the uniform suspension; hence they can induce the presence of coarser particles in the deposits, thereby increasing the value of *Cu*. This position was adopted by Arnaud-Fassetta (1998) in his study of recent flood deposits in the flat bottoms connected to the Little Rhône River in the Camargue delta. Consistently working on the same landforms (levees and proximal floodplain), he emphasized the link between *Cu* variability in uniform suspension and changes in the flood frequency and magnitude, and used this link as a palaeohydrological indicator for late Holocene time. Arnaud-Fassetta (1998) also proposed that the character of the channel banks may explain the preservation of different types of uniform suspensions (from 250 to 290 μm). In theory, uniform suspension should occur in the channel of the Upper Rhône during small floods, but deposits from uniform suspension could not be found in it, because they are scoured by larger flood events or because they cannot be deposited. This question has to be explored further.

The shifting positions of CM *patterns.* In light of this discussion, it may be useful to reconsider the mobility of the complete *CM* pattern in the graph, as reflected by the mobility of the different segments (Fig. 5):

(1) As seen above, the RS segment probably shifts vertically in the *CM* graph, depending on the level of the turbulence in the graded suspension, which in turn controls the highest value of *C* (*Cu*) for uniform suspension.

(2) The length of the RS segment depends on the bottom turbulence of a river, as Passega (1977) found (see above). S shifts to the right when bottom turbulence may influence uniform suspension in the channel. Short RS segments indicate that turbulence remains high in a channel where uniform suspension may be deposited. Also, they indicate that the conditions of sedimentation on a floodplain do not allow a selection of coarse particles to reach remote areas which usually receive uniform suspension with low *M* values. Indeed, the coarse particles have settled out while crossing the alluvial plain.

(3) Bull (1962) was probably the first to propose that the position of a pattern relative to the *CM* line could reflect the degree of sediment sorting, in which proximity would signify good sorting. This was confirmed by Passega (1964), who specified that the distance between segment QR and the *CM* line expresses the degree of sorting of the graded suspension, which is a function of *Cs*.

(4) The position of point Q (QR segment), or *Cs* value, depends on the turbulence of the graded suspension. Its position in the graph is higher in high-energy rivers.

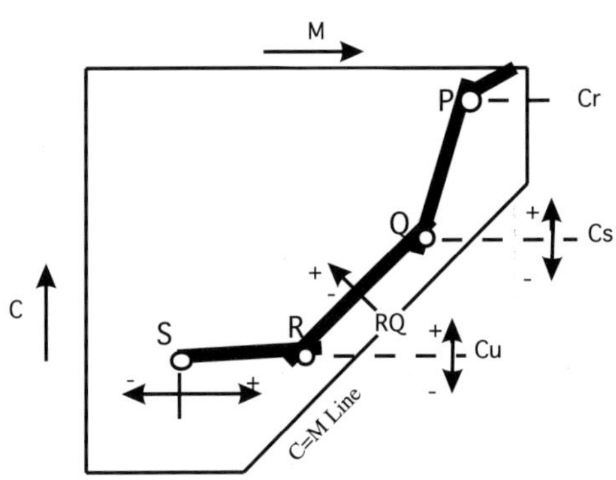

Fig. 5. Latitudes of change of the *CM* pattern from one river to another (after Peiry 1988).

CM *patterns of river environments are indicative of the cohesiveness of floodplain overbank sediments.* Generally speaking, the different zones of a *CM* pattern have their own textural characteristics, which depend on the relative proportion of sand and silt in the sediment. For instance, well-sorted sands from graded suspensions form distinct layers or laminae that have no cohesiveness when compared with sandy silts or silty sands from uniform suspensions, or with silty clays of infilled meanders. To test this, a curve was derived by plotting on two axes the median and the percentage of sediment finer than 50 μm extracted from 120 cumulative curves selected in the Rhône watershed (Fig. 6). Conventionally, particles finer than 50 μm represent the silt–clay content of the plots but this value may be variable. Using an average *CM*, we selected the median *M* of 100 μm as a threshold discriminating between uniform suspension and graded suspension. As shown in Fig. 6, the percentage of sediment finer than 50 μm (or silt–clay content), corresponding to a median of 100 μm, is close to 20% in the fluvial sediments of the Rhône watershed. This distinction is important because uniform suspensions are cohesive whereas graded suspensions are not. A median of 100 μm on the *M* axis of a *CM* graph, converted into a silt–clay content of 20%, provides the threshold value between cohesive and non-cohesive deposits.

Modifying Schumm's (1977) classic graph, we represented the different fluvial pattern types in relation to unit stream power and silt–clay content (Fig. 7). Sinuous and slow-flowing rivers, transporting sand as bedload or as graded suspension close to the bottom, build cohesive floodplains. By contrast, free wandering, gravel-bed rivers or gravel-bed meandering rivers are turbulent enough to transport and to deposit large quantities of graded suspensions on their banks and even on the top of the floodplain. The velocity of flows exceeding bankfull discharge prevents the deposition of large amounts of sediment from uniform suspensions, explaining why these types of rivers build non-cohesive floodplains.

As a consequence, the complete *CM* pattern in a river reach may provide a broad assessment of the

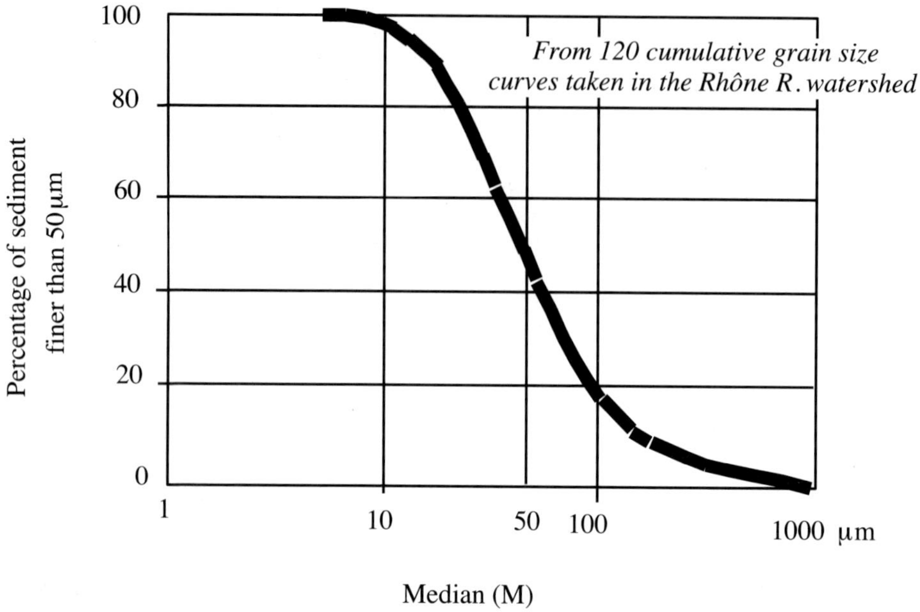

Fig. 6. Diagram showing relationships between median diameter and silt–clay content of alluvial sediments.

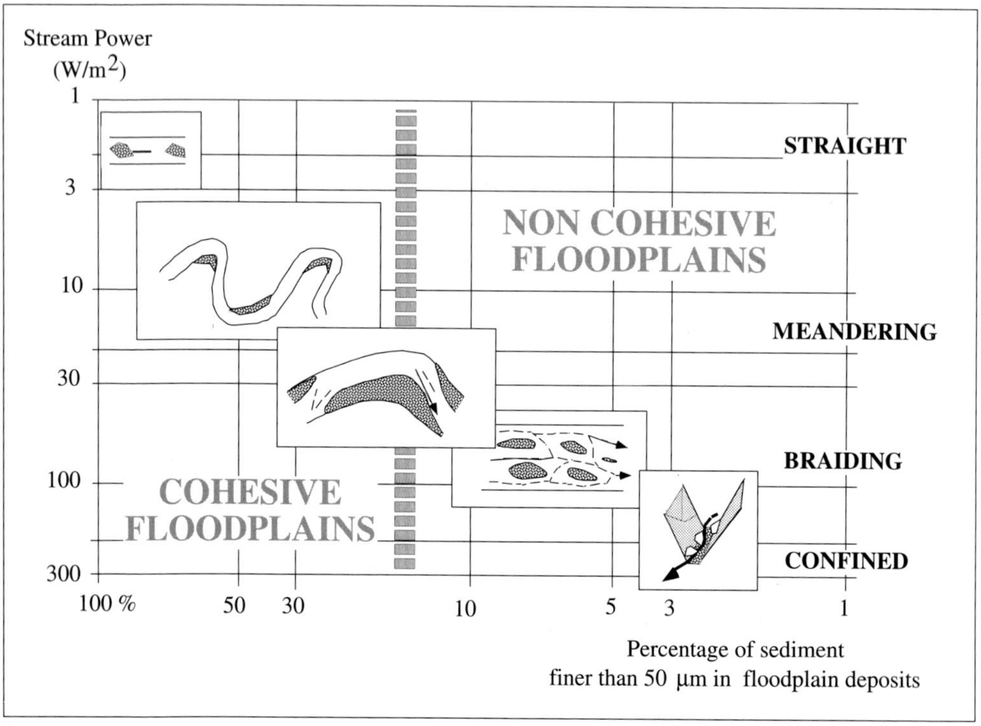

Fig. 7. Relationships between fluvial pattern and floodplain cohesiveness.

cohesiveness of deposits in the channel and across the adjacent floodplain. As the *CM* graph makes it possible to distinguish between different depositional processes and landform units, the task is easy for present-day deposits. However, the erosion of concave river banks cuts into palaeodeposits of different origins, ages, texture and cohesion. As the *CM* pattern of the alluvial plain does not provide the relative contribution of each sediment facies to the general cohesiveness of the plain, further investigations should characterize floodplains and meander banks not only by the occurrence of different sediment facies, but also by their spatial and vertical extension.

The *CM* images in the river continuum: linking fluvial patterns, sedimentary facies and energy

The grain size of overbank deposits along rivers is closely related to energy through the influence of turbulence during floods, which determines the type of suspension deposited on the banks and on the top of the floodplain. This was clearly demonstrated along the Saône River, a tributary of the Rhône, where downstream variation in the distribution of energy (unit stream power) is mainly related to the slope of the water profile, whereas other variables display only slight changes (Astrade 1996; Astrade & Bravard 1999). Therefore, as stated by Nanson & Croke (1992), energy is a controlling variable which influences the geomorphological pattern and the texture of in-channel and overbank deposits.

To complement this approach, we have provided a semi-quantitative assessment of the relationships between the control variables discussed here. Figure 8 displays the theoretical model of downstream change in the *CM* pattern for a complete river continuum, from an outwash plain downstream of the Bossons glacier (Mont Blanc massif) down to the Camargue delta via torrential, braided and meandering patterns selected in the Arve and Rhône valleys. Each *CM* pattern is positioned in relation to energy (represented by the unit stream power) and the related cohesiveness of sediments. The shift of the *CM* pattern across the graph reflects the progressive loss of unit stream power and the downstream fining of sediment texture. The shape of each pattern reflects the degree of sediment sorting, the finer segments corresponding to better sorting. There is some overlap between segments of high-energy rivers. The differences are due to the size of cobbles and to the presence of shelters, which may allow the deposition of sand sheets.

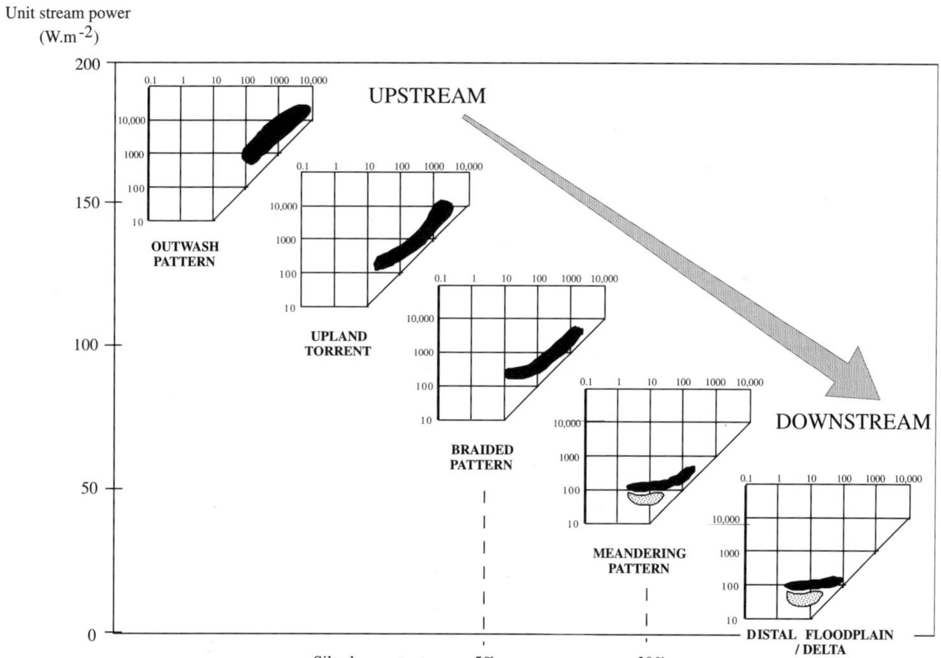

Fig. 8. Upstream--downstream evolution of the *CM* diagram according to changes of the floodplain pattern (the Camargue delta from Arnaud-Fassetta, 1998).

Conclusion

The *CM* pattern technique is useful for precisely describing the sedimentation environments of rivers and associated floodplains. It discriminates well between cohesive and non-cohesive geomorphological units and between floodplain reaches, as uniform suspension is responsible for the deposition of silt-rich sediments whereas graded suspension is sandy and non-cohesive. The construction of a representative *CM* pattern in a valley reach may usefully complement floodplain classification based on energy, on the degree of cohesiveness of alluvial sediments and on geomorphological features (Nanson & Croke 1992), providing a semi-quantitative assessment of the silt-clay content and cohesiveness of the banks prone to lateral erosion.

The authors thank S. B. Marriott, M. Provansal and C. Spencer, who provided helpful comments in reviewing the manuscript.

References

AMOROS, C., ROUX, A.-L., REYGROBELLET, J.-L., BRAVARD, J.-P. & PAUTOU, G. 1987. A method for applied ecological studies of fluvial hydrosystems. *Regulated Rivers*, **1**, 17–36.

ARNAUD-FASSETTA, G. 1998. *Dynamiques fluviales holocènes dans le delta du Rhône*. Thèse de géographie physique, Université de Provence (Aix–Marseille 1).

ASTRADE, L. 1996. *La Saône en crue: dynamique d'un hydrosystème anthropisé*. Thèse de doctorat de géographie, Université Paris IV–Sorbonne.

—— & BRAVARD, J.-P. 1999. Energy gradient and geomorphological processes along a river influenced by neotectonics (the Saône River, France). *Geodinamica Acta*, **12**(1), 1–10.

BRAVARD, J.-P. 1983. Les sédiments fins des plaines d'inondation dans la vallée du Haut-Rhône (approche qualitative et spatiale). *Revue de Géographie Alpine*, **LXXI**, **4**, 363–379.

——, AMOROS, C. & JACQUET, C. 1986. Reconstitution de l'environnement des sites archéologiques fluviaux par une méthode interdisciplinaire associant la géomorphologie, la zoologie et l'écologie. *Revue d'Archéométrie*, **10**, 43–55.

BROWN, A. G. 1985. Traditional and multivariate techniques in the interpretation of floodplain sediment grain size variations. *Earth Surface Processes and Landforms*, **10**, 281–291.

BULL, W. B. 1962. Relation of textural (*CM*) patterns to depositional environments of alluvial-fan deposits. *Journal of Sedimentary Petrology*, **32**, 211–216.

DOEGLAS, D. J. 1946. Interpretation of the results of mechanical analyses. *Journal of Sedimentary Petrology*, **16**,19–40.

FISK, H. N. 1944. *Geological Investigations of the Alluvial Valley of the Lower Mississippi River*. US Army Corps of Engineers, Mississippi River Commission, Vicksburg, MS.

INMAN, D. L. 1949. Sorting of sediment in the light of fluid mechanics. *Journal of Sedimentary Petrology*, **19**, 51–70.

LEWIN, J. 1978. Floodplain geomorphology. *Progress in Physical Geography*, **2**(3), 408–437.

MAGILLIGAN, F. J. 1992. Sedimentology of a fine-grained aggrading floodplain. *Geomorphology*, **4**, 393–408.

NANSON, G. C. & CROKE, J. C. 1992. A genetic classification of floodplains. *Geomorphology*, **4**, 459–486.

PASSEGA, R. 1957. Texture as characteristic of clastic deposition. *Bulletin, American Association of Petroleum Geologists*, **41**(9), 1952–1984.

—— 1964. Grain-size representation by *CM* patterns as a geological tool. *Journal of Sedimentary Petrology*, **34**(4), 830–847.

—— 1977. Significance of *CM* diagrams of sediments deposited by suspensions. *Sedimentology*, **24**, 723–733.

—— & BYRAMJEE, R. 1969. Grain-size image of clastic deposits. *Sedimentology*, **13**, 233–252.

PEIRY, J.-L. 1988. *Approche géographique de la dynamique spatio-temporelle des sédiments. L'exemple de la plaine alluviale de l'Arve (Haute-Savoie)*. Thèse de géographie et aménagement, Université Lyon 3.

ROYSE, C. F. 1968. Recognition of fluvial environments by particle-size characteristics. *Journal of Sedimentary Petrology*, **38**(4), 1171–1178.

SCHUMM, S. A. 1977. *The Fluvial System*. Wiley–Interscience, New York.

VANDENBERGHE, N. 1975. An evaluation of *CM* patterns for grain-size studies of fine grained sediments. *Sedimentology*, **22**, 615–622.

WILLIAMS, P. F. & RUST, B. R. 1969. Sedimentology of a braided river. *Journal of Sedimentary Petrology*, **39**(2), 649–679.

Alluvial architecture of the Mississippi valley: predictions using a 3D simulation model

JOHN S. BRIDGE

Department of Geological Sciences, Binghamton University, Binghamton, NY 13902-6000, USA (e-mail:jbridge@binghamton.edu)

Abstract: The 3D alluvial-architecture model of Mackey & Bridge (1995, *Journal of Sedimentary Resarch*, **B65**, 7–31) is used to simulate the nature of avulsion and alluviation of the Mississippi River in its lower alluvial valley and delta plain during the Holocene period, a time of relative sea-level rise. An important objective of these simulations is to examine the controls on the relative proportion of channel-belt deposits and floodplain deposits within the alluvium. Although the model cannot predict the exact avulsion behaviour of the Mississippi, the timing and location of avulsions are predicted well in a general sense. In the lower Mississippi valley, avulsions commonly originate in an up-valley location near Memphis, and the mean avulsion period is 1000–1500 a. This avulsion frequency, combined with large channel-belt width relative to valley width, has resulted in large channel-deposit proportion along the length of the lower alluvial valley. On the delta plain, the model predicts that avulsions are relatively frequent (intervals of decades to centuries) on active deltas where deposition rate on alluvial ridges is greatest and down-valley slopes are least (related to relative sea-level rise). Avulsions originating from locations on the delta plain more distant from the coast occur at longer time intervals (hundreds to thousands of years), because alluvial-ridge topography develops more slowly in these regions. Such avulsions give rise to discrete delta lobes or subdeltas (e.g. Teche, Lafourche). Channel-deposit proportion is generally low on the delta plain, and decreases towards the Gulf of Mexico. This is primarily due to down-valley increase in floodplain width relative to channel-belt width. However, channel-deposit proportion is locally high near avulsion points. These results indicate that change in alluvial architecture during relative sea-level rise in near-coastal settings is more complicated than suggested by recent sequence-stratigraphic models where the proportion of channel-belt deposits is postulated to be controlled primarily by only two factors: deposition rate and valley width.

Alluvial deposits contain an important record of the morphology, flow and sedimentary processes of the rivers and floodplains of the past. Alluvial deposits are also important aquifers and hydrocarbon reservoirs. It is therefore desirable to be able to interpret the origin of alluvial deposits, and to predict their subsurface characteristics. A particularly important aspect of ancient alluvium is the geometry, proportion and spatial distribution of relatively coarse-grained channel-belt deposits within finer-grained floodplain alluvium (commonly referred to as alluvial architecture). Interpretation of ancient alluvial architecture relies fundamentally on knowledge of the morphology, flow and sedimentary processes, and deposits of modern floodplains and their associated channel belts. However, the stratigraphic architecture of most modern alluvial valleys is not known in detail, and alluvial processes acting over hundreds or thousands of years (e.g. development of alluvial ridges, river avulsions) are poorly understood because they are difficult to study. Thus, there is a great need for more detailed 3D information on modern alluvium (from cores, geophysical profiling, and age determination), but also for development of theoretical, process-based models of alluvial sedimentation, and experimentation with scale models. Although process-based models are in their infancy, they can still aid in the understanding of ancient alluvial architecture.

The purpose of this paper is to use the 3D model of Mackey & Bridge (1995) to simulate the alluvial architecture of the lower Mississippi valley and delta plain. As the processes and deposits of the Mississippi valley and delta plain have been studied extensively (e.g. Fisk 1944, 1947, 1951, 1952; Autin *et al.* 1991; Saucier 1994), the input data required for the simulations are well constrained, and the simulated deposits can be compared closely with observed deposits. The main value of the

From: MARRIOTT, S. B. & ALEXANDER, J. (eds) 1999. *Floodplains: Interdisciplinary Approaches*. Geological Society, London, Special Publications, **163**, 269–278. 1-86239-050-9/99/$15.00

simulations is to suggest hypotheses about the controls on alluvial architecture in the Mississippi valley. However, the simulations also have broader implications for sequence-stratigraphic models that predict that the proportion of channel-belt deposits in ancient alluvium is controlled primarily by variation in only two factors: deposition rate and valley width (e.g. Ross 1990; Shanley & McCabe 1993).

Mackey–Bridge model of alluvial architecture: summary

The first 3D, process-based model of alluvial architecture was developed by Mackey & Bridge (1995). In this model, a floodplain of variable width and length is occupied by a single, active channel belt that contains relatively coarse sediment. Channel-belt width and maximum bankfull depth can be constant or vary in time. Changes in floodplain topography are produced by spatial and temporal variation of channel-belt and floodplain deposition rate, and by compaction and local (within-floodplain) or regional tectonism. The location and timing of avulsions are determined by local changes in cross-valley floodplain slope relative to down-valley channel slope (compare the simple model of Slingerland & Smith (1998)), and by flood magnitude and frequency. The diverted channel follows the locus of maximum floodplain slope. Major differences between this model and the earlier 2D models (e.g. Bridge & Leeder 1979; Bridge & Mackey 1993) are the treatment of avulsion location and period as dependent variables, and constraints on the location of avulsing channels by the points of avulsion and topographic highs on the floodplain. Thus, avulsing channels may be prevented from shifting to the lowest point in a given floodplain section, as was assumed to be the case in the 2D models. The 3D model predicts that the proportion and connectedness of channel-belt deposits vary with distance from points of avulsion and with cross-section position. Thus, relationships between channel-deposit proportion and connectedness derived from 2D models are applicable only to parts of the floodplain located some distance down-valley from avulsion points.

In the 3D model, avulsions occur preferentially where there is a decrease in down-valley channel slope and/or an increase in cross-valley floodplain slope. For example, a down-valley decrease in channel slope and/or a down-valley increase in aggradation rate (associated, for example, with base-level rise) results in high avulsion frequencies in down-valley regions of the floodplain, and possibly increased channel-deposit proportion and connectedness. Clearly, avulsion frequency must be included as a primary control in any model of alluvial architecture. High deposition rates near the apices of alluvial fans tend to increase cross-valley slopes, thus increasing avulsion frequencies in this region. This results in a nodal form of avulsion and relatively high channel-deposit proportion and connectedness in the up-valley parts of fans.

In cases where avulsions tend to occur preferentially in down-valley regions of floodplains, the 3D model predicts avulsion sequences where points of avulsion shift progressively upstream with a corresponding decrease in avulsion period. This is due to continued growth of alluvial ridges (and increasing cross-valley slopes) upstream of avulsion points, whereas new channel-belt segments down-valley from avulsion points have low ridges and a low probability of avulsion. Avulsion sequences are commonly punctuated by abrupt down-valley stepping of avulsion points. Sedimentary successions associated with avulsion sequences may increase upwards in channel-deposit proportion and connectedness, and be capped with overbank deposits with well-developed soils. Such successions may take on the order of 10^3–10^5 a to form, comparable with cycles attributed to tectonism or climate change.

The 3D model predicts marked variations in avulsion period that are related to progressive changes in floodplain relief. During a period of new alluvial ridge growth in an area, a relatively long time (say, on the order of 10^3 a) may elapse before conditions are conducive for an avulsion. However, once some threshold floodplain relief is reached, a series of avulsions occurs with periods on the order of 10^1 a or less. This gives rise to a bimodal or polymodal distribution of avulsion periods. It is expected that relatively mature soils would form during periods when avulsion activity was reduced.

Another important aspect of the 3D model simulations is the influence of the location of new channel belts by prior alluvial ridges. Channel belts may be clustered preferentially on one side of the floodplain and show a distinctive en echelon pattern, possibly giving rise to channel-belt sandstone bodies with high width/thickness. This also means that other parts of the floodplain distant from the active channels experience relatively low overbank deposition rates for extended periods of time, allowing soil development or peat formation.

Tectonic tilting and faulting within the floodplain increase avulsion probabilities in specific areas of the floodplain. Channel belts are generally shifted away from zones of uplift and towards zones of maximum subsidence. However, if channel-belt aggradation keeps pace with fault displacement or tilting, depositional topography causes channels to shift away (off-lap) from areas of maximum subsidence. There is a tendency for channel-deposit proportion and connectedness to increase locally near downthrown areas of the floodplain. However, increases in channel-deposit proportion and connectedness may be caused by other factors. Consequently, it is extremely difficult to determine the location and geometry of tectonic subsidence or uplift from the distribution of channel belts alone.

Many of these predictions of the Mackey–Bridge model agree qualitatively with data from modern rivers. For example, the nodal and up-valley–down-valley shifting modes of avulsion occur in many rivers (details have been given by Mackey & Bridge (1995)). The predicted association of high avulsion frequency and high deposition rate was observed experimentally by Bryant *et al.* (1995). The bimodal or polymodal distribution of avulsion periods has been observed in the Rhine–Meuse delta (Törnqvist 1994). Many features of the avulsion behaviour of the Kosi river over its megafan can be explained using the model, including the spatial distribution of channel-belt deposits (Mackey & Bridge 1995). The 3D model was used to simulate the alluvial architecture of some Miocene Siwalik strata from northern Pakistan (Khan *et al.* 1997). Zones of relatively high channel-deposit proportion and connectedness were explained by increased channel-belt dimensions, deposition rate and avulsion frequency on an ancient megafan comparable with the Kosi megafan.

Despite the qualified success of the Mackey–Bridge model in simulating various aspects of modern-river behaviour, and in generation of new hypotheses to test, further tests of validity and development are clearly required. Rigorous testing of this kind of model is difficult. Mackey & Bridge (1995) concluded: 'There is a critical need for more comprehensive architectural data from modern fluvial systems, especially data related to processes controlling floodplain geometry and channel pattern over periods of thousands of years'. Since the model was originally constructed, a large amount of new 3D field data has become available, particularly from the Rhine–Meuse and Mississippi delta plains, that will allow further model testing and development. Here, the Mackey–Bridge model is tested using data from the Mississippi, and is used primarily to explain how channel-deposit proportion varies in time and space.

Table 1. *Responses of Mississippi valley and delta plain to glacial–interglacial cycles (following Autin* et al. *1991; Saucier 1994)*

Variable	Uplands	Alluvial valley	Coastal plain
Deglaciation–interglacial (Holocene)			
Tectonism	Isostatic uplift	Subsidence increasing towards coast	
Sea level			Rising: transgression
Water supply		Decreasing	Decreasing
Sediment supply		Decreasing	Decreasing
Mean sediment size		Intermediate	Fine
Channel-belt width and depth		High	Low
Channel pattern		Meandering	Low sinuosity
Floodplain width		Low	Very high
Floodplain slope		Intermediate	Low
Deposition or erosion rate	Erosion	Terrace formation up-valley Deposition rate increasing to coast	
Avulsion frequency		Intermediate	High
Channel-deposit proportion		High	Low
Glacial maximum–deglaciation (Pleistocene)			
Tectonism	Crustal loading	Subsidence decreasing towards coast	
Sea level			Low to rising
Water supply		High	High
Sediment supply		High	High
Mean sediment size		High	Intermediate
Channel-belt width and depth		High	High
Channel pattern		Braided	Braided-meandering
Floodplain width		High	Low
Floodplain slope		High	Intermediate
Deposition or erosion rate	Erosion	Maximum deposition to minor erosion	Valley incision to deposition
Avulsion frequency		High, decreasing	Low, increasing
Channel-deposit proportion		High	High, decreasing

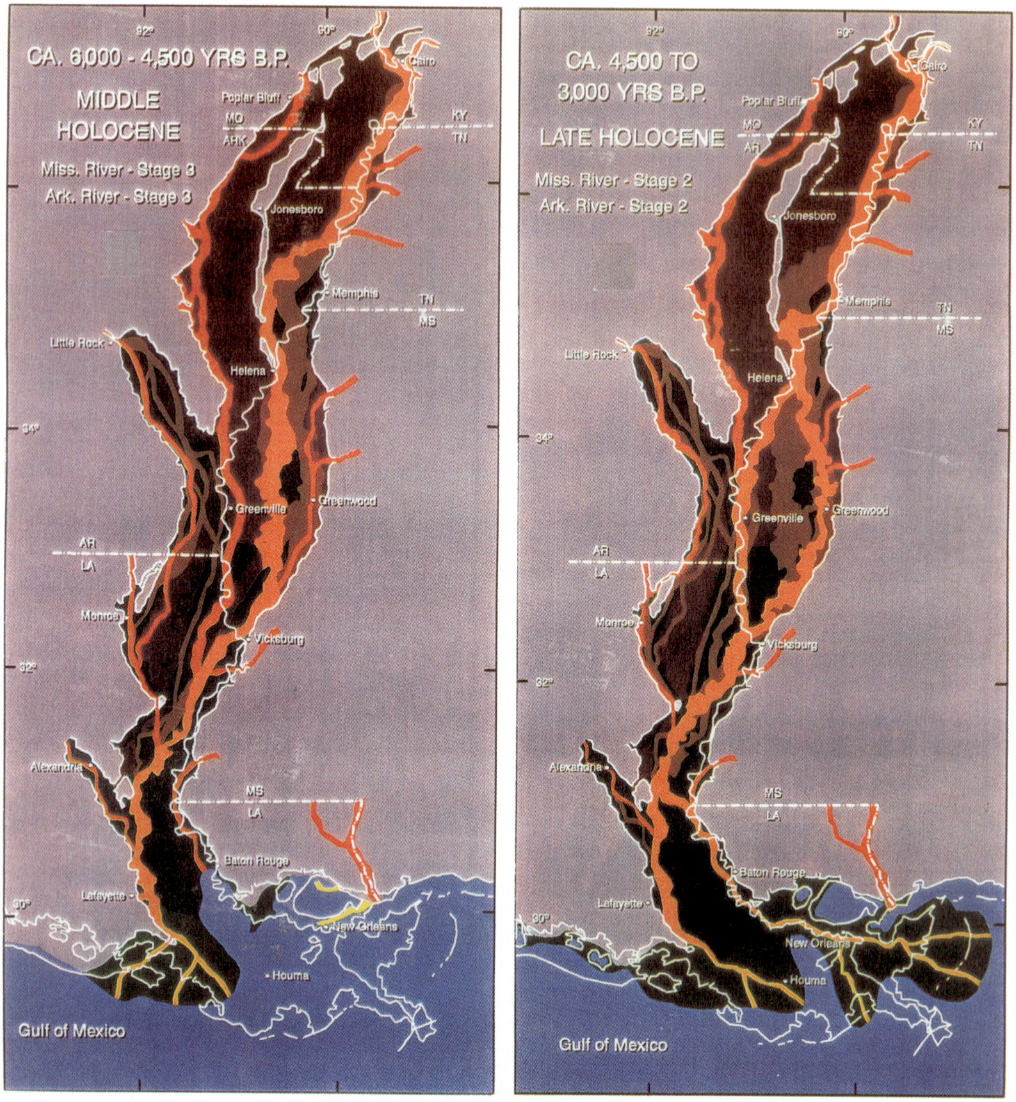

Fig. 1. Maps of Mississippi alluvial valley and delta plain for three periods during Holocene time (from Saucier 1994). Noteworthy features are the avulsion node near Memphis, at the upstream end of the Yazoo Basin, and the points of avulsion near Alexandria, Baton Rouge and New Orleans.

Application to the lower Mississippi valley and delta plain

The primary data sources on Mississippi alluvial processes and deposits are Fisk (1944, 1947, 1951, 1952), Autin *et al.* (1991) and Saucier (1994). The model is appropriate for the Mississippi valley insofar as there is generally a single dominant channel belt that is periodically diverted within a fairly well-defined floodplain area. However, there are variations in time and space of channel-belt and floodplain width, deposition rate, tectonic subsidence and uplift. For example, there are variations in alluvial processes and deposits from the uplands through the alluvial valley to the delta plain that are different for glacial and interglacial periods (Table 1; Autin *et al.* 1991; Saucier 1994). There is also local tectonism in addition to regional, along-valley variations in subsidence or uplift rate (e.g. New Madrid seismic zone: Saucier 1994). The model

was only applied where input variables could be determined with confidence. Thus, simulations using the Mackey–Bridge model were made for the Holocene conditions for the sections of valley between Memphis and Vicksburg (Yazoo Basin), and between the Atchafalaya Basin and the Gulf of Mexico (Fig. 1). Observational data (references above) relevant to these simulations are as follows (see Table 2 for actual input values).

Holocene channel belts and avulsions

Channel-belt width and channel sinuosity vary along the length of the modern Mississippi River and these factors also varied through Holocene time. In the modern alluvial valley, channel belts are 15–25 km wide and their levees may be 3–6 m higher than adjacent floodbasin areas (Fisk 1944). Maximum bankfull depths increase from about 20 m to 40 m from Cairo (at the entrance to the lower Mississippi valley) to Natchez (upstream end of Atchafalaya Basin). On the delta plain, channel-belt widths are 10–15 km, and maximum bankfull depth is up to 50 m (near New Orleans). Channel-belt dimensions clearly varied during the process of avulsion, but channel-belt dimensions were apparently not clearly influenced by climate changes during Holocene time (Saucier 1994).

Table 2. *Input and output variables for runs of the 3D alluvial architecture model*

	Lower Mississippi valley (Yazoo Basin)	Atchafalaya Basin and delta plain
Input variables		
Floodplain width up-valley (km)	48	48
Floodplain width down-valley (km)	60	200
Floodplain length (km)	400	300
Initial upstream floodplain slope	0.00011	0.00012
Slope exponent	–0.1	-0.1
Channel-belt width (km)	16	10
Standard deviation	0.0	0.0
Channel-belt depth (m)	25	40
Upstream channel-belt aggradation rate (m a^{-1})	0.0045	0.006
Downstream channel-belt aggradation rate (m a^{-1})	0.006	
Channel-belt aggradation exponent		–1.0
Overbank aggradation exponent	0.8	0.3
Mean annual maximum flood (cumecs)	50000	50000
Standard deviation	20000	20000
Avulsion discharge exponent	2.0	2.0
Avulsion slope exponent	2.0	2.0
Avulsion slope proportionality constant	0.3	0.3
Tilt rate at down-valley location (m a^{-1})	0.001	0.0038
Output variables		
Mean channel-deposit proportion	0.86	0.42
Mean floodplain aggradation rate (m a^{-1})	0.0036	0.0076
Mean avulsion period (a)	960	789
Range of avulsion periods (a)	200–2500	100–2300
Mean sandstone-body width (km)	25	16
Mean sandstone-body thickness (m)	36	54
Mean thickness of alluvium (m)	42	94

Avulsion occurred commonly in the Memphis region, and avulsion points occur near Alexandria, Baton Rouge and New Orleans (Fig. 1). During the avulsion, river flow may be diverted into or adjacent to parts of abandoned channels, and a new river course may rejoin the original course further down-valley. Channel belts also commonly occupy floodplain margins. The avulsion period averages 1000–1500 a for the lower Mississippi Valley, but is less on the delta plain. According to Fisk (1952), it takes on the order of a hundred years for a former channel belt to be completely abandoned. However, Saucier (1994) believed that some diversions of the Mississippi during Holocene time may have taken several hundred years, during which period the Mississippi discharge was split between two channel belts. However, accurate dating of the activity times of channel belts and subdeltas is difficult, and improvement in dating is required (Saucier 1994; Tornqvist *et al.* 1996). Tectonic uplifts (e.g. Lake County, Monroe) apparently influence the river course and pattern, but not points of avulsion (Saucier 1994). Such local tectonic affects were not included in these simulations, primarily because of a lack of knowledge of them.

Holocene floodplain and alluvium

The modern floodplain slope decreases down-valley from *c.* 0.00015 (at the entrance to the lower valley) to 0.00004 (near Baton Rouge), but there are local, tectonically induced changes in flood-plain slope (not considered here). The slope of the base of the Holocene valley is somewhat larger than the modern valley slope, and it increases slightly down-valley (Fisk 1944, 1947, 1951, 1952; Saucier 1994). Flood-plain width is variable in the alluvial valley (40–80 km), and increases markedly from the Atchalafayala Basin (70 km) to the Gulf of Mexico. Average floodplain deposition rate in the alluvial valley is a few millimetres per year, but increases down-valley, reaching a maximum on the delta plain (Table 2). Subsidence rates also generally increase down-valley and are on the order of millimetres per year on the delta plain. However, subsidence rates are locally variable. The average thickness of Holocene alluvium increases down the lower Mississippi valley from about 40 m to 45 m. Floodbasin deposits are *c.* 10 m thick at Memphis, 13 m thick at Vicksburg, 20 m at Natchez, and 50 m thick at Baton Rouge (Fisk 1944). Channel-deposit proportion decreases slightly from Cairo to

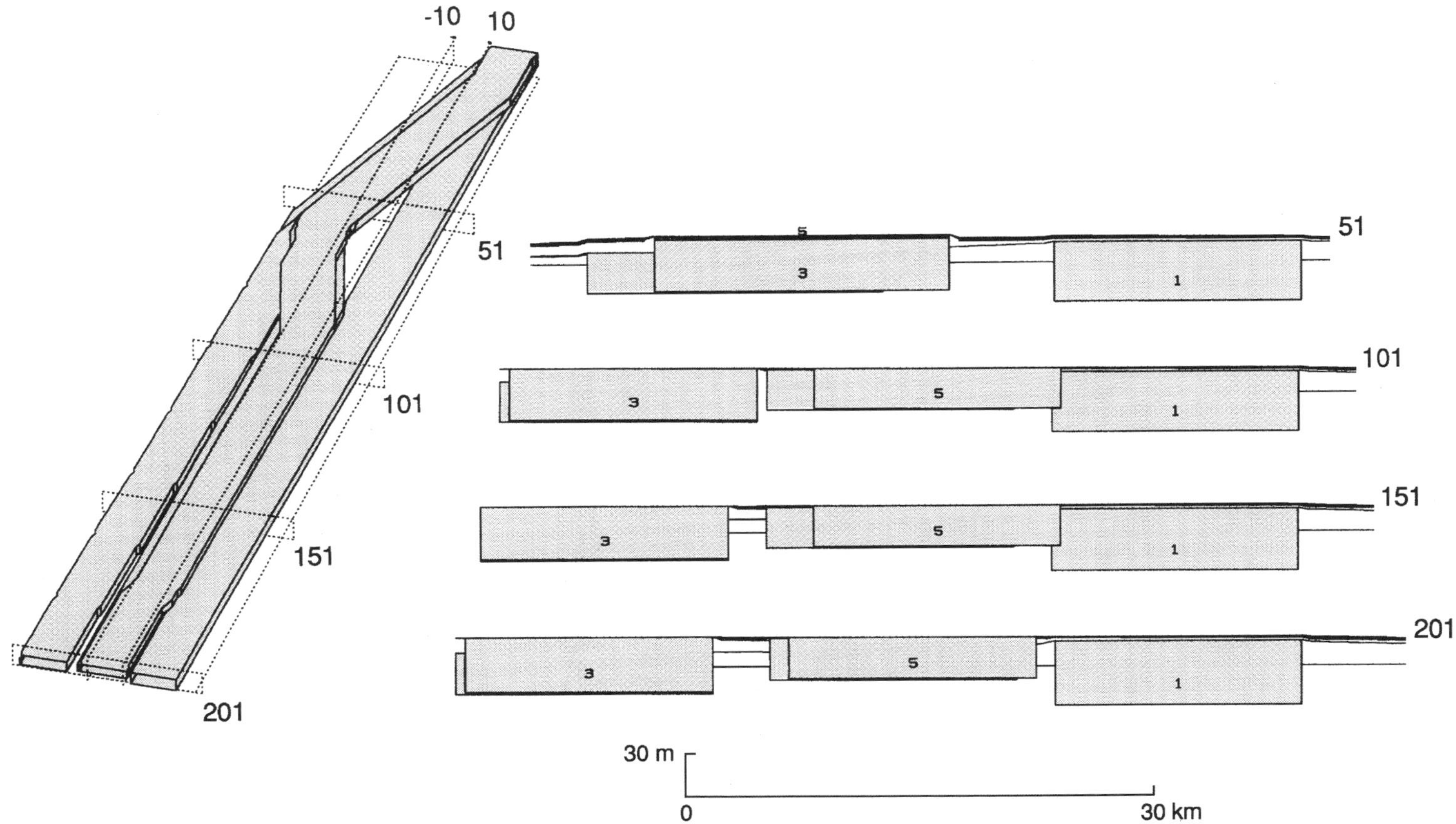

Fig. 2. Simulated alluvial architecture for area comparable with Yazoo Basin, showing 3D perspective diagram and associated cross-sections. Shaded areas are channel belts, numbered in chronological order.

Natchez, and decreases markedly from Natchez to the Gulf of Mexico.

Results and conclusions

The Mackey–Bridge model simulates many, but not all, features of the lower Mississippi valley, and has generated a number of hypotheses to test. The model successfully simulates the observed modes of avulsion and general distribution of channel-belt deposits of the lower Mississippi valley and delta plain during the Holocene relative sea-level rise (Figs 2 and 3). The avulsion points near Memphis are simulated by the model and are possibly due to

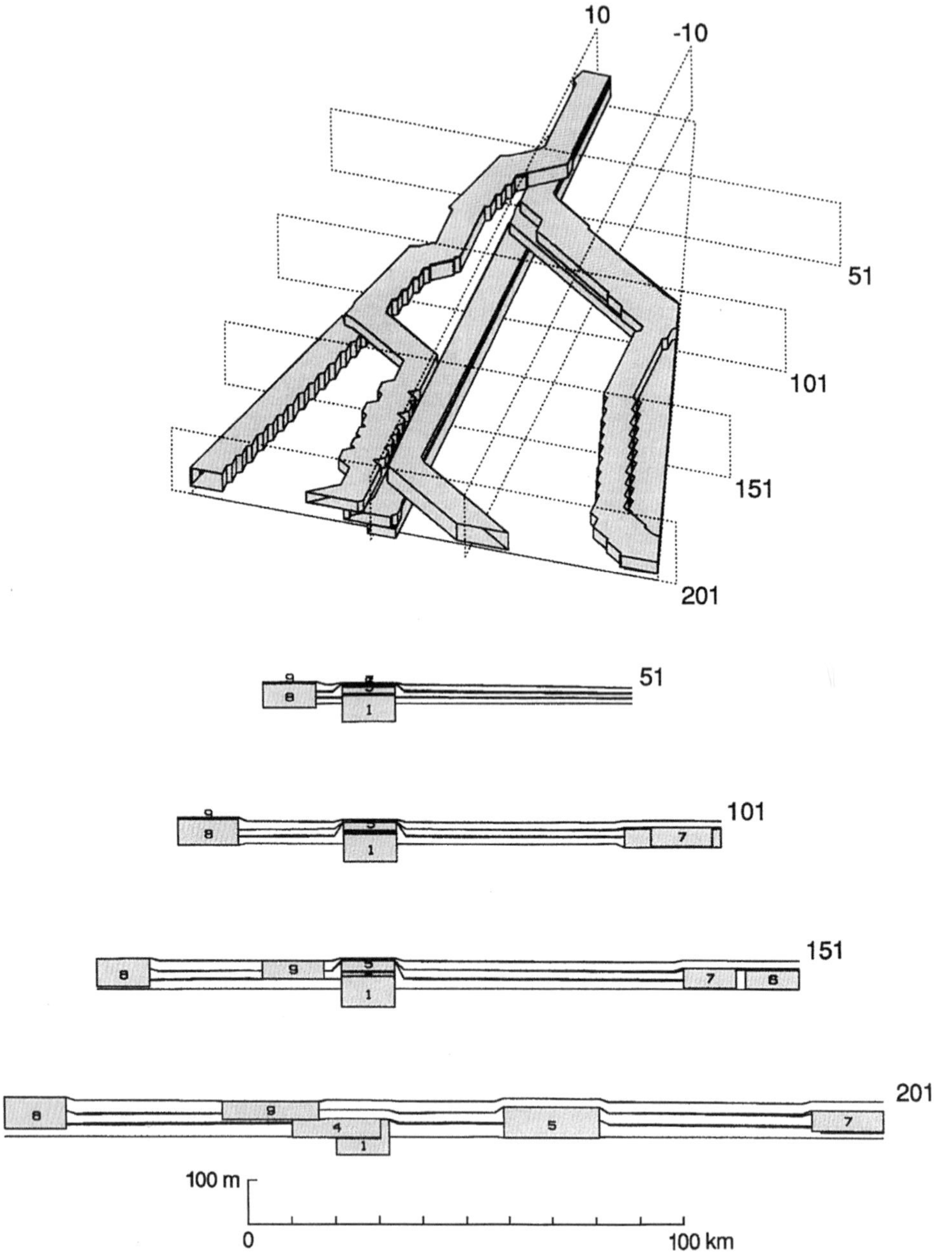

Fig. 3. Simulated alluvial architecture for area comparable with Mississippi delta plain and Atchafalaya Basin.

locally low down-valley slope plus high cross-valley slope as a result of alluvial-ridge deposition. The simulated mean avulsion period of *c.* 1000 a in the lower Mississippi valley is similar to that observed. The high channel-belt proportion along most of the alluvial valley is due primarily to high channel-belt width/floodplain width. Variation in elevation of the base of channel-belt deposits is due to progressive increase in elevation of the channel thalwegs as deposition within the alluvial basin occurs. In reality, the relief of the channel-belt bases will be greater than shown in the simulations, because of the effects of local channel scouring. The simulated average thickness of Holocene alluvium (Table 2) is similar to that observed.

Avulsion periods of decades to centuries near the coast are predicted because the highest deposition rate and lowest down-valley slope occur here. In contrast, avulsion periods on the order of 1000 a are simulated up-valley in the Atchalafaya Basin, because alluvial ridges develop more slowly here. These avulsions are apparently responsible for the development of discrete delta lobes or subdeltas. Channel-deposit proportion is low and decreases down-valley in the Atchalafaya Basin owing primarily to low channel-belt/floodplain width.

Along-valley variation in channel-deposit proportion is strongly controlled by channel-belt width/floodplain width, deposition rate and avulsion frequency, and differential subsidence. On the delta plain at least, the last of these controls is intimately related to rising base level, and subsidence and deposition rates that increase towards the Gulf of Mexico. Avulsions of channels within delta lobes, and diversion of the Mississippi to form a new delta lobe, can be explained in terms of spatial and temporal development of alluvial ridges. New delta lobes are initiated when relief on alluvial ridges is sufficient to initiate an avulsion in an up-valley part of the delta plain. These results have important implications for sequence-stratigraphic models (e.g. Shanley & McCabe 1993, 1994; Wright & Marriott 1993; Blum 1994; Aitken & Flint 1995). If an alluvial sequence-stratigraphic model does not deal with the way in which climate, eustatic sea level, and local and regional tectonism influence temporal and spatial changes in aggradation or degradation rate, avulsion frequency and location, and channel-belt dimensions relative to floodplain dimensions, it is likely to be over simplified.

To put the results of this study in perspective, it must be appreciated that the Mackey–Bridge model is only an initial approach. There are some specific deficiencies that need to be addressed in future. For example, the model is not capable of simulating diversion of channels into pre-existing channels, an important influence on the location of avulsing channels (Fisk 1952; Coleman 1969; Wells & Dorr 1987; Smith *et al.* 1998). Also, the gradual transfer of discharge into a new channel belt, and the gradual reduction in discharge and deposition rate in abandoned channel belts, cannot be simulated at present. The model does not simulate gradual degradation of the channel and floodplain (valley incision and terrace formation). It is particularly desirable to be able to simulate the effects on alluvial architecture of cyclic variations in aggradation and degradation related to base-level and shoreline changes, so as to comprehensively assess the predictions of extant sequence-stratigraphic models. Finally, future models need to be based more on physical principles than on empiricism, but we are a long way from this ideal.

References

AITKEN, J. F. & FLINT, S. S. 1995. The application of high-resolution sequence stratigraphy to fluvial systems: a case study from the Upper Carboniferous Breathitt Group, eastern Kentucky, USA. *Sedimentology*, **42**, 3–30.

AUTIN, W. J., BURNS, S. F., MILLER, R. T., SAUCIER, R. T. & SNEAD, J. I. 1991. Quaternary Geology of the lower Mississippi valley. *In*: MORRISON, R. B. (ed.) *Quaternary Nonglacial Geology: Conterminous US.* The Geology of North America, Geological Society of America, **K-2**, 547–582.

BLUM, M. D. 1994. Genesis and architecture of incised valley fill sequences: a late Quaternary example from the Colorado River, Gulf Coastal Plain of Texas. *In*: WEIMER, P. & POSAMENTIER, H. W. (eds) *Siliciclastic Sequence Stratigraphy*. Memoir, American Association of Petroleum Geologists, **58**, 259–283.

BRIDGE, J. S. & LEEDER, M. R. 1979. A simulation model of alluvial stratigraphy. *Sedimentology*, **26**, 617–644.

—— & MACKEY, S. D. 1993. A revised alluvial stratigraphy model. *In*: MARZO, M. & PUIDEFABREGAS, C. (eds) *Alluvial Sedimentation*. International Association of Sedimentologists, Special Publications, **17**, 319–337.

BRYANT, M., FALK, P. & PAOLA, C. 1995. Experimental study of avulsion frequency and rate of deposition. *Geology*, **23**, 365–369.

COLEMAN, J. M. 1969. Brahmaputra River: channel processes and sedimentation. *Sedimentary Geology*, **3**, 129–239.

FISK, H. N. 1944. *Geological Investigation of the Alluvial Valley of the Lower Mississippi River*. US Army Corps of Engineers, Mississippi River Commission, Vicksburg, MS.

—— 1947. *Fine Grained Alluvial Deposits and their Effects on Mississippi River Activity*. US Army Corps of Engineers, Mississippi River Commission, Vicksburg, MS.

—— 1951. Mississippi River valley geology related to river regime. *American Society of Civil Engineers Transactions*, **117**, 667–682.

—— 1952. *Geological Investigation of the Atchafalaya Basin and the Problem of Mississippi River Diversion.* Mississippi River Commission, Waterways Experiment Station, Vicksburg, MS.

KHAN, I. A., BRIDGE, J. S., KAPPELMAN, J. & WILSON, R. 1997. Evolution of Miocene fluvial environments, eastern Potwar plateau, northern Pakistan. *Sedimentology*, **44**, 221–251.

MACKEY, S. D. & BRIDGE, J. S. 1995. Three-dimensional model of alluvial stratigraphy: theory and application. *Journal of Sedimentary Research*, **B65**, 7-31.

ROSS, W. C. 1990. Modeling base-level dynamics as a control on basin-fill geometries and facies distribution: a conceptual framework. *In*: CROSS, T. A. (ed.) *Quantitative Dynamic Stratigraphy*. Prentice–Hall, Englewood Cliffs, NJ, 387–399.

SAUCIER, R. T. 1994. *Geomorphology and Quaternary Geologic History of the Lower Mississippi Valley.* Mississippi River Commission, Vicksburg, MS.

SHANLEY, K. W. & MCCABE, P. J. 1993. Alluvial architecture in a sequence stratigraphic framework: a case history from the Upper Cretaceous of southern Utah, USA. *In*: FLINT, S. & BRYANT, I. D. (eds) *The Geological Modeling of Hydrocarbon Reservoirs.* International Association of Sedimentologists, Special Publications, **15**, 21–56.

—— & —— 1994. Perspectives on the sequence stratigraphy of continental strata. *Bulletin, American Association of Petroleum Geologists*, **78**, 544–568.

SLINGERLAND, R. & SMITH, N. D. 1998. Necessary conditions for meandering-river avulsion. *Geology*, **26**, 435–438.

SMITH, N. D., SLINGERLAND, R. L., PEREZ-ARLUCEA, M. & MOROZOVA, G. S. 1998. The 1870s avulsion of the Saskatchewan River. *Canadian Journal of Earth Sciences*, **35**, 453–466.

TÖRNQVIST, T. E. 1994. Middle and late Holocene avulsion history of the River Rhine (Rhine–Meuse Delta, Netherlands). *Geology*, **22**, 711–714.

——, KIDDER, T. R., AUTIN, W. J. *ET AL.* 1996. A revised chronology of Mississippi River subdeltas. *Science*, **273**, 1693–1696.

WELLS, N. A. & DORR, J. A. 1987. Shifting of the Kosi River, northern India. *Geology*, **15**, 204–207.

WRIGHT, V. P. & MARRIOTT, S. B. 1993. The sequence stratigraphy of fluvial depositional systems: the role of floodplain sediment storage. *Sedimentary Geology*, **86**, 203–210.

Assessing flood duration gradients and fine-scale environmental change on ancient floodplains

V. P. WRIGHT

Department of Earth Sciences, Cardiff University, Cardiff CF1 3YE, UK; and BG Exploration and Production, 100 Thames Valley Park Drive, Reading RG6 1PT, UK

Abstract: Existing approaches to studying ancient floodplain deposits involve integrating lithofacies and palaeopedological data to identify medium- to long-term controls on deposition. However, deciphering short-term, annual or even seasonal events such as flooding duration and hydroperiod is rarely attempted, yet it is these that control the ecology of floodplains. An approach is presented that provides insights into the patterns of flood duration in ancient floodplain successions. The ability to identify criteria indicating frequent wetting and drying is critical and this is most easily done in successions with carbonate-dominated sediments (palustrine, hard-water systems) or where Vertisol palaeosols are present. Some ancient floodplain deposits contain metre-scale units displaying drying-out trends that reflect aggradation of the floodplain surface. Evidence of hydromorphism and Fe reduction and translocation is also critical for identifying hydromorphism related to flooding, but recognizing such effects in pre-Devonian fluvial successions will be difficult because of the low levels of organic matter in soils of that interval.

In recent years much research has been conducted on ancient floodplain and flood basin deposits, where previously almost all research was focused on the coarser, channel deposits. Of critical importance to this new work was the appreciation of the abundance and usefulness of palaeosols (fossil soils) in such successions. Allen (1974) was one of the first researchers to recognize the potential of integrating lithofacies and palaeosols to assess the broad-scale controls on floodplain deposition. Other studies followed (Retallack 1986; Bown & Kraus 1987; Kraus 1987, 1996; Platt & Keller 1992; Kraus & Aslan 1993; Marriott & Wright 1993; Wright & Marriott 1993; Willis & Behrensmeyer 1994; Bestland 1997; Kraus & Gwinn 1997), which elaborated palaeosol relationships on ancient floodplains and led to the development of concepts such as pedofacies (Bown & Kraus 1987; Kraus 1997), and the identification of avulsion belt deposits (Kraus & Aslan 1993; Kraus 1996) and of catenary relationships (e.g Platt & Keller 1992).

These studies have commonly concentrated on evaluating the likely controls on soil development over medium- to long-term intervals, that is, over many thousands of years or longer. Deciphering short-term changes on a floodplain will be difficult, as such processes may not leave a preservable signature in the sediments; low deposition rates lead to time averaging of events. In this paper, examples are presented of small (metre-scale) units in ancient floodplain successions, which are most readily interpreted as the result of variations in flooding duration. They are crudely analogous to the small-scale 'parasequences' that compose large parts of the record of marine carbonate deposition. It is not suggested that such units are as important in the record of floodplain deposits but this paper may lead others to identify similar units and to assess whether they reflect intrinsic processes, such as floodplain aggradation, or extrinsic factors such as short-term variations in flood magnitude. It is suggested that the highly developed nature of these units might indicate the former presence of seasonal floodplains, which are widespread in tropical and subtropical regions at the present time but have a very poor geological history. To illustrate the possible importance of these floodplain units two types will be reviewed, one based on well-documented palustrine limestones, and the other on a recently identified non-carbonate example. Before presenting these examples some background concepts will be presented, followed by a review of criteria for the recognition of variability in flood duration on ancient floodplains.

From: MARRIOTT, S. B. & ALEXANDER, J. (eds) 1999. *Floodplains: Interdisciplinary Approaches*. Geological Society, London, Special Publications, **163**, 279–287. 1-86239-050-9/99/$15.00

Floodplain palaeosols

The term floodplain is used here in the sense of Junk & Welcomme (1990) to refer to 'areas of low lying land that are subject to inundation by the lateral overflow of waters from rivers or lakes with which they are associated'. Variations in drainage conditions on ancient floodplains have been identified by the presence of well-drained and poorly drained palaeosols, and this has led to the recognition of palaeo-catenas. Catenas are groups of soils that developed under similar climatic conditions and on similar parent materials, but differ because of local drainage conditions, related to relief. On floodplains, back-swamps and cut-offs are typically the sites of poor drainage conditions, whereas the alluvial ridges (levees) have more free-draining soils. Similar associations have been recognized in the geological record, with the different palaeosols occurring in different lithofacies associations, reflecting the different geomorphological settings (e.g. Platt & Keller 1992). Such palaeosols are reflecting differences not in flood frequency but in what might be regarded as perennial drainage situations. Similarly, the associations of poorly- and well-drained palaeosols noted by Kraus & Aslan (1993) and Kraus (1996) in Eocene fluvial successions in Wyoming have been related to periods of avulsion belt development and 'normal' floodplain accretion, respectively. In both catenary associations and avulsion belt–floodplain alternations, the changes in soil–palaeosol types are associated with facies changes and different environments on the floodplain. The types of drainage changes discussed in this paper occur in units without such facies changes and are regarded as the result of changes in flood duration taking place on a local scale.

Flood duration

To understand the specific environmental conditions on an ancient floodplain it would be necessary to be able to assess the specific flooding regime for that site during deposition. The annual pattern of flood inundation of a wetland such as a floodplain is referred to as the hydroperiod (Duever 1990). Flooding may be permanent, as in a perennial pond, or the wetland may be seasonally or intermittently flooded to intermittently exposed (Mitsch & Gosselink 1993). Identifying the hydroperiod for an ancient floodplain may not be possible; indeed, even recognizing clear evidence of wetting and drying is not simple. Within a floodplain the length of time a site is flooded will also depend on the local topography. Figure 1 shows a hypothetical hydrograph for a seasonal floodplain, with three sites. For the first (A), the flood duration is 6 months, but it is 4 months for site B and only 2 months for site C. An analogy can be drawn with tidal flat settings, where the hydroperiod is typically semi-diurnal. In low-lying areas below the low tide mark the tidal flood duration is permanent, but in areas above normal high tide and affected only by spring tides or storm surges, flooding is intermittent. The intermediate sites experience a gradation between these two extremes. These different flooding and exposure regimes can be quantified to produce an exposure index for a tidal flat (Ginsburg *et al.* 1977) and as the different degrees of exposure lead to the formation of a variety of sedimentary structures, it has been possible to assign ranges for exposure time to different features. Thus it is possible to identify flooding regimes in ancient tidal flat sediments, but no such system exists for floodplains, although Platt & Wright (1992) offered a crude system for palustrine wetlands.

Whereas many modern floodplains have flood durations that are relatively short, many large floodplains experience extended periods of flooding on a regular, seasonal basis. This is particularly true of higher-order rivers, e.g fifth order, which can be influenced by seasonal rainfall patterns in large watersheds (Mitsch & Gosselink 1993). Such rivers can experience flooding for periods of 2–6 months a year (Welcomme, in Duever 1990). In tropical and subtropical regions today there are huge areas that experience seasonal flooding with prolonged durations separated by dry periods. Such areas include the Everglades of Florida, the Llanos of Venezuela, the Igapo and Varzea flooded forests of the Amazon, the Pantanal

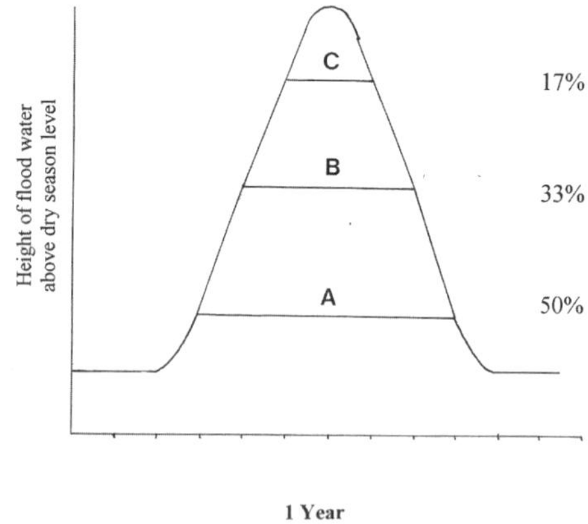

Fig. 1. Effect of small-scale topographic differences on the hydroperiods of three sites (A, B and C) on a hypothetical floodplain, with exposure percentages for each site.

of Brazil, the Okavango of Botswana and the Kakadu of Australia (Dugan 1993). With a regime fluctuating annually between the extremes of flooding and desiccation, it is interesting to consider what the preservation potential would be of features forming during these extremes, and how would we recognize the sedimentary product of such a system?

Hydrological facies gradients on floodplains: palustrine carbonates

The hydrological gradient created across a floodplain, from low-lying areas with a long flood period, to topographically higher areas with a short flood period, will be reflected in the sediments and soils. Following Walther's Law, this gradient should be reflected in a vertical succession during the aggradation of the depositional surface. Such drying-upwards or drying-out units are well known in many continental Mesozoic and Cenozoic successions in Europe in palustrine carbonates. These are limestones that possess a complex mixture of features indicating different degrees of exposure. They typically occur in metre-scale cyclothems, which pass from mud-grade carbonate with molluscan and charophytic plant remains and cyanobacterial or algal structures, through to limestones with features indicating prolonged degrees of exposure including calcrete crusts and related soil features. They were elegantly described in great detail by Freytet & Plaziat (1982), from upper Cretaceous and lower Tertiary successions in France, where they represent the marginal facies of carbonate-rich (hard-water) lakes. There are now many descriptions in the literature (reviewed by Armenteros *et al.* 1997), and such limestones are known also from fluvial deposits, without any association with permanent lakes, and from coastal plain settings. Platt & Wright (1992) made comparisons between palustrine facies and the interior prairies of the Everglades in Florida. They later (Wright & Platt 1995) expanded their model to suggest that palustrine deposits actually represented seasonal carbonate wetlands and that the cyclothems represented transitions from deposits with very low exposure indices to ones with total exposure. They therefore represent a type of catenary succession, but whereas in a typical catena the different component sites are separate from one another and require some geomorphological change to cause a change of drainage, Wright & Platt (1995) argued that the small cyclothems represented a catenary change caused by a localized change in the flood duration at a single site. They suggested that this may be caused by local changes in the hydroperiod as a result of minor differences in topography or of variations in flood height. To distinguish this type of flood duration catena from one related to 'static' topographic differences, Wright & Platt (1995) suggested the term dynamic catena. Of critical importance in this model is the extensive grainification that takes place in such limestones. This process, recently discussed in detail by Armenteros & Daley (1998), involves frequent wetting and drying cycles that cause fine-grained carbonate sediment to form nodular–brecciated fabrics. These later develop into peloidal fabrics to produce the distinctive secondary grainstones characterizing palustrine limestones. Thus the palustrine cyclothem displays the sequence from subaqueous to subaerial with a transition zone characterized by wetting and drying, which could be most easily explained by the aggradation of the floodplain surface leading to its passage through a range of hydroperiods (Fig. 2).

As such wetland carbonate cyclothems are relatively common in some continental and even marginal marine successions (Armenteros *et al.* 1997), what would similar dynamic catenas look like in non-carbonate successions? On the basis of the approach illustrated in Fig. 2, a model is presented here for a non-carbonate (non-hard-water) system, and a possible example presented from the Lower Cretaceous succession of the Isle of Wight.

Dynamic catena model for a non-hard-water floodplain

In distilling a model for such a system a key component would be to identify a preservable feature that indicates wetting and drying cycles. Vertic soil structures are examples of such features that are common in the geological record (e.g Allen 1986; Gustavson 1991, and references therein), and occur in Vertisols, which develop on parent materials that have a high shrink–swell potential (typically rich in smectite) and experience a marked (typically seasonal) contrast in soil moisture (Wilding & Tessier 1988; Coulombe *et al.* 1996). The soils shrink in the dry season and expand in the wet season. Differential stresses create deformation features such as slickensided, curved slip surfaces defining 'pseudo-anticlines', separating distinctive soil peds. The wetting and drying cycles can also create granular fabrics similar to those found in palustrine limestones, but these have a low preservation potential in the soil, typically disappearing as a result of compaction after only shallow burial in sediment (Rust & Nanson 1989). Pseudo-anticlines are, however, widely recognized in ancient floodplain deposits and are included in the model (Fig. 3). This does not imply that all

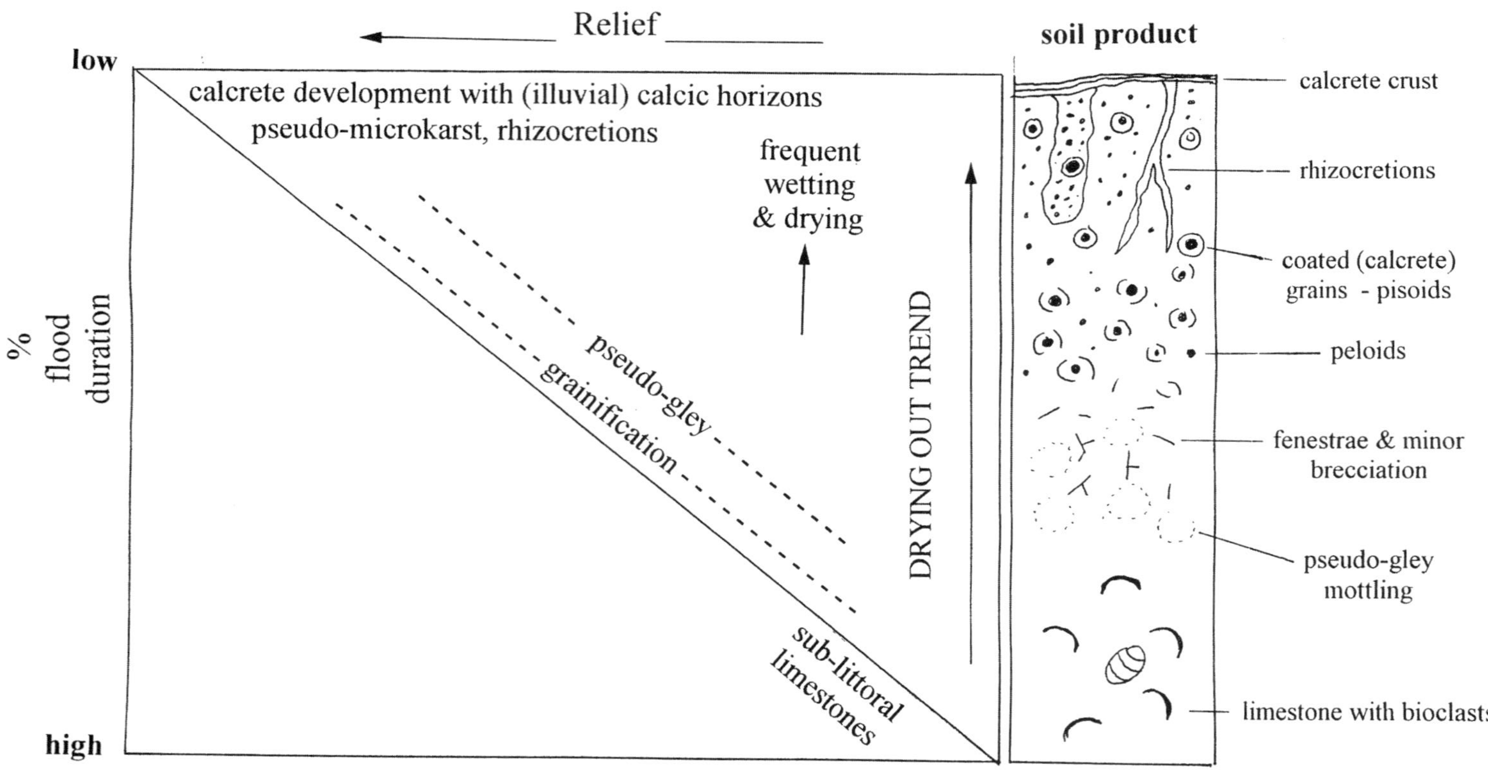

Fig. 2. Sedimentary features in palustrine carbonates produced during drying out, caused by a decrease in flooding frequency as the floodplain surface aggraded (based on model developed by Platt & Wright (1992) and Wright & Platt (1995)). Grainification, which causes textural inversion of matrix-rich carbonates into peloidal, grain-supported textures, is produced by numerous wetting and drying cycles. Metre-scale units such as this are common in many Mesozoic and Cenozoic continental successions in Europe A list of case studies has been provided by Armenteros *et al.*, (1997).

occurrences of vertic features in the stratigraphic record indicate seasonal flood regimes, but suggests that these features, with those discussed below, could signal the sites of such deposystems.

Another key component is the presence of flood-related hydromorphism in the sediments or palaeosols, specifically the presence of pseudo-gley or surface-water gley features. These will be represented by distinctive mottling patterns with areas of reduced colour set within a mainly non-reduced groundmass (Van Breeman 1988). Recently, PiPujol & Buurman (1994, 1997) have provided detailed discussions of surface-water gley soils and palaeosols, and of criteria to recognize hydrological processes in mottled palaeosols. Soils that are subject to alternating prolonged desiccation and hydromorphism, such as those of wet monsoonal regions, can undergo a process known as ferrolysis. This involves the degradation and formation of Al-hydroxyl compounds, which become incorporated into clay lattices to produce pedogenic chlorites (Brinkman 1977).

At the lowest point in the catenary setting the soils will have the characteristics of gleys, with low chroma values reflecting the predominance of reducing conditions, although the saturated interval may have been less than 50% of the annual cycle (Daniels *et al.* 1971). The least flooded areas will be characterized by well-drained soils, which may show evidence of eluvial–illuvial horizons.

Soil mineralogy may also reflect the catenary relationships (Fig. 3). In soils with a long duration of flooding reducing conditions may result in the formation, in suitable conditions, of siderite (Duchaufour 1982). Hematite and goethite may occur in the higher parts of the catenas, with the latter in soils that have a higher moisture content (Schwertmann 1988).

An example of a metre-scale unit that could represent deposition in this type of catena is shown in Fig. 4, from the Lower Cretaceous 'Wealden' succession of the Isle of Wight (Wright *et al.* 2000). These occur in thick, floodplain successions associated with high-sinuosity river channel sands (Stewart 1981). Fine-grained lithofacies include prominent clay plugs (Stewart 1981), and muddy, point-bar deposits (Stewart 1983). The floodplain units contain numerous thin sheet sandstones (crevasse deposits) and a range of palaeosol types (Insole & Hutt 1994; Wright *et al.* 2000).

The author is not aware of any other examples of such gley-to-pseudo-gley-to-Vertisol units in the literature. This raises the question as to why the examples from hard-water systems appear to be more common? This may reflect a preservation potential difference. Features of non-carbonate systems may be prone to overprinting, and mottle patterns may be lost by later wetting or drying phases. The carbonate features clearly have a higher preservation potential as shown by their widespread occurrence. The non-hard-water model offered above requires the presence of vertic features to indicate wetting and drying phases, but these will not develop in parent materials lacking clays with a high shrink–swell potential.

A further complication relates to the recognition of hydromorphic features in some of the most ancient floodplain deposits. For gley and pseudo-gley mottle patterns to develop, reduction, leading to iron mobilization and colour changes, must take place. This reduction is a result of depletion in oxygen caused by microbial decay of organic matter. If the organic matter content is low, even under hydromorphic conditions, the degree of iron reduction might be low. Pre-mid-Devonian soils are likely to have had a low organic content (Algeo & Scheckler 1998), and as a result it may be difficult to detect former hydromorphic conditions in palaeosols older than that interval. Vertic palaeosols are widespread in the Siluro-Devonian Old Red Sandstone fluvial successions of southern Britain, yet despite the evidence for wetting and drying cycles there are no records of gleyed or pseudo-gleyed palaeosol profiles (Allen 1986). Their absence could simply reflect a prevailing semi-arid climate with topographically uniform floodplains or diagenetic overprinting, which has removed ferrous iron, or perhaps the original soils lacked enough organic matter to trigger iron reduction.

Origins of dynamic catenary units

If these small-scale units are present in ancient floodplain successions, an explanation must be offered as to why they occur. Two possibilities exist. First, the drying-out units represent the filling of topographic lows on the floodplain. Such lows will retain flood waters and may also be influenced by ground-water levels. Such depressions tend to be sites of higher sedimentation rates (Walling & He 1998) because they pond flood water at greater depths than surrounding areas and so trap larger amounts of suspended sediment. As the sediment surface aggrades it will pass through a decreasing frequency of flooding related to the local hydro-period. On a larger scale, the whole floodplain might aggrade, so producing a drying-out unit. If a floodplain and levee complex aggrade without any rise in the river flood height, the whole system will undergo a reduced flood frequency. In both situations the shallowing of flood-water depth will both reduce the deposition rate from suspended sediment and produce a drying-out trend. Hence the unit will show an increase in the degree of soil development upwards towards a level at which

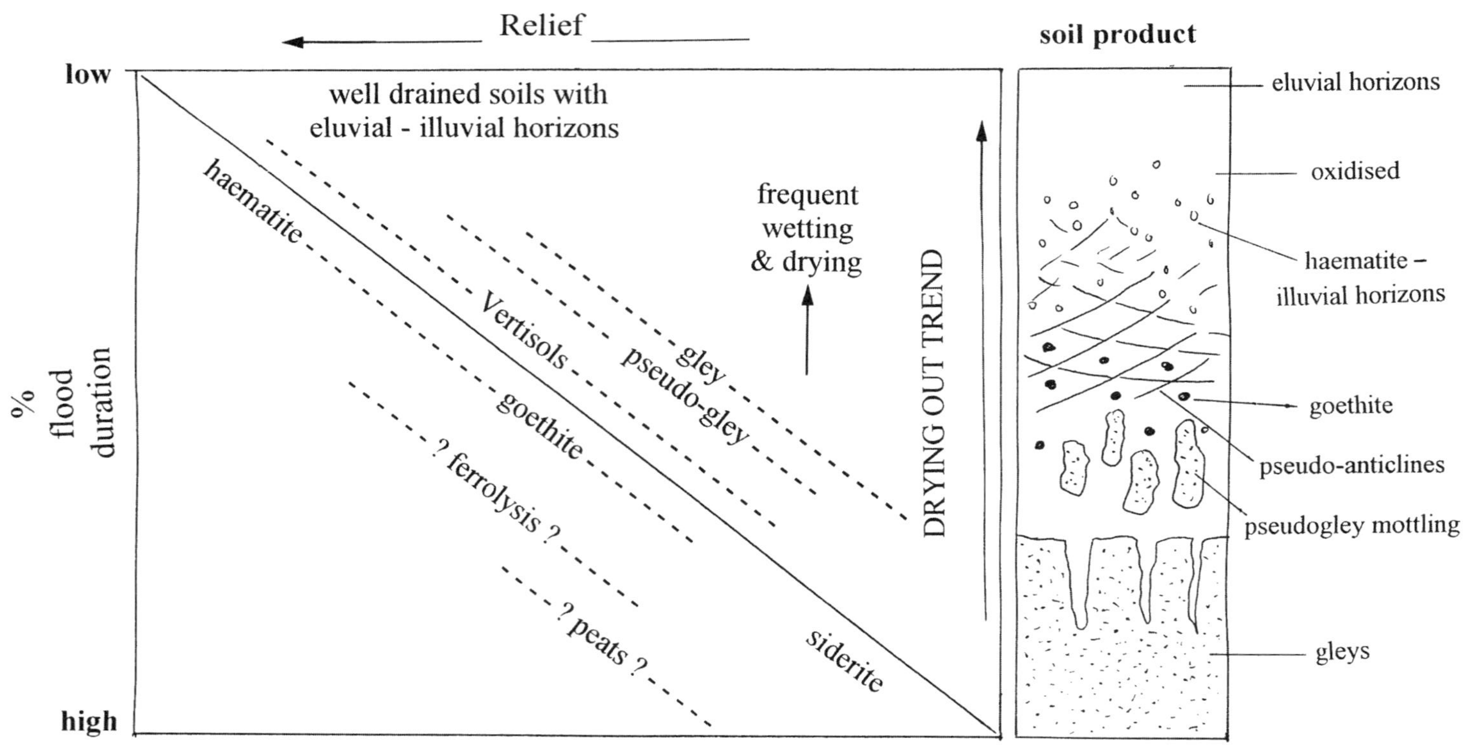

Fig. 3. Prediction of the distributions of key features related to a non-carbonate floodplain wetland where the clays have a high shrink–swell potential (typically smectitic). The drying-out trend reflects the aggradation of the floodplain surface. Evidence of numerous wetting and drying cycles is reflected in the development of vertic features such as pseudo-anticlines and pseudo-gleying.

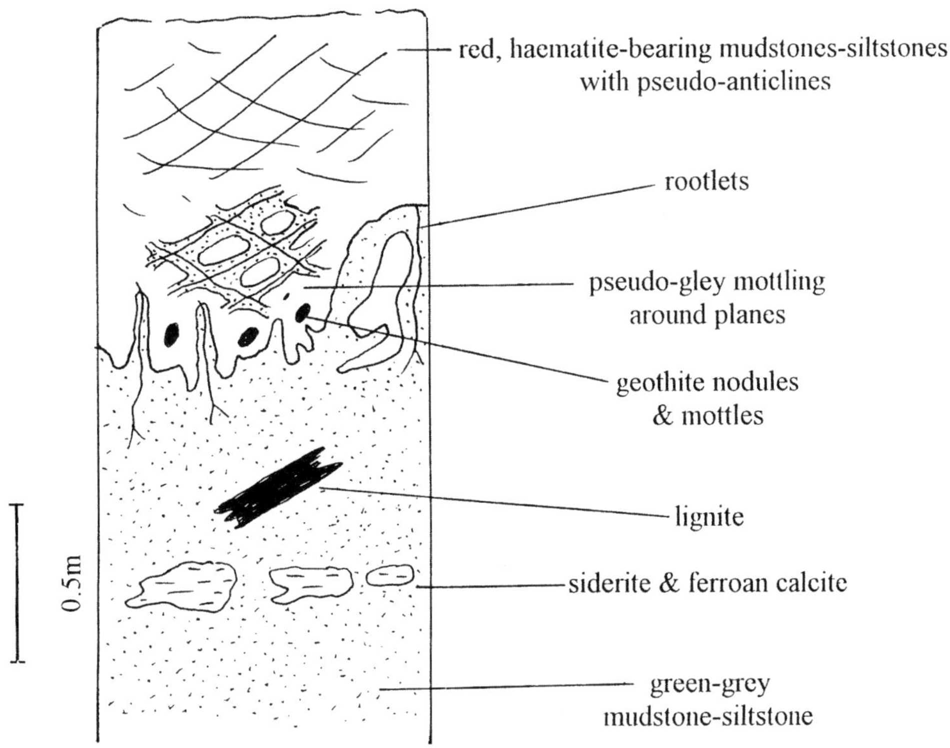

Fig. 4. Example of drying-out unit from the Wealden succession of the Isle of Wight, from 'plant debris bed' at Chilton Chine. See Insole & Hutt (1994) and Wright *et al.* (2000) for details.

aggradation effectively ceases (the regime level of aggradation of Hayward (1985)), the degree of soil development being the inverse of the rate of sedimentation (Marriott & Wright 1993). Thus the prominent soil caps to palustrine cyclothems may indicate decreased deposition rates caused by a decrease in flooding as well as more prolonged exposure. However, the appearance of more strongly developed soil features at the 'drier' tops of these cyclothems might also reflect the fact that hydromorphic conditions retard soil development (Kraus 1997). In summary, it is suggested that ancient floodplains can show small-scale drying-out units, which reflect aggradation in local topographic lows and across larger areas.

Conclusions

Studies of floodplain palaeosols have concentrated on larger-scale relationships but the types of soils that develop on floodplains reflect finer-scale differences in topography and in flood patterns. This is especially true of seasonal floodplain wetlands, which are characteristic of larger river systems. Gradients in flooding frequency occur on floodplains related to topography and flood height, and in seasonal floodplain wetlands these produce features reflecting the high frequency of wetting and drying. During floodplain aggradation, drying-out units can develop, and examples of metre-scale units in ancient floodplain successions that may be attributable to this effect are presented. These units also display evidence of frequent flood–desiccation cycles implying formation in seasonal floodplain wetland settings. Prolonged flooding is typically manifested in the development of hydromorphic features related to iron reduction triggered by microbial decay of organic matter. In pre-mid-Devonian floodplain sediments it may be difficult to recognize such effects because the organic matter contents of these soils would have been low, lowering the potential for hydromorphic reduction and mobilization of Fe.

This study has benefited from discussions with S. B. Marriott on Old Red Sandstone floodplain palaeosols. I thank the two reviewers, P. Friend and K. Taylor, for their suggestions for improving the text.

References

ALGEO, T. J. & SCHECKLER, S. E. 1998.Terrestrial–marine teleconnections in the Devonian: links between the evolution of land plants, weathering processes, and marine anoxic events. *Philosophical Transactions of the Royal Society of London, Series B*, **353**, 113–130.

ALLEN, J. R. L. 1974. Studies in fluviatile sedimentation: implications of pedogenic carbonate units, Lower Old Red Sandstone, Anglo-Welsh outcrop. *Geological Journal*, **9,** 181–208.

—— 1986. Pedogenic calcretes in the Old Red Sandstone facies (late Silurian–early Carboniferous) of the Anglo-Welsh area, southern Britain. *In*: WRIGHT, V. P. (ed.) *Paleosols: their Recognition and Interpretation.* Blackwell, Oxford, 58–86.

ARMENTEROS, I. & DALEY, B. 1998. Pedogenic modification and structure evolution in palustrine facies as exemplified by the Bembridge Limestone (Late Eocene) of the Isle of Wight, southern England. *Sedimentary Geology*, **119**, 275–295.

——, —— & GARCIA, E. 1997. Lacustrine and palustrine facies in the Bembridge Limestone (late Eocene, Hampshire Basin) of the Isle of Wight, southern England. *Palaeogeography, Palaeoclimatology, Palaeoecology*, **128**, 111–132.

BESTLAND, E. A. 1997. Alluvial terraces and paleosols as indicators of early Oligocene climate change (John Day Formation, Oregon). *Journal of Sedimentary Research*, **67**, 840–855.

BOWN, T. M. & KRAUS, M. J. 1987. Integration of channel and floodplain suites in aggrading fluvial systems. I. Developmental sequence and lateral relations of Lower Eocene alluvial paleosols, Willwood Formation, Bighorn Basin, Wyoming. *Journal of Sedimentary Petrology*, **57**, 587–601.

BRINKMAN, R. 1977. Surface water gley soils in Bangladesh: genesis. *Geoderma*, **17**, 111–144.

COULOMBE, C. E., WILDING, L. P. & DIXON, J. B. 1996. Overview of Vertisols: characteristics and impacts on society. *Advances in Agronomy*, **57**, 289–375.

DANIELS, R. B., GAMBLE, E. E. & NELSON, L. A. 1971. Relationships between soil morphology and water-table levels on a dissected North Carolina coastal plain surface. *Soil Science Society of America, Proceedings*, **35,** 157–175.

DUCHAUFOUR, P. 1982. *Pedology, Pedogenesis and Classification.* Allen & Unwin, London

DUEVER, M. J. 1990. Hydrology. *In*: PATTEN, B. C. (ed.) *Wetlands and Shallow Continental Water Bodies, Vol. 1.* SPB Academic, The Hague, 61–89.

DUGAN, P. 1993. *Wetlands in Danger.* Mitchell Beazley & IUCN, London.

FREYTET, P. & PLAZIAT, J. C. 1982. *Continental Carbonate Sedimentation and Pedogenesis—Late Cretaceous and Early Tertiary of Southern France.* Contributions to Sedimentology, **12**.

GINSBURG, R. N., HARDIE, L. A., BRICKER, O. P., GARRETT, P. & WANLESS, H. R. 1977. Exposure index: a quantitative approach to defining position within a tidal zone. *In*: HARDIE, L. A. (ed.) *Sedimentation on the Modern Carbonate Tidal Flats of Northwest Andros Island, Bahamas.* Johns Hopkins University Press, Baltimore, MD, 7–11.

GUSTAVSON, T. C. 1991. Buried Vertisols in lacustrine facies of the Pliocene Fort Hancock Formation, Hueco Bolson, West Texas, and Chihuahua, Mexico. *Geological Society of America Bulletin*, **103**, 448–460.

HAYWARD, M. 1985. Soil development in Flandrian floodplains: River Severn case-study. *In*: BOARDMAN, J. (ed.) *Soils and Quaternary Landscape Evolution.* Wiley, Chichester, 281–299.

INSOLE, A. N. & HUTT, S. 1994. The palaeoecology of the dinosaurs of the Wessex Formation (Wealden Group), Isle of Wight, Southern England. *Zoological Journal of the Linnaean Society*, **112,** 197–215.

JUNK, W. J. & WELCOMME, R. L. 1990. Floodplains. *In*: PATTEN, B. C. (ed.) *Wetlands and Shallow Continental Water Bodies, Vol. 1.* SPB Academic, The Hague, 491–524.

KRAUS, M. J. 1987. Integration of channel and floodplain suites II. Lateral relations of alluvial paleosols. *Journal of Sedimentary Petrology*, **57,** 602–612.

—— 1996. Avulsion deposits in Lower Eocene alluvial rocks, Bighorn Basin, Wyoming. *Journal of Sedimentary Research,* **66**, 354–363.

—— 1997. Lower Eocene alluvial paleosols: pedogenic development, stratigraphic relationships, and paleosol/landscape associations. *Palaeogeography, Palaeoclimatology, Palaeoecology*, **129**, 387–406.

—— & ASLAN, A. 1993. Eocene hydromorphic paleosols: significance for interpreting ancient floodplain processes. *Journal of Sedimentary Petrology*, **63,** 453–463.

—— & GWINN, B. 1997. Facies and facies architecture of Paleogene floodplain deposits, Willwood Formation, Bighorn Basin, Wyoming, USA. *Sedimentary Geology,* **114,** 33–54.

MARRIOTT, S. B. & WRIGHT, V. P. 1993. Palaeosols as indicators of geomorphic stability in two Old Red Sandstone alluvial suites, South Wales. *Journal of the Geological Society of London*, **150,** 1109–1120.

MITSCH, W. J. & GOSSELINK, J. G. 1993. *Wetlands.* Van Nostrand, New York.

PIPUJOL, M. D. & BUURMAN, P. 1994. The distinction between ground-water and surface-water gley phenomena in Tertiary paleosols of the Ebro Basin, NE Spain. *Palaeogeography, Palaeoclimatology, Palaeoecology*, **110**, 103–113.

—— & —— 1997. Dynamics of iron and calcium carbonate redistribution and palaeohydrology in middle Eocene alluvial paleosols of the southeast Ebro Basin margin (Catalonia, northeast Spain). *Palaeogeography, Palaeoclimatology, Palaeoecology*, **134,** 87–107.

PLATT, N. H. & KELLER, B. 1992. Distal alluvial deposits in a foreland basin setting—the Lower Freshwater Molasse (Lower Miocene), Switzerland. *Sedimentology*, **39**, 545–565.

—— & WRIGHT, V. P. 1992. The palustrine problem and the Everglades: towards an exposure index for non-marine systems. *Journal of Sedimentary Petrology*, **62**, 1058–1071.

RETALLACK, G. J. 1986. Fossil soils as grounds for interpreting long term controls on ancient rivers. *Journal of Sedimentary Petrology*, **56,** 1–18.

RUST, B. R. & NANSON, G. C. 1989. Bedload transport of

mud as pedogenic aggregates in modern and ancient rivers. *Sedimentology*, **36**, 291–306.

SCHWERTMANN, U. 1988. Occurrence and formation of iron oxides in various pedoenvironments. *In*: STUCKI, J. W., GOODMAN, B. A. & SCHWERTMANN, U. (eds) *Iron in Soils and Clays*. NATO ASI Series. D. Reidel, Dordrecht, 267–308.

STEWART, D. J. 1981. A meander-belt sandstone of the Lower Cretaceous of southern England. *Sedimentology*, **28**, 1–20.

—— 1983. Possible suspended-load channel deposits from the Wealden Group (lower Cretaceous) of southern England. *In*: COLLINSON, J. D. & LEWIN, J. (eds) *Modern and Ancient Fluvial Systems*. International Association of Sedimentologists, Special Publications, **6**, 369–384.

VAN BREEMAN, N. 1988. Long term chemical, mineralogical and morphological effects of iron-redox processes in periodically flooded soils. *In*: STUCKI, J. W., GOODMAN, B. A. & SCHWERTMANN, U. (eds) *Iron in Soils and Clays*. NATO ASI Series. D. Reidel, Dordrecht, 881–823.

WALLING, D. E. & HE, Q. 1998. The spatial variability of overbank sedimentation on river floodplains. *Geomorphology*, **24**, 209–223.

WILDING, L. P. & TESSIER, D. 1988. Genesis of Vertisols: shrink–swell phenomena. *In*: WILDING, L. P. & PUENTES, R. (eds) *Vertisols: their Distribution, Properties, Classification and Management*. Texas A&M Technical Monograph, **18**, 55–81.

WILLIS, B. J. & BEHRENSMEYER, A. K. 1994. Architecture of Miocene overbank deposits in northern Pakistan. *Journal of Sedimentary Research*, **B64**, 60–67.

WRIGHT, V. P. & MARRIOTT, S. B. 1993. The sequence stratigraphy of fluvial depositional systems: the role of floodplain storage. *Sedimentary Geology*, **86**, 203–210.

—— & PLATT, N. H. 1995. Seasonal wetland carbonate sequences and dynamic catenas: a re-appraisal of palustrine limestones. *Sedimentary Geology*, **99**, 65–71.

——, TAYLOR, K. G. & BECK, V. H. 2000. The paleohydrology of Lower Cretaceous, seasonal wetlands, Isle of Wight, southern England. *Journal of Sedimentary Research*, in press.

Floodplain palaeosols of the Cenomanian Dunvegan Formation, Alberta and British Columbia, Canada: Micromorphology, pedogenic processes and palaeoenvironmental implications

PAUL J. McCARTHY[1,2] & A. GUY PLINT[1]

[1]*Department of Earth Sciences, University of Western Ontario, London, Ont. N6A 5B7, Canada*

[2]*Present Address: Department of Geology and Geophysics, and Geophysical Institute, University of Alaska, Fairbanks, AK 99775-5780, USA*

(*e-mail: mccarthy@gi.alaska.edu*)

Abstract: Floodplains represent both sedimentological and pedological entities that record a unique blend of potential environmental indicators in the rock record. Because floodplains develop in aggradational environments with a high preservation potential, they preserve a detailed record of the local pedosedimentary history. Assemblages of micromorphological features present in floodplain palaeosols record, with a high degree of temporal and spatial resolution, specific pedological processes that operated under particular environmental conditions. Detailed micromorphological analyses of floodplain deposits, in conjunction with their regional distribution, are required for accurate palaeoenvironmental reconstruction of these environments.

Micromorphological features are described from floodplain palaeosols in the Upper Cretaceous (Cenomanian) Dunvegan Formation in northwestern Alberta and northeastern British Columbia, Canada. The palaeosols developed on low-lying and generally poorly drained floodplains during a time when the tectonic subsidence rate was very low. High-frequency fluctuations in base level were superimposed upon this tectonic signature, resulting in at least seven phases of valley cutting and the corresponding development of mature interfluve palaeosols. The major pedogenic processes operating were clay illuviation, redox processes, precipitation of soluble salts (primarily $FeCO_3$) and pedoturbation. Palaeopedological features in the Dunvegan Formation are consistent with features developed in modern cool temperate soils with a non-seasonal precipitation distribution, such as Brunisols and Luvisols. The presence of embedded grain argillans, compound silt–clay coatings and rounded aggregates suggests at least the possibility of limited frost penetration in inland soils. Floodplains of the mid-Cretaceous Dunvegan Formation (palaeolatidude *c.* 65°N) record evidence for cool temperate conditions that supports other palaeobotanical and sedimentological evidence for cool Cretaceous palaeoclimates at high latitudes.

Floodplains represent unique environments in which both sedimentary and pedogenic processes are preserved in the pedosedimentary record. Floodplain palaeosols have a high preservation potential owing to their development in an aggradational setting. Palaeosols are now widely recognized in ancient sedimentary successions, and they have proved useful for palaeoclimatic and palaeogeomorphological reconstruction (Bown & Kraus 1987; Fastovsky & McSweeney 1987; Besley & Fielding 1989; Leckie *et al.* 1989; McCarthy & Leckie 1991; McCarthy *et al.* 1997*a*, *b*; McCarthy & Plint 1998).

Palaeoenvironmental studies of modern and ancient floodplain successions have tended to focus on macroscopic field observations and geochemical and mineralogical trends (Mack 1992; Kraus & Aslan 1993; Aslan & Autin 1998). These studies suggest that floodplain settings are best suited for interpreting depositional and hydrological histories of ancient floodplains, or that hydromorphic palaeosols are not good palaeoenvironmental indicators (Mack 1992). However, because floodplain palaeosols develop via complicated interactions between sedimentation and pedogenesis (Marriott & Wright 1993; McCarthy *et al.* 1997*b*; Aslan & Autin 1998), their morphologies commonly reflect at least partial records of the

From: MARRIOTT, S. B. & ALEXANDER, J. (eds) 1999. *Floodplains: Interdisciplinary Approaches*. Geological Society, London, Special Publications, **163**, 289–310. 1-86239-050-9/99/$15.00

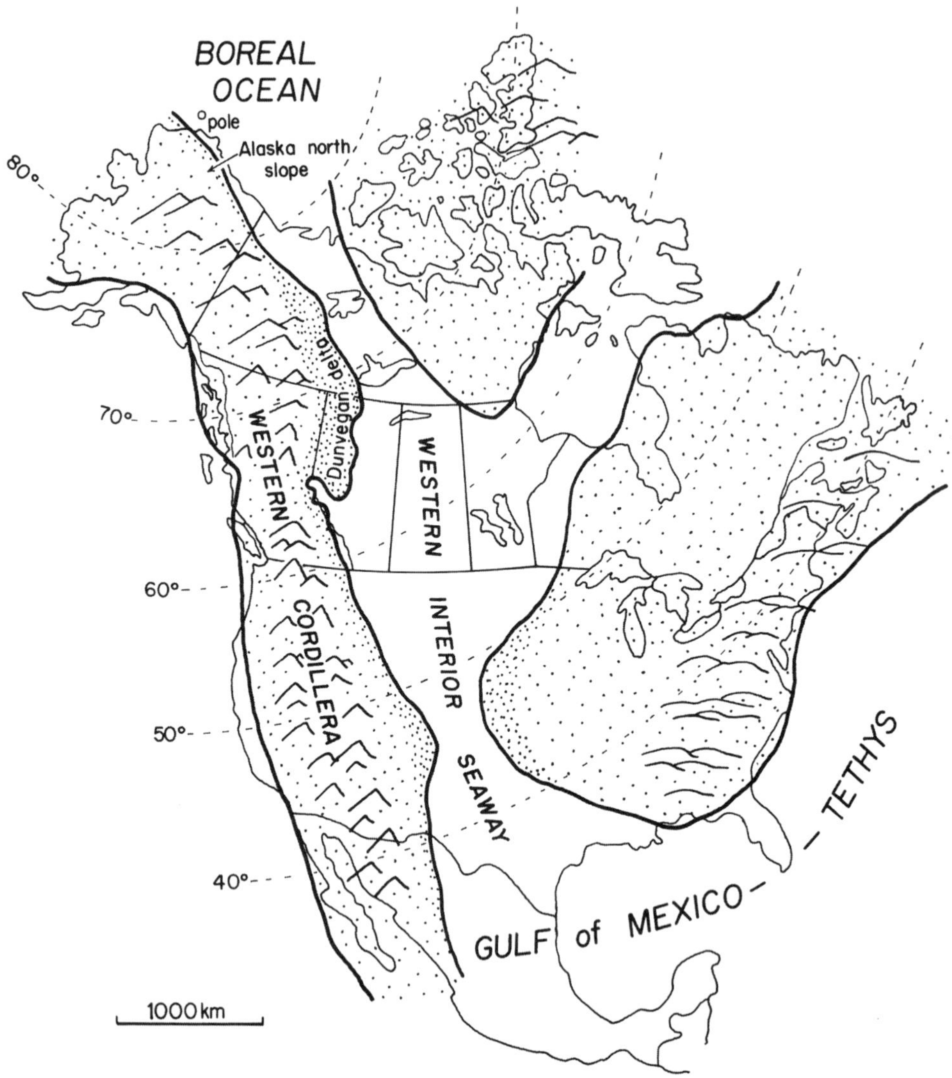

Fig. 1. Palaeogeography of North America in Cenomanian time, showing the location of the Dunvegan Formation floodplain–coastal plain complex in relation to the Western Interior Seaway and the Western Cordillera. Palaeolatitudinal data from Irving *et al.* (1993).

progressive and retrogressive processes involved in their evolution (Johnson & Watson-Stegner 1987). Micromorphological features present in floodplain deposits record evidence of specific pedological processes that operate under particular environmental conditions, yet there remains little detailed documentation of micromorphological features from these environments (McCarthy *et al.* 1998*a*, 1999*a*; McCarthy & Plint 1998). By interpreting assemblages of micromorphological features within the palaeosols, and also their regional distribution, the palaeoenvironment of these ancient floodplains can be understood in some detail.

Micromorphological interpretations are especially important because many pedogenic processes on ancient floodplains can be interpreted from assemblages of features even when the rock successions can be demonstrated to have experienced truncation and/or aggradation (Marriott & Wright 1993; McCarthy *et al.* 1998*a*); conditions which are typical of dynamic floodplain environments. Under these conditions, recognition, classification and

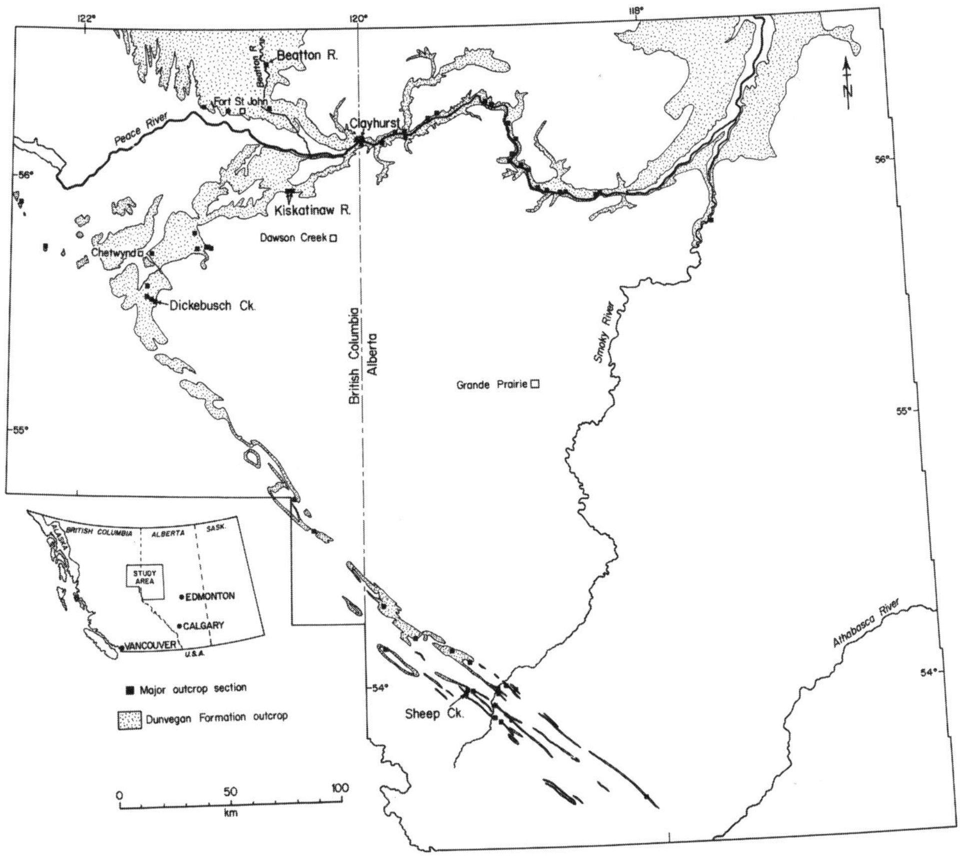

Fig. 2. Location of outcrop stratigraphic sections used to analyse regional palaeosol development in the Dunvegan Formation.

interpretation of typical ABC-type soil profiles can be difficult. The purpose of this paper is, therefore: (1) to describe micromorphological features present in floodplain palaeosols from the Upper Cretaceous (Cenomanian) Dunvegan Formation of Alberta and British Columbia, Canada; (2) to interpret the pedogenic processes that were active in these ancient soils; and (3) to explore the broader implications of these floodplain palaeosols for the understanding of mid-Cretaceous palaeoenvironments.

Geological setting

During Late Cretaceous time, the sediments of the Dunvegan Formation (Early to Mid-Cenomanian) were deposited within the Alberta foreland basin, located in the northwestern part of the Western Interior Seaway, which extended from the Gulf of Mexico to the Arctic Ocean (Caldwell *et al.* 1978; Stott 1982; Caldwell & Kauffman 1993) (Fig. 1). The rising Cordillera to the west provided abundant sediment that was transported to the southeast by rivers flowing parallel to the active basin margin (Stott 1982; Plint & Wadsworth 1997). Within the geographical limits of this study (Fig. 2), the Dunvegan Formation consists primarily of fine- to medium-grained sandstones, siltstones and mudstones deposited in floodplain, coastal plain and deltaic environments (Plint 1996). To date, published work has focused on regional lithostratigraphic and allostratigraphic studies (Fig. 3), and on detailed facies and ichnological analyses of deltaic and marine shoreline deposits (Stott 1982; Bhattacharya & Walker 1991*a*, *b*; Bhattacharya 1993, 1994; Plint 1996; Gingras *et al.* 1998). Incised valleys have been mapped on four discrete sequence boundaries in coastal plain rocks of allomembers E, F, G and H (Plint 1996; Wadsworth & Plint 1997), and corresponding interfluve

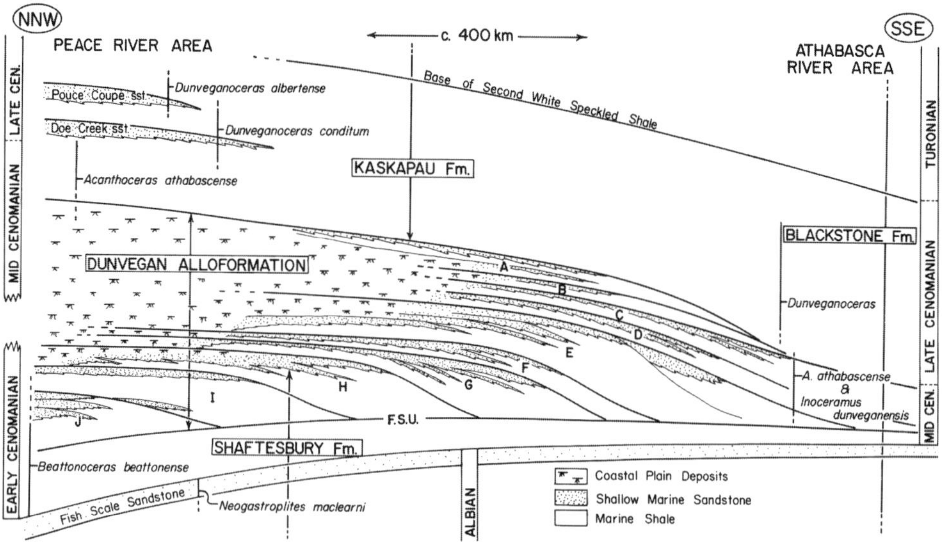

Fig. 3. Summary dip section across regional study area shown in Fig. 2, illustrating lithological and biostratigraphic relationships between the Shaftesbury, Dunvegan, Kaskapau and Blackstone formations. The upper and lower boundaries of the Dunvegan Alloformation are indicated (from Plint (1996); biostratigraphic data from Caldwell *et al.* (1978) and Stott (1963).

palaeosols have been identified (McCarthy & Plint 1998; McCarthy *et al.* 1999*b*).

Maximum burial and subsequent uplift of the Dunvegan Formation occurred in early Eocene time (Kalkreuth & McMechan 1984). Estimated maximum burial temperatures in the Dunvegan Formation were no more than 100–125°C. McCarthy *et al.* (1997*b*, 1998*a*, 1999*a*) detected no significant morphological alteration to pedogenic microfeatures subjected to similar burial temperatures in the Albian Mill Creek Formation in southwestern Alberta, and concluded that pedogenic features could, therefore, be used cautiously as palaeoenvironmental indicators.

Methods

Palaeosol-bearing floodplain deposits of the Dunvegan Formation were examined at outcrop and in the subsurface (Fig. 2). All stratigraphic sections and cores were logged and described according to standard procedures for describing soils (Day 1983). Undisturbed blocks were sampled from selected palaeosols at 10 cm intervals for thin-section preparation. Overlapping thin sections were manufactured from each soil block at 3 cm intervals, resulting in the *de facto* production of a composite mini-monolith throughout each palaeosol. Thin sections were prepared from undisturbed blocks following air-drying and impregnation with a polyester resin (Murphy 1986) and described using the terminology of Brewer (1976) and Bullock *et al.* (1985). More than 500 thin sections and corresponding polished blocks were examined for this study. Polished surfaces of impregnated samples were studied with a binocular microscope under low magnification (up to 10× magnification), and thin sections were subsequently examined under plane- and cross-polarized light using a Zeiss petrographic microscope (10–500× magnification).

Depositional environments and palaeosols

Fluvial channels and valleys

Very fine- to fine-grained, lenticular sandstones up to 15 m thick encased in floodplain mudstones and

Fig. 4. (**A**) Outcrop photograph from Clayhurst, British Columbia, showing typical floodplain deposits in the Dunvegan Formation. Floodplains comprise thin, weakly-developed palaeosols with thick, well developed interfluve palaeosols present at stratigraphic intervals coincident with the presence of incised valleys elsewhere. Person for scale. (**B**) Field view of a thick interfluve palaeosol that crops out in the Kiskatinaw River valley in northeastern British Columbia. Scale bar represents 20 cm. (**C**) Field view of typical floodplain palaeosol from Dickebusch Creek in northeastern British Columbia. (Note the thin coal seam that overlies the 1 m thick palaeosol developed in a fining-upward floodplain succession.) (**D**) Thick, coalified root trace present in 20 cm thick crevasse-splay deposit of the Dunvegan Formation. Hammer for scale.

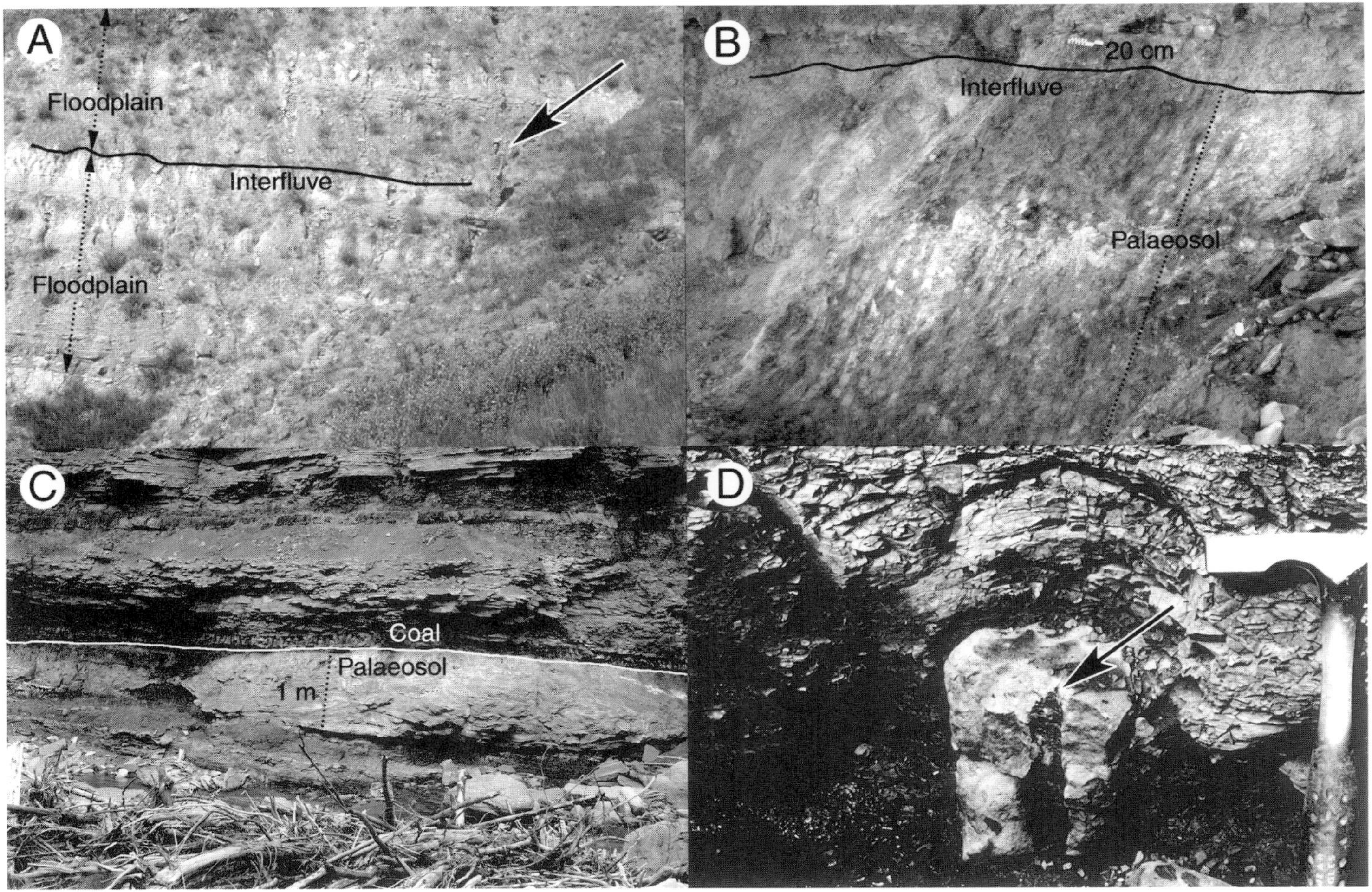
A
Floodplain
Interfluve
Floodplain
B
20 cm
Interfluve
Palaeosol
C
Coal
Palaeosol
1 m
D

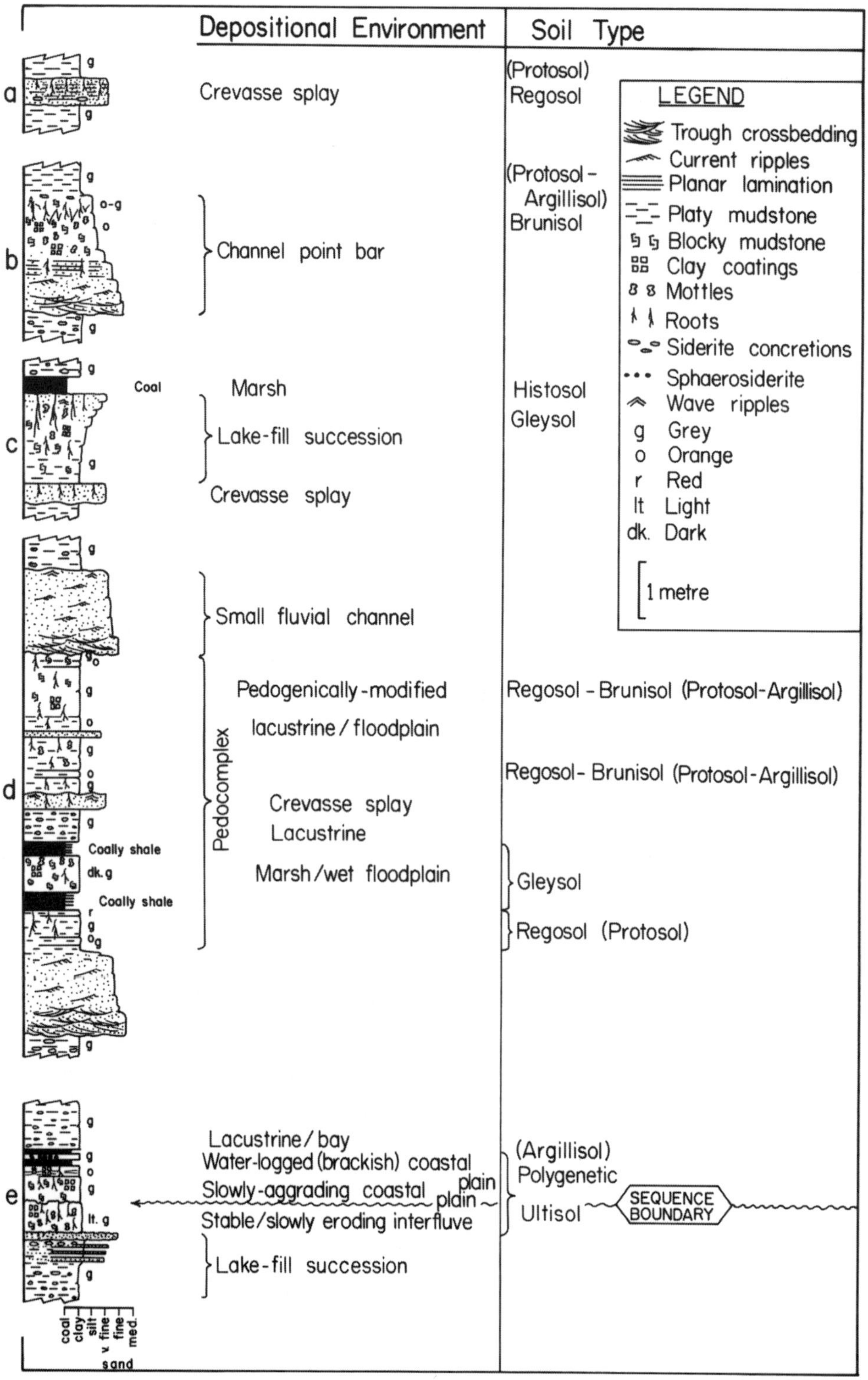
Depositional Environment
Soil Type
a
Crevasse splay
(Protosol)
Regosol
LEGEND
Trough crossbedding
Current ripples
Planar lamination
Platy mudstone
Blocky mudstone
Clay coatings
Mottles
Roots
Siderite concretions
Sphaerosiderite
Wave ripples
g Grey
o Orange
r Red
lt Light
dk. Dark
1 metre
b
Channel point bar
(Protosol-Argillisol)
Brunisol
c
Coal
Marsh
Lake-fill succession
Crevasse splay
Histosol
Gleysol
d
Small fluvial channel
Pedogenically-modified lacustrine / floodplain
Regosol - Brunisol (Protosol-Argillisol)
Regosol - Brunisol (Protosol-Argillisol)
Pedocomplex
Crevasse splay
Lacustrine
Marsh / wet floodplain
Coally shale
dk. g
Coally shale
Gleysol
Regosol (Protosol)
e
Lacustrine / bay
Water-logged (brackish) coastal plain
Slowly-aggrading coastal plain
Stable / slowly eroding interfluve
Lake-fill succession
(Argillisol)
Polygenetic
Ultisol
SEQUENCE BOUNDARY
lt. g
coal
clay
silt
v fine
fine
med.
sand

sheet sandstones are interpreted as anastomosed river channel deposits (Nadon 1994; McCarthy *et al.* 1999*b*). Fine- to medium-grained, trough cross-bedded, multi-storey sandstones up to 25 m thick with generally southeast-trending palaeocurrents are interpreted as valleys filled by point-bar deposits of meandering rivers (Plint 1996). These cross-bedded sandstones are highly lenticular with steeply incised margins. In places, the sandstones are partitioned by large-scale, low-angle accretion surfaces and they may enclose mudstone lenses (Plint 1996; McCarthy *et al.* 1999*b*). Three-dimensional exposures, or dense well-log control, show that the sandbodies are several hundred metres to 1–3 km wide. Valleys can be mapped for over 250 km updip from coeval shorelines (Wadsworth & Plint 1997).

Floodplains and palaeosols

Most of the floodplain sediments (75–80%) in the Dunvegan Formation constitute successions of very fine-grained sandstones and dark grey to brown, laminated carbonaceous mudstones interbedded with occasional thin coals and rooted horizons interpreted to represent deposits of a low-lying coastal plain (Fig. 4) (Plint 1996). Palaeosols in the Dunvegan Formation developed primarily in floodplain claystones and siltstones, although some sandstones display limited evidence of pedogenic modification (Fig. 4), on poorly drained and sedimentologically active floodplains (Plint 1996; McCarthy *et al.* 1999*b*). Detailed field descriptions and geochemical and mineralogical trends in the palaeosols have been presented elsewhere (McCarthy & Plint 1998; McCarthy *et al.* 1999*b*). Palaeosol development can be organized into five main types, based on depositional environment and degree of pedogenic expression (Fig. 5; McCarthy *et al.* 1999*b*), which are briefly summarized below.

Rapid deposition of sediment and generally wet conditions resulted in a preponderance of weakly developed and poorly drained palaeosols that typify floodplain sedimentation in the Dunvegan Formation (Fig. 5a–c). Thin tabular sandstones may contain a few, large root traces, sometimes preserved as vertical to sub-horizontal, coalified structures or expressed as surface cradle knolls, but the sandstones maintain evidence of primary sedimentary stratification (type 1 palaeosols; Fig. 5a). Alternatively, thin sandstones may be overlain by coals less than 40 cm thick (commonly 5–10 cm thick) and/or thin, grey mudstones (less than 50 cm thick) with weakly developed blocky structure (Figs 4C and 5a). The mudstones commonly contain few, dark carbonaceous or coalified root traces and a few, small orange mottles. Type 1 palaeosols are similar to modern Regosols (Canada Soil Survey Committee 1987).

Type 2 palaeosols (Fig. 5b) develop in upward-fining floodplain successions. Typically, they consist of a basal layer of trough cross-bedded to massive sandstone overlain by up to 1 m of blocky mudstone. The mudstone displays a gradual colour change from orange at the base to dark grey at the top. Macroscopic root traces are present in the upper surface of the mudstone. Type 2 palaeosols are similar to modern Brunisols (Canada Soil Survey Committee 1987).

Type 3 palaeosols developed in metre-scale, sandier-upward successions that are interpreted as prograding lacustrine delta mouth bars or prograding levees (Fig. 5c) (Plint 1996). Sharp-based, decimetre-scale current rippled sandstone beds probably represent crevasse splays, whereas thicker cross-bedded sandstones are interpreted as minor (distributary?) channel fills. Upper surfaces of sandstones commonly contain large, coalified or carbonaceous root traces that extend into the underlying mudstone. Finely disseminated sphaerosiderite is common in these mudstones (McCarthy *et al.* 1999*b*). Thin coals or coaly shales up to 20 cm thick typically cap the succession. Type 3 palaeosols are interpreted as Gleysols (Canada Soil Survey Committee 1987). Coal seams are interpreted as Histosols (Fig. 5c).

Elsewhere, where aggradation rates were lower, because of regional or local changes in subsidence rate or sediment supply, periodic additions of thin layers of overbank sediment led to the cumulative development of floodplain soils with pedological features extending throughout much of the stratigraphic section. Even when relatively more mature soils formed, subsequent sedimentation was insufficiently rapid to isolate the palaeosols from the effects of the next phase of soil formation, resulting in compound and complex palaeosol development, expressed as thick floodplain pedo-

Fig. 5. Sedimentary logs showing characteristic relationships between palaeosols and facies associations in the Dunvegan Formation. (**a**) Crevasse splay containing a few, large vertical root traces. (**b**) Fluvial point bar containing Regosols to Brunisols. (**c**) Coarsening-upward lake-fill deposits overlain by coal (Histosol) and large roots. (**d**) Cumulative succession showing thin fluvial channel sands associated with aggradational floodplain sediments. (**e**) Floodplain succession showing sequence boundary marking well-developed interfluve palaeosol. (Soil types classified according to Canada Soil Survey Committee (1987) and palaeosol classification (in brackets) of Mack *et al.* (1993)).

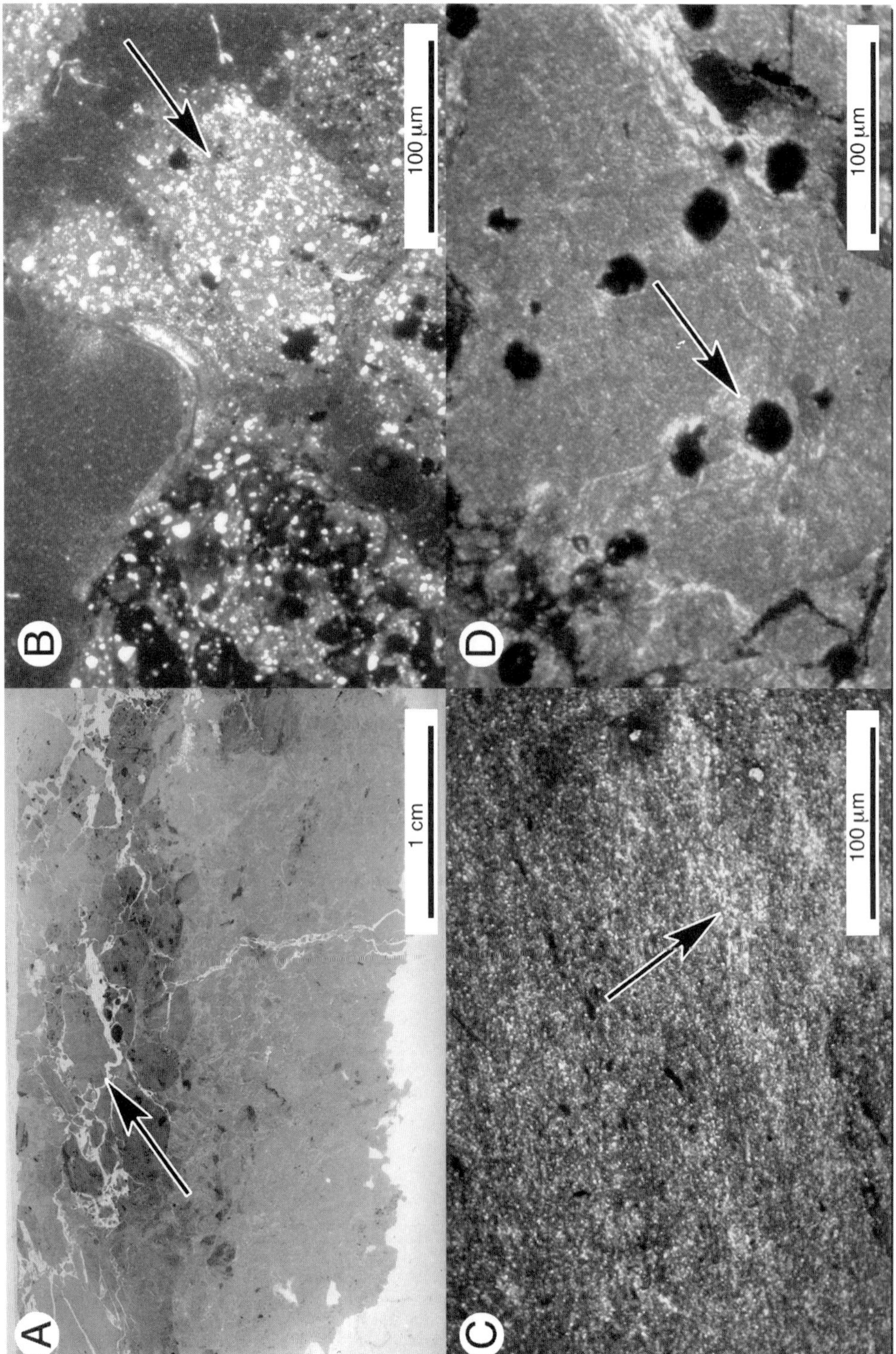
A
1 cm
B
100 µm
C
100 µm
D
100 µm

complexes (Fig. 5d). Although illuvial clay coatings are sometimes identified, they are not abundant enough for designation of Bt horizons (McCarthy *et al.* 1999*b*). Pedocomplexes in the Dunvegan Formation, therefore, attain a maximum development similar to modern Brunisols or, where locally less well-drained, Gleysols (Canada Soil Survey Committee 1987).

During phases of valley cutting and filling, floodplains received no new sediment for prolonged periods, resulting in the development of thick, mature interfluve palaeosols (type 5 palaeosols; Fig. 5e). Interfluve palaeosols consist of light grey, blocky mudstone at least 2 m thick that is rooted throughout. Interfluve palaeosols are characterized by the presence of well-developed Bt horizons containing evidence of translocated clay and are similar to modern Luvisols (Canada Soil Survey Committee 1987). Many of these interfluve palaeosols also display evidence of truncation and erosion (McCarthy & Plint 1998).

The lack of mineralogical and geochemical trends in type 1–4 palaeosols is consistent with weak pedogenesis (Fig. 5a–d). The clay mineralogy of these palaeosols is dominated by illite, with lesser amounts of chlorite, illite–smectite and kaolinite. Variations in SiO_2 and Al_2O_3 closely parallel textural trends (McCarthy *et al.* 1997*c*). In contrast, the clay mineralogy in interfluve palaeosols (type 5 palaeosols, Fig. 5e) is dominated by kaolinite, and chlorite is absent, consistent with prolonged temperate weathering conditions (Allen & Hajek 1989; Bain *et al.* 1993; McCarthy *et al.* 1998*a*, *b*). Weathering surfaces can also be characterized by a marked concentration of resistate elements (e.g. TiO_2, Zr; McCarthy & Plint 1998).

Micromorphological evidence of pedogenic modification

Microstructure and fabric

Macroscopic structural organization in the palaeosols is restricted to the recognition of root traces and blocky peds in outcrop. Laboratory examination indicates that most of the units comprise angular or subangular blocky structures (Fitzpatrick 1993). Peds, of 5–10 mm, have well-accommodated flat or slightly curved surfaces. Voids are continuous to discontinuous, linear in shape and range in size from <50 to 2000 μm. Some layers have composite structures composed of granular and subangular blocky peds (Fig. 6A and B). Small, rounded to sub-rounded granules, 0.2–1 mm in diameter, surround larger, more resistant subangular blocky peds. Within some near-surface layers, zones with striated bow-like or vermiform structures are evident. Both sand- and clay-rich parent materials (floodplain alluvium) retain their primary sedimentary lamination in incipient palaeosols.

Microstructure in the palaeosols is closely related to the nature of the parent material (i.e. sand or clay) and to the intensity of biological activity. The fine-grained palaeosols display predominantly striated birefringence fabrics (b-fabrics; see Bullock *et al.* 1985). Striated b-fabrics are characterized by the presence of elongated birefringent zones in which the individual clay particles show more or less simultaneous extinction. These birefringent zones lack sharp boundaries, and they are difficult to recognize in plane-polarized light. Various types of striated b-fabrics are recognized in the Dunvegan Formation, including porostriated, granostriated, parallel striated, random striated and crescent striated (Fig. 6C and D).

In some mudstones and siltstones, speckled b-fabrics are sometimes observed. These fabrics are characterized by the presence, in the groundmass, of randomly arranged equidimensional speckles of optically oriented clay. Both stipple-speckled b-fabrics, consisting of individual isolated speckles and mosaic speckled fabrics, where the birefringent speckles are in contact with each other, are identified. Voids range in size from <50 μm to 2 mm. Channels, chambers, planes, vugs and packing voids have all been observed (Bullock *et al.* 1985). Preserved sedimentary structures rarely occur in mudstones, but are common in sandstones. Sedimentary microlamination is typically planar and is recognized by subtle grain-size variations.

Interpretation of microfabric

Laminar structures are produced by deposition and subsequent compaction of floodplain sediments. Angular and subangular blocky structures develop through the repeated action of shrink–swell forces generated by wetting and drying processes, and by the action of plant roots and soil organisms (Fanning & Fanning 1989). Composite granular

Fig. 6. (**A**) Thin-section photomicrograph showing well-developed subangular blocky to granular structure (arrow; plane-polarized light). (**B**) Close-up view of subangular blocky pedogenic aggregate (arrow; cross-polarized light). (**C**) Parallel striated birefringence fabric (arrow; cross-polarized light). (**D**) Granostriated birefringence fabric (arrow) surrounding Fe-oxide micronodules (cross-polarized light).

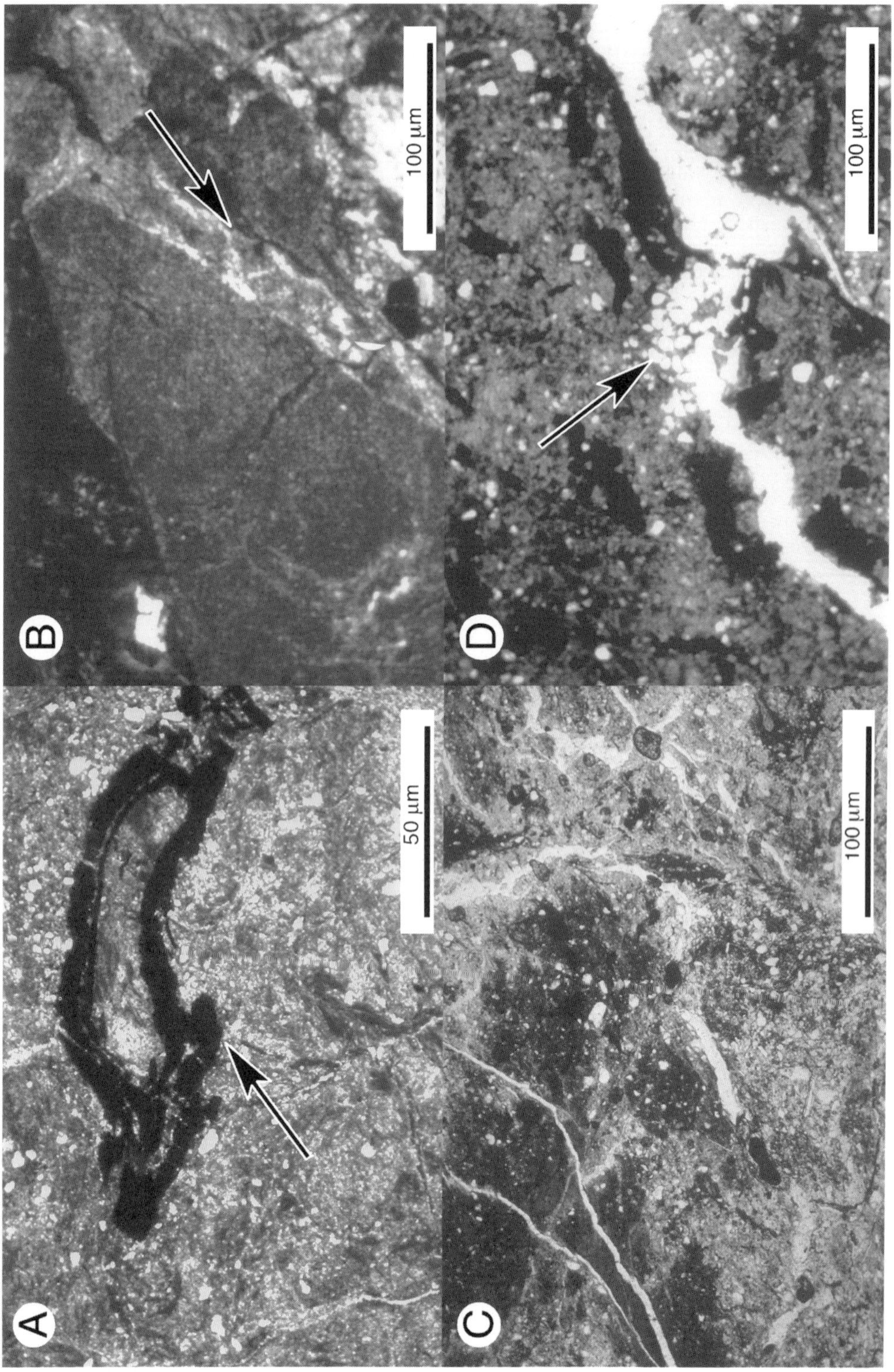
A
50 µm
B
100 µm
C
100 µm
D
100 µm

and subangular blocky structures may result either from wetting and drying of layers containing enhanced concentrations of expanding-layer clays or, alternatively, the small granules may be produced by biological activity with the subangular blocky peds remaining as more resistant aggregates (Fitzpatrick 1993). Another possibility is that the aggregates became rounded during transport (Mucher & Morozova 1982).

The development of palaeosol microfabrics reflects the original sedimentary material, subsequent modification during pedogenesis and, probably, compaction. Realignment of detrital clay flakes into birefringent zones is commonly attributed to the effects of pressure and tension stresses produced by alternate wetting and drying of the soil (Brewer 1976). Speckled b-fabrics are generated by suspension settling or flocculation of fine alluvial material during sedimentation (Bullock *et al.* 1985). The extent of fabric development is related to a number of factors including type of clay minerals, moisture regime of the soil, time and initial composition of the parent material (Kemp 1985). Post-burial compaction is another possible mechanism for the development of strial fabrics, although in the case of the Dunvegan palaeosols, its effects have probably enhanced some of the fabrics rather than produced them outright. Preserved sedimentary microlamination indicates that, at least locally, pedogenesis was not intense or prolonged. Void channel formation is attributed to shrinking and swelling processes and biological activity resulting from plants and organisms.

Biological features and fabrics

Root traces and channels. Fine-grained rocks of the Dunvegan Formation are extensively penetrated by pedotubules (features with a tubular external form and sharp external boundaries; Brewer 1976, p. 236), which are predominantly fine (1–2 mm diameter), vertically oriented, downward tapering and branching features (average. 3–20 cm long, maximum 1.6 m long). Subhorizontal, tubular, branching structures associated with flat-lying accumulations of organic matter also occur. Organically lined root traces commonly have a similar length and thickness and decrease in frequency with depth.

Root traces are preserved in several ways.

(1) Typically, root traces are present as thin carbon films or as carbonaceous and/or coalified material. Sometimes the root trace has a hollow interior, which has subsequently been filled with silica cement or fine-grained material (Fig. 7A).

(2) Clay-filled root traces are less common and are typically coated with multiple, layered, birefringent clay laminae. In some cases, the clays form compound coatings, interlayered with Fe–Mn-oxide coatings and silt coatings. Clay-filled root traces may or may not be associated with organic material.

(3) Agglomerations of sphaerosiderite may pseudomorph root traces and organic matter (e.g. Ludvigson *et al.* 1998).

Biological fabrics. Preserved burrows typically appear as subhorizontal elliptical tubules 0.5–2 mm in diameter (Fig. 7B). In thin section, the tubules have smooth, slightly irregular boundaries and are filled with silty clay or clay material similar to the surrounding matrix. In a few cases, the burrows are lined by thin, oriented clay coatings, rarely associated with some preserved organic material. Commonly, these elliptical tubules have a crescentic or vermiform internal fabric identifiable under plane-polarized light by a subtle yet distinct reorganization of the b-fabric. Locally, several generations of burrowing may be recognized by the superimposition of several tubules that cross-cut each other (McCarthy & Plint 1998).

Other intervals contain evidence of intense bioturbation as indicated by a complete lack of grain-size or colour differentiation and the intimate association of a disrupted, isotropic groundmass, disseminated organic matter, small granular aggregates and larger subangular blocky aggregates (Fig. 7C and D). A few channels with simple and compound fills were also observed.

Organic matter. Thick coal seams do not occur in the Dunvegan Formation (Stott 1982; Plint 1996). Coal is present, however, as thin (maximum 50 cm thick) discontinuous seams interlayered with alluvial mudstones and sandstones. Coals are absent from interfluve palaeosols, but thin layers of coaly shale are present in floodplain deposits. In

Fig. 7. (**A**) Root trace (arrow) infilled with fine-grained material (cross-polarized light). (**B**) Burrow in fine-grained mudstone displaying well-developed bow-shaped fabric (arrow; cross-polarized light). (**C**) Intensely bioturbated silty mudstone showing interdigitation of grain sizes and pedogenic aggregates typical of biological reworking (plane-polarized light). (**D**) Pod of silt-sized quartz grains in mudstone (arrow) probably emplaced by biological activity (plane-polarized light).

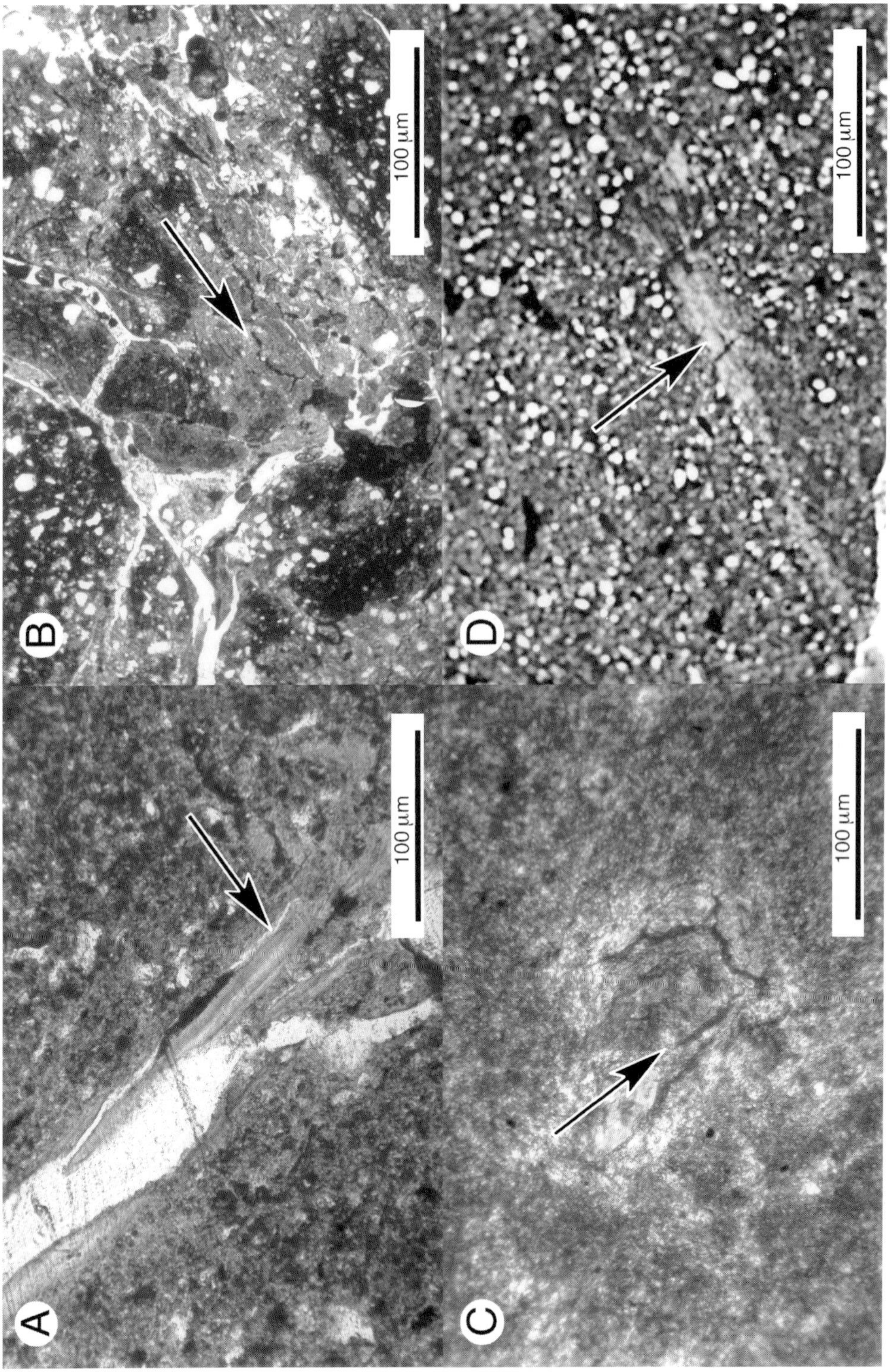
A
100 µm
B
100 µm
C
100 µm
D
100 µm

thin section, organic matter occurs as: (1) small (0.5–3 mm), opaque fragments, occasionally exhibiting some preserved cellular structure; these fragments are typically disrupted or incomplete and are filled with silty–clay matrix material; (2) fine, opaque particles, commonly dispersed throughout the matrix or associated with former bedding planes; organic fragments are more common near heavily rooted and/or bioturbated zones.

Interpretation of biological features

The abundance of root traces and organic material, and the presence of thin coals and coaly shales suggest that these coastal plains were heavily vegetated and poorly drained. Crescentic and vermiform structures resemble features produced in modern soils by the action of burrowing insects and earthworms (Fitzpatrick 1993; Aslan *et al.* 1995). The presence of intensely bioturbated zones suggests that some surfaces were stable for prolonged periods of time. The bioturbated zones closely resemble dense to very dense 'excremental total fabrics' of biological origin that may be found in modern soils (Courty & Fedoroff 1985).

Textural features

Clay coatings and infillings. Simple clay coatings are of two types: (1) microlaminated clay coatings and infillings are present as yellowish orange to pale yellow–grey, laminated continuous void coatings (25–60 μm thick; Fig. 8A); (2) disturbed and degraded clay coatings are common. The latter coatings occur as local clay-rich zones (30–50 μm thick) within a matrix containing weakly birefringent clay in domain-like units, which appear to grade into the surrounding matrix, but can still be distinguished in plane- and cross-polarized light (Fig. 8B). Some of these clay concentrations have a granular extinction pattern resulting from internal reorganization of the clay platelets.

Compound, layered coatings up to 100 μm thick consist of dark yellowish orange clay interbedded with silt-size material, and rarely Fe–Mn-oxides. These coatings are present in the upper parts of some interfluve palaeosols.

Commonly, microlaminated, crescentic clay coatings partially or completely fill pores to form discontinuous to continuous void infillings (Bullock *et al.* 1985) (Fig. 8C). Infillings occasionally consist of fine silt, typically associated with carbonaceous root traces.

Linear and irregular clay concentrations. Linear clay concentrations are intrapedal striations that represent either old ped argillans (clay coatings), the peds no longer being separated, or channels or planar voids totally filled with illuvial clay. Irregular clay concentrations are usually found as equant to prolate intrapedal concentrations. These features are always simple, grey to greyish brown, generally non-laminated, and composed of clay with low birefringence and very weak, wavy extinction (Fig. 8D). In large numbers, they may impart a generally striated appearance to the matrix. They are usually clay rich, but they may also consist of silt and silty clay layers.

Grain coatings. These are coatings of clay either impregnated with or coated by Fe-oxides around skeleton grains in the groundmass (Fig. 9A and B). Grain coatings are present in interfluve palaeosols and uncommon in sandstone and sandy siltstone floodplain palaeosols.

Papules and pedorelicts. Papules are small, subangular to subrounded, clay-rich pellets containing fabrics and/or colours different from the surrounding matrix. Papules have sharp boundaries and are not commonly associated with voids (Fig. 9C). Pedorelicts are subrounded nodules containing matrix fabrics different from the surrounding palaeosol matrix (Fig. 9D).

Interpretation of textural features

Laminated, well-oriented clay coatings in soils provide evidence for illuvial translocation of clay (Bullock & Thompson 1985). Clay illuviation requires that water percolated through the soil to eluviate the clay, and that the soil must have dried out periodically so that the translocated clay was retained (McKeague 1983). The multi-layered nature of many of the coatings suggests that this process was active over multiple events. Microlaminated clay coatings are deposited by evaporating capillary water in soils. Moderately oriented coatings and infillings indicate saturated

Fig. 8. (**A**) Thick, microlaminated clay coating along large void (arrow) characteristic of illuvial clay coatings found in well-drained Luvisols (cross-polarized light). (**B**) Disrupted illuvial clay coatings in silty mudstone (arrow; plane-polarized light). (**C**) Well-developed, complete clay infilling (arrow) formed by complete plugging of a small pore by illuvial clay (cross-polarized light). (**D**) Linear clay concentration present in mudstone matrix (arrow; plane-polarized light).

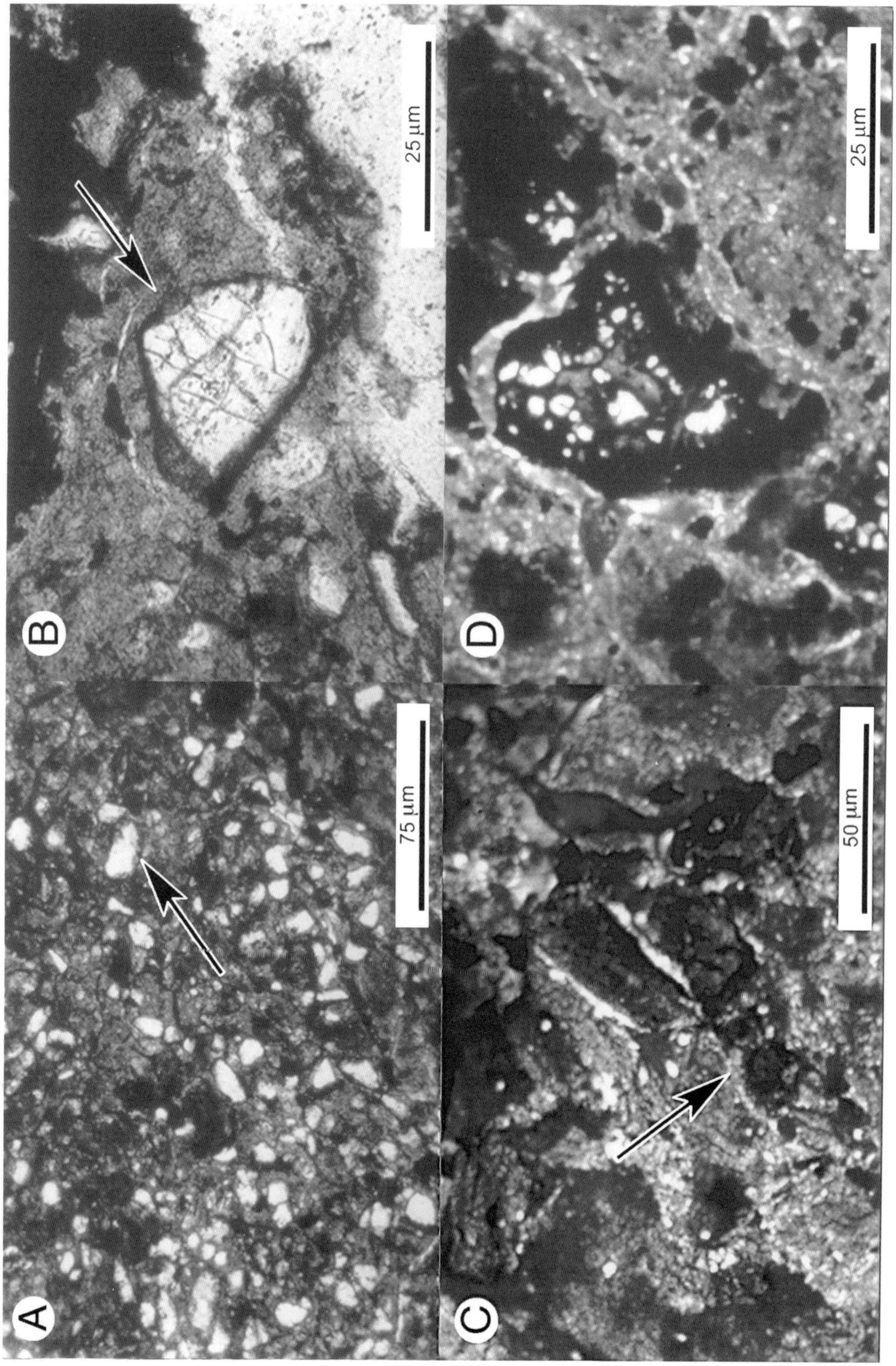
A
75 µm
B
25 µm
C
50 µm
D
25 µm

soil conditions within a stable groundmass (i.e. water is moving through discrete pores). Weakly birefringent clay domains having a granular extinction pattern are indicative of ageing of illuvial clay coatings and subsequent fragmentation and assimilation into the matrix in response to shrinking and swelling and bioturbation (Fitzpatrick 1993; Wang *et al.* 1995). Silt coatings, and silty clay compound coatings, form under higher-energy conditions, and their presence is commonly attributed to wetter, and possibly colder, conditions than those required for clay translocation, or to the presence of lithological discontinuities (Fedoroff *et al.* 1990; Nettleton *et al.* 1994). Several explanations for the formation of grain argillans have been advanced including the following: (1) stress argillans resulting from shrinking and swelling; (2) clay coatings collect and develop preferred orientation during transport; (3) coatings are former free grain argillans that are rounded and intermixed with other material during transport (Bullock & Murphy 1979; Kemp 1985). Embedded grain argillans have, however, been traditionally associated with short-distance transport of frozen soil material (Bullock & Murphy 1979).

Linear and irregular clay concentrations probably originate from compression of void argillans by pedoturbation or from total infilling of irregular voids with illuvial clay (Bullock & Murphy 1979), although some may be intercalations that develop when the soil is saturated and water is moving throughout the whole, unstable groundmass (Fedoroff 1998, pers. comm.). Both papules and pedorelicts formed elsewhere and were transported and redeposited in their present position. Owing to their fragile nature, they probably result from low-energy erosion and short-distance transport by colluvial processes (Brewer 1976). Some of them may also result from intense biological reworking.

Ferruginous features

Coatings and hypocoatings. Local redistribution of iron oxides results in mottled zones in which the groundmass is weakly impregnated with iron oxides. Mottles typically have an amiboidal shape and gradational boundaries with the adjacent matrix. The matrix material is similar in the mottles and adjacent groundmass. Mottles range in size from <10 mm to hundreds of millimetres. Ferruginous void coatings and hypocoatings occur as thin, 0.5–1.5 mm, dark red features on void walls (Fig. 10A) and surrounding root tubules. Rarely, these coatings are associated with coatings and concentrations of manganese oxides.

Nodules. Ferruginous nodules (<0.5 to 10 mm diameter) are present as strongly impregnated, discrete concentration features having regular, subrounded to well-rounded shapes and sharp to distinct boundaries with the adjacent groundmass (Fig. 10B). Nodules may envelope silt- and sand-size mineral grains. Orthic nodules, which are developed in material identical to that of the host sediment, are the most common type but typic (undifferentiated internal fabric), nucleic (foreign core), geodic (hollow interior) and aggregated (formed from several more or less rounded entities) nodules are also observed.

Depletion coatings or surface residues. Outer surfaces of pedogenic aggregates from which iron or clay, or both, have been removed (Fig. 10C) are termed depletion coatings or surface residues (Bullock *et al.* 1985; Fitzpatrick 1993). In the Dunvegan Formation, both clay and Fe-oxides are removed from aggregate surfaces along major soil channels. The surface residue manifests itself as a thin zone (up to 75 μm thick) that is bleached and contains less clay and correspondingly more silt and sand than the interior of the aggregate.

Interpretation of ferruginous features

Ferruginous nodules and void coatings are interpreted to have formed by segregation of Fe oxides as a result of wetting and drying over numerous redox cycles. The morphology of some of the nodules suggests that they may have formed as oxidation products of earlier formed siderite (Driese *et al.* 1995; PiPujol & Buurman 1997). Under conditions of free drainage, percolating water removes Fe-oxides and clay preferentially from the surface of aggregates, resulting in depletion coatings or surface residues. The surfaces of aggregates commonly become progressively more impoverished in Fe-oxides and clay (Fitzpatrick 1993). Fluctuating drainage conditions

Fig. 9. (**A**) Mudstone containing sand- and silt-sized grains coated with illuvial clay (arrow; plane-polarized light). (**B**) Close-up view of grain coating (arrow) showing irregular thickness surrounding the grain and clear distinction between grain coating and clay in the surrounding groundmass (plane-polarized light). (**C**) Dark reddish orange papules (arrow) in bioturbated groundmass (cross-polarized light). (**D**) Large compound pedorelict containing quartz grains cemented by Fe-oxides (plane-polarized light).

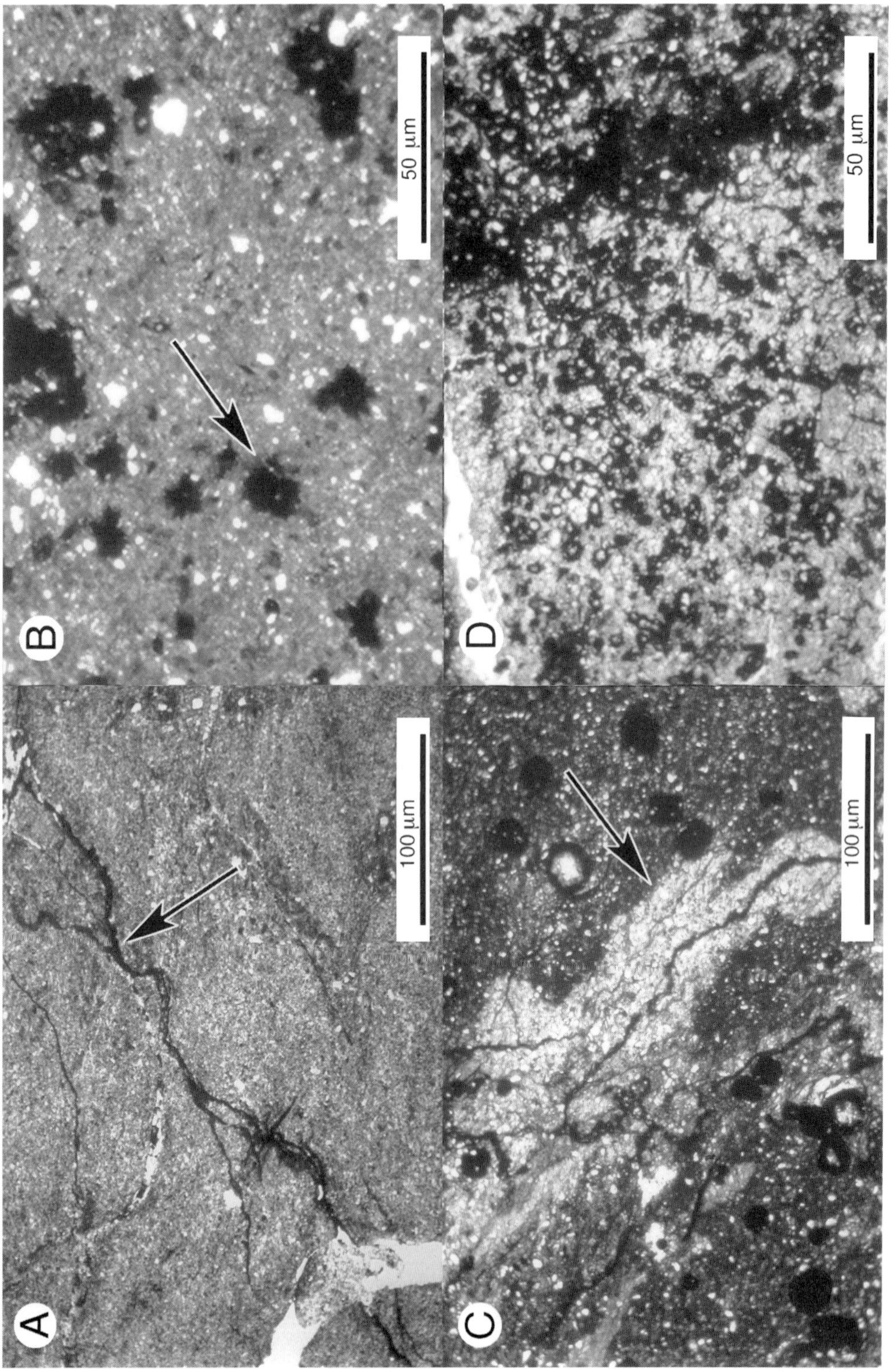
A
100 µm
B
50 µm
C
100 µm
D
50 µm

typically result in spatial variations in redox conditions within clay-rich horizons, and the development of ferruginous segregations as mottles. The common occurrence of ferruginous coatings along the surfaces of large voids suggests that the palaeosols were saturated for at least part of the time (Bouma *et al.* 1990). Iron-rich void coatings would have precipitated as the soils were drying out, when the partial pressure of oxygen was highest in the large voids, which would have dried out first (Vepraskas *et al.* 1994).

Other features

Siderite. Siderite is present as small spherulites 0.5–2 mm in diameter (Fig. 10D). Some spherules have a geodic pattern with small voids in the centre, whereas others are filled with calcite or silica. Siderite spherules may be coated by a thin rim of Fe-oxide, and some occur as agglomerations that pseudomorph root traces and organic fragments. Typically, the siderite is either superimposed on illuviated clay or disseminated throughout the fine-grained matrix.

Sphaerosiderite commonly consists of coarsely crystalline siderite that radiates outward from a central mass. In mudstones, spherules do not enclose detrital grains and they are commonly bounded by compactionally deformed clay laminae, resulting in a locally granostriated birefringence fabric. Large irregular siderite concretions (20–50 mm diameter) formed locally, apparently in association with sandier layers.

Barite. Rare, very fine, euhedral laths of barite have been identified by energy-dispersive X-ray analysis (McCarthy & Plint 1998), associated with large carbonaceous root traces at the tops of some interfluve palaeosols.

Interpretation of other features

Siderite forms in wetland soils (bog-ore) and is commonly associated with organic matter under reducing conditions (Landuydt 1990). Sphaerosiderites precipitate from sulphate-poor, iron-rich, reducing pore fluids within, or immediately above, methane-producing zones (Gonzalez *et al.* 1996). Therefore, we interpret sphaerosiderites in the Dunvegan Formation as having formed below the water table, like their modern counterparts (Ludvigson *et al.* 1998). Siderite spherules superimposed on illuviated clay suggest that drainage conditions in the palaeosols deteriorated. The occurrence of Fe-oxide coatings covering many spherules also suggests periods of oxidation, probably during fluctuations in local water table levels. The dissemination of sphaerosiderite throughout the matrix at certain levels within the palaeosols indicates that they were saturated and below the ground-water table, at least for some of the time. Authigenic barite occurs in hydromorphic soils associated with saline ground-water conditions (Lynn *et al.* 1971; Stoops & Zavaleta 1978) that are coeval with marine transgressions downdip (McCarthy & Plint 1998).

Discussion

Palaeoenvironmental implications of micromorphological features

The major pedogenic processes operating in floodplain palaeosols of the Dunvegan Formation were clay illuviation, redoxymorphic processes, precipitation of sphaerosiderite and pedoturbation. Poorly drained conditions on the floodplains are suggested by the overall dull grey colour, abundance of sphaerosiderite and organic matter in the palaeosols, and by the presence of thin coals. The associated presence of lacustrine shales also indicates that permanent lakes existed on these floodplains (Plint 1996). However, the key factor in the formation of illuvial clay coatings is the presence of alternating wet and dry conditions (Fedoroff *et al.* 1990). Pale yellow, silty clay, microlaminated void coatings suggest illuviation under increasingly poorly drained, and possibly cooler, soil conditions although the microlamination indicates evaporation of capillary water (i.e. the soils dried out periodically; Fedoroff 1997; McCarthy *et al.* 1998*a*). The presence of moderately oriented clay coatings and linear concentrations (intercalations?) within the matrix suggests that some horizons were saturated for at least part of the time (Fedoroff 1998, pers. comm.). Fragmented clay coatings and weakly birefringent linear and irregular clay concentrations within the

Fig. 10. (**A**) Fe-oxides coating large planar void (arrow) in mudstone (cross-polarized light). (**B**) Fe-oxide nodules (arrow) in mudstone. (Note the sharp, yet irregular boundaries of the nodules with the surrounding matrix (cross-polarized light).) (**C**) Fe-depletion residue (arrow) in mudstone (plane-polarized light). (**D**) Sphaerosiderite disseminated throughout the matrix of a mudstone palaeosol (plane-polarized light).

groundmass are attributed to ageing of illuvial clay coatings and their subsequent incorporation into the groundmass through the combined action of bioturbation and physical shrink–swell processes (i.e. wetting and drying; Fitzpatrick 1993; Wang *et al.* 1995).

Evidence for fluctuating redox conditions, attributed to variations in ground-water levels, can be detected by analysing assemblages of micromorphological features. The presence of void coatings, mottles and nodules of Fe-oxides and Fe-oxide depletion coatings adjacent to large voids indicates that ferruginous compounds were redistributed within the palaeosols (Vepraskas *et al.* 1994; Driese *et al.* 1995). Iron oxides coating surfaces of voids and channel margins indicate that the surrounding groundmass was at least periodically saturated, and that the soil also dried out periodically, allowing the dissolved Fe^{2+} to migrate towards the more oxidizing larger pores, where it was precipitated as Fe^{3+} when the soil dried out (Bouma *et al.* 1990; Vepraskas *et al.* 1994).

Iron-oxide depletion coatings indicate that ponded water was present at some times and places on the floodplains whereas the phreatic zone (true ground-water table) was located below the solum (Duchaufour 1982; PiPujol & Buurman 1994; McCarthy *et al.* 1998*a*). As this ponded surface water drained through the vadose zone, Fe^{2+} was removed from the large pores and precipitated in more oxidizing ped interiors and/or transported out of the palaeosol with migrating ground water (Bouma *et al.* 1990; Vepraskas *et al.* 1994; PiPujol & Buurman 1997). Although similar features are known to form diagenetically during early burial (e.g. diffusion gleyans of Retallack 1991), the coexistence of pale orange, microlaminated illuvial clay coatings that must have formed under well-drained conditions (Fedoroff 1997; McCarthy *et al.* 1998*a*) within Fe-depletion zones suggests variable redox conditions at different times (McCarthy *et al.* 1999*a*). This condition is more consistent with pedogenic development within a zone of fluctuating water table conditions than with diagenetic development during continuous burial of the palaeosol beneath the water table (PiPujol & Buurman 1994; McCarthy *et al.* 1999*a*). Similarly, the presence of partially oxidized sphaerosiderite superimposed upon illuvial clay coatings suggests that local water tables fluctuated and that individual floodplain palaeosols were, at different times, saturated and well drained. Although these variable drainage conditions could be related to changes in regional climate patterns, it is more likely that, in the Dunvegan Formation, they reflect changing topography and variable parent materials on the floodplains (McCarthy *et al.* 1999*b*).

Palaeoenvironmental interpretation of Dunvegan floodplains

Taken together, the micromorphological features present in floodplain palaeosols of the Dunvegan Formation are similar to those found in modern Brunisols and Luvisols (Canadian System of Soil Classification; Canada Soil Survey Committee 1987) located in south–central Canada (McKeague *et al.* 1969, 1972; Wang *et al.* 1995; McCarthy & King 1997), and to Pleistocene palaeosols documented from England (Bullock & Murphy 1979; Chartres 1980) and northern Canada (Tarnocai & Smith 1989). By comparison with these modern soils and Pleistocene palaeosols, the Dunvegan palaeosols probably formed under cool temperate conditions with mean annual temperatures in the range of 7–13°C and precipitation amounts from *c.* 375 to >1250 mm a^{-1}. Driese *et al.* (1995) further suggested that redoximorphic features in palaeosols are characteristic of warm to cool temperate palaeoclimates with mean annual temperatures in the range of 5–20°C and seasonal saturation. Although some seasonal saturation is suggested by the redoximorphic pedogenic features in the Dunvegan Formation, it may be related to seasonal snowmelt in the Cordillera to the west and consequent overbank flooding, rather than to any pronounced seasonality in precipitation distribution.

The northern limit of modern Luvisols occurs around 61°N latitude, south of the –3.5°C mean annual temperature (MAT) isotherm with mean annual precipitation (MAP) of at least 375 mm (Tarnocai 1997). Modern Luvisols that develop in temperate climatic zones in northern North America are commonly subjected to periods of sub-zero temperatures, which leave no morphologically distinct features in the soils (Bentley 1978). Cryogenic soil features develop only under cold climate conditions with MAT of –2°C or less (Tarnocai 1997). In the absence of evidence for permafrost conditions (e.g. ice wedge casts, solifluction structures), it is difficult to detect evidence of freezing conditions in palaeosols (Van Vliet-Lanoe 1999). Nevertheless, some of the micromorphological features present in the Dunvegan Formation, including embedded grain argillans, compound silt–clay coatings and rounded pedogenic aggregates, are commonly interpreted as evidence for freezing conditions in Pleistocene palaeosols (Bullock & Murphy 1979; Chartres 1980; Fedoroff & Goldberg 1982; Kemp 1985, 1987; Tarnocai & Smith 1989) and their presence may suggest at least limited freezing conditions on Dunvegan floodplains inland from the seaway (Fedoroff *et al.* 1990).

These palaeoenvironmental interpretations,

based on Dunvegan palaeosols, are consistent with macrofloral data from the Dunvegan Formation (Bell 1963), which suggest a temperate, humid to sub-humid palaeoclimate with a non-seasonal distribution of precipitation (Upchurch & Wolfe 1993). On the basis of palaeobotanical data, Upchurch & Wolfe (1993) suggested a mean annual temperature of about 13°C for the Cenomanian period at around 70°N latitude, with a latitudinal temperature gradient of 0.3°C per degree latitude. The Dunvegan Formation had a Cenomanian palaeolatitude of *c.* 65°N (Irving *et al.* 1993), which suggests a mean annual temperature of about 14.5°C.

Parrish & Spicer (1989) interpreted flora from the Dunvegan Formation as mixed coniferous forest with a diverse angiosperm understorey, similar to the modern low-montane mixed coniferous forest. This physiognomy is similar to that from the Chandler Formation from the North Slope of Alaska and is consistent with forests growing at a mean annual temperature of no more than 13°C (Parrish & Spicer 1988*a*, *b*; Spicer & Parrish 1990). Using the nomogram devised by Wolfe (1979), Spicer & Parrish (1990) interpreted a mean annual temperature range of about 20°C for this low-montane coniferous forest. This implies a possible warm month mean temperature of *c.* 20°C and a cold month mean temperature of *c.* 0°C, with a maximum low temperature of no less than −11°C (Spicer & Parrish 1990; Parrish *et al.* 1998). Therefore, on the basis of both palaeopedological and palaeobotanical data, the Dunvegan floodplains may be interpreted to have formed under a cool, temperate palaeoclimate where short duration frost may have been a frequent occurrence, although a long, continuous winter freeze seems unlikely (Spicer & Parrish 1990).

Conclusions

Micromorphology is a valuable but under-utilized tool for the analysis of pedogenically modified floodplain deposits. Integrating data derived from palaeosols with traditional sedimentological and palaeobotanical data, and placing this information within its proper regional context, greatly enhances the palaeoenvironmental interpretation of floodplain successions. Therefore, floodplains, developed in aggradational settings with a high preservation potential, hold a largely untapped record of their pedosedimentary history.

Floodplain palaeosols from the Cenomanian Dunvegan Formation in northwestern Alberta and northeastern British Columbia, Canada, contain abundant, exceptionally well-preserved micromorphological features that are similar to those found in cool temperate Brunisols and Luvisols in North America today. The major pedogenic processes that operated in the Dunvegan palaeosols (clay illuviation, redox processes, siderite precipitation and pedoturbation) are interpreted from individual features. By analysing assemblages of features within individual palaeosols, the overall floodplain palaeoenvironment can be interpreted. Assemblages of micromorphological features in the Dunvegan Formation indicate cool temperate, humid to sub-humid conditions. Some features (embedded grain argillans, silt–clay compound coatings and rounded aggregates) suggest at least the possibility of limited freezing conditions on inland floodplains. Micromorphological evidence from Dunvegan floodplains, therefore, supports other sedimentological and palaeobotanical evidence for cool temperate palaeoenvironmental conditions at high latitudes during mid-Cretaceous time.

We appreciate the generous financial support provided for this research by Petro-Canada Ltd, Union Pacific Resources, Wascana Energy Inc., Amoco Canada Ltd and Husky Oil Operations; additional funding was provided by an NSERC operating grant to A. G. Plint. We thank G. Wood for thin-section production, I. Craig for assistance with preparation of photomicrographs, and C. McIntosh and M. Lumsdon for help in the field. We are grateful to V. P. Wright, M. J. Kraus and J. Alexander for comments that greatly improved the final version of this paper.

References

ALLEN, B. L. & HAJEK, B. F. 1989. Mineral occurrence in soil environments. *In*: DIXON, J. B. & WEED, S. B. (eds) *Minerals in Soil Environments*. Soil Science Society of America, Madison, WI, 199–278.

ASLAN, A. & AUTIN, W. J. 1998. Holocene flood-plain soil formation in the southern lower Mississippi Valley: implications for interpreting alluvial paleosols. *Geological Society of America Bulletin*, **110**, 433–449.

——, —— & TORNQVIST, T. E. 1995. Holocene to Wisconsinan sedimentation, soil formation and evolution of the Mississippi River flood plain, southern Mississippi valley (LMV). Geological Society of America, Field Trip Guidebook, **3**, 59–93.

BAIN, D. C., MELLOR, A., ROBERTSON-RINTOUL, M. S. E. & BUCKLAND, S. T. 1993. Variations in weathering processes and rates with time in a chronosequence of soils from Glen Feshie, Scotland. *Geoderma*, **57**, 275–294.

BELL, W. A. 1963. *Upper Cretaceous Floras of the Dunvegan, Bad Heart, and Milk River Formations of Western Canada*. Geological Survey of Canada, Bulletin, **94**.

BENTLEY, C. F. 1978. *Photographs and Descriptions of some Canadian Soils*. University of Alberta, Edmonton.

BESLEY, B. M. & FIELDING, C. R. 1989. Palaeosols in Westphalian coal-bearing and red-bed sequences, central and northern England. *Palaeogeography, Palaeoclimatology, Palaeoecology*, **70**, 303–330.

BHATTACHARYA, J. P. 1993. The expression and interpretation of marine flooding surfaces and erosional surfaces in core; examples from the Upper Cretaceous Dunvegan Formation, Alberta foreland basin, Canada. *In*: POSAMENTIER, H. W., SUMMERHAYES, C. P., HAQ, B. U. & ALLEN, G. P. (eds) *Sequence Stratigraphy and Facies Associations*. International Association of Sedimentologists, Special Publications, **18**, 125–160.

—— 1994. Cretaceous Dunvegan Formation of the Western Canada Sedimentary Basin. *In*: MOSSOP, G. D. & SHETSEN, I. (compilers) *Geological Atlas of the Western Canada Sedimentary Basin*. Canadian Society of Petroleum Geologists and Alberta Research Council, Calgary, 335–352.

—— & WALKER, R. G. 1991*a*. Allostratigraphic subdivision of the Upper Cretaceous Dunvegan, Shaftesbury, and Kaskapau formations in the northwestern Alberta subsurface. *Bulletin of Canadian Petroleum Geology*, **39**, 145–164.

—— & —— 1991*b*. River- and wave-dominated depositional systems of the Upper Cretaceous Dunvegan Formation, northwestern Alberta. *Bulletin of Canadian Petroleum Geology*, **39**, 165–191.

BOUMA, J., FOX, C. A. & MIEDEMA, R. 1990. Micromorphology of hydromorphic soils: applications for soil genesis and land evaluation. *In*: DOUGLAS, L. A. (ed.) *Soil Micromorphology: a Basic and Applied Science*. Elsevier, Amsterdam, 257–278.

BOWN, T. M. & KRAUS, M. J. 1987. Integration of channel and floodplain suites. I. Developmental sequence and lateral relations of alluvial palaeosols. *Journal of Sedimentary Petrology*, **57**, 587–601.

BREWER, R. 1976. *Fabric and Mineral Analysis of Soils*. Kriegerg, New York.

BULLOCK, P. & MURPHY, C. P. 1979. Evolution of a palaeo-argillic Brown Earth (Palaeudalf) from Oxfordshire, England. *Geoderma*, **22**, 225–252.

—— & THOMPSON, M. L. 1985. Micromorphology of Alfisols. *In*: DOUGLAS, L. A. & THOMPSON, M. L. (eds) *Soil Micromorphology and Soil Classification*. Soil Science Society of America, Special Publications, **15**, 17–48.

——, FEDOROFF, N., JONGERIUS, A., STOOPS, G. & TURSINA, T. 1985. *Handbook for Soil Thin Section Description*. Waine Research, Wolverhampton.

CALDWELL, W. G. E. & KAUFFMAN, E. G. (eds) 1993. *Evolution of the Western Interior Basin*. Geological Association of Canada, Special Paper, **39**.

——, NORTH, B. R., STELCK, C. R. & WALL, J. R. 1978. A foraminiferal zonal scheme for the Cretaceous System in the interior plains of Canada. *In*: STELCK, C. R. & CHATTERTON, B. D. E. (eds) *Western and Arctic Canadian Biostratigraphy*. Geological Association of Canada, Special Paper, **18**, 495–575.

CANADA SOIL SURVEY COMMITTEE, SUBCOMMITTEE ON SOIL CLASSIFICATION 1987. *The Canadian System of Soil Classification*. Canada Department of Agriculture Publication, 1646.

CHARTRES, C. J. 1980. A Quaternary soil sequence in the Kennet valley, central southern England. *Geoderma*, **23**, 125–146.

COURTY, M. A. & FEDOROFF, N. 1985. Micromorphology of recent and buried soils in a semi-arid region of northwestern India. *Geoderma*, **35**, 287–332.

DAY, J. H. 1983. *The Canadian Soil Information System (CanSIS). Manual for Describing Soils in the Field*, 1982 revised. Expert Committee on Soil Survey, Agriculture Canada, Ottawa, Ont, **LRRI 82-52**.

DRIESE, S. G., SIMPSON, E. L. & ERIKSSON, K. A. 1995. Redoximorphic palaeosols in alluvial and lacustrine deposits, 1.8 Ga Lochness Formation, Mount Isa, Australia: pedogenic processes and implications for palaeoclimate. *Journal of Sedimentary Research*, **A65**, 675–689.

DUCHAUFOUR, P. 1982. *Pedology*. George Allen & Unwin, London.

FANNING, D. S. & FANNING, M. C. B. 1989. *Soil Morphology, Genesis and Classification*. Wiley, New York.

FASTOVSKY, D. E. & MCSWEENEY, K. 1987. Paleosols spanning the Cretaceous–Paleogene transition, eastern Montana and western North Dakota. *Geological Society of America Bulletin*, **99**, 66–77.

FEDOROFF, N. 1997. Clay illuviation in red Mediterranean soils. *Catena*, **28**, 171–189.

—— & GOLDBERG, P. 1982. Comparative micromorphology of two late Pleistocene palaeosols (in the Paris basin). *Catena*, **9**, 227–252.

——, COURTY, M. A. & THOMPSON, M. L. 1990. Micromorphological evidence of palaeoenvironmental change in Pleistocene and Holocene palaeosols. *In*: DOUGLAS, L. A. (ed.) *Soil Micromorphology: a Basic and Applied Science*. Elsevier, Amsterdam, 653–666.

FITZPATRICK, E. A. 1993. *Soil Microscopy and Micromorphology*. Wiley, New York.

GINGRAS, M. K., MACEACHERN, J. A. & PEMBERTON, S. G. 1998. A comparative analysis of the ichnology of wave- and river-dominated allomembers of the Upper Cretaceous Dunvegan Formation. *Bulletin of Canadian Petroleum Geology*, **46**, 51–73.

GONZALEZ, L. A., LUDVIGSON, G. A., METZGER, R. A., WITZKE, B. J. & BRENNER, R. L. 1996. Meteoric sphaerosiderite lines: a tool for the study of paleohydrology and paleoclimates in humid continental environments. *Geological Society of America*, Abstracts with Programs, **28**, A36.

IRVING, E., WYNNE, P. J. & GLOBERMAN, B. R. 1993. Cretaceous paleolatitudes and overprints of North American craton. *In*: CALDWELL, W. G. E. & KAUFFMAN, E. G. (eds) *Evolution of the Western Interior Basin*. Geological Association of Canada, Special Paper, **39**, 91–96.

JOHNSON, D. L. & WATSON-STEGNER, D. 1987. Evolution model of pedogenesis. *Soil Science*, **143**, 349–366.

KALKREUTH, W. D. & MCMECHAN, M. E. 1984. Regional pattern of thermal maturation as determined from coal rank studies, Rocky Mountain Foothills and Front Ranges north of Grande Cache, Alberta—implications for petroleum exploration. *Bulletin of Canadian Petroleum Geology*, **35**, 263–295.

KEMP, R. A. 1985. *Soil Micromorphology*. Quaternary Research Association, Technical Guide, **2**.

—— 1987. Genesis and environmental significance of a buried middle Pleistocene soil in eastern England. *Geoderma*, **41**, 49–77.

KRAUS, M. J. & ASLAN, A. 1993. Eocene hydromorphic palaeosols: significance for interpreting ancient floodplain processes. *Journal of Sedimentary Petrology*, **63**, 453–463.

LANDUYDT, C. J. 1990. Micromorphology of iron minerals from bog ores of the Belgian Campine area. *In*: DOUGLAS, L. A. (ed.) *Soil Micromorphology: A Basic and Applied Science*. Elsevier, Amsterdam, 289–294.

LECKIE, D. A., FOX, C. A. & TARNOCAI, C. 1989. Multiple palaeosols of the Late Albian Boulder Creek Formation, British Columbia, Canada. *Sedimentology*, **36**, 307–323.

LUDVIGSON, G. A., GONZALEZ, L. A., METZGER, R. A., WITZKE, B. J., BRENNER, R. L., MURILLO, A. P. & WHITE, T. S. 1998. Meteoric sphaerosiderite lines and their use for paleohydrology and paleoclimatology. *Geology*, **26**, 1039–1042.

LYNN, W. C., TU, H. Y. & FRANZMEIER, D. P. 1971. Authigenic barite in soils. *Soil Science Society of America, Proceedings*, **33**, 160–161.

MACK, G. H. 1992. Palaeosols as indicators of climatic change at the Early–Late Cretaceous boundary, southwestern New Mexico. *Journal of Sedimentary Petrology*, **62**, 483–494.

——, JAMES, W. C. & MONGER, H. C. 1993. Classification of paleosols. *Geological Society of America Bulletin*, **105**, 129–136.

MARRIOTT, S. B. & WRIGHT, V. P. 1993. Palaeosols as indicators of geomorphic stability in two Old Red Sandstone alluvial suites, South Wales. *Journal of the Geological Society of London*, **150**, 1109–1120.

McCarthy, P.J. & King, R.H. 1997. Characteristics and genesis of two Gray Brown Luvisols, southwestern Ontario, Canada. *The Great Lakes Geographer*, **4**, 7-24.

—— & LECKIE, D. A. 1991. Preliminary observations on Lower Cretaceous (Albian) paleosols in the Mill Creek Formation of southwestern Alberta. *In*: *Current Research, Part B*. Geological Survey of Canada, Paper, **91-1B**, 51–58.

—— & PLINT, A. G. 1998. Recognition of interfluve sequence boundaries: integrating paleopedology and sequence stratigraphy. *Geology*, **26**, 387–390.

——, MARTINI, I. P. & LECKIE, D. A. 1997*a*. Anatomy and evolution of a Lower Cretaceous alluvial plain: sedimentology and palaeosols of the upper Blairmore Group, southwestern Alberta, Canada. *Sedimentology*, **44**, 197–220.

——, PLINT, A. G. & LONGSTAFFE, F. J. 1997*c*. Recognition and interpretation of interfluve sequence boundaries, Upper Cretaceous Dunvegan Formation, NE British Columbia. Canadian Society of Petroleum Geologists—Society for Sedimentary Geology (SEPM) Joint Annual Convention, 1–6 June, Calgary, AB., Abstract volume, 186.

——, MARTINI, I. P. & LECKIE, D. A 1997*b*. Pedosedimentary history and floodplain dynamics of the Lower Cretaceous upper Blairmore Group, southwestern Alberta, Canada. *Canadian Journal of Earth Sciences*, **34**, 598–617.

——, —— & —— 1998*a*. Use of micromorphology for interpretation of complex alluvial palaeosols: examples from the Mill Creek Formation (Albian), southwestern Alberta, Canada. *Palaeogeography, Palaeoclimatology, Palaeoecology*, **143**, 87-110.

——, FACCINI, U. F. & PLINT, A. G. 1998*b*. Evolution of an ancient floodplain: palaeosols and alluvial architecture in a sequence stratigraphic framework, Cenomanian Dunvegan Formation, NE British Columbia, Canada. *Floodplains '98*, 24–26 June, Norwich, UK, Abstract volume, 29.

——, MARTINI, I. P. & LECKIE, D. A 1999*a*. Pedogenic and diagenetic influences on void coating formation in Lower Cretaceous palaeosols of the Mill Creek Formation, southwestern Alberta, Canada. *Geoderma*, **87**, 209–237.

——, FACCINI, U. F & —— 1999*b*. Evolution of an ancient coastal plain: palaeosols, interfluves and alluvial architecture in a sequence stratigraphic framework, Cenomanian Dunvegan Formation, NE British Columbia, Canada. *Sedimentology*, **46**, in press.

MCKEAGUE, J. A. 1983. Clay skins and the argillic horizon. *In*: BULLOCK, P. & MURPHY, C. P. (eds) *Soil Micromorphology, Volume 2. Soil Genesis*. AB Academic, Berkhamsted, 367–388.

——, MACDOUGALL, J. I., LANGMAID, K. K. & BOURBEAU, G. A. 1969. Macro- and micromorphology of ten reddish-brown soils from the Atlantic provinces. *Canadian Journal of Soil Science*, **49**, 53–63.

——, MILES, N. M., PETERS, T. W. & HOFFMAN, D. W. 1972. A comparison of Luvisolic soils from three regions in Canada. *Geoderma*, **7**, 49–69.

MUCHER, H. J. & MOROZOVA, T. D. 1982. The application of soil micromorphology in Quaternary geology and geomorphology. *In*: BULLOCK, P. & MURPHY, C. P. (eds) *Soil Micromorphology*. AB Academic, Berkhamsted, 151–194.

MURPHY, C. P. 1986. *Thin Section Preparation of Soils and Sediments*. Pudoc, Wageningen.

NADON, G. C. 1994. The genesis and recognition of anastomosed fluvial deposits: data from the St. Mary River Formation, southwestern Alberta, Canada. *Journal of Sedimentary Research*, **B64**, 451–463.

NETTLETON, W. D., BRASHER, B. R., BAUMER, O. W. & DARMODY, R. G. 1994. Silt flow in soils. *In*: RINGROSE-VOASE, A. J. & HUMPHREYS, G. S. (eds) *Soil Micromorphology: Studies in Management and Genesis*. Elsevier, Amsterdam, 361–372.

PARRISH, J. T. & SPICER, R. A. 1988*a*. Late Cretaceous terrestrial vegetation: a near-polar temperature curve. *Geology*, **16**, 22–25.

—— & —— 1988*b*. Middle Cretaceous wood from the Nanushuk Group, central North Slope, Alaska. *Palaeontology*, **31**, 19–34.

—— & —— 1989. Flora of the Dunvegan Formation (Cenomanian, British Columbia) and comparison with the flora of the Niakogon tongue of the Chandler Formation (Cenomanian, Alaska).

Geological Society of America, Abstracts with Programs, **21**, A147.

——, DANIEL, I. L., KENNEDY, E. M. & SPICER, R. A. 1998. Palaeoclimatic significance of mid-Cretaceous floras from the Middle Clarence Valley, New Zealand. *Palaios*, **13**, 149–159.

PIPUJOL, M. D. & BUURMAN, P. 1994. The distinction between ground-water gley and surface-water gley phenomena in Tertiary palaeosols of the Ebro basin, NE Spain. *Palaeogeography, Palaeoclimatology, Palaeoecology*, **110**, 103–113.

—— & —— 1997. Dynamics of iron and calcium carbonate redistribution and palaeohydrology in middle Eocene alluvial palaeosols of the southeast Ebro Basin margin (Catalonia, northeast Spain). *Palaeogeography, Palaeoclimatology, Palaeoecology*, **134**, 87–107.

PLINT, A. G. 1996. Marine and nonmarine systems tracts in fourth-order sequences in the Early–Middle Cenomanian Dunvegan Alloformation, northeastern British Columbia, Canada. *In*: HOWELL, J. A. & AITKEN, J. F. (eds) *High Resolution Sequence Stratigraphy: Innovations and Applications.* Geological Society, London, Special Publications, **104**, 159–191.

—— & WADSWORTH, J. 1997. Interacting tectonic and eustatic controls on regional palaeogeography, incised valley systems and sequence geometry: Cenomanian Dunvegan Formation, Alberta basin, Canada. International Association of Sedimentologists, *18th Regional European Meeting of Sedimentology*, 2–4 September, Heidelberg, Germany, Abstract volume, 278.

RETALLACK, G. J. 1991. Untangling the effects of burial alteration and ancient soil formation. *Annual Review of Earth and Planetary Sciences*, **19**, 183–206.

SPICER, R. A. & PARRISH, J. T. 1990. Late Cretaceous–early Tertiary palaeoclimates of northern high latitudes: a quantitative view. *Journal of the Geological Society of London*, **147**, 329–341.

STOOPS, G. J. & ZAVALETA, A. 1978. Micromorphological evidence of barite neoformation in soils. *Geoderma*, **20**, 63–70.

STOTT, D. F. 1963. *The Cretaceous Alberta Group and Equivalent Rocks, Rocky Mountain Foothills, Alberta.* Geological Survey of Canada, Memoir, **317**.

—— 1982. *Lower Cretaceous Fort St. John Group and Upper Cretaceous Dunvegan Formation of the Foothills and Plains of Alberta, British Columbia, District of Mackenzie and Yukon Territory.* Geological Survey of Canada, Bulletin, **328**.

TARNOCAI, C. 1997. Palaeosols of the northern part of North America: their features and significance as indicators of past climates. *In*: MARTINI, I. P. (ed.) *Late Glacial and Post-Glacial Environmental Changes.* Oxford University Press, Oxford, 1–24.

—— & SMITH, C. A. S. 1989. Micromorphology and development of some central Yukon palaeosols, Canada. *Geoderma*, **45**, 145–162.

UPCHURCH, G. R. & WOLFE, J. A. 1993. Cretaceous vegetation of the Western Interior and adjacent regions of North America. *In*: CALDWELL, W. G. E. & KAUFFMAN, E. G. (eds) *Evolution of the Western Interior Basin.* Geological Association of Canada, Special Paper, **39**, 243–282.

VAN VLIET-LANOE, B. 1999. Frost and soils: implications for palaeosols, palaeoclimates and stratigraphy. *Catena*, **34**, 157–183.

VEPRASKAS, M. J., WILDING, L. P. & DREES, L. R. 1994. Aquic conditions for soil taxonomy: concepts, soil morphology and micromorphology. *In*: RINGROSE-VOASE, A. J. & HUMPHREYS, G. S. (eds) *Soil Micromorphology: Studies in Management and Genesis.* Elsevier, Amsterdam, 117–131.

WADSWORTH, J. A. & PLINT, A. G. 1997. Distribution and character of incised valley deposits, Cenomanian Dunvegan Formation, Alberta basin. Canadian Society of Petroleum Geologists–Society for Sedimentary Geology (SEPM) Joint Annual Convention, 1–6 June, Calgary, AB., Abstract volume, 286.

WANG, C., BREWSTER, G. R. & WEBB, K. T. 1995. Micromorphological evidence of pedogenetic pathway of a Podzolic Gray Luvisol (Falmouth series) in Nova Scotia. *Canadian Journal of Soil Science*, **75**, 491–496.

WOLFE, J. A. 1979. *Temperature Parameters of Humid to Mesic Forests of Eastern Asia and Relation to Forests of other Regions of the Northern Hemisphere and Australasia.* US Geological Survey, Professional Paper, **1106**.

Nature and distribution of heavy minerals in the Natal Group, South Africa

K. W. LIU

Department of Geology, University of Durban–Westville. Private Bag X54001, Durban 4000, South Africa (e-mail:kwliu@pixie.udw.ac.za)

Abstract: The author provides an example of heavy minerals concentrated in floodplain sediments. Nine types of heavy mineral have been found in fluvial sediments of the Ordovician Natal Group, South Africa, among which ilmenite is the most abundant. These heavy minerals are concentrated mainly in the Eshowe and Inanda Formations of the succession, and vary in content from 1.6 to 20.1% (average contents of 3.86% and 4.18%, respectively). This is in contrast to the shallow-marine Kranskloof and Dassenhoek Members, with average contents of only 1.14% and 1.27%, respectively. Host rocks of the heavy minerals are mainly arkose and lithic arkose, with subordinate subarkose. Grain-size analysis shows that heavy mineral rich beds are composed mostly of fine to medium size sand and are better sorted than the host rocks. The main depositional environments for the heavy minerals are branched or braided channels, rather than the main channel, as the moderate hydrodynamic regime is more favourable for the accumulation of fine to medium arenaceous heavy minerals. Palaeogeography and mineral assemblages confirm that the main provenance of the heavy minerals lay to the north of the depositional basin, namely, the Archaean Kaapvaal Craton, with the Natal Structural and Metamorphic Province as a supplementary source.

Floodplain sediments are widely distributed in modern and ancient sedimentary environments as a result of flooding events. Most of these events are catastrophic disasters that cause heavy loss to properties and lives. However, on the other hand, individual events may construct large fertile areas suitable for agriculture and fishery, and also accumulate mineral deposits. In the case of the Ordovician sequence of the Natal Group (*c*. 490 Ma), South Africa, heavy minerals including ilmenite, magnetite, rutile, titanite, zircon, hematite, garnet, apatite and hornblende have been widely deposited in a floodplain environment, and have locally reached economic potential. This paper aims to describe the mineralogical and petrological characteristics, and the geological occurrences, of the heavy minerals, and further to discuss the sedimentary environments and provenance of the heavy minerals in the Natal Group.

Stratigraphy and distribution

The Ordovician Natal Group, distributed along the east coast of South Africa, is predominantly a fluvial sequence (Visser 1974; Kingsley 1975; SACS (South Africa Committee for Stratigraphy) 1980; Thomas *et al.* 1993*a*; Marshall 1989, 1994; Liu 1997). Geographically, it extends from Hlabisa (Zululand) in the north to Hibberdene in the south. The sediments were deposited mainly in a floodplain environment covering an area over 280 km in length in a NE–SW direction and over 70 km in width (Fig. 1). Stratigraphically, the Natal Group has been subdivided into five formations (SACS 1980): the Eshowe, Mkunya, Mlazi, Inanda and Hibberdene Formations (Fig. 2). The Eshowe Formation is predominantly a reddish brown arkose and subarkose unit with subordinate reddish purple mudstone. Heavy minerals are disseminated in this unit and the distribution areas reach north to Ulundi and south to Hibberdene. The Mkunya Formation, in the middle of the sequence (Fig. 2), consists of the Kranskloof, Situndu and Dassenhoek Members and is composed chiefly of quartz arenite with lesser amounts of subarkose. Both the Kranskloof and Dassenhoek Members have been described as shallow marine sediments, including tidalites (Roberts 1981, 1990; Liu & Cooper 1998), in which heavy minerals are also present, but less abundant. The Situndu Member, separating the Kranskloof and Dassenhoek Members, represents a return to fluvial sedimentation during a period of marine regression. It comprises mainly reddish fine conglomerate and coarse sandstone, with sparse heavy minerals present in the strata. The Mlazi

From: MARRIOTT, S. B. & ALEXANDER, J. (eds) 1999. *Floodplains: Interdisciplinary Approaches*. Geological Society, London, Special Publications, **163**, 311–325. 1-86239-050-9/99/$15.00

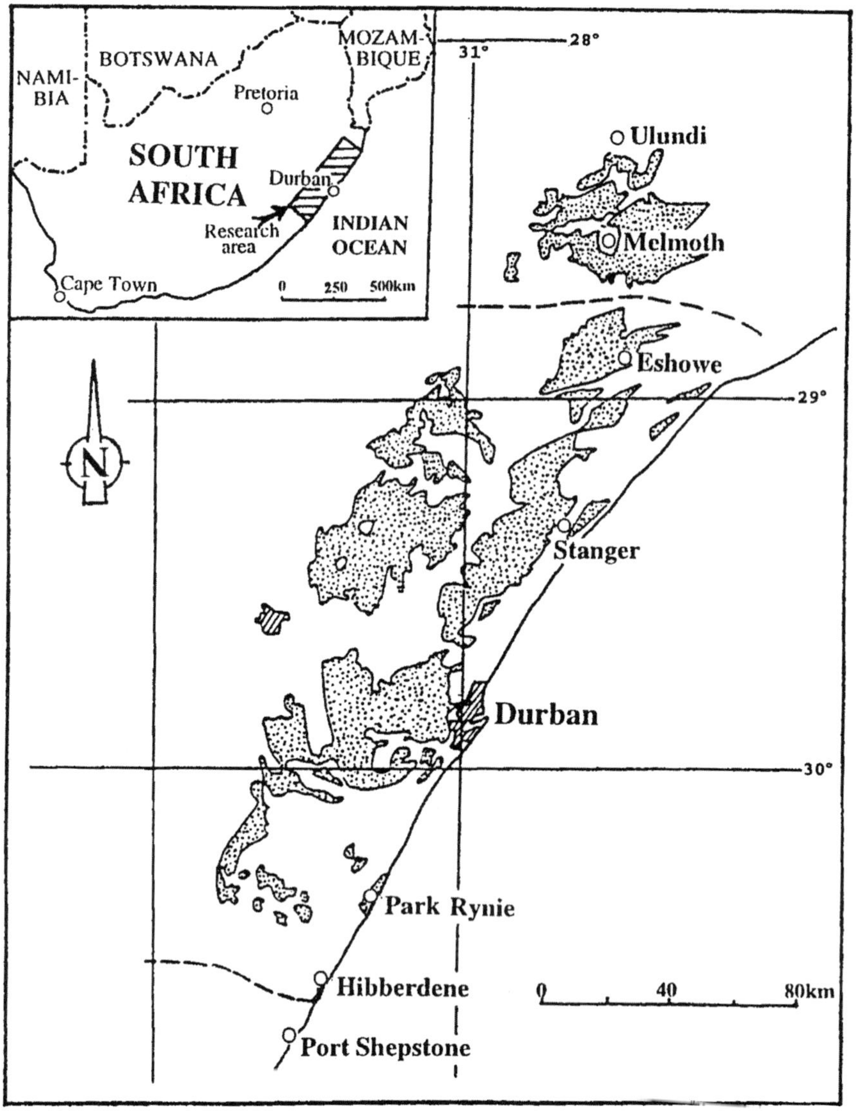

Fig. 1. Distribution of the Natal Group.

Formation, similar to the Situndu Member, is also a brownish fine conglomerate and coarse sandstone unit of fluvial origin. Its deposition was a result of rejuvenation of topographic relief in the source area to the north of the depositional basin. Heavy minerals, however, are less concentrated in this unit. The Inanda Formation is the thickest and most widely distributed unit in the Natal Group, comprising brownish red sandstone and mudstone, and representing floodplain sedimentation in a wide rift basin (Hobday & von Brunn 1979; Marshall 1994). Heavy minerals are very common in the Inanda Formation and are distributed throughout the unit. The Hibberdene Formation at the top of the Natal Group is a shallow marine unit that is totally different from the underlying fluvial Natal Group. From its stratigraphy and sedimentology, it is clear that it belongs to another depositional cycle and it has much in common with the Msikaba Formation (Visser 1974). Therefore, this unit will not be discussed here. In any case, few heavy minerals are present in the Hibberdene Formation.

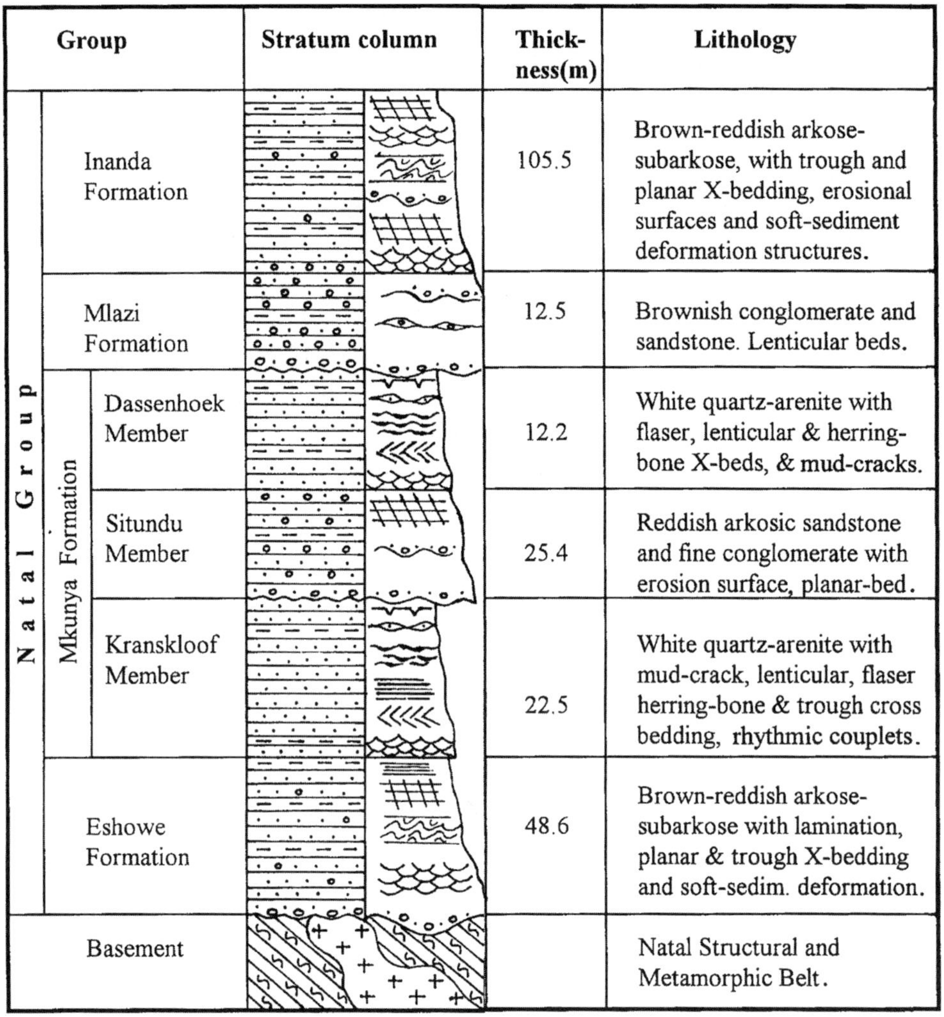

Fig. 2. Stratigraphic subdivision of the Natal Group.

Field occurrences

Field investigations show that heavy minerals mainly occur in the Eshowe and Inanda Formations; they are less concentrated in the Kranskloof and Dassenhoek Members, and are sparse in the Situndu Member and Mlazi Formation. Four types of field occurrence of heavy mineral have been recorded in the Natal Group: thin beds, laminations, concentrated pockets, and disseminated grains scattered throughout the host rocks.

Thin, heavy mineral beds are usually 100–300 mm thick and dark coloured, extending laterally more than 20 m without disruption. Small-scale cross-bedding, ripple marks and lamination structures are found commonly in the heavy mineral beds (Fig. 3B). In some strata, convolute bedding is also developed (Fig. 3A), indicating soft-sediment deformation before consolidation (Matthews 1961; Rhodes & Leith 1967; Mathew 1971). This type of heavy mineral occurrence is found only in the Eshowe and Inanda Formations, which are the two richest heavy mineral bearing units in the Natal Group. In the north of Eshowe, at least three thin beds (total thickness 0.75 m) of heavy mineral layers have been found in an exposure 15.4 m thick.

Heavy mineral laminae are dark coloured and usually <10 mm in thickness, alternating with light

Fig. 3. Field photographs of thin-bedded heavy minerals. (**A**) Thin-bedded heavy minerals with convoluted laminae at Park Rynie beach. (**B**) Thin-bedded heavy minerals showing small-scale cross-bedding and ripple marks, north Eshowe. Pencil in the middle for scale. It should be noted that soft-sediment deformation was a post-depositional phenomenon, unrelated to heavy mineral concentration.

coloured laminae made up of detrital quartz and feldspar (see Fig. 5D, below). Individual laminae extend laterally in the field for only 0.5–6 m, and then pinch out or amalgamate with light quartzo-feldspathic layers.

Heavy mineral pockets vary in size from 10 to 50 mm in diameter and are patchily or irregularly shaped, often accumulating behind obstacles, such as small pebbles, as a result of a reduction in hydrodynamic energy in the lee of such obstacles. A single pocket can concentrate hundreds of heavy mineral grains mixed with other detrital light minerals (see Fig. 5B and C).

Disseminated grains are the most common type of heavy mineral occurrence in the Natal Group, and are distributed throughout the strata (see Fig. 5A), particularly the Eshowe and Inanda Formations.

There is a general tendency for heavy minerals to be concentrated in the lower part of each formation in the vertical profile, and also a tendency for them to be concentrated in the middle to lower reaches, rather than the upper reaches in the palaeostream system.

Petrological characteristics

A Swift automatic point counter attached to a Nikon polarizing microscope was used for modal composition analysis, with >550 point counts for each sample. Results showed that the different stratigraphic units of the Natal Group have distinct mineral compositions and, particularly, heavy mineral concentrations (Table 1). The Eshowe and Inanda Formations have the highest heavy mineral content, and the Situndu Member and Mlazi Formation the lowest, whereas the Kranskloof and Dassenhoek Members have intermediate heavy mineral contents. These differences are strongly linked to the lithologies and depositional environ-

Table 1. *Modal mineral compositions (%) of different stratigraphic units in the Natal Group*

Stratum unit	Slides	Qtz+Cht	Feldspar	Lithics	Matrix	Heavies	Average
Inanda Formation	12	51.8–70.4	13.6–28.7	3.1–15.4	Common	1.6–20.1	4.18
Mlazi Formation	4	50.3–68.8	13.3–27.6	4.0–14.1	Common	0.3–3.2	0.66
Dassenhoek Member	6	76.2–82.6	4.8–11.2	6.3–10.4	Rare	0.6–4.3	1.14
Situndu Member	7	59.5–70.3	16.3–30.5	4.2–11.7	Common	0.4–3.5	0.85
Kranskloof Member	6	77.4–86.6	3.8–8.8	5.1–10.3	Rare	0.5–4.8	1.27
Eshowe Formation	11	52.6–70.4	12.2–29.3	4.8–15.1	Common	1.8–19.2	3.86

Slides, statistical numbers of thin sections; Qtz, quartz; Cht, Chert; Heavies, percentage of heavy minerals; Average, average percentage of heavy minerals.

ments, as the Eshowe and Inanda Formations are fluvial in origin, whereas the Kranskloof and Dassenhoek Members are shallow marine. Although the Situndu Member and Mlazi Formation are also fluvial units, their lithologies are mainly conglomeratic and are too coarse to concentrate heavy minerals that are mainly medium to fine sand sized.

Ternary classification (Fig. 4) shows that the samples from the Kranskloof and Dassenhoek Members fall mainly in the quartz arenite field, with a few in the subarkose field; samples from the Eshowe and Inanda Formations are mainly arkose and lithic arkose, with a few falling in the subarkose field. It is clear that heavy mineral concentration is strongly linked to arkosic and lithic–arkosic lithologies, rather than quartz arenites. Petrological study also shows that grain roundness and sorting of both the Eshowe and Inanda Formations is much poorer than that of the Kranskloof and Dassenhoek Members. It is likely, therefore, that texturally and compositionally immature rocks of fluvial origin are more favourable for hosting heavy minerals in the Natal Group.

Petrographic study in thin sections shows that heavy minerals occur also as disseminated grains (Fig. 5A), concentration pockets (Fig. 5B and C) and laminae (Fig. 5D), replicating their macroscopic occurrences in the field. It can be seen that

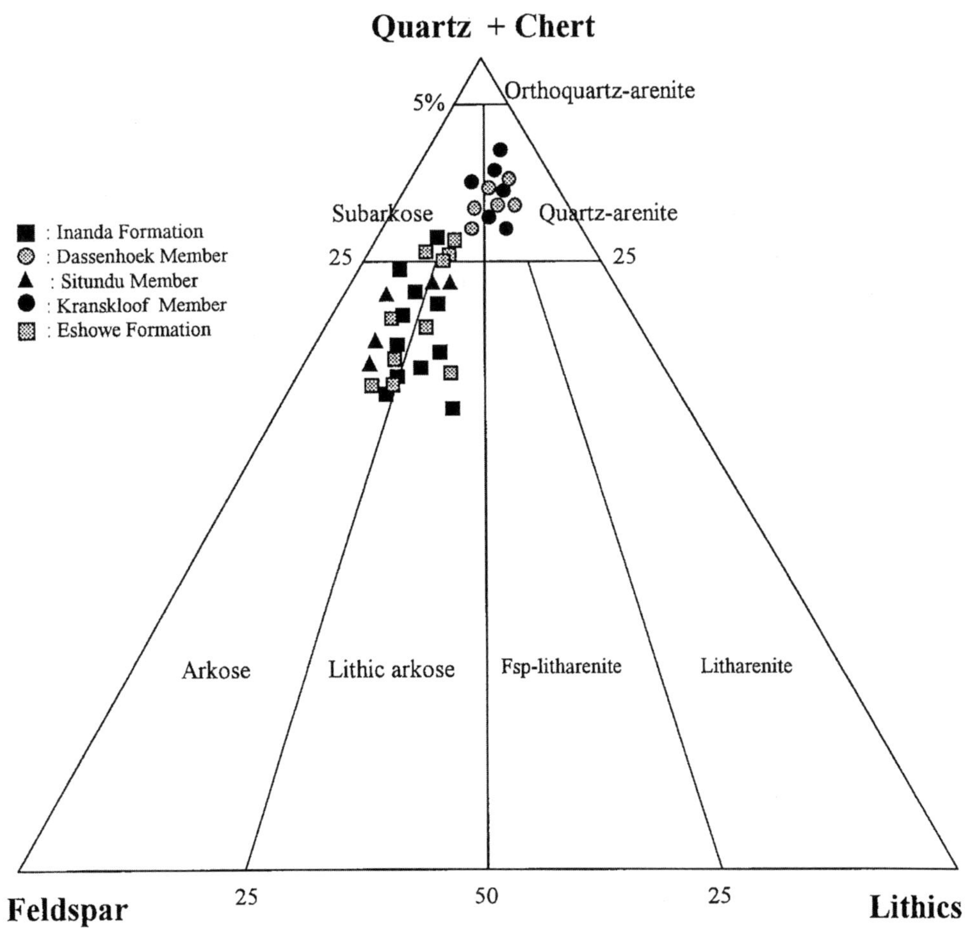

Fig. 4. Ternary classification diagram for the Natal Group sandstones. Heavy minerals occur mainly in the Eshowe and Inanda Formations, which are fluvial units dominated by arkose and lithic arkose; whereas the Kranskloof and Dassenhoek Members are shallow marine sequences characterized by quartz arenite and subarkose, and are poorer in heavy minerals.

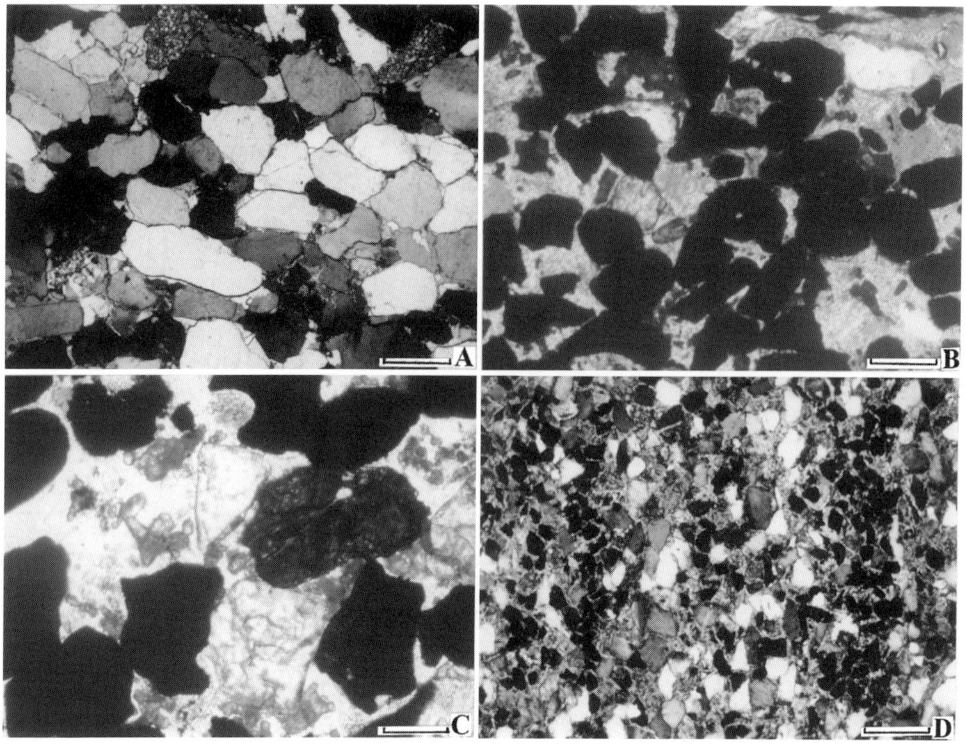

Fig. 5. Microscopic occurrences of heavy minerals. All photographs are under cross-polarized light. (**A**) Ilmenite (black) disseminated among detrital quartz and feldspar (white and grey). The scale bar represents 0.30 mm. (**B**) A concentrated pocket of ilmenite and magnetite (black), in which some of the grains have been welded together by diagenetic processes. It is difficult to distinguish magnetite from ilmenite under transmitted light. The scale bar represents 0.30 mm. (**C**) A concentrated pocket of ilmenite (black) and rutile (middle right, red interference colour). The scale bar represents 0.2 mm. (**D**) Heavy mineral laminae of ilmenite and magnetite (black). It should be noted that stratigraphical top is to the left. The scale bar represents 1.2 mm.

most of the heavy minerals are subrounded to rounded in shape, and have uniform grain size in each individual layer, indicating good sorting and textural maturity that is better than that of the quartzo-feldspathic grains. As a result of diagenetic modification, some of the heavy mineral grains have greatly changed from their original shapes and have been welded together (Fig. 5B).

Mineralogy of heavy minerals

Microscopy and scanning electron microscopy (SEM; JEOL Model JSM-6100) with energy dispersive spectrometry (EDS) were employed for mineralogical study of the heavy minerals.

Nine types of heavy mineral were found in the Natal Group: zircon, ilmenite, apatite, garnet, rutile, magnetite, titanite, hornblende and hematite. Among them, ilmenite is the most abundant.

Rutile (TiO_2)

Rutile is one of the common heavy minerals present in the Natal Group. In reflected light with oil-immersion lenses, rutile is bluish grey or brownish in colour and shows weak bireflectance and pleochroism. Its high anisotropism is often masked by strong colourful internal reflections. The shape of the rutile grains is rounded to well rounded, with high sphericity. Grain sizes range between 0.15 and 0.4 mm, which is usually coarser than the para-genetic magnetite and ilmenite (Fig. 5C), probably as a result of its lower special gravity. SEM-EDS diagram shows a strong peak of Ti with no other peak present, and the TiO_2 content is >97% (Fig. 6, Table 2).

Magnetite (Fe_3O_4)

Magnetite occurs as opaque, iron-black, rounded to subrounded grains and shows magnetic character,

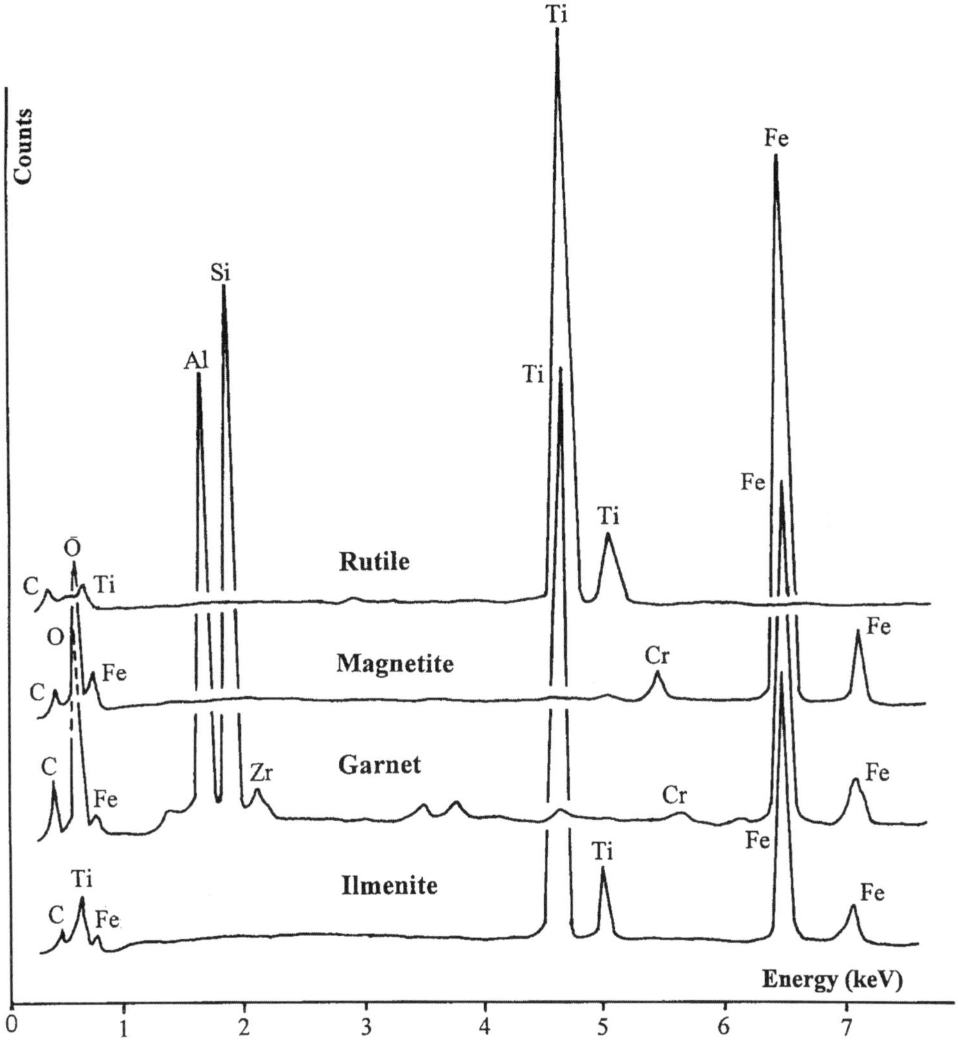

Fig. 6. Representative SEM-EDS patterns of the heavy minerals rutile, magnetite, garnet and ilmenite.

with grain sizes commonly in the range of 0.1–0.25 mm in diameter. Because of diagenetic modification, some of the grains have been welded together and become irregular (Fig. 5). Typical chemical compositions of magnetite by SEM-EDS analysis show that the dominant peak is iron (FeO=93.45%), with some small peaks of Cr, V and Ti, which probably occur as replacement elements (Fig. 6, Table 2).

Garnet ($Fe_3Al_2(SiO_4)_3$)

Garnet occurs as rounded, medium sand-sized grains and is only a minor heavy mineral in the Natal Group. It is translucent, brownish red in parallel light and isometric in polarized light, with a high refractive index. Garnet is a little coarser than the paragenetic magnetite and ilmenite in the thin sections and, sometimes, shows internal circular zoning. SEM-EDS analysis shows three strong peaks of Si, Al and Fe, with SiO_2=30.33%, Al_2O_3=23.29% and FeO=37.52% (Fig. 6, Table 2); therefore, it belongs to the Fe–Al series of garnet.

Ilmenite ($FeTiO_3$)

Iron–titanium oxide minerals are resistant to weathering and, therefore, remain in the tail of lag

Table 2. *Representative chemical compositions (wt %) of heavy minerals by SEM-EDS analysis*

	Na_2O	MgO	Al_2O_3	SiO_2	P_2O_5	S	K_2O	CaO	TiO_2	V_2O_5	Cr_2O_3	MnO	FeO	ZnO	ZrO_2	O
Rutile (TiO_2)																
wt %	0.08	0.00	0.00	0.16	0.00	0.00	0.01	0.05	97.04	0.53	0.04	0.10	1.11	0.50	0.38	–
Stoic.	0.02	0.00	0.00	0.03	0.00	0.00	0.00	0.01	11.68	0.06	0.01	0.01	0.14	0.06	0.03	24
Magnetite (Fe_3O_4)																
wt %	0.02	0.08	0.07	0.07	0.00	0.00	0.03	0.05	1.02	1.14	4.02	0.00	93.45	0.00	0.05	–
Stoic.	0.01	0.03	0.02	0.02	0.00	0.00	0.01	0.01	0.20	0.21	0.87	0.00	21.32	0.00	0.01	24
Garnet ($Fe_3Al_2(SiO_4)_3$)																
wt %	0.94	0.65	23.29	30.33	0.00	0.00	0.67	0.93	1.01	0.25	1.00	0.63	37.54	0.88	1.88	–
Stoic.	0.31	0.16	4.42	5.10	0.00	0.00	0.14	0.17	1.13	0.03	0.13	0.09	5.24	0.11	0.15	24
Ilmenite ($FeTiO_3$)																
wt %	0.61	0.00	0.08	0.40	0.00	0.00	0.05	0.20	53.36	1.50	0.31	0.35	43.01	0.13	0.00	–
Stoic.	0.23	0.00	0.02	0.08	0.00	0.00	0.01	0.04	7.77	0.19	0.05	0.06	6.97	0.02	0.00	24
Zircon ($ZrSiO_4$)																
wt %	0.43	0.16	0.62	31.03	0.00	0.00	0.00	1.12	0.59	0.61	0.08	0.09	1.72	0.79	62.75	–
Stoic.	0.15	0.04	0.14	5.72	0.00	0.00	0.00	0.22	0.08	0.07	0.01	0.02	0.26	0.11	5.64	24
Apatite ($Ca_5(PO_4)_3F$)																
wt %	0.28	0.01	0.12	0.52	37.84	0.00	0.03	60.59	0.34	0.00	0.00	0.00	0.27	0.00	0.00	–
Stoic.	0.03	0.00	0.02	0.09	5.22	0.00	0.01	10.58	0.04	0.00	0.00	0.00	0.04	0.00	0.00	24
Hornblende ($(Ca,Mg,Al,Fe)(Si_4O_{11})_2(OH)_2$)																
wt %	1.04	10.57	13.84	34.93	0.22	0.05	4.91	6.33	2.23	0.22	0.08	0.75	24.75	0.07	0.00	–
Stoic.	0.34	2.63	2.72	5.82	0.03	0.02	1.04	1.13	0.28	0.02	0.01	0.11	3.45	0.01	0.00	24
Titanite ($CaTiSiO_4(O,F)$)																
wt %	0.10	0.00	0.18	24.16	0.00	0.00	0.11	20.10	54.61	0.00	0.06	0.00	0.69	0.00	0.00	–
Stoic.	0.03	0.00	0.03	3.68	0.00	0.00	0.02	3.47	6.25	0.00	0.01	0.00	0.08	0.00	0.00	24
Hematite (Fe_2O_3)																
wt %	1.12	0.10	1.62	1.48	0.00	0.00	0.85	0.08	0.52	1.47	0.85	1.56	90.26	0.00	0.08	–
Stoic.	0.60	0.04	0.51	0.13	0.00	0.00	0.18	0.02	0.11	0.27	0.18	0.22	20.59	0.00	0.01	24

Stoic, stoichiometry; the results are based on 24 oxygen atoms. All iron reported as FeO.

deposits to form heavy mineral deposits in the Natal Group. In reflected light with oil immersion, ilmenite is dark pinkish to dark brownish with moderate bireflectance and pleochroism, but high anisotropism. It is fine to medium sand-sized, with moderate roundness and sorting. Ilmenite always occurs with magnetite and they are very difficult to distinguish both with the naked eye and even with a microscope. X-ray diffraction also does not satisfactorily distinguish them, but SEM-EDS and electron microprobe are useful tools for discrimination of these two minerals. There are two strong peaks of Ti and Fe in the SEM-EDS figures, and both the TiO_2 and FeO are over 40% in the Natal Group samples (Fig. 6, Table 2).

Zircon ($ZrSiO_4$)

Zircon in the Natal Group occurs as rounded or slightly prismatic grains of fine to medium sand-size, and is slightly reddish, brownish or colourless in parallel light. Some zircons are inhomogeneous and display circular zoning parallel to grain boundaries. Zircon shows a high order of interference colours of bright pink, green and yellow, and also shows parallel extinction characteristics. SEM-EDS analysis shows two strong peaks of Zr and Si, with ZrO_2=62.75% and SiO_2=31.03%, which is near the perfect stoichiometric molecular formula for zircon (Fig. 7, Table 2).

Apatite ($Ca_5(PO_4)_3F$)

Apatite occurs as very fine sand-sized grains and is only a minor heavy mineral in the Natal Group. It occurs as transparent, colourless or light green, subprismatic or rounded grains with low interference colour. Apatite shows parallel extinction and is always dispersed among other heavy minerals, namely, ilmenite and rutile. SEM-EDS analysis shows two strong peaks of Ca and P, with P_2O_5=37.84% and CaO=60.59%, which perfectly

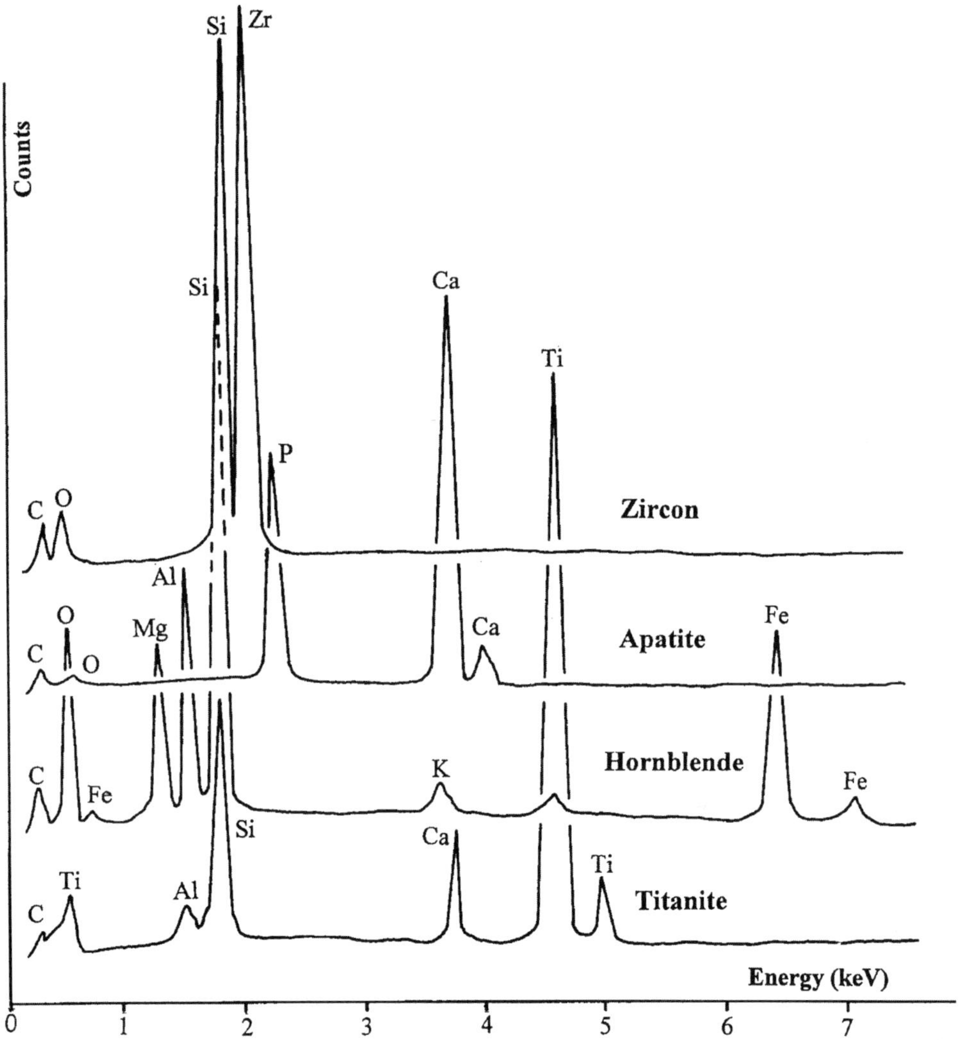

Fig. 7. Representative SEM-EDS patterns of the heavy minerals zircon, apatite, hornblende and titanite.

matches the stoichiometric molecular formula of apatite (Fig. 7, Table 2).

Hornblende $(Ca,Mg,Al,Fe)(Si_4O_{11})_2(OH)_2)$

Hornblende occurs in the Natal Group as medium sand-sized grains with low to medium roundness and sphericity, and is greenish in parallel light and pleochroic in polarized light, with characteristic rhombic cleavage. ESM-EDS analysis shows strong Si, Al, Mg and Fe peaks, with MgO=10.57%, Al_2O_3=13.84%, FeO (total)=24.75%, SiO_2=34.93% and CaO=6.33% (Fig. 7, Table 2); therefore, it is an Fe–Mg-rich variety. Hornblende is a rare heavy mineral in the Natal Group, probably reflecting a dry palaeoclimate and rapid depositional history of the sediments.

Titanite $(CaTiSiO_4(O,F))$

Titanite is less common in the Natal Group, and is a light brownish or colourless mineral in parallel light with high interference colour in polarized light and strong pleochroism. It is characterized by wedge-shaped crystals and {110} cleavage. SEM-EDS analysis shows a strong Ti peak and less strong Ca and Si peaks, with TiO_2=54.61% (Fig. 7, Table 2).

Hematite (Fe_2O_3)

Hematite occurs as rounded grains of fine sand-size with high sphericity, and is normally opaque but with a slight red brownish colour in very thin sections in transmitted light, which serves to distinguish it from magnetite. The SEM-EDS pattern of hematite is very similar to that of magnetite, with a dominant iron peak and FeO (total iron) >90% (Table 2). Hematite is only a minor heavy mineral in the Natal Group.

The above-mentioned nine types of heavy minerals are not equally distributed in the strata. On the basis of the field investigation and statistics of eight thin sections, ilmenite is the most abundant heavy mineral in the Natal Group, followed by magnetite and rutile. Calculating the total heavy mineral content as 100%, an average content of the different types of heavy minerals is as follows: ilmenite 57.8%, magnetite 13.5%, rutile 8.7%, titanite 5.6%, zircon 4.4%, garnet 3.7%, hematite 3.2%, apatite 1.8% and hornblende 1.3%.

Grain-size distribution

Grain-size analysis was undertaken with a Nikon polarizing microscope with >550 point counts for each sample. The standard method of long (*a*) axis measurement was employed and the apparent dimensions of thin-section grain sizes were converted to normal sieve size before calculation (Friedman 1958, 1962). The counted grains included quartz, feldspar, lithics and heavy minerals, and the finest grain size counted was 5ϕ, which is satisfactory for common use (Visher 1969; Dai & Chen 1978; Johnson 1994; Liu & Greyling 1996).

Results of grain-size analysis (Fig. 8) show that there are major differences between heavy mineral beds and their host strata. Grain-size distribution of the heavy mineral beds is finer, more uniform and better sorted than that of the host rocks. Heavy mineral beds have grain sizes ranging from 1.5 to 5ϕ, with dominant sizes between 0.1 and 0.3 mm, whereas the host rocks have a grain-size distribution ranging from –1 to 5ϕ, that is, very coarse to very fine sands, and thus poorly sorted. In a single bed, light minerals, such as quartz and feldspar, are always coarser than the heavy minerals. The cumulative frequency percentage curves of the heavy mineral beds are steeper, and consist of only two segments, suspension and saltation loads, lacking traction loads. On the other hand, the host rocks show three segments, i.e. suspension, saltation and traction loads, and the cumulative frequency percentage curves are much flatter, implying poorer sorting. These differences between heavy mineral beds and host rocks are also found in other heavy mineral deposits in the world (Force & Stone 1990; Force 1991), and the reasons are believed to be different specific gravity and settling velocities of different minerals during transportation and deposition, that is, coarser light minerals and finer, heavy minerals in proportion to their densities (Slingerland 1977, 1984; Slingerland & Smith 1986).

Field investigation shows that grain size of the sediments in the Natal Group generally fines southward, which corresponds to the palaeoslope direction (Hobday & von Brunn 1979), and changes from conglomerate to sandstone and then sand–siltstone with mudstone. Sandstone is the most widely distributed lithofacies in the middle and lower reaches of the palaeostream system. The sorting, textural and mineralogical maturity become steadily better from the upper stream dominated environment to the lower floodplain environment.

Provenance and sedimentary environment

Heavy mineral concentration is controlled by complex interactions, among which provenance, hydrodynamic condition and depositional environment are the most important factors.

The proportion and assemblage of heavy minerals in the Natal Group indicate that most of the minerals are typical of metamorphic and mafic igneous origin, particularly the ilmenite, magnetite, hornblende and garnet. It is possible that felsic igneous rocks and sedimentary rocks may have also provided additional sources for the heavy mineral assemblages if we consider the concentration of rutile, zircon and hematite. The evidence from microscopy and field studies supports this estimation; there are abundant metamorphic and igneous rock lithics present in the rocks, and banded iron formation pebbles also occur in the strata.

Previous studies show that the Natal Group was deposited in a NE–SW oriented rift basin, and the major palaeocurrent direction was from NE to SW (Hobday & von Brunn 1979; Roberts 1981; Marshall 1988, 1994; Liu 1997). Thus, the main source area probably lay to the north of the Natal Group basin. The local geology demonstrates that the Archaean Kaapvaal Craton (2.65–3.5 Ga) lies to the north of the Natal Group basin, and is a greenstone–granitoid terrane with a volcano-sedimentary cover sequence that includes banded iron formations (Pongola Supergroup) (Tankard *et al.* 1982; De Wit *et al.* 1992; Gold 1993; Thomas *et al.* 1993*b*, 1994; Hilliard 1997; De Wit 1998). Therefore, the Kaapvaal Craton undoubtedly acted as a major source for the heavy mineral supply. On

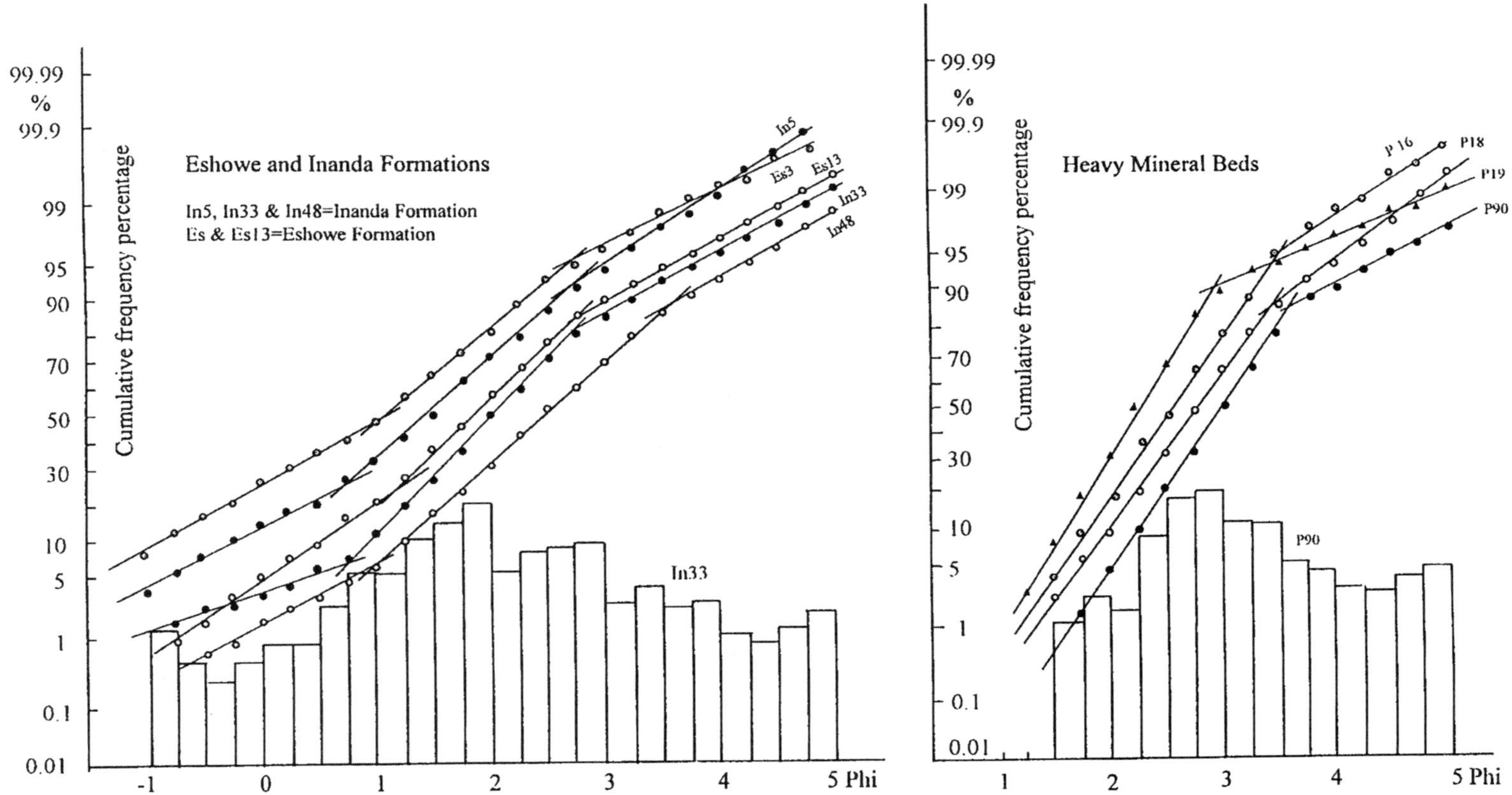

Fig. 8. Grain-size analysis of heavy minerals and their host rocks of the Eshowe and Inanda Formations

the other hand, local sources, namely, the Natal Structural and Metamorphic Province, provided an additional supply (Du Toit 1931; Kent 1938), but this was not an important source according to the present evidence.

The heavy minerals of the Natal Group are chiefly hosted in the fluvial dominated units of the Eshowe and Inanda Formations, not in the shallow marine units of the Kranskloof and Dassenhoek Members. It is clear that sedimentary facies and depositional environment played a major role in the accumulation of the heavy minerals. Detailed sedimentary facies and environment study of the Natal Group shows that the heavy minerals were deposited on a wide floodplain environment in the middle and lower reaches of the Ordovician fluvial system, rather than the upper reaches (Fig. 9). The richest heavy mineral concentrations were coincident with the branched or braided side-channels, rather than the main fluvial channel, although the main channel was dominant in the middle and lower reaches. This phenomenon is explained as being controlled by hydrodynamic conditions: in the upper reaches, relief was steep and fluvial channels were narrow, the current was swift and hydrodynamic energy was too strong to allow heavy minerals to settle. In the middle and lower reaches of the floodplain, shallower gradients

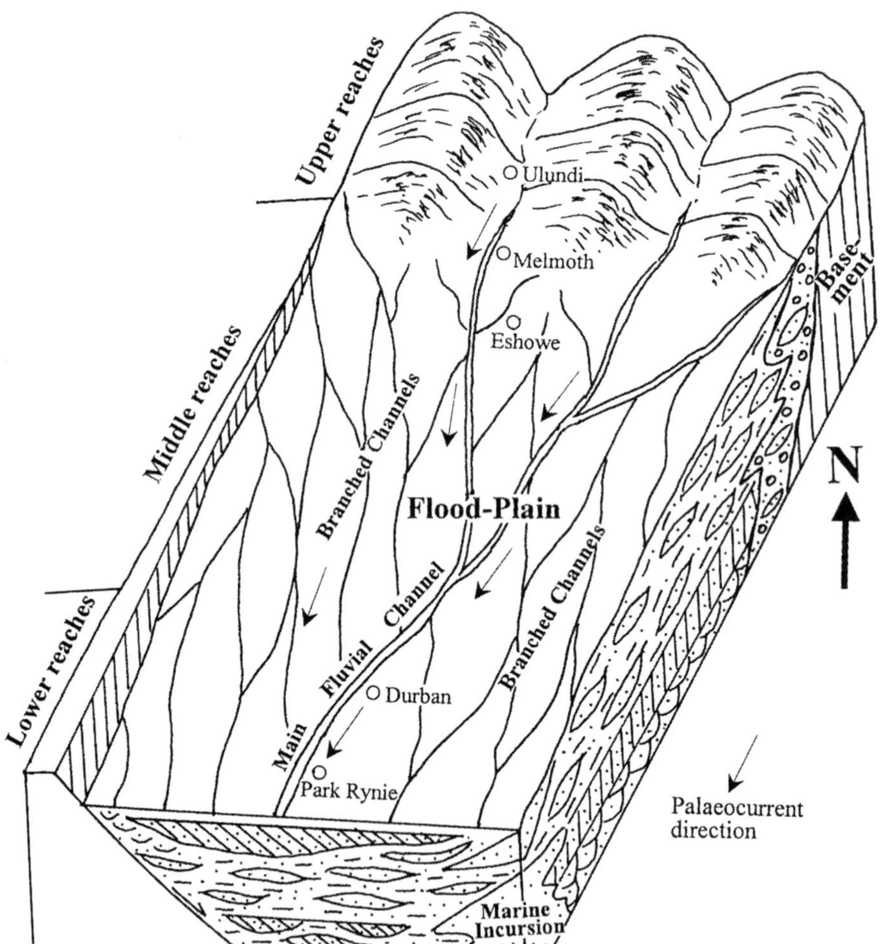

Fig. 9. Sketch diagram illustrating the depositional model for heavy mineral accumulations in the Natal Group. It should be noted that heavy minerals are deposited mainly in the lower and middle reaches, rather than the upper reaches of the palaeostream system. The branched or braided side-channels are the favourable sedimentary environment for heavy mineral accumulation.

greatly reduced current velocity, and allowed heavy minerals to be deposited. Owing to coarser light grains moving faster than finer dense grains, hydrodynamic differentiation led to concentrations of heavy minerals (Slingerland 1977, 1984; Slingerland & Smith 1986). As the hydrodynamic energy in the branched or braided side-channels was, in most cases, weaker than that of the main channel, the richest heavy mineral concentration, therefore, was deposited in the side-channels.

In general, branched or braided side-channels had the most favourable environment for the accumulation of fine to medium sand-sized heavy minerals, where hydrodynamic energy was moderate (Komar & Wang 1984; Komar *et al.* 1984).

Evidence shows that there is a general tendency for the heavy mineral accumulation in the Natal Group to become gradually richer toward the lower reaches. However, this tendency is uneven, and the two richest areas are localized in the Natal Group, one in the area between Melmoth and Eshowe, and the other between Durban and Park Rynie, where thin-bedded heavy minerals are common.

Comparison with other heavy mineral deposits

Three types of placer heavy mineral deposits are known in the geological record, fluvial, glacio-lacustrine and marine shoreline deposits; among which, shoreline deposits are the most economically viable.

Marine shoreline deposits of heavy minerals are currently mined at Trail Ridge and Green Cove Springs, along the coastal areas of northeastern Florida and southeastern Georgia. The heavy mineral assemblage consists of ilmenite (*c.* 50%), zircon (15%) and staurolite (15%), with minor sillimanite, tourmaline and rutile (Pirkle *et al.* 1974, 1977). The average content of total heavy minerals is about 4%. Along the east coast of Australia, from south Sydney to Brisbane, there is the world's most valuable heavy mineral resource. The mineral assemblage consists of rutile, zircon and ilmenite, in which rutile is the most abundant. Ore reserves along the east coast of Australia are very large, but ore grade is <1%, which is much lower than the similar sands from the east coast of the USA (Whitworth 1959; McKellar 1975). In Western Australia, the Geographe Bay and Eneabba deposits are dominated by ilmenite, with less zircon and rutile. Heavy mineral contents >10% are common (Collins & Baxter 1984; Force 1991). At Richards Bay, South Africa, heavy minerals have been mined since 1967. The main heavy mineral is also ilmenite, with minor zircon, rutile, monazite, hornblende, magnetite and garnet. Heavy mineral contents ranging from 10 to 14% are common, and an average economic content is 5.9% (Hammerbeck 1976; Fockema 1986).

All the above-mentioned major shoreline heavy mineral deposits are hosted in unconsolidated sands of Pleistocene to Holocene age, and deposited by tidal–wave interactions aided by coastal aeolian processes. Besides quartz and feldspar, a frequent gangue mineral is carbonate. These characteristics contrast with the heavy mineral deposits in the Natal Group. Apart from the rutile-dominated deposits of eastern Australia, all the other areas are ilmenite-dominant deposits. It is interesting that the grain sizes of heavy minerals vary between 0.1 and 0.3 mm, similar to that of the Natal Group. As the hydraulic regime of a shoreline is different from that of the fluvial system, the ores and host sediments in shorelines are better sorted than those of the Natal Group. Further, shoreline ores are larger, laterally more persistent, and the grades are higher than those of the Natal Group, with the exception of the ores in the Australian east coast where the average heavy mineral contents are <1%. The source of heavy minerals in the Richards Bay deposits is believed to be sedimentary rocks, particularly the Ordovician Natal Group, the Late Palaeozoic Karoo Supergroup and the Cenozoic Port Durnford Formation. However, the ultimate sources are probably igneous and metamorphic suites.

The Pleistocene Gbangbama deposits of Sierra Leone were mined as rutile ores, and also contained ilmenite, garnet and minor pyroxene and zircon. The ores are mainly sand-sized and poorly sorted, showing common characteristics of fluvial sediments. Unlike the deposits of the Natal Group, the sediments are not consolidated and were deposited directly on a weathered bedrock of amphibolites and granulites (Force 1991). The contents of rutile ranges from 0.5 to 2%, with total heavy mineral contents between 1 and 5%. Similar to the Natal Group, the sediments are of fluvial origin, and fluvial channels are the favourable depositional environments for the heavy minerals (Raufuss 1973; Force 1991).

Heavy mineral deposits have been found at the Port Leyden delta, New York. The deposits consist largely of sandstones with an average heavy mineral content of 3.5%, locally 4–10%. The ores contain ilmenite, pyroxene, garnet, zircon, sillimanite and minor rutile (Stone & Force 1980; Force & Stone, 1990). Unlike the Natal Group, the deposits are glaciolacustrine in origin, and are better sorted than fluvial sediments.

In general, there are many similarities between the heavy mineral deposits of the Natal Group and those in other areas, reflecting the common characteristics of placer deposits globally, particularly in terms of mineral assemblages, grain-size distributions and source pre-requirements. Differences, on the other hand, reflect peculiarities of the Natal Group, particularly with respect to geological occurrence, ore grade and age.

Conclusions

Nine types of heavy minerals have been found in the Natal Group: ilmenite, magnetite, rutile, titanite, zircon, garnet, hematite, apatite and hornblende. Among these, ilmenite and magnetite are the most abundant, whereas apatite and hornblende are rare. Heavy minerals are mostly confined to fluvial sediments of the Eshowe and Inanda Formations, with average contents of 3.86% and 4.18%, respectively. Heavy minerals are less frequent in the Kranskloof and Dassenhoek Members of shallow marine origin, with average contents of 1.14–1.27% only. Host rocks of the heavy minerals are mainly arkose and lithic arkose, with a few subarkose; pure quartz arenite is unfavourable for the concentration of heavy minerals. The heavy minerals in the Natal Group occur as thin beds, laminae, concentrated pockets and disseminated grains; the thin-bedded heavy minerals hold the richest concentrations and are thus of the most economic significance. Grain-size analysis shows that most of the heavy minerals are fine to medium sand-sized and are better sorted than the host rocks. The main depositional environments for heavy mineral accumulation are branched or braided side-channels on floodplains, rather than main channels or overbank environments. Two rich areas of heavy mineral concentration have been localized, one between Eshowe and Melmoth, and the other between Durban and Park Rynie. Mineral characteristics and assemblages confirm that the provenance of the heavy minerals lay to the north of the depositional basin and was, most probably, the Archaean Kaapvaal Craton, where banded iron formations, basaltic lavas, greenstone and igneous rock suites are present. Locally, the Proterozoic Natal Structural and Metamorphic Province may have acted as a supplementary source area. More work is needed to ascertain the ore-scale, space distribution and economic potential. This study provides an example of floodplain-hosted mineral deposits.

The author sincerely thanks A. Sudamah for the preparation of thin sections, and A. Rajh for assistance with the photography. M. R. Cooper is gratefully acknowledged for reading the manuscript and improving the English text. The author highly appreciates the effort and contribution made by S. B. Marriott and J. Alexander for the organization of the Floodplains '98 Conference and for editing the conference proceeding. B. Turner is thanked for reviewing and improving the manuscript. This research project was funded by grants from the University of Durban–Westville and the Foundation for Research and Development (FRD), South Africa.

References

Collins, L. B. & Baxter, J. L. 1984. Heavy mineral-bearing strandline deposits associated with high-energy beach environments, southern Perth Basin, Western Australia. *Australian Journal of Earth Sciences*, **31**, 287–292.

Dai, D. L. & Chen, R. X. 1978. *Grain-size Analysis of Sediments (Rocks) and its Applications*. Geological Publishing House, Beijing, 1–147.

De Wit, M. J. 1998. On Archaean granites, greenstones, cratons and tectonics: does the evidence demand a verdict? *Precambrian Research*, **91**, 181–226.

——, Roering, C., Hart, R. J. *et al.* 1992. Formation of an Archaean continent. *Nature*, **357**, 553–562.

Du Toit, A. L. 1931. *The Geology of the Country Surrounding Nkandla, Natal*. Explanation Sheet 109 (Nkandla), Geological Survey of South Africa, 1–111.

Fockema, P. D. 1986. The heavy mineral deposits north of Richards Bay. *In*: Anhaeusser, C. R. (ed.) *Mineral Deposits of Southern Africa*. Geological Society of South Africa, Johannesburg, 2301–2307.

Force, E. R. 1991. *Geology of Titanium-Mineral Deposits*. Geological Society of America, Special Papers, **259**, 1–112.

—— & Stone, B. D. 1990. Heavy mineral dispersal and deposition in sandy deltas of glacial Lake Quinebaug, Connecticut. *US Geological Survey Bulletin*, **1874**, 1–21.

Friedman, G. M. 1958. Determination of sieve-size distribution from thin-section data for sedimentary petrological studies. *Journal of Geology*, **66**, 394–416.

—— 1962. Comparison of moment measures for sieving and thin-section data for sedimentary petrological studies. *Journal of Sedimentary Petrology*, **32**, 15–25.

Gold, D. J. C. 1993. *The geological evolution of a part of the Pongola Basin, South-eastern Kaapvaal Craton*. PhD thesis, University of Natal, Pietermaritzburg.

Hammerbeck, E. C. I. 1976. Titanium. *In*: Coetzee, C. B. (ed.) *Mineral Resources of the Republic of South Africa*. South African Geological Survey Handbook, **7**, 221–226.

Hilliard, P. 1997. *Structural evolution and tectonostratigraphy of the Kheis Orogen and its relationship to the south-western margin of the Kaapvaal Craton*. PhD thesis, University of Durban–Westville.

Hobday, D. K. & von Brunn, V. 1979. Fluvial sedimentation and paleogeography of an early Palaeozoic failed rift, southeastern margin of Africa. *Palaeogeography, Palaeoclimatology, Palaeoecology*, **28**, 169–184.

Johnson, M. R. 1994. Thin section grain size analysis revisited. *Sedimentology*, **41**, 985–999.

Kent, L. E. 1938. The geology of a portion of Victoria

County, Natal. *Transactions of the Geological Society of South Africa*, **41**, 1–36.

KINGSLEY, C. S. 1975. A new stratigraphic classification implying a lithofacies change in the Table Mountain Sandstone in southern Natal. *Transactions of the Geological Society of South Africa*, **78**. 43–55.

KOMAR, P. D. & WANG, C. 1984. Processes of selective grain transport and the formation of placers on beaches. *Journal of Geology*, **92**, 637–655.

——, BABA, J. & CUI, B. Q. 1984. Grain-size analyses of mica within sediments and the hydraulic equivalence of mica and quartz. *Journal of Sedimentary Petrol*ogy, **54**, 1379–1391.

LIU, K. W. 1997. Stratigraphy and depositional environments of the Natal Group at Park Rynie Beach, KwaZulu–Natal Province. *South African Journal of Geol*ogy, **100**, 137–150.

—— & COOPER, M. R. 1998. Tidalites in the Natal Group. *South African Journal of Geology,* **101**, 307–312.

—— & GREYLING, E. H. 1996. Grain-size distribution and cementation of the Cretaceous Mzamba Formation of Eastern Cape, South Africa: a case study of a storm-influenced sediment. *Sedimentary Geology,* **107**, 83–97.

MARSHALL, C. G. A. 1988. Some aspects of the Natal Group between Durban and Eshowe. *Geocongress '88, Extended Abstracts*, Geological Society of South Africa, Durban, 379–383.

—— 1989. *Stratigraphy and Sedimentology of the Natal Group in the Melmoth and Hlabisa Areas*. Annual Technical Report of the Geological Survey of South Africa, 101–103.

—— 1994. *The stratigraphy of the Natal Group*. MSc thesis, University of Natal, Pietermaritzburg.

MATHEW, D. 1971. A palaeoenvironmental study of the basal zone Table Mountain Sandstone at Kloof Gorge. *Petros*, **3**, 6–17.

MATTHEWS, P. E. 1961. Slump structures in the Table Mountain Series of Natal. *Transactions of the Geological Society of South Africa*, **64**, 55–69.

MCKELLAR, J. B. 1975. The eastern Australia rutile province. *In*: KNIGHT, C. L. (ed.) *Economic Geology of Australia and Papua New Guinea*. Australasian Institute of Mining and Metallurgy Monograph, **5**, 1055–1061.

PIRKLE, E. C., PIRKLE, W. A. & YOHO, W. H. 1974. The Green Cove Springs and Boulougne heavy-mineral sand deposits of Florida. *Economic Geology*, **67**, 1129–1137.

——, —— & —— 1977. *The Highland Heavy-mineral Sand Deposit on Trial Ridge in Northern Peninsular Florida*. Florida Bureau of Geology Report of Investigation, **84**, 1–50.

RAUFUSS, W. 1973. Structur, Schwermineralfuhrung, Genese, and Bergbau der sedimentaren Rutil Lagerstatten in Sierra Leone (Westafrika). *Geologische Jahrbuch, Reihe D*, **5**, 3–52.

RHODES, R. C. & LEITH, M. J. 1967. Lithostratigraphic zones in the Table Mountain Series of Natal. *Transactions of the Geological Society of South Africa*, **70**, 15–28.

ROBERTS, D. L. 1981. *The stratigraphy and sedimentology of the Natal Group in the Durban area*. MSc thesis, University of Natal, Durban.

—— 1990. Marine incursions in the Natal Group, eustasy or tectonism? *Extended Abstracts, Geocongress '90*. Geological Society of South Africa, Cape Town, 467–470.

SACS (SOUTH AFRICAN COMMITTEE FOR STRATIGRAPHY) 1980. *Stratigraphy of South Africa. Part 1: Lithostratigraphy of the Republic of South Africa, South West Africa/Namibia, and the Republics of Bophuthatswana, Transkei, and Venda*. Handbook of the Geological Survey of South Africa, **8**, 1–690.

SLINGERLAND, R. L. 1977. The effects of entrainment on the hydraulic equivalence relationships of light and heavy minerals in sands. *Journal of Sedimentary Petrology*, **47**, 753–770.

—— 1984. Role of hydraulic sorting in the origin of fluvial placers. *Journal of Sedimentary Petrology*, **54,** 137–150.

—— & SMITH, N. D. 1986. Occurrence and formation of water–land placers. *Annual Review of Earth and Planetary Science*, **14**, 113–147.

STONE, B. D. & FORCE, E. R. 1980. The Port Leyden, New York, heavy mineral deposit. *New York State Museum Bulletin,* **436**, 57–61.

TANKARD, A. J., JACKSON, M. P. A., ERIKSSON, K. A., HOBDAY, D. K., HUNTER, D. R. & MINTER, W. E. L. 1982. *Crustal Evolution of Southern Africa*. Springer, New York, 1–523.

THOMAS, R. J., CORNELL, D. H., MOORE, J. M. & JACOBS, J. 1994. Crustal evolution of the Namaqua–Natal Metamorphic Province, southern Africa. *South African Journal of Geology,* **97,** 8–14.

——, MARSHALL, C. G. A., WATKEYS, M. K., FITCH, F. J. & MILLER, J. A. 1993*a*. K–Ar and $^{40}Ar/^{39}Ar$ dating of the Natal Group, Southeast Africa: a post Pan-African molasse? *Journal of African Earth Science*, **15**, 453–471.

——, VON VEH, M. W. & MCCOURT, S. 1993*b*. The tectonic evolution of southern Africa: an overview. *Journal of African Earth Sci*ence, **16,** 5–24.

VISHER, G. S. 1969. Grain size distribution and depositional processes. *Journal of Sedimentary Petrology*, **39**(3), 1074–1106.

VISSER, J. N. J. 1974. The Table Mountain Group: a study in the deposition of quartz arenites on a stable shelf. *Transactions of the Geological Society of South Africa,* **77**, 229–237.

WHITWORTH, H. F. 1959. *The Zircon–Rutile Deposits on the Beaches of the East Coast of Australia with Special Reference to their Mode of Occurrence and the Origin of the Minerals.* New South Wales Department of Mines Technical Report, **4**, 7–60.

Index

Page numbers in *italics* refer to Figures and page numbers in **bold** refer to Tables